Turbulent Jets and Plumes

Joseph H. W. Lee · Vincent H. Chu ·
Adrian C. H. Lai

Turbulent Jets and Plumes

A Lagrangian Approach

Second Edition

Springer

Joseph H. W. Lee
Faculty of Innovation Engineering
Macau University of Science
and Technology
Macao, China

Vincent H. Chu
Department of Civil Engineering
and Applied Mechanics
McGill University
Montreal, QC, Canada

Adrian C. H. Lai
Coastal Engineering and Management
Surbana Jurong (SJ Group)
Singapore, Singapore

ISBN 978-3-032-01186-2 ISBN 978-3-032-01187-9 (eBook)
https://doi.org/10.1007/978-3-032-01187-9

This Springer imprint is published by the registered company Springer Nature Switzerland AG
The registered company address is: Gewerbestrasse 11, 6330 Cham, Switzerland

To Winnie, Alice, and Clara

Preface

Turbulent jets and plumes are turbulent flows produced by momentum and buoyant forces, and are ubiquitous in nature and engineered environments. Turbulent mixing and transport is also relevant to the sustainable development of modern societies. Numerous examples of jets and plumes can be found: cooling tower and smokestack emissions; forest fires; volcano eruptions; deep sea vents; atmospheric thermals; marine outfall sewage discharges; thermal discharges from nuclear power stations; sediment laden river plumes; brine discharges from desalination, heat dissipation convective plumes on nanoscale micro-electronics chips; feeding of bivalves (shellfish) on the sea bed; and space vehicle landing on Mars. The prediction of the path and mixing of jets and plumes is important for sound environmental control and impact assessment. The study of this important turbulent shear flow is a key to understanding a variety of environmental and industrial mixing problems. Jet and plume modeling is also at the core of real-time forecast systems for water management in smart cities—for example, daily beach water quality forecasts to protect public health.

Many technical papers have been written on this subject. For a variety of reasons, the newcomer often finds it difficult to sieve through the enormous literature on the subject. Most of the published literature is concerned with laboratory measurements near the source. Without proper interpretation, the laboratory results are not suitable for engineering application. The user manuals of mathematical models used in industrial practice also do not cover adequately the fundamentals of the basic principles. A great variety of turbulent closures, ranging from simple entrainment hypothesis to fully three-dimensional Reynolds-stress-transport models, have been introduced. Often the student fails to see the correct underlying physical concepts, and is sometimes misled by the common mistakes found in the literature.

This book is an outgrowth of two decades of research and teaching by the authors on the subject. The objective of this simple text is to introduce the fundamental concepts of the mixing of turbulent buoyant flows as well as incorporate recent advances. Jets and plumes are introduced as slender turbulent shear flows. We give a consistent and unified development of a Lagrangian approach for prediction of free jets and plumes, including effects of crossflow and ambient density stratification. The connection of Lagrangian and Eulerian formulations is discussed. Regardless of the detailed formulation, it is our view that the conceptual understanding of the process is greatly facilitated by a Lagrangian thinking or interpretation. This theme persists throughout the text, starting from simple cases and building towards the practically important case of a turbulent buoyant jet in a density-stratified crossflow. Basic ideas are illustrated by ample use of flow visualization using the laser-induced fluorescence and video imaging techniques. The robustness of the unified integral method is illustrated in a variety of complex engineering applications. An extensively validated Lagrangian computer code for the computation of buoyant jets is also described. A practical modeling tool in the form of an interactive virtual reality software, VISJET, is offered to the reader to apply the techniques to solve his or her own problems.

The First Edition was written based on the belief that study of this type of boundary layer flows can benefit from a clear physical understanding coupled with a solid mastery of integral modeling. However, with the recent rapid advances in Computational Fluid Dynamics (CFD) and Artificial Intelligence (AI)-based modeling approaches, a paradigm shift to problem solving that does not require physical understanding appears to be taking place. At this juncture of the Fourth Industrial Revolution, one may question if CFD and AI-based modeling approaches will soon replace 'simple' integral modeling? This has motivated the authors in writing this Second Edition, in which we attempt to address the question implicitly.

The advances in computing power and CFD software over the past decades have been very excitng. Compared to 20 years ago when the First Edition was published, CFD is no longer the privilege of universities and research institutes that have access to high-performance computing clusters and domain experts; CFD is becoming much more accessible to consultant firms and individuals. It has increasingly become the norm in engineering communities to include CFD as a tool in delivering consultancy projects; and independent CFD enthusiasts are seen actively exchanging resources and discussion in improving setups and showcasing individual works.

The trend of increasing CFD usage will almost certainly continue in future, but one should also be cautioned not to over-rely on CFD results, because CFD solvers all have their own assumptions and limitations. Correct CFD model setup is also important in generating reliable results that make physical sense; a minor mistake in choosing the boundary conditions can lead to completely wrong results. The key to avoiding all these pitfalls is, of course, the physical understanding of the problem—which is the focus of this book. Through the theory, worked examples, problems and mathematical models developed progressively in this book, we hope the readers can gain a solid understanding of the physical processes of jets and plumes, and have the knowledge in critically interpreting outputs generated from process-based numerical models or the very recent AI models.

New Features

This Second Edition includes the following additional original features to enhance physical understanding of the subject:

- Many worked examples and problems have been added to demonstrate the application of jet and plume theory to real life problems for easier understanding.
- In response to queries from many readers, a complete solution manual to the extensive problems at the end of each chapter is given, which also encourages readers to attempt these problems and test their own understanding.
- The subject can best be understood by observation conducted from simple experiments. Hence, many experimental visualization videos and computer animation have been incorporated, both for developing better understanding of the problems through observations, and for having a sense of realistic buoyant jet flows in a world that is full of numerically or AI-generated images and videos.

When used correctly, CFD is certainly a very powerful tool that everyone needs to embrace. To illustrate this, this new edition also gives:

- Examples that combine the use of CFD with jet and plume theory to provide innovative solutions to complex engineering problems.
- A section to highlight important points in CFD simulations of buoyant jets which is especially useful for beginners of CFD.

Engineering Application of Jet and Plume Models

This brings us back to the question of whether integral modeling or first-order theory is still needed? For this, we have added an entirely new chapter on:

- Modeling of multiple jet and plume interaction; and the coupling between near-field plume models with far field shallow water circulation models.

These are considered the most complex and commonly encountered situations in practice that can be handled by integral modeling. Such practical problems are still very challenging for even the most advanced CFD tools to handle reliably, and we believe that simple first-order models with analytical clarity will always retain its importance in solving complex engineering problems. We encourage the readers to attempt these problems using both the new version of VISJET provided and the CFD approach to draw their own conclusions.

Who Should Read This Book?

The text is suitable for senior undergraduate or first-year postgraduate students as part of a curriculum in Water and Environmental Engineering. The book also complements standard fluid mechanics textbooks in that it covers the fundamental environmental fluid mechanics concepts not normally covered in these texts—such as turbulent mixing, buoyancy, and density stratification. The book can serve as an introduction to turbulent shear flow in a course on environmental fluid mechanics, and may be used as a reference by students in physics, mechanical and chemical engineering, and earth and environmental sciences. CFD professionals and enthusiasts will also find the basic experimental data and theory useful for testing and validating their CFD models.

Macao, China — Joseph H. W. Lee
Montreal, Canada — Vincent H. Chu
Singapore — Adrian C. H. Lai

Acknowledgments

We would like to acknowledge the notable contributions and continual support of our past students and associates, in particular Reverend Dr. Valiant Cheung, Dr. Paul Chu, and Prof. Chen Guoqian, who have worked productively and maintained an interest in this subject. We would also like to express our special appreciation for the stimulating exchanges and discussions with our colleagues Prof. Gerhard Jirka and Prof. Ian Wood. We thank Prof. Wenping Wang for the development of advanced computer graphics and visualization tools in VISJET. The assistance of Prof. C. P. Kuang, Dr. H. C. Chan, and Dr. Daeyoung Yu on the preparation of the manuscript is well-appreciated. Preparation of the Second Edition has been greatly facilitated by discussions with Dr. David K. W. Choi, Dr. Tree S. N. Chan, and Dr. Chris C. K. Lai.

The research embodied in this work has been supported at various times by the National Research Council of Canada, the Croucher Foundation, the Alexander von Humboldt Foundation, the Hong Kong Research Grants Council, and the Macao Science and Technology Development Fund. The VISJET development has been supported by the Hong Kong Innovation and Technology (ITF) Fund. This book has been typeset using LaTeX 2_ε. We would also like to thank the Hong Kong Jockey Club Charities Trust Fund that supported the WATERMAN coastal water quality forecast system project during 2009–2012. Finally, our deep appreciation goes to our families for their unfailing support for our intellectual pursuits.

Contents

1 Introduction 1

2 Turbulent Jets 25
1 Plane Jet 30
1.1 Governing Equations 32
1.2 Integral Equations 33
1.3 Eulerian Integral Model 34
1.4 Entrainment Hypothesis 37
2 Round Jet 38
2.1 Mean Flow Structure 38
2.2 Additional Remarks on 3D Jet 41
3 Theory Versus Experiment 45
3.1 Mean Properties 45
3.2 Turbulence Properties 48
4 The Top-Hat Profile 51
5 Prediction of Potential Core Length 54
6 Summary 54

3 Turbulent Buoyant Plumes 63
1 Buoyancy and Reduced Gravity 64
2 Turbulent Round Plume 69
2.1 Dimensional Considerations 69
2.2 Eulerian Integral Model 72
2.3 Effect of Initial Momentum: Vertical Buoyant Jet 83
2.4 Buoyancy Reduction Due to Density-Temperature Nonlinearity 86
3 Lagrangian Approach for Plume Modelling 88
4 Negatively Buoyant Jets 93
5 Turbulent Line Plume 96
6 Summary 100

4 Inclined Buoyant Jet in Stagnant Environment 111
1 Lagrangian Model for Buoyant Jet in Stagnant Fluid 114
1.1 Zone of Established Flow (ZEF) 116
1.2 The Potential Core (ZFE) 117
2 Numerical Solution 121
2.1 Jet Trajectory and Potential Core Development 121
2.2 Dilution 124
2.3 Boundary Effects 128
3 Application Examples 130
4 Summary 137

5 Density Stratification 145
1 Buoyancy Variation 146
1.1 Salinity Equation 147
1.2 Temperature Equation 148
1.3 Stratification Frequency 151
2 Thermals in Stratified Fluid 153
2.1 Round Thermals 154
2.2 Line Thermal 160
3 Plumes in Stratified Fluid 162
3.1 Round Plume in Linearly Stratified Environment 164
3.2 Plane Plume in Linearly Stratified Environment 167
3.3 Plumes in Arbitrary Density Stratification 174
4 Plume in a Container 176
5 Summary 180

6 Turbulent Round Jet in Coflow 189
1 Summary of Experimental Observations 189
1.1 Cross-sectional Images 190
1.2 Gaussian Profiles 191
2 Integral Model 193
2.1 The Natural but Incorrect Formulation 194
2.2 Jet Spreading Hypothesis 196
2.3 Governing Equations 198
2.4 Approximate Initial Conditions 199
2.5 Prediction of Potential Core Length 200
2.6 Alternative Definition of Characteristic Velocity 200
3 Asymptotic Solutions: Strong and Weak-Jet 201
4 Comparison of Theory with Experimental Data 203
4.1 Jet Spreading Rate 203
4.2 Centerline Dilution 203
4.3 Centerline Excess Velocity Decay 205
4.4 The Entrainment Coefficient 206
5 Correlation of Model Results with Experiments 207
5.1 Visual Boundary 207
5.2 Flow-Weighted Average Dilution 209

5.3 Summary of Experimental Data 211
5.4 Ambient Turbulence 213
6 Summary 213

7 Jet in Crossflow: Advected Line Puffs 219
1 Length Scales and Regimes 221
1.1 Line Puff Analogy for MDFF 223
1.2 Similarity Variables for the Line Puff 224
2 1D Model of Line Puff 226
3 2D Model of Line Puff 227
3.1 Numerical Simulation of Line Puffs 228
4 3D Model of Jet in Crossflow 233
4.1 The Advected Line Puff 233
4.2 3D Model of Advected Line Puff 233
5 Measurements in Advected Line Puffs 237
5.1 Time-Averaged Properties 240
5.2 Turbulence Properties 245
6 Practical Application 246
7 Summary 251

8 Plume in Crossflow: Advected Line Thermals 253
1 Length Scales and Regimes 256
1.1 Line-Thermal Analogy for BDFF 256
1.2 Similarity Variables for the Line Thermal 258
2 1D Model of Line Thermal 260
3 2D Model of Line Thermal 262
3.1 Numerical Simulation of Line Thermals 262
4 3D Model of Plume in Crossflow 270
4.1 The Advected Line Thermal 270
4.2 3D Model of Advected Line Thermal 272
5 Measurements in Line Thermals 276
5.1 Experiments on Advected Line Thermals 276
5.2 Concentration Measurements 277
5.3 Summary of Experimental Data 280
6 Buoyant Jet in Crossflow 283
6.1 Line Thermals and Puffs 283
6.2 Application Example 284
6.3 Case Study 288
7 Summary 290

9 General Lagrangian Formulation 297
1 Elemental Volume 297
2 Method of Excesses 299
3 Spreading Hypothesis 301
3.1 Projected Area Entrainment 302
3.2 Surface Area Entrainment 304

3.3 Shear Entrainment 305
3.4 Summary 306
4 Puffs and Thermals 306
5 Buoyant Jet in Crossflow 308
5.1 Zone of Established Flow 309
5.2 Potential Core in the ZEF 311
6 Summary 318

10 Numerical Modelling and Field Application 321
1 Initial Dilution of Buoyant Plumes in a Current: The BDNF and BDFF 322
1.1 The BDNF-BDFF Transition 328
2 JETLAG—A Lagrangian Buoyant Jet Model 329
2.1 Overview of Model 330
2.2 Basic Concepts of Lagrangian Model 331
2.3 Model Formulation 332
2.4 Shear and Vortex Entrainment 334
2.5 Formulation for Near-Far Field Transition 337
2.6 Comparison of Model Predictions with Laboratory Data 341
3 Field Application and Verification 342
3.1 Post-Operation Monitoring of Sydney Outfall 342
3.2 Field Verification at North West New Territories Outfall, Urmston Road, Hong Kong 347
3.3 Environmental Impact Assessment of the Hong Kong SSDS Ocean Outfall 349
3.4 VISJET Interactive Virtual Reality Model 350
4 Other Specialized Applications 353
5 Role of Computational Fluid Dynamics (CFD) 355
5.1 Near-Field CFD Simulation 355
5.2 Far-Field Circulation Model 357

11 Mixing of Multiple Buoyant Jets 365
1 Multiple Jets in Stagnant Ambient 365
1.1 Momentum Superposition 366
2 Multiple Plumes in Stagnant Ambient 369
2.1 Kinetic Energy Superposition 370
2.2 Experiments and Numerical Simulations 374
3 Multiple Buoyant Jets in Stagnant Ambient 375
3.1 Dynamic Interaction of Multiple Buoyant Jets 376
3.2 Theory Versus Experiments 384
4 Multiple Buoyant Jets in a Crossflow 387
4.1 Effects of a Crossflow on Dynamic Interaction 387
4.2 Composite Dilution of a Buoyant Jet Group 389
4.3 VISJET Application in Multiple Buoyant Jet Groups 393

4.4 Remark on Dynamic Interaction of Jets in Crossflow 394
5 Dynamic Coupling of Near and Far Field Models in Predicting Mixing and Transport 396
5.1 Motivation 396
5.2 The Distributed Entrainment Sink Approach (DESA) 397
5.3 Validation of DESA by Laboratory Experiments 401
5.4 Case Studies 406
6 Summary 412

Appendix A: Density of Seawater 419

Appendix B: Notation 425

References 427

Index 441

Chapter 1
Introduction

Turbulent jets and plumes are a class of turbulent flows produced by momentum and buoyancy sources. A jet is the flow produced by a continuous source of momentum (Fig. 1) while a plume by a continuous source of buoyancy (Fig. 2). The appearance of the jet is similar to that of the plume although the turbulent motions in the two flows are produced by different mechanisms. Mixing in the jet is directly related to the inertia of the turbulent eddies. In the plume, the buoyant force produces the inertia, which leads to mixing.

Besides a problem of basic interest in fluid mechanics, the study of jets and plumes has applications in many practical problems. Examples of jets and plumes can be readily found. A cigarette smoker observes (maybe unknowingly!) a plume break up into eddies many times a day. Smoke plumes from factories and fossil-fueled electric power stations carry undesirable products of combustion such as sulphur dioxide and particulates which can cause severe pollution under unfavourable meteorological conditions. The supercritical high-velocity flow in many hydraulic structures and flood protection works can be considered a jet. Fire plumes have also been extensively studied to improve the design of smoke detectors and sprinkler systems in buildings and atriums of large shopping plazas. In many coastal cities, pre-treated sewage are discharged as buoyant jets through submarine outfalls located on the sea bed. Similarly, large flows of condenser cooling water from steam-electric power generation are often discharged at an elevated temperature of 5–10 °C above the natural background value (thermal effluent) through this type of submerged outlets.

A good grasp of jets and plumes greatly facilitates the understanding of water quality control and quantitative environmental impact and risk assessment. Environmental regulations and water quality objectives are mostly set in terms of pollutant concentrations in receiving waters. It is important to be able to predict the

Supplementary Information The online version contains supplementary material available at https://doi.org/10.1007/978-3-032-01187-9_1.

J. H. W. Lee et al., *Turbulent Jets and Plumes*,
https://doi.org/10.1007/978-3-032-01187-9_1

Fig. 1 High Reynolds number turbulent jet produced by the test of a rocket (three million pound thrust) by Lockheed in the hills of Redland, California (from *Los Angeles*, Sunset-Lane, Menlo Park, California 1968. Courtesy of Lockheed Martin)

Fig. 2 Thermal plume generated by the spectacular fire that destroyed Barcelona's Gran Teatro del Liceo (from *Time*, February 14, 1994. Courtesy of Gamma Imaginechina)

concentration distribution in the vicinity (the near field) of a discharge for a given discharge design and location, waste load, and environmental conditions. This is required for defining mixing zones and to minimize the impact of discharges on sensitive receivers (e.g. nearby beaches for amenities, wetland reserve, or fisheries). As we will see, the proper design of a submarine outfall—a hydraulic structure—can have a profound effect on the water quality actually observed near the discharge.

Jets and plumes are also related to many natural environmental processes. Hot "black smokers" released from deep sea hydrothermal vents, volcanic eruptions and forest fires, and sinking brine plumes due to freezing of openings on polar ice cap, are some examples. The local natural environment can also be altered by jets and plumes for the benefit of mankind—dissolved oxygen level in the hypolimnion of a lake can be raised through jet mixing with the oxygen rich upper layer waters. On the other hand, nutrient-rich bottom waters in some fjords can be effectively brought up to the surface layer by plume action to increase the productivity of fisheries.

Environmental discharges as turbulent jets and plumes form a class of 'active' designs for which the engineer can control the amount of mixing, and hence the degree of dilution, near the discharge point. Within this near field, the discharge itself generates the flow field and mixing. The velocity and concentration fields are hence coupled and have to be solved simultaneously. An important engineering parameter is the dilution. The dilution at a point is defined as the ratio of the discharge concentration (or concentration excess above the ambient) to the concentration (excess) at the point. This is a measure of the mixing capacity of the discharge.

Jets and plumes are often analysed as momentum and buoyancy sources. Since velocity is momentum per unit mass, any means of producing velocity in a fluid is a momentum source. Under the earth's gravitational attraction, the buoyancy force is associated with the density variation in the fluid (Archimedes' Principle). Many physical processes can lead to changes in fluid density and hence sources of buoyancy, e.g. heating by radiation and conduction, diffusion of solute, or simply the discharging of fluids and mixtures into a receiving environment of different density. With buoyancy, momentum would be generated at a rate equal to the buoyancy force (Newton's second law of motion). The Euler equation governing this momentum generation process is

$$\rho \frac{D\mathbf{u}}{dt} = -\nabla p + \rho g' \hat{\mathbf{k}} \tag{1}$$

where ρ = fluid density, $\mathbf{u}$ = velocity = momentum per unit mass of the fluid, ∇p = pressure gradient, $\rho g'$ = buoyancy force per unit volume of the fluid, $g' = g(\rho_a - \rho)/\rho$ = reduced gravity = buoyancy force per unit mass of the fluid (subscript a denotes reference ambient values), and $\hat{\mathbf{k}}$ = unit vector in the +z-direction (opposite to direction of gravity). The effect of viscosity is often ignored in the formulation of the turbulent jets and plumes at high Reynolds numbers.

The hot gas discharging from a rocket engine as shown in Fig. 1 is a continuous momentum source. In this example, the momentum is generated by the rocket engine at a rate equal to the thrust of the rocket engine, which is three million pounds! Since the initial momentum flux is huge, the role of buoyancy in the momentum generation is insignificant. Figure 2 shows the turbulent motion of a plume of smoke generated by the fire that destroyed most of Barcelona's Gran Teatro del Liceo. The source of the buoyancy in this plume is the heat produced by the fire. The initial momentum flux of the fire on the ground is negligible. The subsequent generation of momentum in the plume is determined by the buoyancy flux of the plume. If both momentum and buoyancy are present at the source, momentum would be the dominant effect in the region near the source. Buoyancy becomes significant only when the momentum produced by the buoyancy force has exceeded the initial momentum at the source. Figure 3 is another example of turbulent motion produced by momentum and buoyancy. The steam, smoke and ash are generated by the eruption of Mount St. Helens. The sudden release of momentum has produced essentially a *puff*. The flow may eventually become like a *thermal* as buoyancy become the dominant effect far from the source. Puffs and thermals are produced by instantaneous sources of momentum and buoyancy respectively, while jets and plumes by continuous sources (see Table 1 and Example 1.2).

Motions of enormous magnitude can be generated by minute density differences if the buoyancy force acts on the fluid for a long period of time. Examples can be found in the atmosphere, lakes, and oceans where buoyancy forces act over a column of fluid of large vertical extent. The updraft produced by the action of thermals in the atmosphere is an example. The initial density difference due to uneven heating of the ground by solar radiation is quite small. However, the motion of the thermals produced by this density difference is massive; glider pilots depend on these to keep the aircraft afloat. Another example of utilizing the buoyancy force can be found in the offshore disposal of wastewater through a diffuser located at the floor of the ocean. The mixing of the effluent close to the diffuser is controlled by the action of the jet. With a 2–3% initial density difference between the wastewater (essentially freshwater) and the sea water, the dilution of the wastewater is chiefly determined by the action of the plume. The dilution of a wastewater effluent in large depth is, therefore, primarily a function of the buoyancy flux of the discharge.

Figure 4 shows a plume rise in a stably stratified atmosphere. In this more complex example, the plume is affected by the velocity of the cross wind and the density variation in the atmosphere. The plume is bent over by the crossflow as the momentum and buoyancy of the surrounding fluid is entrained into the plume. The entrainment or mixing process causes the momentum and buoyancy in the plume to approach ambient values; the plume eventually levels off as the relative buoyancy is reduced to zero. The development of the plume in this example is therefore a problem of a concentrated source of buoyancy being modified by the momentum and buoyancy of its surrounding environment.

Many predictive methods have been developed over past decades to study the turbulent mixing processes in jets and plumes. Experimental data have been accumulated for flows with different source and ambient conditions. It is the purpose of this book to provide a systematic review of these findings and the various aspects of the phenomena as momentum and buoyancy interaction problems. The book is divided into eleven chapters. The basic concepts of turbulent mixing and entrainment and techniques of analysis are treated at length in Chaps. 2 and 3 for straight jets and

Fig. 3 Mammoth fists of steam, smoke, and ash—some thrusting up to 6000 m—rose from the peak of Mount St. Helens near Portland, Oregon, in a volcanic eruption (from *Life*, May 1980. Courtesy of Associated Press)

plumes. Inclined buoyant jets in stagnant environment are considered in Chap. 4 as a prelude to the general integral formulation, which includes the treatment of the initial potential core. The effect of ambient density stratification is examined in Chap. 5. In Chaps. 6–8 we look at several important basic flows necessary for understanding the complicated and practically important problem of a turbulent buoyant jet in a stratified crossflow: coflowing jets, puffs and thermals. In Chap. 9, a general Lagrangian formulation of the jet in crossflow problem is discussed. In Chap. 10, an extensively

Table 1 Turbulent shear flows generated by continuous and instantaneous sources of momentum and buoyancy in a stationary environment of uniform density. M_o, F_o are the source momentum and buoyancy fluxes for a jet and plume respectively. M_{po}, B_o are the source momentum and buoyancy force for the puff and thermal respectively. D, d_o are the dimensions of the point and line sources, and w_o is the source velocity. z = vertical co-ordinate above source; t = time after release. The expressions for volume, V, or volume flux, Q, are deduced by dimensional analysis

Flows	Description	Source strength and dimensions	Volume flux or Volume
Round jet	Point source of momentum flux	$\frac{M_o}{\rho_o} = \frac{\pi}{4} D^2 w_o^2 \sim \frac{L^4}{T^2}$	$Q \sim z\sqrt{\frac{M_o}{\rho_o}}$
Plane jet	Line source of momentum flux	$\frac{M_o}{\rho_o} = d_o w_o^2 \sim \frac{L^3}{T^2}$	$Q \sim \sqrt{\frac{M_o z}{\rho_o}}$
Round puff	Instantaneous point source of momentum	$\frac{M_{po}}{\rho_o} = w_o \frac{\pi}{6} D^3 \sim \frac{L^4}{T}$	$V \sim \left[\frac{M_{po}}{\rho_o}\right]^{\frac{3}{4}} t^{\frac{3}{4}}$
Line puff	Instantaneous line source of momentum	$\frac{M_{po}}{\rho_o} = w_o \frac{\pi}{4} d_o^2 \sim \frac{L^3}{T}$	$V \sim \left[\frac{M_{po}}{\rho_o}\right]^{\frac{2}{3}} t^{\frac{2}{3}}$
Round plume	Point source of buoyancy flux	$\frac{F_o}{\rho_o} = g'_o \frac{\pi}{4} D^2 w_o \sim \frac{L^4}{T^3}$	$Q \sim \left[\frac{F_o}{\rho_o}\right]^{\frac{1}{3}} z^{\frac{5}{3}}$
Plane plume	Line source of buoyancy flux	$\frac{F_o}{\rho_o} = g'_o d_o w_o \sim \frac{L^3}{T^3}$	$Q \sim \left[\frac{F_o}{\rho_o}\right]^{\frac{1}{3}} z$
Round thermal	Instantaneous point source of buoyancy force	$\frac{B_o}{\rho_o} = g'_o \frac{\pi}{6} D^3 \sim \frac{L^4}{T^2}$	$V \sim \left[\frac{B_o}{\rho_o}\right]^{\frac{3}{4}} t^{\frac{3}{2}}$
Line thermal	Instantaneous line source of buoyancy force	$\frac{B_o}{\rho_o} = g'_o \frac{\pi}{4} d_o^2 \sim \frac{L^3}{T^2}$	$V \sim \left[\frac{B_o}{\rho_o}\right]^{\frac{2}{3}} t^{\frac{4}{3}}$

validated Lagrangian model based on the projected area entrainment hypothesis is illustrated by a number of practical example applications. In Chap. 11, the buoyant jet theory is extended to multiple source discharges, and a novel approach is presented to couple the near-field jet model to a far-field circulation model.

Throughout this text we aim at developing a solid 1D analytical framework in analysing jet and plume problems, and demonstrate the advantages of using a Lagrangian approach. A general Lagrangian spreading hypothesis can be used in the formulation of turbulent entrainment, and this can be interpreted using the concept of dominant eddies. The Lagrangian method is also simpler and more convenient in its implementation because the important results can be obtained without actually specifying the velocity and concentration profiles, which are not generally known for complex flow situations. In many instances, more meaningful physical interpretation of the model predictions can be obtained using this approach.

Numerical examples and problems are provided in each chapter for the reader to acquire a quantitative appreciation of the material. Throughout the text, the problems

Fig. 4 Thermal plume in a stratified cross wind. The plume from the Lamma Power Station, Hong Kong, is observed to be bent over by the crossflow and level off at an elevation where the density of the plume is the same as the cross wind (photo taken in August 1982)

at the end of the chapters are occasionally referred to by the abbreviation 'Prob.xx'. For example, Prob. 1.7 and Prob. 3.14 refer to the specific problems at the end of this chapter and Chap. 3 respectively. Results of computer simulations are often included to illustrate the need for both mathematical models and order of magnitude estimates based on a physical understanding of the problem. An interactive virtual reality Lagrangian model, VISJET, is provided to assist the reader to obtain a better physical understanding of jets and plumes. We end this chapter with four examples to show how turbulent motions are generated by momentum and buoyancy forces, and how useful engineering estimates can be obtained using simple models without the need to consider the details of the turbulent motion.

Example 1.1 *Sand dumping from a dredger*
A 100 kg of sand is dumped overboard from a dredging vessel into a lake. A sand-water mixture is formed as the cloud of sand mixes with its surrounding. The mixture is a heavy fluid element which descends down the water body by virtue of its buoyancy. Find the buoyancy force associated with the thermal. Assuming that the radius of the sand cloud, R, spreads at a rate proportional to the vertical velocity, $w = dz/dt$,

$$\frac{dR}{dt} = \beta \frac{dz}{dt}, \tag{2}$$

find the velocity and radius of the sand cloud after it has reached a depth of 5 m below the water surface. Calculate the depth when the sand particles begin to separate from the cloud. Assume a spreading rate of $\beta = 0.34$ *and an average fall velocity of* w_s=5 *cm/s for the sand particles (relative density 2.6).*

Solution: This is an example of a round thermal (an instantaneous source of buoyancy) in a stagnant environment. The heavier sand particles are the source of buoyancy which generates the turbulence and causes the sand to mix with the surrounding water.

The downward buoyancy force associated with the 100 kg of sand in water is the weight of the sand minus the weight of the displaced water (Archimedes Principle). The weight of the sand is $(\rho_s \mathcal{V}_o)g = 100\,\text{kg} \times 9.81\,\text{m/s}^2$. The weight of the displaced water is 2.6 times lighter. Hence, the buoyancy force

$$B = \rho_s g \mathcal{V}_o - \rho_w g \mathcal{V}_o = 100 \times 9.81 \times \left(1 - \frac{1}{2.6}\right) = \underline{603.7}\ \text{N}$$

where ρ_s = density of sand, ρ_w = density of water, ρ_s/ρ_w = specific gravity of the sand $\simeq$2.6, $\mathcal{V}_o$ = volume of water displaced by the sand, $\rho_s \mathcal{V}_o$ = mass of sand = 100 kg. As the thermal (cloud of sand-water mixture) sinks, it mixes with the surrounding water by turbulent entrainment; the radius of the thermal R increases with the distance z from the source. According to Eq. 2,

$$R = R_o + \beta z$$

(see Chap. 9 for the hypothesis leading to Eq. 2). The volume $\mathcal{V}$ and mass $\mathcal{M}$ of the thermal also increase correspondingly as (approximating the cloud as a sphere with characteristic radius R):

$$\mathcal{V} = \frac{4\pi}{3} R^3$$

$$\mathcal{M} = \mathcal{M}_o + \rho_w(\mathcal{V} - \mathcal{V}_o)$$

where R_o and $\mathcal{M}_o$ are the initial radius and initial mass of the thermal at the source, respectively. If the density of the surrounding water is uniform, the downward buoyancy force associated with the thermal (603.7 N in this example) would stay constant. The sand-water cloud (thermal) moves collectively with a velocity that is much greater than the sand particle fall velocity.

According to Newton's second law of motion, the momentum of the thermal, $(1 + k)\mathcal{M}w$, increases with time in proportional to the buoyancy force, B, as follows (see also Eq. 23 in Chap. 5):

$$(1 + k)\mathcal{M}w = Bt$$

where $w = dz/dt$ is downward velocity and k an *added mass* coefficient associated with the motion outside of the thermal. For a sphere accelerating in an irrotational fluid, $k = 0.5$. For large z, the initial mass $\mathcal{M}_o$ and initial volume $\mathcal{V}_o$ may be ignored as a first approximation. Therefore,

$$\mathcal{M} \simeq 4\rho_w \pi R^3/3 = 4\rho_w \pi \beta^3 z^3/3$$

and the momentum equation becomes

$$\frac{4\pi}{3}(1+k)\rho_w \beta^3 z^3 \frac{dz}{dt} = Bt.$$

Integrating this equation gives the trajectory of the thermal with respect to time, $z(t)$:

$$z = \left[\frac{3B}{2\pi\beta^3 \rho_w (1+k)}\right]^{\frac{1}{4}} t^{\frac{1}{2}}.$$

The velocity (momentum per unit mass) of the thermal is then:

$$w = \frac{dz}{dt} = \left[\frac{3B}{32\pi\beta^3 \rho_w (1+k)}\right]^{\frac{1}{4}} t^{-\frac{1}{2}} = \left[\frac{3B}{8\pi\beta^3 \rho_w (1+k)}\right]^{\frac{1}{2}} z^{-1} \quad (3)$$

At a depth of 5 m below the water surface, the sand-cloud mixture has a velocity of

$$w_{5\text{m}} = \sqrt{\frac{3 \times 603.7}{8\pi \times 0.34^3 \times 1.5 \times 1000}} \frac{1}{5} = \underline{0.22} \text{ m/s}.$$

which is still four times larger than the fall velocity of the constituent sand particles. The corresponding radius of the cloud is $R = 0.34 \times 5 = \underline{1.7}$ m, with a Reynolds number (based on the diameter) of $Re = Vd/\nu = 0.22 \times 3.4/10^{-6} = 7.5 \times 10^5$. Rather significant turbulent motions have been generated by the buoyancy. At this depth of 5 m, the sand and water stay together in the thermal as a mixture since the fall velocity $w_s \ll w$. The sand particles do not begin to separate from the thermal until the thermal velocity is reduced to a velocity comparable to the fall velocity at a depth of about 22 m below the water surface.

The above result is based on the 'point source' assumption that $\mathcal{V}_o$ and $\mathcal{M}_o$ are negligible. This assumption has led to the unrealistic infinite velocity at the point source (see Eq. 3). The realistic solution as shown in Fig. 5 is obtained by numerical integration of the equation of motion including the finite-source effects of $\mathcal{V}_o$ and $\mathcal{M}_o$ (Prob. 1.4). Round puffs or thermals also have other important applications, such as the fluid mechanics of virus spread, sneezes and coughs, as illustrated in Problems 1.5 and 1.6).

Example 1.2 *Sand slurry plume in crossflow*
A sand-water slurry is pumped continuously from a 0.4 m diameter jet nozzle at the bottom of a barge into a lake. The jet velocity is 3 m/s and the waste material has a relative density of 1.03. The discharge forms a plume in crossflow. A line thermal of negatively buoyant material is formed as the vessel is moving with a velocity of 0.5 m/s relative to the lake water. Find the buoyancy force per unit length of the line thermal. Assuming that the radius of thermal, R, spreads at a rate proportional to

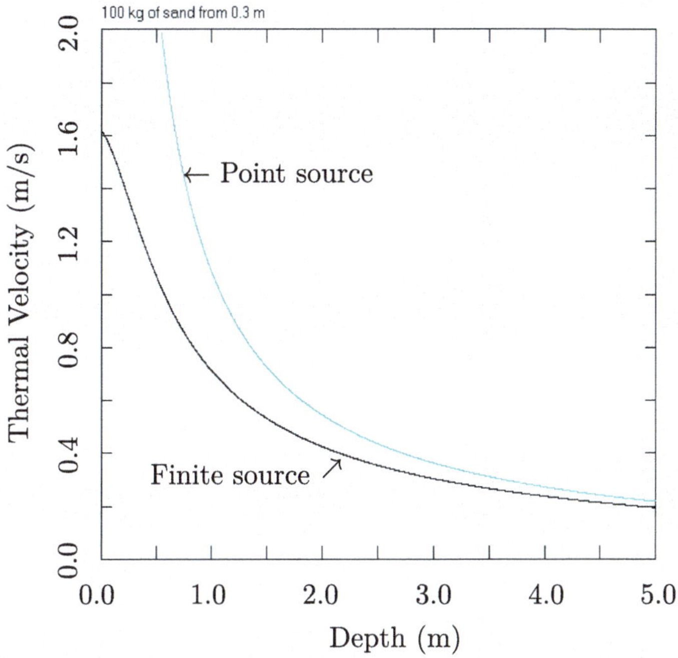

Fig. 5 The thermal velocity produced by the dumping of 100 kg of sand at an elevation of 0.3 m above the water surface. Dashed line is the point source approximation

the vertical velocity, $w = dz/dt$,

$$\frac{dR}{dt} = \beta \frac{dz}{dt}, \tag{4}$$

find the velocity and radius of the cloud after the waste has reached a depth of 20 m below the water surface. Assume a spreading rate of $\beta = 0.4$ *and an average fall velocity of 5 cm/s for the sand particles.*

Solution: The path of the waste plume from the moving vessel forms a line thermal as shown in Fig. 6a. Calculation of the thermal velocity is made on an elemental volume AB produced by the release of the heavy material over a period of one second. Over a period of one second, the vessel moves forward a distance of 0.5 m. The heavy material is distributed over this same distance to form the elemental volume. Therefore the length of the elemental volume is 0.5 m. The jet discharge flow is $Q_o = 3 \times \pi \times 0.4^2/4 = 0.378\,\text{m}^3/\text{s}$. The buoyancy flux is $F_o = (\rho_o - \rho_a)gQ_o = (1030 - 1000) \times 9.81 \times 0.38 = 111$ N/s. Therefore, the volume per unit length of this element is $V_o = 0.38/0.5 = 0.76\,\text{m}^3/\text{m}$ and the buoyancy force per unit length of this element is

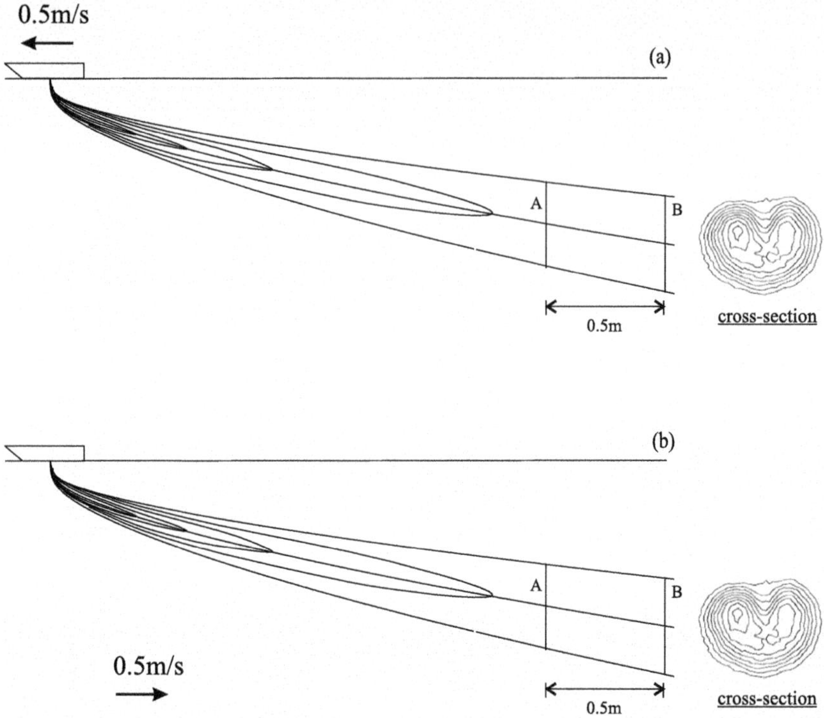

Fig. 6 **a** Line thermal produced by discharging a sand-water slurry from a dredging vessel moving at a velocity of 0.5 m/s. **b** The equivalent plume-in-crossflow problem. The elemental volume AB is an element along the path of the buoyant jet produced by the buoyant flux at the source for a period of one second. A buoyant force of 222 N is associated with the heavy slurry in the elemental volume AB. The length of the volume is 0.5 m

$$B = (\rho_s - \rho_a) g \mathcal{V}_o = (1030 - 1000) \times 9.81 \times 0.76 = \underline{222}\ \mathrm{N/m}$$

The calculations for the thermal velocity follows the same procedure as the previous example. As before, the momentum of the thermal, $(1 + k)\mathcal{M}w$, increases with time in proportional to the buoyancy force, B, as follows:

$$(1 + k)\mathcal{M}w = Bt$$

where $w = dz/dt$ is downward velocity and k the *added mass* coefficient associated with the irrotational motion outside of the thermal. The radius of the thermal R increases linearly with distance from the source. Ignoring the source size,

$$R \simeq \beta z,$$

We have assumed the distance to be z as a first approximation by ignoring the horizontal movement. The justification of this assumption will become clear as we examine the problem of advection line thermal in Chap. 8. The volume $\mathcal{V}$ per unit length of the cylindrical line thermal is $\mathcal{V} = \pi R^2$. The mass per unit length of the same is $\mathcal{M} = \rho_w \pi R^2 = \rho_w \pi \beta^2 z^2$. The momentum equation now become

$$\pi(1+k)\rho_w \beta^2 z^2 \frac{dz}{dt} = Bt.$$

The velocity of the thermal can then be obtained as (Eq. 23):

$$w = \frac{dz}{dt} = \frac{2}{3}\left[\frac{3B}{2\pi\beta^2\rho_w(1+k)}\right]^{\frac{1}{3}} t^{-\frac{1}{3}} = \left[\frac{2B}{3\pi\beta^2\rho_w(1+k)}\right]^{\frac{1}{2}} z^{-\frac{1}{2}} \tag{5}$$

For a circular cylinder accelerating in an irrotational fluid, $k = 1.0$ (see Chaps. 7 and 8 for this added mass coefficient). At a depth of 20 m below the water surface, this velocity is

$$w_{20\text{m}} = \sqrt{\frac{2 \times 222}{3\pi \times 0.4^2 \times 2.0 \times 1000 \times 20.0}} = \underline{0.086}\ \text{m/s}.$$

At this depth, the sand and water may not stay together in the thermal as a mixture since the fall velocity of the sand (5 cm/s) is now comparable to the velocity of the thermal (8.5 cm/s).

The thermal velocities in the previous two examples are determined by two parameters: the added-mass coefficient k and the spreading coefficient β. The added mass coefficient is to account for the irrotational motion surrounding the thermal. The spreading coefficient on the other hand is the parameter that characterizes the turbulent motion within the thermal. Similar parameterization is employed to obtain one-dimensional solutions for a variety of jet-and-plume problems examined in this monograph. The value of the added mass coefficient has been determined using two- and three-dimensional turbulence models for line puffs and thermals in Chaps. 7 and 8.

Example 1.3 *Horizontal buoyant jet for sewage disposal*
In a small coastal town, an innovative sewage disposal strategy calls for discharging the fine-screened wastewater through a bored hole of 5 cm diameter (bored through a rocky cliff) into the sea in a horizontal offshore direction at the rate of 0.04 m³/s. The density of sewage is 2.5% smaller than the density of ambient sea water. Calculate the momentum flux and buoyancy flux of the discharge at the source. Define a length scale based on these source fluxes. Assuming that buoyancy effect is negligible, find the jet dilution at the distance of $x = 5$ m from the source. Estimate the ammonia nitrogen concentration at this distance if the effluent concentration is 10 mg/L.

Solution: This is another example of a continuous source since the wastewater (with momentum and buoyancy) is discharged from the pipe continuously at a rate of $Q_o = 0.04\ \text{m}^3/\text{s}$. The rate of supply of momentum by the source is the momentum *flux* (momentum per unit time). The rate of supply of buoyancy is the buoyancy flux. First, the mass *flux* (mass per unit time) is calculated:

$$\frac{\text{mass}}{\text{time}} = \rho Q_o$$

As momentum = mass × velocity, the momentum *flux* (momentum per unit time), M_o, is given by:

$$M_o = \frac{\text{momentum}}{\text{time}} = \frac{\text{mass}}{\text{time}} \times \text{velocity} = \rho Q_o V_o \tag{6}$$

For this example, $\rho = 1000\,\text{kg/m}^3$, $A_o = \pi D^2/4 = \pi(0.05)^2/4 = 1.964 \times 10^{-3}\,\text{m}^2$, and $Q_o = 0.04\ \text{m}^3/\text{s}$. The discharge velocity is $V_o = Q_o/A_o = 20.37\ \text{m/s}$. The sewage is directed at this high velocity away from the shore. Hence, the specific momentum flux (momentum flux per unit mass) is

$$\frac{M_o}{\rho} = Q_o V_o = \underline{0.815}\ \text{m}^4/\text{s}^2$$

As the buoyancy force per unit mass is $\frac{\Delta\rho_o g \mathcal{V}}{\rho \mathcal{V}} = g'_o$, the buoyancy flux, F_o, is given by:

$$\frac{\text{buoyancy force}}{\text{time}} = \frac{\text{mass}}{\text{time}} \times \frac{\text{buoyancy force}}{\text{mass}} = \rho Q_o g'_o$$

The specific buoyancy flux (buoyancy flux per unit mass) is:

$$\frac{F_o}{\rho} = Q_o g'_o = 0.04 \times 0.025 \times 9.81 = \underline{9.81 \times 10^{-3}}\ \text{m}^4/\text{s}^3$$

We see from the above calculations that the dimensions of the specific momentum flux and buoyancy flux are

$$\left[\frac{M_o}{\rho}\right] = \frac{L^4}{T^2}, \quad \left[\frac{F_o}{\rho}\right] = \frac{L^4}{T^3}.$$

Based on these fluxes, a length scale can be defined by eliminating the dependence on time:

$$l_s = \frac{(M_o/\rho)^{\frac{3}{4}}}{(F_o/\rho)^{\frac{1}{2}}}$$

For $M_o/\rho = 0.815\ \text{m}^4/\text{s}^2$ and $F_o/\rho = 9.81 \times 10^{-3}\,\text{m}^4/\text{s}^3$,

$$l_s = \underline{8.7}\,\text{m}.$$

This length scale of around 9 m defines the relative effect of momentum and buoyancy. Momentum is significant close to the discharge. Buoyancy does not become significant until the distance from the source has exceeded this length scale (see Chaps. 3 and 4).

At $x = 5$ m, since $x < l_s$, we may neglect the effect of buoyancy as a first approximation, and assume that the jet is straight and that the momentum flux in the jet is constant within the first 5 m. By virtue of the jet momentum, the wastewater is mixed with the ambient fluid. The jet volume flux at a distance x downstream is given by (Chap. 2):

$$Q = \sqrt{\pi}\beta\sqrt{\frac{M_o}{\rho}}\,x,$$

where $\beta = 0.17$ is the jet spread rate. Therefore, at $x = 5$ m,

$$Q = \sqrt{\pi} \times 0.17 \times \sqrt{0.815} \times 5 = \underline{1.36}\ \mathrm{m^3/s}.$$

The jet is diluted by the fluid entrained from its surrounding environment. The dilution ratio is

$$S = \frac{Q}{Q_o} = \frac{1.36}{0.04} = \underline{34}.$$

Due to mixing of the jet with the entrained fluid, the tracer concentration is reduced on average by a factor equal to this dilution ratio. To calculate the concentration of ammonia nitrogen (which in its un-ionised form can be toxic to marine life), a simple mass balance with the source nitrogen flux gives:

$$c_o Q_o = C Q$$

where c_o and C are the ammonia nitrogen concentration of the discharge and at $x = 5$ m respectively. If the source concentration is $c_o = 10$ mg/L, the ammonia nitrogen concentration at $x = 5$ m would be

$$C = 10/34 = \underline{0.29}\ \mathrm{mg/L}$$

The calculations made in this example are based on the assumption that the jet is straight and that the buoyancy effect is negligible; average jet properties have been adopted. The ambient tracer concentration has also been neglected (Prob. 1.2). Further details concerning velocity and concentration distribution and the important role of buoyancy in turbulent mixing are given in Chaps. 2 and 3.

Example 1.4 *Safe landing of space vehicle on Mars*

The interaction of exhaust jets and plumes from a space landing vehicle with the ground surface of a planet is an important problem in space exploration. In a recent Mars mission, an unmanned rover vehicle is deployed to collect long term data. A Mars lander engine (MLE) is necessary to ensure safe landing of the rover vehicle.

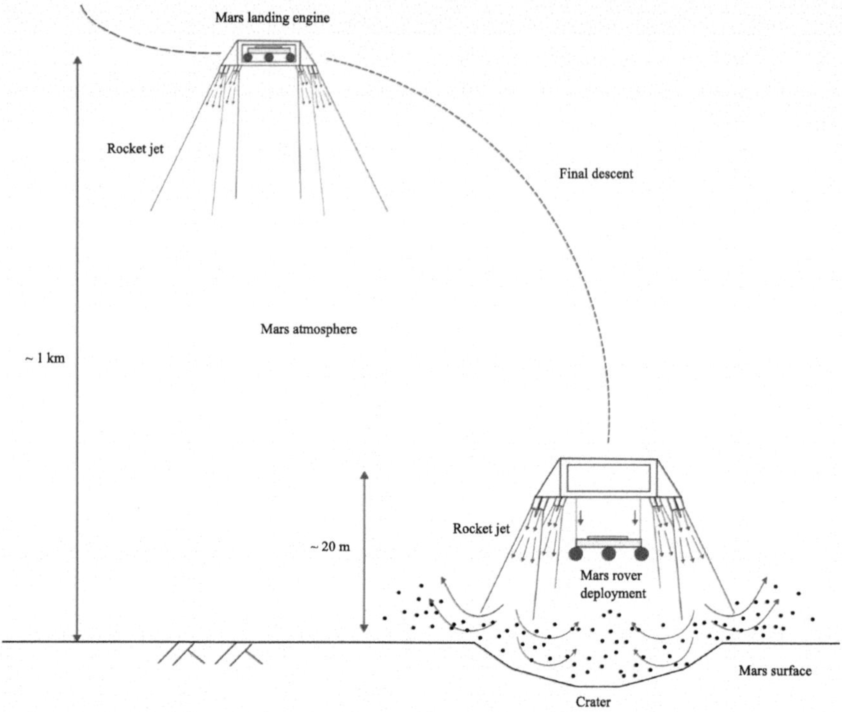

Fig. 7 Safe landing of a Mars rover vehicle using rocket jets (adapted from NASA/JPL-Caltech's illustration)

The MLE carried 400 kg landing propellant, a 600 kg descent spacecraft, and a rover vehicle of mass 1000 kg. High velocity jets are ejected from rocket thrusters in order to provide sufficient deceleration during the final descent stage (Fig. 7).

(a) At 1 km from the Martian surface, suppose the MLE decelerated from 100 to 30 m/s in 5 s, find the thrust required. Also, given that the total mass flow rate of the rocket propellant $\dot{m} = 8$ *kg/s, determine the initial velocity of the rocket jet.*

(b) At 20 m from the Martian surface, the MLE has to descend slowly using rocket jets, and the rover vehicle is gently lowered down to the Martian surface. Assuming the rocket jet initial velocity is constant during the entire descent process, determine the jet velocity u_i *and radius* r_i *when it impacts the Martian surface if the rocket jet is ejected from: (i) a single nozzle with diameter* $D = 0.57$ *m; and (ii) eight nozzles each with diameter* $D = 0.20$ *m (to keep the same total nozzle area). The acceleration due to gravity on Mars is* $g = 3.73$ *m/s*^{2}*. The ambient and exhaust gas density are given as* $\rho_a = 0.015$ *kg/m*^{3} *and* $\rho_o = 0.030$ *kg/m*^{3} *respectively. Assume the spreading rate of the jet as* $\beta = 0.2$*. As a first order estimate, ignore compressibility effects and assume a vertical jet discharge.*

Solution: (a) The mass of the MLE is $M = (400 + 600 + 1000) = 2000$ kg. To determine the thrust F_{th} required to decelerate the MLE, we consider:

$$\text{Thrust} = \frac{\Delta\text{momentum}}{\Delta\text{time}}$$

$$F_{th} = \frac{\Delta(Mu)}{\Delta t} = \frac{2000(30 - 100)}{5} = \underline{-20{,}560 \text{ N}}$$

Assuming that the jet is ideally expanded with $P_o \approx P_a$. With given total mass flow rate of the rocket propellant and the thrust required, the initial velocity can be estimated from:

$$\text{Force} = \Delta(\text{momentum flux}) = \Delta(\text{mass flux} \times \text{velocity})$$

and hence:

$$u_o = \frac{F_{th}}{\dot{m}} = \frac{20{,}560 \text{ N}}{8 \text{ kg/s}} = \underline{2570 \text{ m/s}}$$

(b) To determine the jet velocity and radius, we first determine the relative effect of momentum and buoyancy of the rocket jet discharge. Using the single nozzle configuration as an example, the volume flux Q_o, specific momentum flux M_o/ρ_o and specific buoyancy flux F_o/ρ_o are computed as:

$$Q_o = [(\pi D^2)/4]u_o = 647 \text{ m}^3/\text{s}$$

$$M_o/\rho_o = Q_o u_o = 647 \times 2570 = 1.66 \times 10^6 \text{ m}^4/\text{s}^2$$

$$F_o/\rho_o = Q_o g' = 647 \times 3.73 \times (0.03 - 0.01)/0.01 = 4.8 \times 10^3 \text{ m}^4/\text{s}^3$$

The momentum length scale $l_s = (M_o)^{4/3}/(F_o)^{1/2} = 2.8 \times 10^6$ m. It can be seen the influence of momentum is significantly larger than buoyancy at a height of 20 m, so buoyancy effects can be neglected as a first approximation. The same conclusion can be reached with a 8-nozzle configuration.

At 20 m from the planet surface, (i) for single nozzle configuration, the expanded jet radius can be determined using the given spreading rate as:

$$r_i = \frac{D}{2} + \beta z = \frac{0.57}{2} + 0.2 \times 20 = \underline{4.3 \text{ m}} \tag{7}$$

By momentum conservation and assuming the jet velocity is uniform across the cross-section,

$$\rho_o \pi (D/2)^2 u_o^2 = \rho_i \pi r_i^2 u_i^2 \tag{8}$$

With the rapid mixing between the jet and the ambient (a dilution in the order of 30), it is expected that the jet density will become close to that of the ambient at 20 m from the discharge and $\rho_i \approx \rho_a$. Hence: The velocity when impacting the Martian surface is given as:

$$u_i = \sqrt{u_o^2\left(\frac{\rho_o}{\rho_a}\right)\left(\frac{D}{2r_i}\right)^2} = \sqrt{2570^2 \times \left(\frac{0.030}{0.015}\right) \times \left(\frac{0.57}{2 \times 4.3}\right)^2} = \underline{241\,\text{m/s}}$$

(ii) for 8-nozzle configuration, the expanded jet radius can be determined as:

$$r_i = \frac{D}{2} + \beta z = \frac{0.20}{2} + 0.2 \times 20 = \underline{4.1\text{ m}} \tag{9}$$

$$u_i = \sqrt{u_o^2\left(\frac{\rho_o}{\rho_a}\right)\left(\frac{D}{2r_i}\right)^2} = \sqrt{2570^2 \times \left(\frac{0.030}{0.015}\right) \times \left(\frac{0.20}{2 \times 4.1}\right)^2} = \underline{89\,\text{m/s}}$$

From (i) and (ii), it can be seen at 20 m from the source, the jet radius is governed by the jet spreading and less dependent on the initial diameter. At large distances the jet mixing is relatively independent of the source geometry. It is also seen that for a single nozzle configuration, the rocket jet velocity impacting the ground can be very large, which can lead to crater formation and cause problems to the stability and safety of the lander (Sonar and Katsuragi 2024). On the other hand, by using 8-nozzle configuration with the same total nozzle area (and hence mass and momentum flux), the jet mixing will be enhanced because there is a larger area for turbulent entrainment with the ambient fluid, resulting in a smaller velocity on impact. The predictions of jet mixing will be further explained in Chap. 2, and the topic of multiple jet interaction will be discussed in Chap. 11. w Note that the physics of a rocket jet in space is highly complex; more detailed analysis relating to landing of space vehicles can be found in Metzger et al. (2009) and Zhang et al. (2022). Different from the Earth and Mars, the Moon can be considered practically vacuum, and the physics of rocket jets discharge will be markedly different.

Problems

1.1 Consider a round submerged outlet (diameter D) in depth H and discharging buoyant effluent at an angle ϕ_o to the horizontal in otherwise stagnant fluid. The discharge velocity, density, and tracer (or pollutant) concentration are u_o, ρ_o, c_o; the ambient current u_a is assumed to be uniform and horizontal, and of uniform density ρ_a. The kinematic viscosity of the discharge and ambient fluid are both ν. g is the acceleration due to gravity (Fig. 8). Of interest is the dilution $S = \dfrac{c_o}{c(x, y, z)}$, where $c(x, y, z)$ = concentration.

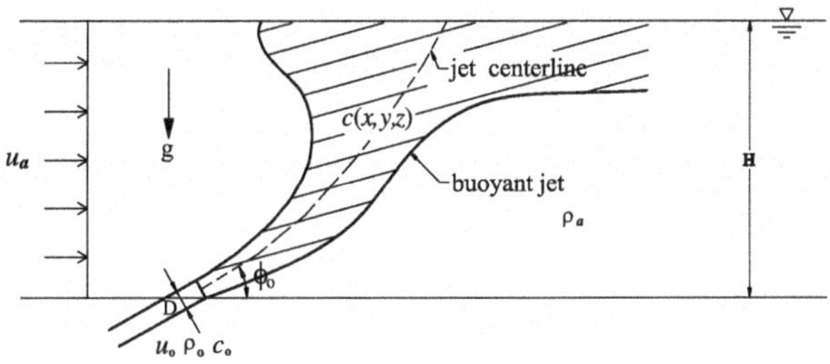

Fig. 8 A submerged buoyant jet

Based on physical considerations, it is reasonable to postulate a dependence of $c(x, y, z)$ on the discharge and ambient parameters, and the vertical location z, in the following manner:

$$fn\left(c, c_o, u_o, \frac{\Delta\rho_o}{\rho_a}g, D, z, \phi_o, \nu, u_a\right) = 0$$

Show that the following dimensionless relation can be obtained by dimensional analysis.

$$\frac{c_o}{c} = fn\left(\frac{u_o}{\sqrt{g\frac{\Delta\rho_o}{\rho_a}D}}, \frac{z}{D}, \frac{u_o D}{\nu}, \phi_o, \frac{u_o}{u_a}\right)$$

The dilution $S = c_o/c_c$ is governed by a jet densimetric Froude number $Fr = \frac{u_o}{\sqrt{g\frac{\Delta\rho_o}{\rho_a}D}}$, the discharge angle ϕ_o, the relative depth $\frac{z}{D}$, the jet Reynolds number $Re = \frac{u_o D}{\nu}$, and $\frac{u_o}{u_a}$, the ratio of discharge to ambient velocity. For a discharge in stagnant fluid, $u_a = 0$, the dilution is determined by Fr, z/D, and ϕ_o. The densimetric Froude number is a measure of the ratio of the inertia force to buoyancy force per unit mass; it is an important buoyant jet parameter, especially in cases when the source dimension cannot be neglected. $Fr \to \infty$ corresponds to the limiting case of a pure momentum jet ($\Delta\rho \to 0$); high values of Fr can be achieved also by high jet velocity, $u_o \to \infty$, or small source dimensions, $D \to 0$. Similarly, the limiting case of a plume is given by $Fr \to 0$. Plume behaviour is enhanced by low velocities, and relatively large source buoyancy and dimension.

1.2 For the buoyant jet in Example 1.3, determine the dissolved oxygen concentration at 5 m from the discharge if the dissolved oxygen concentration in the source and ambient fluid are 1 mg/L and 7 mg/L respectively.

1.3 The round puff is produced by a 'point instantaneous source' of momentum. The eruption of the Mount St. Helens as shown in Fig. 3 is an example. The impulse produced by the eruption is a product of the initial mass of the eruption and the corresponding initial velocity; i.e., M_o = mass × velocity $= \rho \frac{\pi}{6} D^3 w_o$. As fluid from the surrounding is drawn into the round puff through a process known as turbulent entrainment, the volume of the eruption, V, increases with time t following a functional relation that is dependent on the initial impulse and time, $V = \text{fn}(M_o, t, \rho)$.

Using dimensional analysis, derive the following relation (see Table 1):

$$V \sim \left(\frac{M_o}{\rho}\right)^{\frac{3}{4}} t^{\frac{3}{4}} \tag{10}$$

A similar analysis can be made for the line puff (an instantaneous line source of momentum), with $V \sim (M_o/\rho)^{\frac{2}{3}} t^{\frac{2}{3}}$, where V and M_o now represent the volume and momentum per unit length of the line puff respectively (see Table 1 and Chap. 7).

1.4 For the sand dumping problem in Example 1.1, if the lump of sand is released at 0.3 m above the water surface, determine (a) the free fall velocity of the sand particles just before impinging the free surface; and (b) the initial velocity and momentum of the sand-cloud mixture in the water. State any assumptions made (see also Chap. 5 and Probs. 5.5 and 8.3).

1.5 Spread of infectious disease by moist air in housing estate

In a densely populated housing estate in a city, residential flats are vertically arranged in 30 storey-buildings (Fig. 9). The high rise block is separated from an adjacent block by a narrow re-entrant space (light well). The air movement in such confined space (with plan dimensions of only 1.5 m wide by 6 m) is very limited. A possible way to spread an infectious disease (such as a deadly virus) in these poorly-ventilated spaces is through the transport by moist air released from the bathroom of flats (which typically face the re-entrant space) after a hot shower of an infected patient(s).

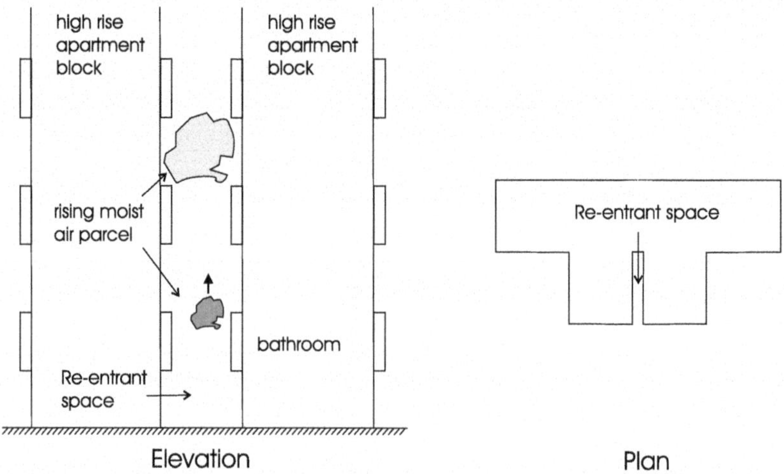

Fig. 9 Air movement in a confined re-entrant space of a residential building

Assume the virus is carried in a volume of moist air of initial volume $V_o = 5\,\text{m}^3$, relative density difference $\Delta\rho_o/\rho_a \approx 0.001$. Assume a linear spreading hypothesis for the round thermal (moist air parcel), $R = \beta z$, where R=radius of the thermal, and $\beta = 0.34$. Neglecting lateral boundaries, use Eq. 3 to find the dilution (inverse of relative concentration) and velocity of the virus-laden moist air after rising a vertical distance of $z = 3$ m (one floor), and $z = 30$ m (10 floors). Comment on your results.

1.6 "Safe social distance" in Covid-19 crisis

Violent respiratory events such as coughs and sneezes play a key role in transmitting respiratory diseases between infectious and susceptible individuals (Bourouiba et al. 2014). A cough or sneeze by an individual infected with an epidemic disease (such as SARS or COVID-19) is in essence a violent release of buoyant hot moist air at high velocity and carrying pathogens in the form of droplets. Experiments have shown that the initial velocity of such events can be as high as 10–20 m/s! In its simplest form, a coughing or sneezing event can be modeled as a turbulent round puff—an instantaneous source of momentum. The turbulent puff mixes with the environment and grows in size by entrainment of the surrounding fluid; the virus concentration is hence diluted and the forward velocity reduced. Any pathogen-carrying droplet drifting with the puff will follow a trajectory governed by the puff velocity $U(t)$ and the droplet vertical settling velocity w_s. It is of interest to estimate the "safe social distance"—i.e. the appropriate separation distance to avoid being infected.

Assume the volume expired during a cough or sneezing event is V_o and the initial velocity is U_o. Hence the initial momentum of the turbulent puff is given by $M_o = V_o U_o$ (Fig. 10). The evolution of the droplet laden turbulent cloud can be described in terms of its volume $V(t) = \frac{4}{3}\pi R^3$, radius $R(t)$, and velocity $U(t)$:

(a) Show by a momentum balance that:

$$\frac{d}{dt}(1+k)M(t) = 0$$

Assuming linear spreading of the round puff, $R = \beta x$, with spreading coefficient $\beta = 0.34$, show that:

$$x = \left(\frac{3M_o}{\pi(1+k)\beta^3}\right)^{1/4} = A^{1/4}t^{1/4}$$

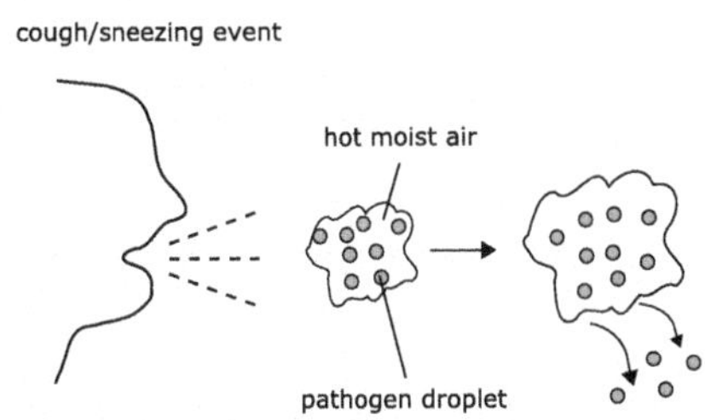

Fig. 10 A violent respiratory event

where $A = \frac{3M_o}{\pi(1+k)\beta^3}$. (b) As the horizontal puff velocity, $U(t) = dx/dt$, show that:

$$U = \frac{dx}{dt} = \frac{A}{4x^3}$$

(c) The tiny droplets (diameter of the order of 10 μm) can be considered to drift with the puff while settling at its own settling velocity w_s. The droplet trajectory is then given by $dz/dx = w_s/U$. Show that the droplet trajectory is given by $z = \frac{w_s}{A}x^4$, and hence the horizontal travel for a droplet to fall a distance of z can be estimated by:

$$x = \left(\frac{Az}{w_s}\right)^{1/4}$$

Using typical values of extreme sneezing events: $V_o = 1$ L; $U_o = 10$ m/s; $w_s = 6.5$ mm/s, and assuming a vertical fall of $z = 1$ m, estimate the "safe social distance" to avoid contamination from the violet respiratory event. Comment on your results and discuss the assumptions.

1.7 Rocket thruster jet
The space vehicle in Example 1.4 ejects a gas mixture at a total mass rate of $\dot{m} = 8$ kg/s from the combustion of liquid hydrazine (Fig. 11), which produces a thrust of $F_{th} \approx 20{,}560$ N. The average atmospheric condition of Mars are given as pressure $P_a = 610$ Pa, temperature $T_a = 210$ K, and specific gas constant $R_a = 190$ J/(kg K). The jet at the exit plane (after leaving the highly pressurized and hot reaction chamber) of the nozzle has a temperature $T_o = 250$ K, and specific gas constant of the gas mixture with complete combustion is $R_o = 390$ J/(kg K). Given that the exit pressure

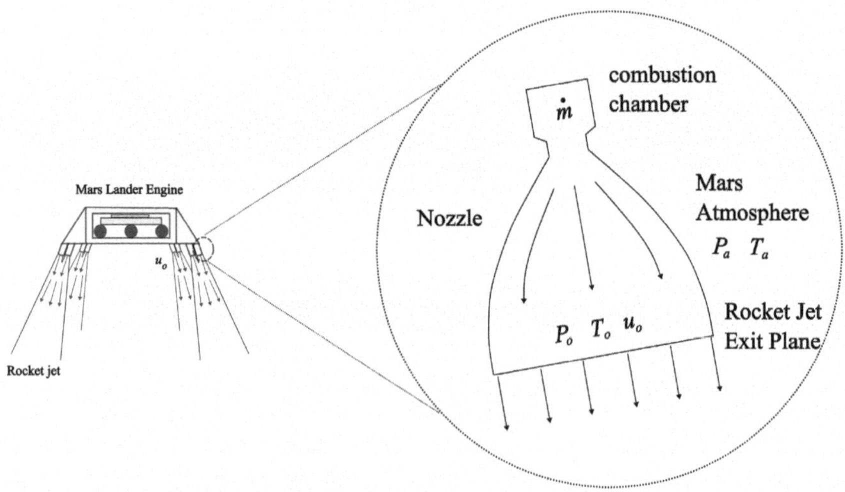

Fig. 11 High velocity jet generated by a rocket thruster

$P_o = 2900$ Pa, determine the initial velocity and density of the rocket thruster jet, and the density of the Mars atmosphere. Assume an 8-nozzle configuration, each nozzle with diameter $D = 0.20$ m.

Chapter 2
Turbulent Jets

A jet is the flow generated by a continuous source of momentum. The Reynolds number of a jet can be conveniently defined as $Re = \frac{u_o D}{\nu}$ where u_o = jet velocity; D = jet diameter; ν = kinematic viscosity of the fluid. Experiments have shown that if Re exceeds about 2000, the jet flow will be turbulent. In most problems of concern to the engineer, the Reynolds number would be large and the jet flow will be turbulent. The round jet shown in Fig. 1 in Chap. 1 is an example. Turbulent eddies of many sizes and shapes are observed due to the presence of the smoke as a tracer of the turbulent motion. There is a general tendency for the length scale of the turbulent motion to increase as the eddies move along the jet. Induced by the motion of the eddies, fluid from the surrounding environment is drawn into the jet through a process called *turbulent entrainment.* The extent of the turbulent zone increases and the concentration of the tracer decreases with distance from the source as the source fluid is diluted by the fluid entrained from the surroundings. The jet in this example produces a thrust (three million pounds) for the rocket engine. In many engineering applications, the jet is a device for mixing of source fluid with the ambient fluid. Figure 1 is a turbulent water jet produced in the laboratory from a high velocity discharge into a weak coflowing current. Rhodamine dye is introduced into the jet as a tracer. The turbulent structure in the jet is revealed by the fluorescence image, which is induced by a sheet of laser light through the center plane of the jet. Despite the enormous difference in the size, and the absence of the fine-scale turbulent structure, the laboratory jet is observed to spread at the same rate as the jet produced by the rocket engine. Apparently, the entrainment process and the spreading rate of a turbulent jet is determined by the large and dominant eddies extending across the full width of the jet. The ultimate mixing of the entrained fluid with the source fluid is carried out by the small eddies that circulate around the dominant eddies. It can also be seen that the initial development of the jet is confined to a mixing layer at

Supplementary Information The online version contains supplementary material available at https://doi.org/10.1007/978-3-032-01187-9_2.

J. H. W. Lee et al., *Turbulent Jets and Plumes*,
https://doi.org/10.1007/978-3-032-01187-9_2

Fig. 1 Potential core of a turbulent jet in a coflow; $u_o = 22.1\,\text{cm/s}$, $D = 2\,\text{cm}$; coflow velocity $u_a = 1.37\,\text{cm/s}$

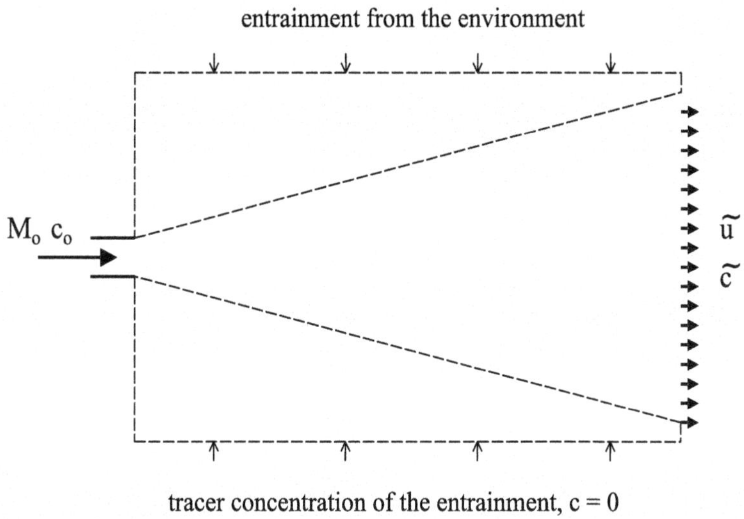

Fig. 2 Control volume and top-hat profile for mass and momentum fluxes in a round jet. Tracer concentration of the entrained ambient fluid is zero

the edge of the jet. There is a core of essentially irrotational fluid close to the source, about $6D$ in length, that is not affected by the jet diffusion. Beyond this potential core, all the source fluid is mixed with the ambient fluid, and the mean flow is fully established.

Example 2.1 *Waste water is discharged into a lake at a rate of 1.6* m^3/s *through a 1 m diameter pipe at the outfall. Assuming momentum flux conservation, find the volume flux of the jet at a distance of 10 m from the source. Assume that the jet radius,* $\tilde{b}$*, increases with the distance from the source, x, at a rate* $d\tilde{b}/dx = 0.17$*.*

Assuming tracer mass flux conservation, find the tracer concentration at the same location if the concentration at the source is $c_o = 100$ *mg/L, and the ambient concentration is zero (Fig. 2).*

SOLUTION: The cross sectional area of the jet at the source is

$$A_o = \frac{\pi}{4} D^2 = \frac{\pi}{4} \times 1^2 = 0.785\,\text{m}^2$$

and the initial jet volume flux (discharge) is $Q_o = 1.6\ \text{m}^3/\text{s}$. The jet velocity at the source (i.e. the momentum per unit mass of fluid) is

$$V_o = \frac{Q_o}{A_o} = \frac{1.6}{0.7853} = 2.038\,\text{m/s}$$

With this velocity, the jet is discharging its momentum at a rate equal to

$$M_o = V_o \rho Q_o = \text{momentum flux} = \frac{\text{momentum}}{\text{mass}} \times \frac{\text{mass}}{\text{volume}} \times \frac{\text{volume}}{\text{time}}$$

Dividing this momentum flux of the jet by the fluid density gives

$$\text{specific momentum flux} = \frac{M_o}{\rho} = V_o Q_o = 2.038 \times 1.6 = 3.261\ \text{m}^4/\text{s}^2$$

Due to its momentum, the jet entrains ambient fluid outside the jet boundary into the main turbulent stream thus increasing the volume flux of the jet. The mixing results in a change in both the jet velocity and width. The radius of the jet $\tilde{b}$ is expected to depend on the momentum flux of the jet M_o, the fluid density ρ and distance from the source x; i.e.,

$$\tilde{b} = \text{fn}(M_o, \rho, x)$$

The only dimensionless parameter in this functional relation is $\tilde{b}/x$, which must be a constant. According to this dimensional consideration, the width of the jet increases linearly with distance from the source. The rate is $d\tilde{b}/dx = 0.17$, which is obtained from experimental observation. The momentum flux of the jet is constant. Everything else is not: the volume flux increases and the energy flux decreases with distance from the source. Turbulence is generated as energy is dissipated in the jet. At a distance of $x = 10$ m, the jet properties are:

$$\text{jet radius} = \tilde{b} = 0.5 + 0.17 \times 10 = 2.2\,\text{m}$$

$$\text{cross sectional area} = \tilde{A} = \pi \times 2.2^2 = 15.21\ \text{m}^2$$

Momentum flux: In an environment of negligible pressure gradient, the momentum flux of the turbulent jet is constant. Equating the momentum flux at $x = 10$ m to the momentum flux at the source gives

$$\frac{M}{\rho} = \tilde{u}^2 \tilde{A} = 3.261\ \text{m}^4/\text{s}^2$$

The average velocity of the jet at 10 m distance from the source is

$$\tilde{u} = \left[\frac{3.261}{15.21}\right]^{\frac{1}{2}} = 0.463\,\text{m/s}.$$

The volume flux at the same location is

$$Q = \tilde{u}\tilde{A} = 0.463 \times 15.21 = \underline{7.04}\,\text{m}^3/\text{s}.$$

The jet is diluted. The dilution ratio is

$$S = \frac{Q}{Q_o} = \frac{7.04}{1.6} = 4.4.$$

Over a distance of 10 m, the volume flux and jet radius are both increased by 4.4 times through jet mixing.

Tracer mass flux: The mass flux of a conservative tracer (e.g. dye) is constant:

$$\text{tracer mass flux} = \frac{\text{tracer mass}}{\text{volume}} \times \frac{\text{volume}}{\text{time}} = c_o Q_o = \tilde{c} Q$$

Therefore, the tracer concentration in the jet is

$$\tilde{c} = c_o \frac{Q_o}{Q} = \frac{100}{4.4} = \underline{22.7}\ \text{mg/L}$$

Energy Flux: At 10 m, the specific kinetic energy flux is

$$\frac{E}{\rho} = \frac{1}{2}\tilde{u}^2 Q = \frac{1}{2} \times (0.463)^2 \times 7.042 = 0.755\ \text{m}^5/\text{s}^3$$

This is compared with the flux at the source, which is

$$\frac{E_o}{\rho} = \frac{1}{2} V_o^2 Q_o = \frac{1}{2} \times (2.038)^2 \times 1.6 = 3.323\ \text{m}^5/\text{s}^3.$$

The kinetic energy flux of the jet decreases with distance from the source. Nearly 77% of the initial kinetic energy is dissipated within 10 m. The dissipated energy is utilized by the jet to generate the turbulence, which induces the mixing and entrainment of surrounding fluid into the turbulent jet. The fluid power needed to produce the turbulent jet is

$$E_o = \frac{1}{2}\rho V_o^2 Q_o = 1000 \times 3.323 = 3323\ \text{W}$$

Top-hat Profile: For convenience, we have carried out the calculations in this introductory example using a 'top-hat' profile (i.e. assuming constant velocity distribution across the jet). The velocity $\tilde{u}$ and concentration $\tilde{c}$ of the top-hat profile are assumed to be the average velocity and concentration of the dominant eddies - the large eddies occupying the jet cross section that are responsible for the overall transport of mass and momentum. Small eddies circulating within the dominant eddies are assumed to have no net effect on the overall transport, so that the mass and momentum fluxes are calculated based on the top-hat profiles.

Gaussian Profile: The top-hat profile is convenient in a Lagrangian formulation of the jet using a reference frame following the motion of the dominant eddies. In a fixed

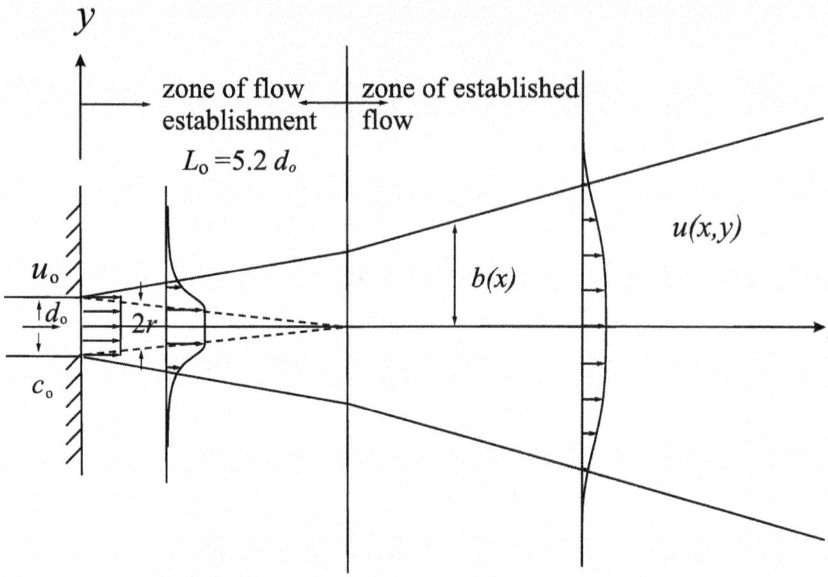

Fig. 3 Turbulent slot jet showing the Zone of Flow Establishment (ZFE) and Zone of Established Flow (ZEF)

reference frame, and from an Eulerian point of view, the velocity and concentration profiles across the jet are Gaussian. Mass and momentum in the turbulent jet move back and forth, and left and right, by the random action of the turbulent motion. The result of this random advection is an essentially Gaussian probability density function for the tracer concentration. The profile for the velocity is Gaussian as well, since momentum transport is analogous to mass transport, and velocity is momentum per unit mass of the fluid. Typical mean velocity profile in the fully developed region of the jet is Gaussian as shown in Fig. 3.

The Lagrangian formulation of jets and plumes using the top-hat profile is convenient. A justification for using the top-hat profile is the dominant-eddy hypothesis. The hypothesis, and the connection between the top-hat and the Gaussian profiles, will be further discussed in this and later chapters.

1 Plane Jet

We begin the study of turbulent jets using the conventional procedure. First the analytical framework for the Eulerian formulation is developed for the plane jet, which sometimes is referred to as the slot jet. Figure 3 shows the mean flow of the slot jet on the x- and y-plane. At the source the velocity and concentration is u_o and c_o, respectively. The turbulent flow in the jet is produced by discharging the source fluid from a slender slot. The development of the jet is divided into two zones: Zone of Flow Establishment (ZFE) and Zone of Established Flow (ZEF). The velocity and concentration in the potential core of the ZFE are constant. Surrounding the potential core is the mixing layer. The exchange of momentum between the core and the quiescent fluid across the mixing layer leads to the profiles in the mixing layer as follows:

$$u(x,y) = u_o \exp\left[-\frac{(y-r)^2}{b^2}\right], \quad c(x,y) = c_o \exp\left[-\frac{(y-r)^2}{(\lambda b)^2}\right]; \; y > r$$

$$u(x,y) = u_o, \qquad c(x,y) = c_o; \qquad y \leq r \tag{1}$$

where $r(x) =$ half-width of the potential core, $b(x) =$ width of the mixing layer, $u(x, y) =$ velocity, $c(x, y) =$ concentration, and $y =$ lateral distance from the center line. The parameter λ is introduced to account for the difference between the diffusion of mass and diffusion of momentum. The width of the concentration profile $\lambda b(x)$ is generally wider than the width of the velocity profiles $b(x)$. At the end of the potential core, $r \to 0$ by definition.

In the Zone of Established Flow (ZEF), $x > 5.2d_o$, when the turbulence has penetrated to the centerline, the velocity and concentration distributions are *self-similar* (this means the profiles at different x all look similar in shape, so that if the profiles at different x are scaled properly, all these should collapse onto one curve), and can be well-approximated by Gaussian distributions:

$$u(x,y) = u_m \exp\left[-\frac{y^2}{b^2}\right], \quad c(x,y) = c_m \exp\left[-\frac{y^2}{(\lambda b)^2}\right]; \; x > 5.2d_o \tag{2}$$

where $u_m(x)$ and $c_m(x)$ are the velocity and concentration maxima along the centerline. The width of the jet, $b(x)$, is defined at a lateral location where the x-component of the velocity is equal to $1/e$ of the centerline value. Experimental measurements of plane jet by Albertson et al. (1950), Miller and Comings (1957), Bradbury (1965) have found the jet spreads linearly with a growth rate

$$\frac{db}{dx} \simeq 0.1 \tag{3}$$

Since the width of the jet is small compared with the longitudinal length scale, the turbulent jet is a slender shear flow and boundary-layer type of approximation is applicable.

1.1 Governing Equations

The governing equations for the steady incompressible turbulent-mean flow of the jet are the continuity equation

$$\frac{\partial u}{\partial x} + \frac{\partial v}{\partial y} = 0, \tag{4}$$

the momentum equations

$$\rho u \frac{\partial u}{\partial x} + \rho v \frac{\partial u}{\partial y} = -\frac{\partial p}{\partial x} - \frac{\partial \rho \overline{u'^2}}{\partial x} - \frac{\partial \rho \overline{u'v'}}{\partial y}, \tag{5}$$

$$\rho u \frac{\partial v}{\partial x} + \rho v \frac{\partial v}{\partial y} = -\frac{\partial p}{\partial y} - \frac{\partial \rho \overline{u'v'}}{\partial x} - \frac{\partial \rho \overline{v'^2}}{\partial y}, \tag{6}$$

and the tracer-mass conservation equation

$$u \frac{\partial c}{\partial x} + v \frac{\partial c}{\partial y} = -\frac{\partial \overline{u'c'}}{\partial x} - \frac{\partial \overline{v'c'}}{\partial y}. \tag{7}$$

where $(u, v) =$ mean velocity, $c =$ mean concentration (tracer-mass per unit volume), $p =$ pressure, and $\rho =$ fluid density; the prime denote the turbulent fluctuations and overbar the time average of the fluctuations. The boundary conditions for a free jet are: $u \to 0$, $\overline{u'v'} \to 0$, and $\overline{u'c'} \to 0$ as $y \to \pm\infty$. With these boundary conditions the jet is assumed to be free from solid boundary and ambient current effects.

Furthermore, since $b \ll x$, by continuity $v \ll u$, and $\frac{\partial}{\partial x} \ll \frac{\partial}{\partial y}$. It can also be shown that $\frac{\partial v^2}{\partial y} \ll \frac{\partial \overline{v'^2}}{\partial y}$ (see Problem 2.6). Hence, based on boundary-layer approximation, the y-momentum equation (Eq. 6) is reduced to $\frac{\partial (p + \rho \overline{v'^2})}{\partial y} = 0$, and hence

$$p + \rho \overline{v'^2} \simeq p_\infty \tag{8}$$

In the x-momentum equation, the pressure p has been shown to be approximately constant so that $\frac{\partial p}{\partial x} \approx 0$ (Abramovich 1963). Other equations, according to the boundary-layer approximation, are then the continuity equation

$$\frac{\partial u}{\partial x} + \frac{\partial v}{\partial y} = 0, \tag{9}$$

the x-momentum equation

$$\rho u \frac{\partial u}{\partial x} + \rho v \frac{\partial u}{\partial y} + \frac{\partial \rho \overline{u'^2}}{\partial x} - \frac{\partial \rho \overline{v'^2}}{\partial x} = -\frac{\partial \rho \overline{u'v'}}{\partial y}, \tag{10}$$

and the tracer-mass conservation equation

$$u\frac{\partial c}{\partial x} + v\frac{\partial c}{\partial y} + \frac{\partial \overline{u'c'}}{\partial x} = -\frac{\partial \overline{v'c'}}{\partial y}, \tag{11}$$

subjected to the same boundary conditions as specified before. Thus we have 3 variables of the mean flow, u, v, c, and three governing equations. But the turbulent covariances $\overline{u'v'}$, $\overline{v'c'}$ are the additional unknowns due to the separation of the flow into mean and fluctuation parts. This is the problem of the *closure*. A *turbulence model* must be introduced to relate these covariances with the mean flow.

1.2 Integral Equations

A one-dimensional procedure to achieve the turbulent closure of the problem is to integrate across the turbulent jet. First, the x-momentum equation is re-written to a conservative form. Since,

$$\rho u\frac{\partial u}{\partial x} + \rho v\frac{\partial u}{\partial y} = \frac{\partial \rho u^2}{\partial x} + \frac{\partial \rho u v}{\partial y} - \rho u\left(\frac{\partial u}{\partial x} + \frac{\partial v}{\partial y}\right) \tag{12}$$

the x-momentum equation (Eq. 5) becomes (on invoking Eq. 8; see Problem 2.5)

$$\frac{\partial \rho u^2}{\partial x} + \frac{\partial \rho u v}{\partial y} + \frac{\partial \rho \overline{u'^2}}{\partial x} - \frac{\partial \rho \overline{v'^2}}{\partial x} = -\frac{\partial \rho \overline{u'v'}}{\partial y} \tag{13}$$

Using the boundary condition and integrating across the jet, from $y = -\infty$ to $y = +\infty$, we have for the left hand side of the equation

$$\frac{d}{dx}\int_{-\infty}^{\infty} \rho u^2 + \rho(\overline{u'^2} - \overline{v'^2})dy + [\rho u v]_{-\infty}^{\infty}$$

and for the right hand side

$$\int_{-\infty}^{\infty}(-\frac{\partial \rho \overline{u'v'}}{\partial y})dy = [-\rho\overline{u'v'}]_{-\infty}^{\infty} = 0.$$

Therefore,

$$\frac{d}{dx}\int_{-\infty}^{\infty}[\rho u^2 + \rho(\overline{u'^2} - \overline{v'^2})]dy = 0. \tag{14}$$

The momentum flux is preserved. Thus, the slot jet is a line source of momentum flux. If M_o is this momentum flux per unit length of the slot at the source,

$$M = \int_{-\infty}^{\infty} [\rho u^2 + \rho(\overline{u'^2} - \overline{v'^2})]\, dy = M_o \tag{15}$$

Similarly, we have for tracer-mass conservation,

$$u\frac{\partial c}{\partial x} + v\frac{\partial c}{\partial y} = \frac{\partial uc}{\partial x} + \frac{\partial vc}{\partial y} = -\frac{\partial \overline{u'c'}}{\partial x} - \frac{\partial \overline{v'c'}}{\partial y} \tag{16}$$

Integrating across the jet as before,

$$\frac{d}{dx}\int_{-\infty}^{\infty} [uc + \overline{u'c'}]\, dy = \left[-\overline{v'c'}\right]_{-\infty}^{\infty}. \tag{17}$$

The right hand side of this equation would be zero if the concentration c is the concentration excess above its value in the ambient. Hence, the excess-mass flux is constant. If Γ_o is this excess-mass flux per unit length of the slot,

$$\Gamma = \int_{-\infty}^{\infty} [uc + \overline{u'c'}]\, dy = \Gamma_o. \tag{18}$$

Measurements show that the two turbulence quantities in Eq. 14 are of the same order; a detailed evaluation using the data of Miller and Comings (1957) shows that the contribution of the turbulence terms to the momentum flux integral is only around 4 percent (Prob. 2.7). If the longitudinal fluxes due to the turbulent advection is ignored,

$$M \simeq \int_{-\infty}^{\infty} \rho u^2\, dy \simeq M_o, \tag{19}$$

$$\Gamma \simeq \int_{-\infty}^{\infty} uc\, dy \simeq \Gamma_o. \tag{20}$$

These are the *approximated* form of the equations since the parts of fluxes due to turbulent advection have been ignored. The Eulerian integral model of the jet to be described in the next section is based on these approximate equations.

1.3 Eulerian Integral Model

The conventional integral model of the turbulent jet is based on the assumption of the Gaussian profiles (Eqs. 1 and 2). To demonstrate the idea, the calculations for the ZEF is considered in this section. With the velocity and concentration profiles given by Eq. 2, the specific or kinematic momentum flux M/ρ (consisting of only kinematic dimensions of length and time) and the tracer-mass flux are functions of the width of the jet, b, the maximum velocity, u_m, and the maximum concentration,

c_m, as follows

$$\frac{M}{\rho} = \int_{-\infty}^{\infty} u^2\, dy = \sqrt{\frac{\pi}{2}} u_m^2 b \tag{21}$$

$$\Gamma = \int_{-\infty}^{\infty} uc\, dy = \sqrt{\frac{\pi \lambda^2}{1+\lambda^2}} u_m c_m b \tag{22}$$

In most cases we will be working with the kinematic fluxes; henceforth the specific momentum flux will be denoted by M unless otherwise stated.

Equating the expressions in the above equations to the flux at the source, $M_o = {u_o}^2 d_o$ and $\Gamma_o = u_o c_o d_o$, and assuming that $b = \beta_G x$,

$$\frac{u_m}{u_o} = \left[\sqrt{\frac{2}{\pi}} \frac{1}{\beta_G}\right]^{\frac{1}{2}} \left(\frac{x}{d_o}\right)^{-\frac{1}{2}}, \tag{23}$$

$$\frac{c_m}{c_o} = \left[\frac{1+\lambda^2}{\lambda^2 \beta_G} \frac{1}{\sqrt{2\pi}}\right]^{\frac{1}{2}} \left(\frac{x}{d_o}\right)^{-\frac{1}{2}}. \tag{24}$$

The consequence of momentum conservation is the jet dilution. The volume flux (per unit length of the slot) Q at a distance x from the source is greater than the volume flux, Q_o, at the source, since

$$Q = \int_{-\infty}^{\infty} u\, dy = \sqrt{\pi} u_m b = \left[\sqrt{2\pi} \beta_G\right]^{\frac{1}{2}} \left(\frac{x}{d_o}\right)^{\frac{1}{2}} u_o d_o \tag{25}$$

The average dilution ratio,

$$S = \frac{Q}{Q_o} = \left(\sqrt{2\pi} \beta_G\right)^{\frac{1}{2}} \left(\frac{x}{d_o}\right)^{\frac{1}{2}} \tag{26}$$

The jet spread rate β_G, defined based on the Gaussian velocity profile, has been determined experimentally by Albertson et al. (1950) to be $\beta_G = 0.154$. This is somewhat larger than values of 0.13 from measurements of Bradbury (1965), 0.116 from the data of Kotsovinos (1976), and 0.119 from Miller and Comings (1957). Kotsovinos's experiments also give a λ value of 1.35. Adopting a value of $\beta_G = 0.12$ and $\lambda = 1.35$ we have the solution summarized in Table 1.

Since $u_m \sim x^{-\frac{1}{2}}, b \sim x$, then the kinetic energy flux, which can be defined as $E = \int_{-\infty}^{\infty} \frac{1}{2} u^3 dy$, varies as $\sim {u_m}^3 b \sim x^{-\frac{1}{2}}$ and $\rightarrow 0$ as $x \rightarrow \infty$. The kinetic energy in the jet is continually converted into the energy of turbulence, and the latter steadily decaying through viscous shear. Eventually, far from the source, all the initial kinetic energy is dissipated.

Since

$$d_o = \frac{{u_o}^2 d_o^2}{{u_o}^2 d_o} = \frac{{Q_o}^2}{M_o}, \tag{27}$$

the solution can alternatively be expressed in terms of the initial volume and momentum fluxes only. e.g.

$$\text{average dilution} = \overline{S} = 0.55\ \frac{{M_o}^{\frac{1}{2}} x^{\frac{1}{2}}}{Q_o} \tag{28}$$

or

$$Q = 0.55\ M_o^{\frac{1}{2}} x^{\frac{1}{2}} \tag{29}$$

At a given x, the larger the initial momentum M_o, the greater the entrained flow in the jet up to x.

Despite the approximation made to the fluxes, the expression obtained from the integral prediction of volume flux, $Q = 0.55\ M_o^{1/2} x^{1/2}$, is in excellent agreement with the velocity measurements by Kotsovinos and List (1977), which gives a coefficient of $Q/(\ M_o^{1/2} x^{1/2}) = 0.54$. This type of engineering analysis, in which the functional form of the velocity and concentration profiles are specified, has proven to give rather accurate predictions of jet properties. The functional dependence of the jet properties is independent of the exact form of the profile assumed. The value of the experimental constant (the spreading rate β) is however a function of the chosen profile.

Example 2.2 *A slot turbulent jet discharges liquid at a mean velocity 3 m/s into a liquid of the same density. Find the maximum time-averaged velocity, tracer concentration, and the average dilution at a distance of 20 m from the jet slot. Slot width = 0.1 m, and the initial concentration of the tracer = 1000 mg/L.*

Solution: The dimensionless parameter $x/d_o = 20/0.1 = 200$. Substituting x/d_o and $c_o = 1000$ mg/L into the formulae in Table 1, the maximum velocity (u_m), maximum concentration (c_m), and the average dilution ($\overline{S}$) are found to be 0.55 m/s, 161 mg/L and 7.8 respectively.

The solution given above are valid only in the zone of established flow (ZEF). The solution for the zone of flow establishment (ZFE) is actually implied in the stated formulation; the thickness of the potential core $r(x)$ varies linearly as $2r/d_o =$

Table 1 Plane jet formulae based on slot opening d_o and source velocity u_o for $x \geq 5.2 d_o$

Jet width	$b = 0.12\ x$
Centerline velocity	$u_m/u_o = 2.58\ (x/d_o)^{-\frac{1}{2}}$
Centerline concentration	$c_m/c_o = 2.27\ (x/d_o)^{-\frac{1}{2}}$
Centerline dilution	$S = 0.44\ (x/d_o)^{\frac{1}{2}}$
Average dilution ratio	$\overline{S} = 0.55\ (x/d_o)^{\frac{1}{2}}$

$1 - x/L_o$ (Prob. 2.4). In most practical applications, $x \gg d_o$, and the zone of flow establishment can be ignored altogether.

1.4 Entrainment Hypothesis

By examining the continuity equation, we can study the closure problem again in a different way. Integrating the continuity equation across the jet, we have

$$\frac{dQ}{dx} = \frac{d}{dx}\int_{-\infty}^{\infty} u dy = [-v]_{-\infty}^{\infty} = 2v_e \tag{30}$$

where $v_e = |v|_{y=\pm\infty}$ = entrainment velocity. Thus the closure problem amounts to saying something about the entrainment velocity and its relationship with the local jet characteristics. We can get some insight (or hindsight!) about this relationship by using the solution for $Q(x)$, Eq. (25), to back calculate dQ/dx

$$\frac{dQ}{dx} = \frac{(\sqrt{2\pi}\beta_G)^{\frac{1}{2}}}{2} u_o \left(\frac{x}{d_o}\right)^{-\frac{1}{2}} = 2\left(\frac{\sqrt{\pi}\beta_G}{4}\right) u_m \tag{31}$$

Comparing Eqs. (30) and (31), we have $v_e \propto u_m$: the entrainment velocity $v_e(x)$ is proportional to the local jet centerline velocity. This means that, alternatively, we could have solved the same problem by beginning from an assumption that $v_e = \alpha u_m$ (where α is an 'entrainment coefficient'), and solving the continuity, momentum, and mass conservation equations. In many buoyant jet problems, turbulent closure can be circumvented by this *entrainment hypothesis* proposed by Morton, Taylor and Turner (1956) to describe the broad mechanics of the turbulent flow rather than the detailed mechanisms. The entrainment velocity is assumed to be proportional to a local characteristic velocity (the centerline velocity) via the entrainment coefficient which can be determined experimentally. The continuity and momentum equations then become:

$$\frac{dQ}{dx} = \frac{d}{dx}(\sqrt{\pi} u_m b) = 2\alpha u_m \tag{32}$$

$$\frac{dM}{dx} = \frac{d}{dx}(\sqrt{\pi/2}\, u_m^2 b) = 0 \tag{33}$$

Noting that $M = M_o$ and $u_m = \sqrt{2}M/Q$, $b = Q^2/(\sqrt{2\pi}M)$, the continuity equation can be solved to give (neglecting source flows for high dilutions, $Q/Q_o \gg 1$) $Q^2 = 4\sqrt{2}\alpha M_o\, x$. A linear jet spread of $b = (\frac{4}{\sqrt{\pi}}\alpha_G)x$ is hence predicted by this entrainment assumption. The experimental constants in the spreading and entrainment hypotheses are hence related by

$$\beta_G = \frac{4}{\sqrt{\pi}} \alpha_G \qquad (2D\ jet) \tag{34}$$

where the subscript 'G' denotes the Gaussian profile.

2 Round Jet

2.1 Mean Flow Structure

The general experimental features described above for a slot jet also apply for the case of the flow from a circular orifice (diameter D) (Fig. 4). The diffusion thickness spreads linearly; static pressure is approximately constant; the same type of boundary layer approximations can be made. For this axisymmetric case, the length of the potential core is 6.2D; the mean axial velocity and concentration profiles are found to attain self-similarity beyond the potential core:

1. In the zone of flow establishment (ZFE), $x \leq 6.2D$;

$$u = u_o, c = c_o; \qquad r \leq R \tag{35}$$

$$u = u_o \exp\left[-\frac{(r-R)^2}{b^2}\right], \; c = c_o \exp\left[-\frac{(r-R)^2}{\lambda^2 b^2}\right]; \; r \geq R \tag{36}$$

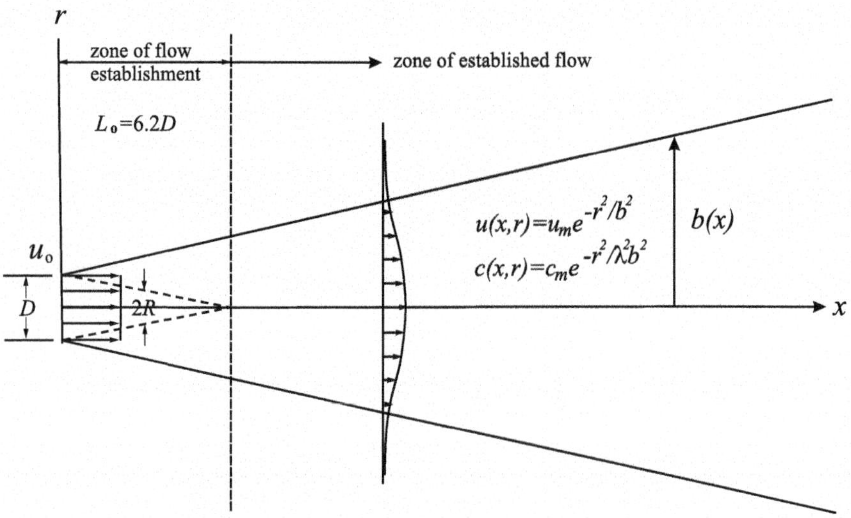

Fig. 4 A round turbulent jet

2. In the zone of established flow (ZEF), $x \geq 6.2D$, the axial velocity and tracer concentration profiles are self-similar and Gaussian:

$$u = u_m \exp\left[-\left(\frac{r}{b}\right)^2\right], c = c_m \exp\left[-\left(\frac{r}{\lambda b}\right)^2\right] \tag{37}$$

where (x, r) = streamwise and radial co-ordinates, and $u_m(x)$ and $c_m(x)$ are the centerline maximum velocity and concentration respectively.
3. The turbulent round jet spreads linearly, with

$$b = \beta x \tag{38}$$

The above jet structure, qualitatively similar to those of the slot jet, has been extensively confirmed by experiments (Albertson et al. 1950; Ricou and Spalding 1961; Papanicolaou and List 1988; Chu 1996). However, the experimental constants (β, λ, α) are in general different from those of the slot jet.

The governing equations for the steady flow of a round jet in the (x, r) co-ordinate system are the continuity equation

$$\rho\frac{\partial u}{\partial x} + \rho\frac{1}{r}\frac{\partial}{\partial r}(rv) = 0 \tag{39}$$

the x-momentum equation

$$\rho u\frac{\partial u}{\partial x} + \rho v\frac{\partial u}{\partial r} = -\frac{1}{r}\frac{\partial}{\partial r}(r\rho\overline{u'v'}) \tag{40}$$

the tracer-mass conservation equation

$$u\frac{\partial c}{\partial x} + v\frac{\partial c}{\partial r} = -\frac{1}{r}\frac{\partial}{\partial r}(r\overline{v'c'}) \tag{41}$$

The x-momentum equation can be written as:

$$\begin{aligned} \rho u\frac{\partial u}{\partial x} + \rho v\frac{\partial u}{\partial r} &= \rho u\frac{\partial u}{\partial x} + \rho v\frac{\partial u}{\partial r} + \rho u\left(\frac{\partial u}{\partial x} + \rho\frac{1}{r}\frac{\partial}{\partial r}(rv)\right) \\ &= -\frac{1}{r}\frac{\partial}{\partial r}(r\rho\overline{u'v'}) \end{aligned} \tag{42}$$

Multiplying this equation by $2\pi r$ and then integrating from $r = 0$ to $r = \infty$ we have

$$\frac{d}{dx}\int_0^\infty \rho u^2 2\pi r dr = 0 \tag{43}$$

Table 2 Round jet formulae based on nozzle diameter D and source velocity u_o, and the formulae based on jet momentum flux M_o, for $x \geq 6.2D$

Jet width	$b = 0.114x$	
Centerline velocity	$u_m/u_o = 6.2\,(x/D)^{-1}$	$u_m = 7.0M_o^{1/2}x^{-1}$
Centerline concentration	$c_m/c_o = 5.26/(x/D)^{-1}$	
Centerline dilution	$S = 0.19\,x/D$	
Average dilution ratio	$\overline{S} = 0.32\,x/D$	$Q = 0.286\,M_o^{\frac{1}{2}}\,x$

The momentum flux in a turbulent round jet is again preserved. Similarly, the mass flux of tracer across the jet is constant.

$$\frac{d}{dx}\int_0^\infty uc2\pi r dr = 0 \tag{44}$$

Substituting the assumed profiles into Eqs. 43 and 44 we get the kinematic momentum flux

$$M(x) = \int_0^\infty u^2 2\pi r dr = \frac{\pi}{2}({u_m}^2 b^2) = u_o^2 \frac{\pi D^2}{4} \tag{45}$$

and the mass flux

$$\int_0^\infty uc2\pi r dr = \frac{\pi\lambda^2}{1+\lambda^2}(u_m c_m b^2) = u_o c_o \frac{\pi D^2}{4} \tag{46}$$

Assuming a linear spreading

$$b(x) = \beta\, x, \tag{47}$$

the solution is then given by:

$$\frac{u_m}{u_o} = \frac{1}{\sqrt{2}\beta}\left(\frac{x}{D}\right)^{-1} \tag{48}$$

$$\frac{c_m}{c_o} = \frac{1+\lambda^2}{2\sqrt{2}\lambda^2\beta}\left(\frac{x}{D}\right)^{-1} \tag{49}$$

The extensive measurements by Albertson et al. (1950) and Wygnanski and Fiedler (1969) in round jets have suggested a jet spread rate of $\beta_G = \frac{db}{dx} = 0.114$. This is close to the mean value of 0.107 from different investigators as given by Fischer et al. (1979). The more recent measurements of Papanicolaou and List (1988) give $\beta_G = 0.108$ and $\lambda = 1.2$. Adopting $\beta_G = 0.114$, $\lambda = 1.2$, the jet properties is summarised in Table 2.

2.2 Additional Remarks on 3D Jet

1. In a turbulent round jet, it is seen that the velocity is inversely proportional to the distances from source, while the volume flux increases linearly with distance. We have $Q \sim x, u_m \sim x^{-1}, M =$ constant, and the kinetic energy flux $E \sim u_m^3 b^2 \sim x^{-1}$; all the energy (velocity head) is ultimately dissipated (this is the discharge from an orifice into an infinite reservoir—see also Prob. 2.10). Note that the total amount of entrained flow depends only on the momentum flux M_o and x. It is interesting to note that the local Reynolds number $R = \frac{u_m b}{\nu} \sim u_m b$ is equal to a constant in a round jet.

2. The exact momentum conservation
Similar to the 2D case, in the above we have made some bold assumptions and have dropped all the turbulence terms $\overline{u'^2}$, $\overline{v'^2}$ other than turbulent shear. A more precise analysis retaining the turbulence terms would have resulted in the following global x-momentum conservation (Pratte and Keffer 1972):

$$\frac{d}{dx}\int (u^2 + \overline{u'^2} - \frac{(\overline{v'^2} + \overline{w'^2})}{2})dr = 0 \tag{50}$$

where $\overline{w'}$ is the turbulent fluctuation in the azimuthal direction. Measurements by Wygnanski and Fiedler (1969) show that the magnitude of $v' \approx w'$, and hence Eq. 50 is often written as:

$$\frac{d}{dx}\int (u^2 + \overline{u'^2} - \overline{v'^2})dr = 0 \tag{51}$$

As the two turbulence quantities are of the same order and their distributions similar in shape, their net contribution to the momentum flux may be assumed negligible (approximately 10 % measured by Wang and Law 2002); the final statement of global momentum conservation reverts to Eq. 43.

3. The Entrainment Hypothesis
By examining the continuity equation, we can study the closure problem again in a different way. Integrating the continuity equation across the jet, it can be shown that:

$$\frac{dQ}{dx} = \frac{d}{dx}(\pi u_m b^2) = 2\pi(rv) = Q_e \tag{52}$$

where the right hand side represents the local entrainment flux into the jet. Defining the inflowing entrainment velocity at $r = b$ as the entrainment velocity and invoking the entrainment hypothesis, we have:

$$v_e = |v|_{r=b} = \alpha u_m \tag{53}$$

where the entrainment velocity is assumed to be proportional to the centerline velocity via the entrainment coefficient α. The problem can then be closed as:

$$\frac{d}{dx}(\pi u_m b^2) = 2\pi(\alpha u_m)b \tag{54}$$

Similar to the 2D case, it can be shown that (Prob. 2.2)

$$\beta_G = 2\alpha_G \qquad (3D\ jet) \tag{55}$$

for the Gaussian profile.

4. Effects of ambient turbulence

In many field or prototype scale situations, significant turbulence may be present in the ambient flow. The presence of ambient turbulence will generally affect the mixing characteristics of the jet. It is well known that turbulence consists of a cascade of eddy sizes. The large eddies of the ambient flow will lead to meandering of the jet and increase its absolute spread rate, while the small eddies with size comparable to the jet length scale can alter or even destroy the jet structure if the eddy velocity scales are also comparable (Kiya et al. 1986).

Lai et al. (2019) showed that the $x-$momentum conservation Eq. 51 remains valid for an isotropic and homogeneous ambient turbulence, but the net contribution of the turbulence quantities is no longer negligible in the equation as in stagnant ambient. As the jet mean flow decays downstream, the turbulence quantities become increasingly dominant, eventually causing the jet to break up. An ambient turbulence length scale l_c has been developed by dimensional analysis to compare the relative importance of the jet momentum and ambient turbulence:

$$l_c = \left(\frac{M_o^3}{\epsilon_a^2}\right)^{1/8} \tag{56}$$

where ϵ_a is the turbulent kinetic energy dissipation rate of the ambient turbulence. When $x \gg l_c$, the jet mixing is only minimally affected; however when $x \ll l_c$, the jet mixing is noticeably affected and the jet can even be broken up (Fig. 5). Lai et al. (2019) observed that the jet is broken up when $x \approx 1.5 l_c$. Before the jet breaks up, the jet mixing characteristics can be predicted by a second-order model that solves Eq. 51 with the turbulence terms accounted for.

For more complicated cases with ambient turbulence present in a coflow or cross-flow, its lumped effects on the jet can be reasonably accounted for using a modified spreading hypothesis (Chap. 6). The importance of ambient turbulence in field conditions can be illustrated with a case study (Chap. 8).

Example 2.3 *In some cases the turbulent free jet solution can be used to estimate the initial dilution for wastewater discharges with negligible density differences,* $Fr \to \infty$ *(e.g. sewage disposal in fresh water). Consider a submerged vertical round port sewage discharge (diameter* 0.15 *m) at the bottom of a river of depth* 14 *m. The*

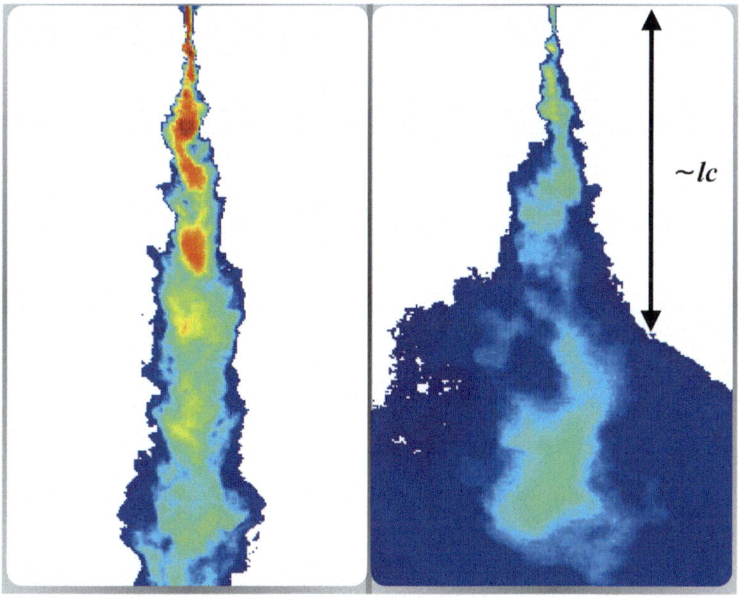

Fig. 5 False colour snapshot of a particle plume discharging vertically downwards into a stagnant ambient (left) and an ambient with turbulence (right)

density difference is practically zero and the initial velocity and coliform bacteria concentration are 1 *m/s and* 10^8*counts/mL respectively. Estimate the centerline velocity, average dilution, and bacteria concentration near the surface.*

Solution: The Reynolds number of this jet discharge is $Re = \frac{1 \times 0.15}{10^{-6}} = 1.5 \times 10^6$. Using the formulae in Table 2 the average dilution, the centerline velocity, and the centerline bacteria concentration are $\overline{S} = 0.32(14/0.15) = \underline{30}$; $u_m = 1.0(6.2)(14/0.15)^{-1} = \underline{0.066}$ m/s; and $c_m = 10^8 \times 5.26(14/0.15)^{-1} = \underline{5.64 \times 10^6}$ counts/mL.

For these typical depth and port dimensions it is seen that rather modest dilutions are achieved; note that the action of jet diffusion results in a rather rapid velocity decay to 1/15 of the initial value at discharge point. The region of practical interest is well into the zone of established flow (ZEF), with $x/D \sim 90$.

Example 2.4 *Jet in ambient turbulence*
In a coastal city, wastewater is discharged through a submerged vertical round port with diameter $D = 0.25$ *m into a water depth of* $H = 15$ *m. Given that the jet initial velocity* $u_o = 2$ *m/s, and turbulent kinetic energy dissipation rate of the coastal water* $\epsilon_a = 10^{-6}$ m^2/s^3. *Discuss the significance of the ambient turbulence.*

Solution: The jet initial specific momentum flux is computed as:

$$M_o = Q_o u_o = \frac{\pi D^2}{4} u_o^2 = 0.196\ \mathrm{m^4/s^2}$$

The ambient turbulence length scale can then be computed:

$$l_c = \left(\frac{M_o^3}{\epsilon_a^2}\right)^{1/8} = \left[\frac{0.196^3}{(10^{-6})^2}\right]^{1/8} = 17.2\ \mathrm{m}$$

Since $H/l_c = 15/17.2 = 0.87 < 1$, it is expected that the ambient turbulence will not significantly affect the jet structure, but might slightly affect the jet mixing (increase in spreading and dilution) comparing to the case of completely stagnant ambient.

Example 2.5 *Bivalve jets and nature-based solution*
Benthic bivalves, such as clams and mussels, can regulate ambient phytoplankton by their siphon system. The bivalves consume the phytoplankton intake, and discharge the filtered water as a jet with much lower phytoplankton concentration (Fig. 6). The bivalves' natural filtering capability has often been considered as an option for nature based solution to improve water quality in estuaries and coastal waters. Consider a single bivalve in Fig. 6, the intake ambient phytoplankton concentration (as chlorophyll-a) is given as $C_a = 10\ \mu g/L$*, and the jet initial velocity and concentration are* $U_o = 10$ *cm/s and* $C_o = 1\ \mu g/L$ *respectively. The diameter of the jet discharge is* $D = 2$ *mm. Determine the distance of the jet when its maximum concentration deficit* $\Delta C_m = C_a - C_m$ *has dropped to 5% of the initial concentration deficit* ΔC_o.

Solution: The change in concentration of a jet with distance x is given in Table 3 as $\Delta C_m = 5.94 Q_o \Delta C_o M_o^{-\frac{1}{2}} x^{-1}$. The volume and momentum flux of the jet discharge are computed as:

$$Q_o = \frac{\pi D^2}{4} U_o = \frac{\pi (0.002)^2}{4} \times 0.1 = 3.14 \times 10^{-7}\ \mathrm{m^3/s}$$

$$M_o = Q_o U_o = 3.14 \times 10^{-7} \times 0.1 = 3.14 \times 10^{-8}\ \mathrm{m^4/s^2}$$

The initial concentration deficit $\Delta C_o = C_a - C_o = 10 - 1 = 9\ \mu g/L$, and when $\Delta C_m = 0.05 \Delta C_o = 0.45\ \mu g/L$, the distance of the jet is:

$$x = \frac{5.94 Q_o \Delta C_o M_o^{-\frac{1}{2}}}{\Delta C_m} = \underline{0.21\ \mathrm{m}}$$

It can be seen that there is significant phytoplankton concentration decay close to the bed. Field observations show that there is generally a thin concentration boundary layer of the order $O(0.1\mathrm{m})$ which is phytoplankton depleted caused by group of benthic bivalves (Monismith et al. 1990; O'Riordan et al. 1995). In practice, the jet discharge will be under the effect of an ambient crossflow and will be discharged in a jet group (see Chaps. 7 and 11 respectively).

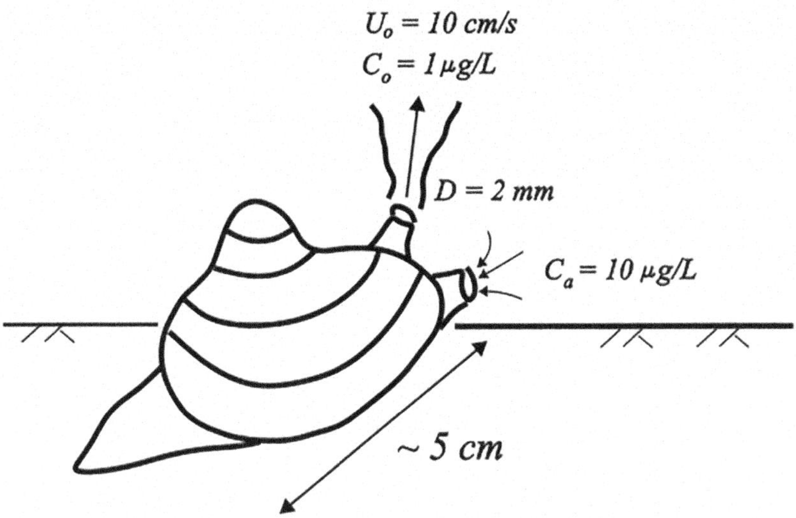

Fig. 6 A benthic bivalve intaking phytoplankton and discharging the filtered jet through a siphon system

3 Theory Versus Experiment

We have developed a semi-empirical theory for turbulent jets, which enables us to estimate the important jet characteristics to a degree of accuracy sufficient for many environmental hydraulics or engineering calculations. A summary of the jet properties is given in Table 3. These properties have been extensively validated by experiments.

3.1 Mean Properties

Figure 7 shows the measured radial profile of the normalized time-mean axial ($x-$) velocity at various downstream locations for a turbulent round jet in stagnant fluid. The measurements were made by laser doppler anemometry (LDA) for a water jet with diameter $D = 1$ cm and discharge velocity $u_o = 1.0, 0.9, 0.8, 0.7$ m/s (corresponding to Runs 1 to 4 respectively). It is seen that the streamwise velocity profile is self-similar and can be well-approximated by a Gaussian distribution. In Fig. 8, the derived centerline velocity variation for a round turbulent jet (Eq. 48) is plotted against the measured values; the classical data of a round air jet by Albertson et al. (1950) is also shown. The experimental results show clearly the existence of a potential core for about six diameters from the source, and the predicted variation in the ZEF is well-confirmed; the same result has been obtained by many different

Table 3 Summary of jet properties (Based on Gaussian profiles)

Parameter	Round jet	Plane jet
Jet volume flux $Q = \int u dA$	Dimensions L^3T^{-1} $Q = \pi u_m b^2$	L^2T^{-1} $Q = \sqrt{\pi} u_m b$
Specific momentum flux $M = \int u^2 dA = M_o$	L^4T^{-2} $M = \frac{\pi}{2} u_m^2 b^2$	L^3T^{-2} $M = \sqrt{\frac{\pi}{2}} u_m^2 b$
Maximum time-averaged velocity u_m	$u_m = 7.0 M_o^{\frac{1}{2}} x^{-1}$	$u_m = 2.58 M_o^{\frac{1}{2}} x^{-\frac{1}{2}}$
Jet width b	$b = 0.114x$	$b = 0.12x$
Maximum time-averaged concentration excess c_m	$c_m = 5.94 Q_o C_o M_o^{-\frac{1}{2}} x^{-1}$	$c_m = 2.27 Q_o C_o M_o^{-\frac{1}{2}} x^{-\frac{1}{2}}$
Average dilution $\overline{S} = \frac{Q}{Q_o}$	$\overline{S} = 0.29 M_o^{\frac{1}{2}} x Q_o^{-1}$	$\overline{S} = 0.55 M_o^{\frac{1}{2}} x^{\frac{1}{2}} Q_o^{-1}$
Jet spreading angle β	$\beta_G = 0.114\ (\beta = 2\alpha)$	$\beta_G = 0.12\ (\beta = \frac{4}{\sqrt{\pi}}\alpha)$
Entrainment coefficient α	$\alpha_G = 0.057$	$\alpha_G = 0.053$
Ratio of concentration to velocity width, λ	$\lambda = 1.2$	$\lambda = 1.35$

investigators (e.g. see Fischer et al. 1979). Similar statements can be made for the two-dimensional slot jet.

Figure 9 shows some recent measurements of the tracer concentration (scalar) field in the jet cross-section using the Laser-induced Fluorescence (LIF) technique (Chu 1996). The contours of the time-averaged concentration in the jet cross-section and the radial profile are shown. The measured concentration c normalized by the center-line maximum c_m is plotted against the dimensionless radial distance r/b_{gc}, where $b_{gc} = \lambda b_g$ is the concentration half-width defined by the radial location at which $C = e^{-1} C_m$. It can be seen the radial concentration variation is self-similar; the data at different sections collapse nicely onto one curve and can be well-approximated by the Gaussian distribution. The radial symmetry and self-similar Gaussian concentration distributions are demonstrated. The predicted linear spread of concentration width ($b_{gc} = \lambda b_g$) and linear increase of centerline dilution with distance downstream are also well-supported by data (Fig. 10). The width of the Gaussian concentration profile, b_{gc}, is generally wider than that of the velocity profile, b_g. Measured jet spread rate by various investigators are quite close: 0.114 (Albertson et al. 1950), and 0.104 by Papanicolaou and List (1988) and Chu (1996) whose data also revealed that $db_{gc}/dx = 0.125$, and $\lambda = b_{gc}/b_g = 1.2$. The predicted linear increase of volume flux (or average dilution), $Q = 0.286 M_o^{1/2} x$ (Table 2), is also in excellent agreement with the formula $Q = 0.282 M_o^{1/2} x$ obtained by Ricou and Spalding (1961) from direct volume flux measurement in a round jet and the more recent value of 0.284 given by Papanicolaou and List (1988).

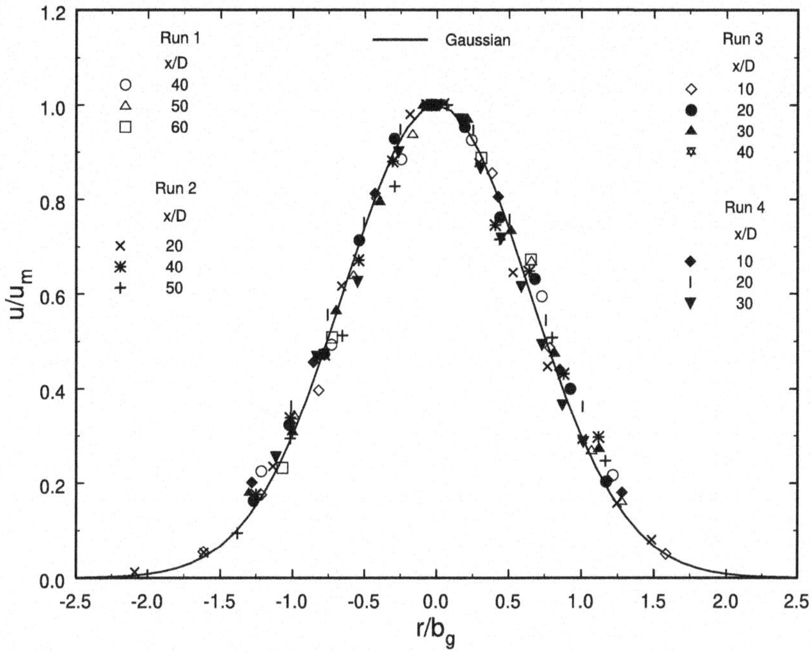

Fig. 7 Measured radial profile of time-averaged velocity in a round jet

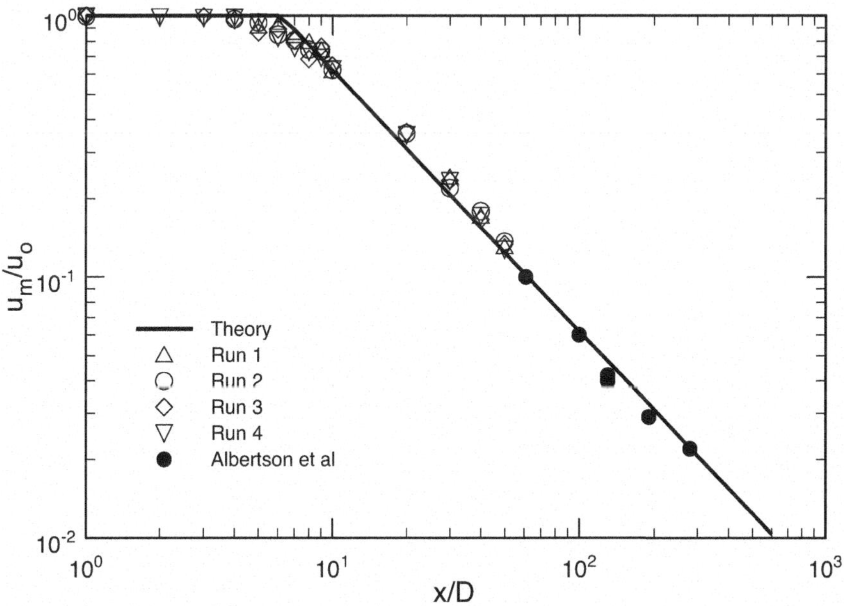

Fig. 8 Centerline velocity variation of a turbulent round jet. In the zone of established flow, $x \geq 6.2D$, the experimental data follow closely the integral model prediction (solid line)

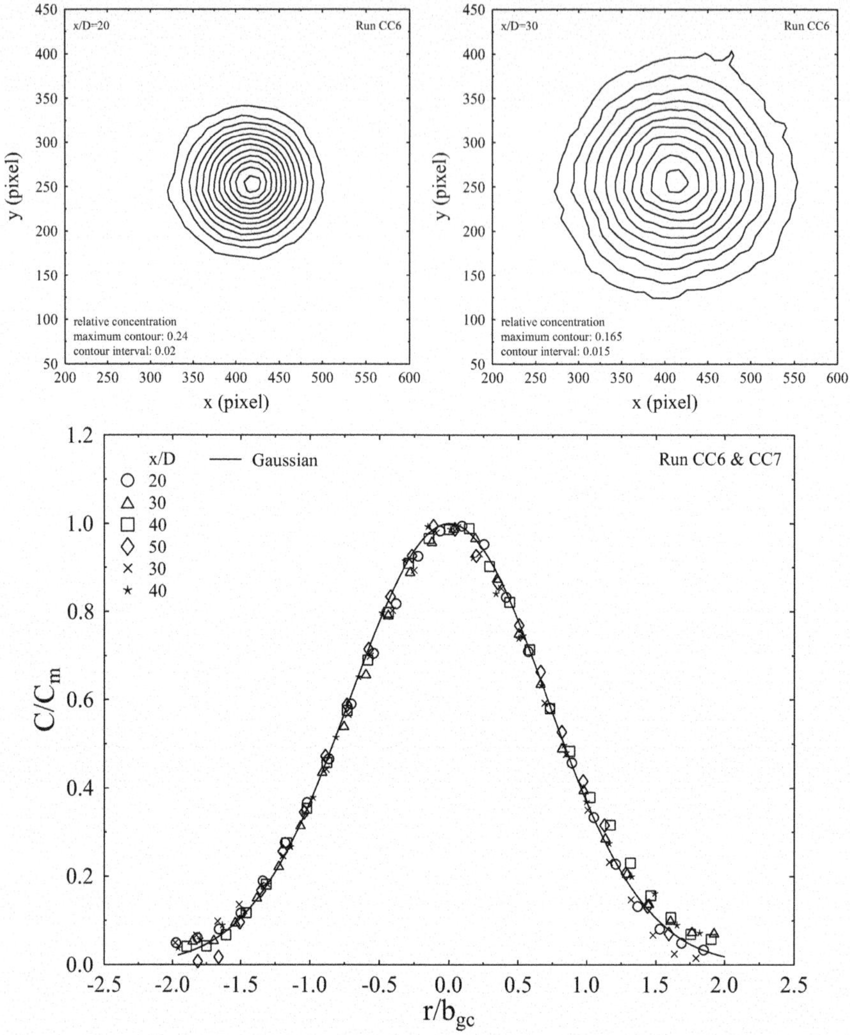

Fig. 9 Measured cross-section tracer concentration field in a round jet (Chu 1996)

3.2 *Turbulence Properties*

The foregoing results can be used to predict time-averaged quantities only. The instantaneous velocity or concentration at any point, however, can differ substantially from the time-mean value. As the eddies move around in a random fashion, a given position may be occupied by either turbulent jet flow or non-turbulent irrotaional fluid. The intermittency factor, γ, is the probability (percentage of time) when turbulent motion is detected at a fixed position. Instrumentation such as the hot-film

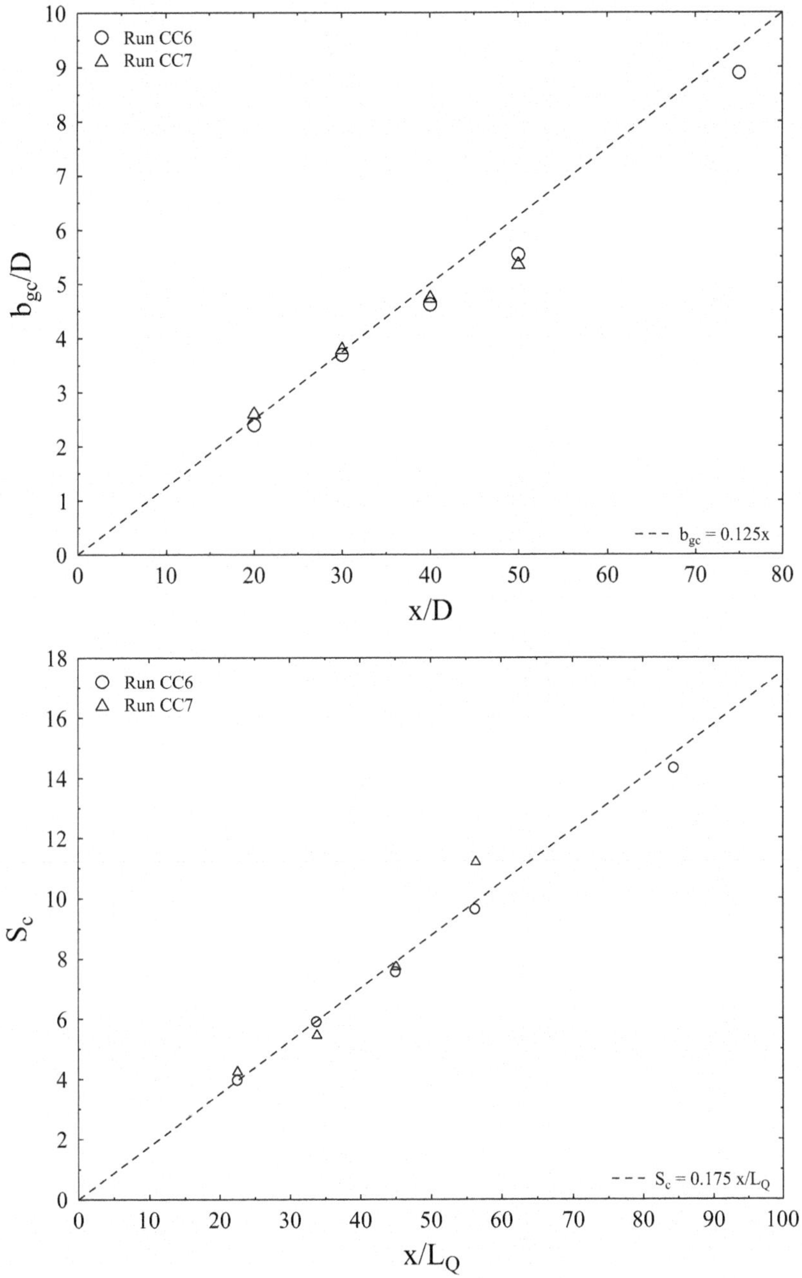

Fig. 10 Concentration half-width and centerline dilution in a round jet (Chu 1996); $L_Q = (\pi/4)^{1/2} D$

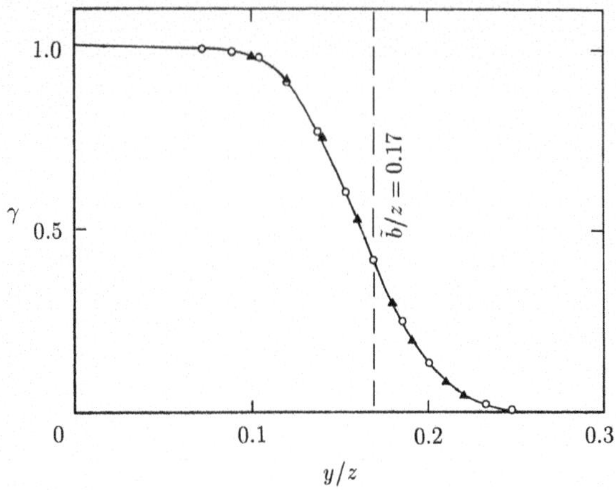

Fig. 11 The intermittency-factor profile of a round jet. Data are reproduced from Corrsin and Kistler (1954) and Wygnanski and Fiedler (1969)

anemometer can be trained to distinguish turbulent motion from the motion of its irrotational surrounding. Vorticity is measured by the anemometer and the part of the fluid motion with a vorticity level exceeding certain threshold is considered turbulent. The part of the motion with a vorticity below the threshold is irrotational. Figure 11 show the γ-profile for the velocity of a turbulent round jet obtained by Corrsin and Kistler (1954) and Wygnanski and Fiedler (1969). Note that if $\gamma = 0.5$ is taken to be the edge of the jet, then the jet boundary is located when $r/x \approx 0.162$, or when $r \approx \sqrt{2}b_g$.

The large turbulent fluctuations are illustrated in Fig. 12, which shows measurements of the instantaneous maximum and minimum concentration at different sections of the ZEF in a round turbulent jet (Chu 1996). The ratio of the instantaneous to the mean centerline concentration is plotted against $\frac{r}{b_{gc}}$. It is seen that the peak concentration can exceed the mean by as much as 50%. Figure 13 shows the radial profile of concentration intermittency (γ = fraction of time a given position is occupied by the jet fluid). It can be seen there is a coherent central core of turbulent fluid ($\gamma = 1$). The jet edge can be defined by the radial location at which $\gamma = 0.5$; it is located at around $r \approx 1.2b_{gc} = \sqrt{2}b_g$. In general, although self-similarity of the mean flow is observed beyond the potential core, the turbulence properties (turbulent stresses, intermittency) reach a self-similar state after a longer period of jet development. For example, for a slot jet, the flow only becomes self-preserving (self-similarity of both the mean flow and turbulence quantities) beyond $x/B \approx 30$ (Bradbury 1965). For a round jet, self-preservation is reached only after $x/D \approx 50$ (Wygnanski and Fiedler 1969, Papanicolaou and List 1989, Chu 1996).

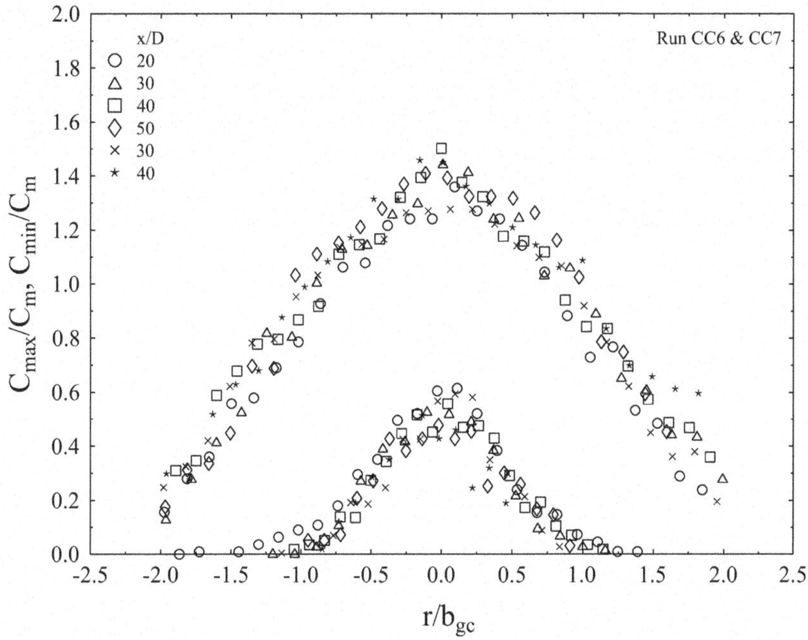

Fig. 12 Radial profile of instantaneous maximum and minimum concentration in a round jet (Chu 1996)

4 The Top-Hat Profile

It is now apparent that the key physical variable controlling jet mixing is the **jet momentum flux**, and the governing equations can be written solely in terms of the integral fluxes. In fact, based on this physical insight (or hindsight), all the characteristic properties could have been deduced by dimensional reasoning alone to within a constant (Prob. 2.3). It is sometimes useful to represent the mass (volume) and momentum fluxes by a *'top-hat'* profile. The momentum and mass flux integrals can be evaluated by assuming a simple velocity profile:

$$u = \begin{cases} U \text{ if} r \leq B \\ 0 \text{ otherwise} \end{cases} \tag{57}$$

where U, and B are the velocity and half-width of an equivalent jet with a sharp boundary and uniform velocity, U, carrying the same mass flow and momentum flux as the actual jet. By equivalence of mass and momentum fluxes, the following relations between the two profiles can be obtained (Fig. 14):

$$\pi U B^2 = \pi u_m b_g^2 \tag{58}$$

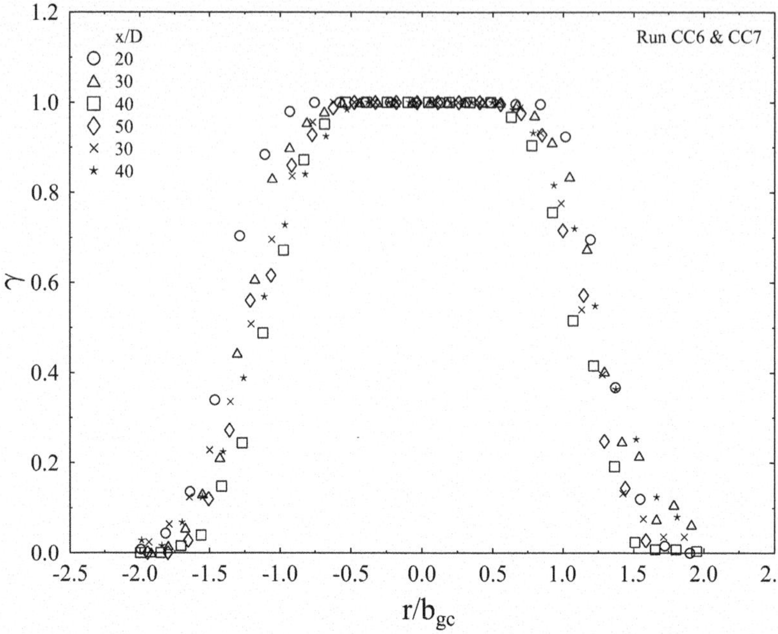

Fig. 13 Radial profile of concentration intermittency in a round jet

$$\pi U^2 B^2 = \frac{\pi}{2} u_m^2 \, b_g^2 \tag{59}$$

Thus, we have,

$$U = \frac{u_m}{2} \tag{60}$$

$$B = \sqrt{2}\, b_g \tag{61}$$

After U and B are obtained, the centerline or average concentration (dilution) can be deduced from tracer mass conservation (Chu et al. 1999). Similarly, the top-hat profile variables for a 2D slot jet is uniquely related to the Gaussian equivalents by $U = u_m/\sqrt{2}$, $B = \sqrt{\frac{\pi}{2}} b_g$.

The use of top-hat profiles not only offers analytical simplicity. As discussed above, the definition of top-hat width matches with the point of 50 percent velocity intermittency. The concentration defined by $r = B$ also seems to correspond to the visual edge of the jet. Figure 15 shows a LIF image of the instantaneous concentration field of the jet cross-section; the visual edge of the jet cross-section corresponds to the concentration contour defined by $c(r = \sqrt{2} b_g) \approx 0.25 c_m$. Physically, the use of top-hat profile can be justified using the concept of dominant eddy (see later discussion). For many practical problems, a simple Gaussian profile simply does not exist. The use of top-hat profile avoids the ambiguity in profile assumption in

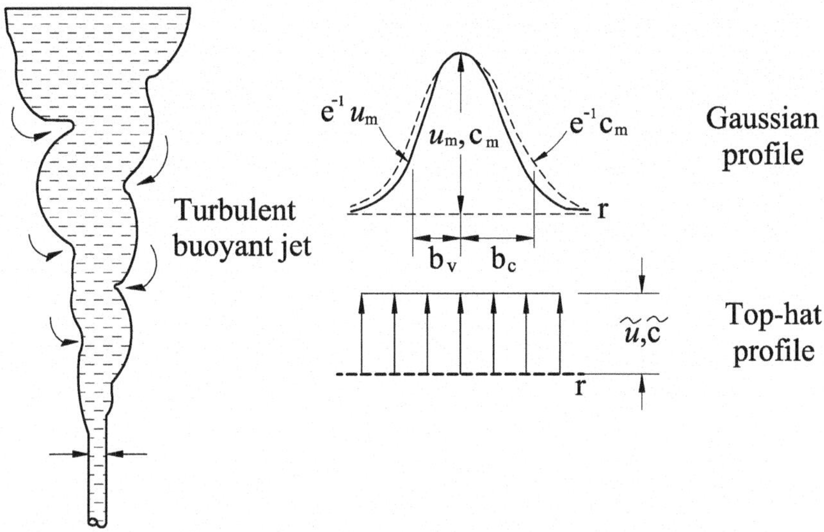

Fig. 14 Turbulent buoyant jet in stagnant fluid (turbulent mean velocity and concentration distributions described by Gaussian profiles; path averaged jet properties defined by top-hat profiles)

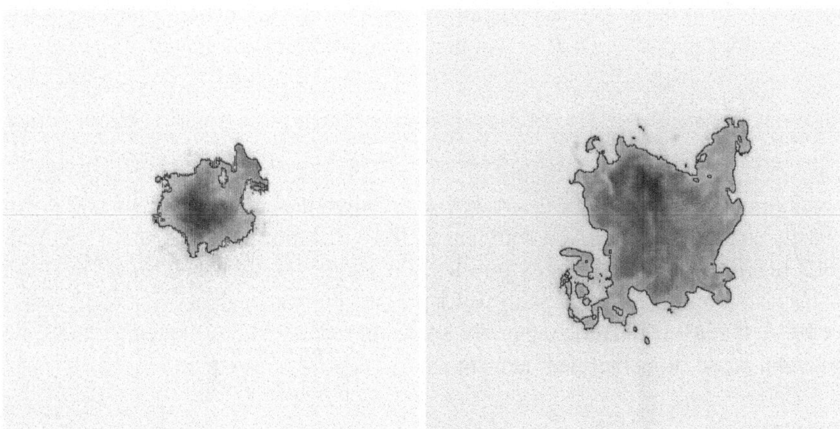

Fig. 15 LIF image of instantaneous concentration field in round jet cross-section

transition between flow regimes while allowing simplicity in mathematical derivations. It has tremendous advantages when the challenging problem of a crossflow is considered—in fact two of the much-tested models that we will describe both use top-hat profiles.

5 Prediction of Potential Core Length

We illustrate the ease with which the top-hat profile can be used with the linear jet spread assumption to yield a prediction of the potential core length. As an example, consider the round jet shown in Fig. 4. In the potential core, when turbulent diffusion has not penetrated into the centerline axis, we denote the half-width of the potential core by R, and the half-width up to the jet boundary B. In general $R(x)$ is not necessarily linear. The width of the shear/mixing layer is $(B - R)$. At the source, $x = 0$, $B = R = D/2$, while at the end of the potential core, $x = x_e$, we have $u_m = u_o$, and $U = u_o/2$ respectively (as within the core the x- velocity is u_o). Adopting a linear jet spread hypothesis and invoking momentum conservation between $x = 0$ and $x = x_e$, we then have:

$$\frac{d}{dx}(B - R) = \beta$$

$$\pi(u_o/2)^2 B^2 = \pi u_o^2 D^2/4 \tag{62}$$

where $\beta = \sqrt{2}\beta_G$ refers to the spread rate defined by the top-hat half-width. The above momentum conservation equation gives $B(x_e) = D$. Integrating from the source, $x = 0$, to the end of the potential core, $x = x_e$, and invoking the conditions $B = R = D/2$ and $B = D$; $R = 0$ at the respective limits, we have

$$\frac{x_e}{D} = \frac{1}{\beta} \tag{63}$$

Substituting, $\beta = \sqrt{2}\beta_G = \sqrt{2} \times 0.114 = 0.16$ into the equation gives $x_e/D = 6.2$ - exactly equal to the observed core length. Note that the spreading rate is independently obtained from the velocity profiles in the zone of established flow. The result is most encouraging; a similar analysis for the two-dimensional jet gives $x_e/d_o = 5.2$ (for $\beta_G = 0.154$). In a later chapter, the same approach will be adopted for the more demanding case of an inclined buoyant jet.

6 Summary

The streamwise momentum flux of a turbulent jet is conserved. In this chapter, we have presented the basic analytical framework for integral modelling of turbulent free jets in a stagnant ambient fluid. The governing equations based on the boundary layer approximations are developed. It is shown that turbulence closure can be achieved either by a jet spreading or an entrainment hypothesis. The characteristic properties of the self-similar jet are derived and the predictions (Table 3) are shown to be in excellent agreement with experimental data.

Recent advances in data-driven methods and analysis have allowed better understanding and visualization of the jet turbulence structures, which effectively complements the traditional statistical description. Lai et al. (2009) presented an early attempt in using Proper Orthogonal Decomposition (POD) to analyze a jet flow, and reconstructed the cross-sectional concentration field of a jet. Dynamic mode decomposition (DMD) developed by Schmid (2010) has recently become the dominant tool; its advantage over POD is the ability to decompose the flow into 'modes' with fixed frequency and varying amplitude—which is intuitively akin to the observation that turbulence consists of a range of eddy sizes, each with corresponding eddy velocity and time scale. Interested readers are encouraged to explore further with readily available online materials of DMD references and tools (e.g. Brunton and Kutz 2019).

Problems

2.1 (a) By assuming self-similar Gaussian velocity and concentration profiles for a round momentum jet, derive the relations between the jet volume flux Q, kinematic momentum flux M, tracer mass flux Γ, and the centerline velocity u_m, half-width b, and concentration c_m. Determine the ratio of average to centerline dilution if the ratio of concentration to velocity width, λ, is 1.16.

(b) Use the entrainment hypothesis to deduce the variation of the jet characteristic properties with downstream distance x in terms of the source fluxes and the entrainment coefficient α. Neglect the initial source flow and the zone of flow establishment.

2.2 (a) Explain what is meant by (i) the 'Entrainment Hypothesis', and (ii) the 'Spreading Hypothesis'.

(b) Assume the structure of the flow and tracer concentration in a two-dimensional (slot) turbulent jet with source kinematic momentum flux M_o (in the x-direction) and tracer mass flux Γ_o is schematized by 'top-hat' profiles:

$$\begin{aligned} u(x, y) &= U(x) \\ c(x, y) &= C(x) \qquad y \leq B(x) \\ u(x, y) &= 0 \\ c(x, y) &= 0 \qquad y > B(x) \end{aligned}$$

where $U(x)$, $B(x)$, $C(x)$ are the jet average velocity, half-width, and concentration as described by the top-hat profile. Use the entrainment assumption to deduce the characteristic jet properties $U(x)$, $B(x)$, $C(x)$ in terms of M_o and Γ_o, and the entrainment coefficient α. Neglect the initial source flow and the zone of flow establishment.

(c) Solve the same problem by using the spreading hypothesis for turbulent closure. Establish the relation between the entrainment coefficient α and the spreading coefficient β for the 2D jet for this profile. Will the same relation hold if Gaussian profiles are assumed instead of top-hat profiles?

2.3 (a) By applying a momentum balance to a suitable control volume and invoking the boundary layer approximations, show that the x-momentum equation for a two-dimensional momentum jet (oriented in the x-direction) is:

$$u\frac{\partial u}{\partial x} + v\frac{\partial u}{\partial y} = \frac{1}{\rho}\frac{\partial \tau}{\partial y}$$

where $\tau = \tau_{yx}$ is the shear stress, and the other terms have their usual meaning. Assume the static pressure to be nearly constant throughout the zone of jet diffusion.

Give a physical interpretation to τ for the case of a laminar and a turbulent jet. Show that the jet momentum flux $M(x) = \int_{-\infty}^{\infty} \rho\, u^2\, dy$ is a constant for either case.

(b) Using dimensional analysis, deduce the variation of the centerline velocity u_m and average dilution Q/Q_o in a 2D turbulent jet with the downstream distance x, to within a proportionality constant that can be determined by basic experiments.

(c) At a distance of 100 m, determine the centerline dilution in a non-buoyant turbulent round jet of initial diameter 1 m.

2.4 Adopting the velocity profiles in the zone of flow establishment (ZFE) of a 2D jet, Eq. 1 and Fig. 3, show by momentum conservation that $x_o/d_o = (\sqrt{\pi/2\beta})^{-1}$, where $x_o = L_o$ is the length of ZFE, and $\beta = db/dx$ is the jet spread rate. Show that within the ZFE, the inner boundary of the diffusion layer (edge of potential core) is given by $2r/d_o = 1 - x/x_o$. The experimental value of β ranged from 0.154 (Albertson et al.1950) to 0.116 (Kotsovinos 1976).

2.5 For a 2D plane jet, experimental measurements (Miller and Comings 1957) show that in the y-momentum equation, the dominant terms in the mixing field of the jet cross-section are the pressure and the transverse turbulent stress; hence the y-momentum equation reduces to $p + \rho\overline{v'^2} \approx p_\infty$, where p_∞= pressure outside the jet flow. Show that the x-momentum equation becomes:

$$u\frac{\partial u}{\partial x} + v\frac{\partial u}{\partial y} = -\frac{\partial \overline{u'^2}}{\partial x} - \frac{\partial \overline{u'v'}}{\partial y} - \frac{1}{\rho}\frac{\partial (p_\infty - \rho\overline{v'^2})}{\partial x}$$

Hence derive the following global x-momentum conservation:

$$\rho\frac{d}{dx}\int_{-\infty}^{\infty}(u^2 + (\overline{u'^2} - \overline{v'^2}))dy = -\int_{-\infty}^{\infty}\frac{\partial p_\infty}{\partial x}dy$$

In the case of a uniform external pressure p_∞, and when $\overline{u'^2} \sim \overline{v'^2}$, the momentum balance reduces to the simple form obtained by neglecting all pressure gradients.

$$\frac{d}{dx}\int_{-\infty}^{\infty} u^2\, dy = 0$$

2.6 For the plane jet above, show that the y-momentum equation is given (in conservative form) by:

$$\frac{\partial uv}{\partial x} + \frac{\partial v^2}{\partial y} = -\frac{\partial \overline{u'v'}}{\partial x} - \frac{\partial \overline{v'^2}}{\partial y} + \frac{1}{\rho}\frac{\partial (p_\infty - p)}{\partial y}$$

where $p - p_\infty$ = dynamic pressure. Each of the terms on the left hand side (LHS) of the equation represents convective y-acceleration of the mean flow; its order of magnitude can be estimated by $\sim V^2/b$, where V, b are y-velocity and width scales. On the other hand, the dominant turbulent stress term (second term on the RHS), can be estimated by $\frac{\partial \overline{v'^2}}{\partial y} \sim v_*^2/b$ where v_* is a characteristic scale of the turbulent velocity fluctuation. If experiments show that $v_*^2/u_m^2 \approx 0.06$ near the centre of the jet (u_m = centerline velocity), show that $\frac{\partial v^2}{\partial y} \ll \frac{\partial \overline{v'^2}}{\partial y}$, and hence the approximate y-momentum equation reduces to $(p - p_\infty + \rho\overline{v'^2}) = 0$. (Note that the transverse mean velocity can be estimated by $V \sim u_m \frac{db}{dx} \approx 0.1 u_m$).

2.7 Experimental measurements for a plane turbulent jet show that the transverse distribution of turbulent stresses can be well-approximated by the following (Miller and Comings 1957):

$$\frac{\overline{u'^2}}{u_m^2} = 0.17e^{-1.29\eta^2} - 0.10e^{-2.90\eta^2}$$

$$\frac{\overline{v'^2}}{u_m^2} = 0.057e^{-0.76\eta^2}$$

where $\eta = y/b$ and b is the Gaussian half-width based on the $1/e$ velocity point.

Determine the percentage contribution of the turbulence quantities in the momentum flux integral of Prob. 2.5.

2.8 Estimate the position of maximum turbulent shear in the cross-section of (i) a plane turbulent jet, and (ii) a round turbulent jet. For a plane jet the shear stress can be written as $\tau = -\rho\overline{u'v'} = \rho\nu_t \frac{\partial u}{\partial y}$, where ν_t is the eddy or turbulent kinematic viscosity. The eddy viscosity can be assumed to be constant in the jet cross section and related to the characteristic velocity $u_m(x)$ and length scale $b(x)$. (This is known as Prandtl's free shear layer hypothesis for turbulence closure). Compare your result with the experimental measurements of Bradbury (1965) and Wygnanski and Fiedler (1969) which show the location of maximum shear at $y = 0.72\, b_g$ and $r = 0.71\, b_g$ for a 2D and 3D jet respectively.

2.9 In the rocket jet shown in Fig. 1 in Chap. 1, estimate the jet Reynolds number assuming a thrust of three million pounds (13.4×10^6 N), density and viscosity of air = 1.2 kg/m^3 and 2×10^{-5} m^2/s respectively.

2.10 The kinetic energy of a turbulent round jet is eventually dissipated into the energy of turbulence. The kinetic energy flux is defined as $E = \int \frac{1}{2}\rho u^3 dA$. As the jet spreads with distance from the source, the kinetic energy flux decreases as x^{-1} due to the conversion of the mean flow energy to turbulent motion. Using the round jet solution, evaluate dE/dx and show that the rate of energy conversion per unit mass of the fluid in a volume equal to $\pi b^2 dx$ is:

$$P = \frac{\beta}{6}\frac{u_m^3}{b}$$

where u_m and b are the centerline velocity and Gaussian radius of the round jet respectively.

The production of the turbulent energy is initially associated with the large scale turbulent motion. Through the nonlinear process of the energy cascade, this energy is passed onto the turbulent motion of smaller and smaller eddies, and finally dissipated at the molecular level as heat.

2.11 Use the jet solution developed in Chap. 2 (Table 3) to derive an expression for the stream function $\psi(x, y)$ for the 2D jet, where $\frac{\partial \psi}{\partial y} = u(x, y)$, and $\psi(x, 0) = 0$. Sketch a few streamlines to indicate the general flow pattern.

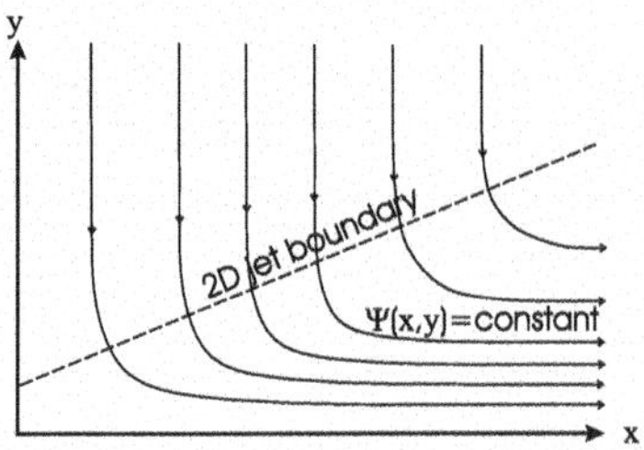

2.12 *Jets for Saving Lakes*:
The hypolimnion of a lake is often low in dissolved oxygen in the summer due to eutrophication and pollution. Sometimes it is desired to raise the oxygen level in the anoxic lower layer while maintaining the stratification (e.g. to keep the low temperature water for fisheries). One way is to inject a highly-oxygenated stream of water into the hypolimnion. The attached figure shows a flow of 1 m^3/s of sidestream withdrawn from a 8 m deep lower layer of a small lake 5 km in length. The sidestream is oxygenated with commercial oxygen by a special device to a DO concentration of 100 mg/L; it is then reintroduced into the hypolimnion and distributed through a number of jet nozzles mounted at 4 m from the bottom. As a reference this system can add around 8 mg/L of DO to a natural water flow of 12 m^3/s.

(a) If the jet-induced velocity at the distance where the jet spreads to the bottom, $x = x_1$, is not to exceed 0.1 m/s (to prevent re-suspension of polluted bottom sediment), determine the minimum nozzle diameter and the number of jet nozzles required. Energy considerations dictate that the initial jet velocity should not exceed

6 m/s. Assume the jets act independently of each other and neglect the ambient velocity.

(b) For your design estimate the ratio of entrained flow to the jet flow at $x = x_1$. Determine also the distance required for the DO concentration in the jet to drop down to 5 mg/L. Assume zero ambient DO concentration.

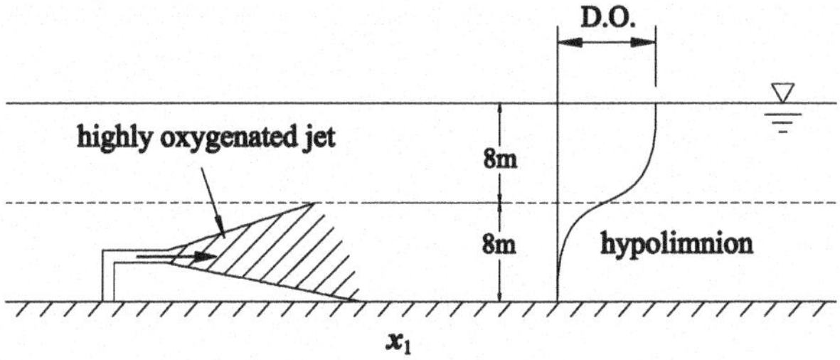

2.13 *Jet Interference*: (a) Consider two identical turbulent free jets located on the y-axis and discharging in proximity to each other; the initial jet velocity and x-momentum of each jet are u_o and M_o respectively. According to Reichardt's inductive theory of turbulence, the jet interaction can be accounted for by superimposing u^2 (where u = turbulent mean velocity), a measure of the x-momentum. In the plane containing the two jets (the $x - y$ plane), the x-momentum is given by:

$$u^2(x, y) = u_m^2(x) \sum_i e^{-2(y-y_i)^2/b^2}$$

where y_i is the location of the jet on the y-axis, and $u_m(x)$, $b(x)$ are the centerline velocity and width of the single jet as given by Table 3.

At the location when the two jets merge, when $2b(x) = s$, find the x-velocity at the point M indicated on the figure (mid-way between the two jets). How will your result change if jet velocities (instead of momentum) are additive.

This result can be generalized to predict the velocity for an array of interfering jets (located at mid-depth of hypolimnion) for Prob. 2.12.

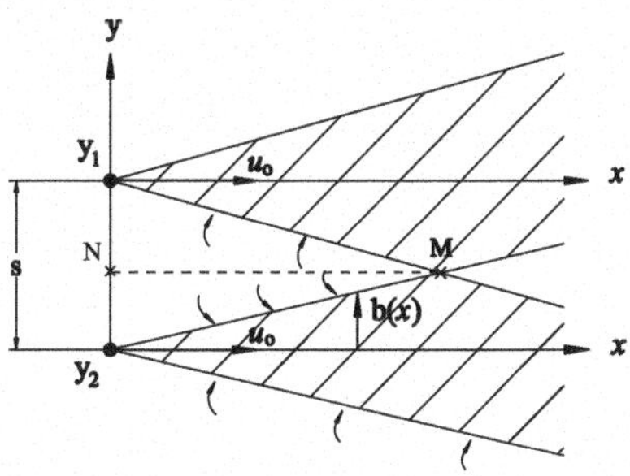

(b) Consider the governing x-momentum equation for an axisymmetric jet:

$$\frac{\partial r U^2}{\partial x} + \frac{\partial r U V}{\partial r} = 0$$

where U, V are instantaneous velocities in the x- and radial direction in a cylindrical co-ordinate system. Based on experiments, Reichardt assumes that the rate of transfer of x-momentum in the transverse (radial) is proportional to the transverse gradient in x-momentum:

$$\overline{UV} = -\Lambda(x)\frac{\partial U^2}{\partial r}$$

where $\Lambda(x)$ is a momentum transfer length that is only a function of x. By neglecting turbulent normal stress terms, we then have

$$\frac{\partial u^2}{\partial x} = \frac{\Lambda}{r}\frac{\partial}{\partial r}\left(r\frac{\partial u^2}{\partial r}\right)$$

Based on self-similarity, show that $M \sim u_m^2 b^2$, and u^2 must have a form of $u^2 = \frac{K}{b^2} f(r/b)$, where K= constant, and $b \sim x$. Hence show that if $\Lambda = \frac{b}{2}\frac{db}{dx}$, the solution to the above equation is

$$u^2 = \frac{K}{b^2} e^{-(r/b)^2}$$

If $\Lambda = kx$ where k=constant, the momentum equation can be transformed to a linear "diffusion" equation in terms of u^2 and a stretched x-coordinate $X = kx^2/2$:

$$\frac{\partial u^2}{\partial X} = \frac{1}{r}\frac{\partial}{\partial r}\left(r\frac{\partial u^2}{\partial r}\right)$$

Thus at the same X (or x), i.e. if all jets have the same x-dependence for $\Lambda(x)$, the interaction in the transverse $(y - z)$ plane can be obtained by super-imposing u^2.

2.14 *Dynamic interaction of multiple jets*:
The external flow induced by a jet is irrotational and can be solved by potential flow theory. Taylor (1958) showed that the jet induced flow can be solved by representing the jet as a line sink. Consider a single round turbulent jet discharging in the $+x$ direction in otherwise stagnant fluid. The initial jet momentum flux is M_o. It can be shown that the jet volume flux $Q \sim M_o^{1/2} x$ increases linearly with x; the jet entrainment flow (dQ/dx) is hence constant. The action of the jet can then be modelled by a line sink of strength dQ/dx (m^2/s) per unit x-length. If the jet is discretized along its axis into N elements each of length $\Delta\xi$, the jet-induced flow can be obtained by summing the flow due to a distribution of N closely spaced point sinks (Fig. 16) (Lai and Lee 2012). The velocity potential $\phi_i(x, y, z)$ due to a point sink located at ξ_i is given by (e.g. Milne-Thomson 1968):

$$\phi_i = \frac{m}{4\pi r}\Delta\xi$$

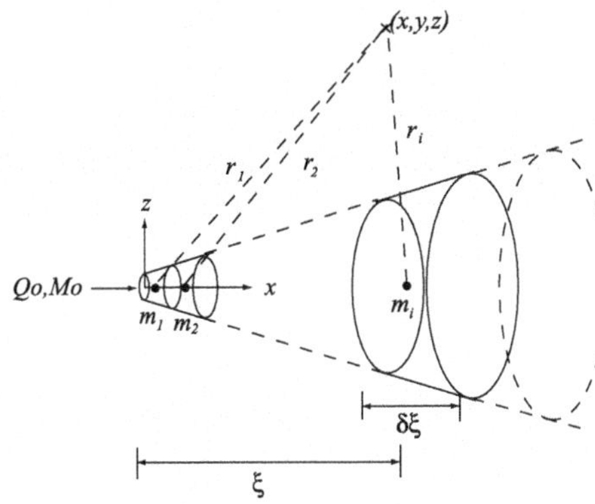

Fig. 16 A horizontal pure jet

where $m = -dQ/dx.r = \sqrt{(x-\xi)^2 + R^2}$ is the distance of the point from the sink; x is coordinate in the jet-wise direction, and $R = \sqrt{y^2 + z^2}$ is the radial direction in cylindrical coordinates.

(a) The velocity induced by a point sink is given by:

$$(u_x)_i = -\frac{\partial \phi_i}{\partial x}; \quad (u_R)_i = -\frac{\partial \phi_i}{\partial R}.$$

Show by integration along ξ that the velocity at a point (x, R) induced by a infinitely long round jet is:

$$u_x = -\frac{m}{4\pi}\frac{1}{\sqrt{x^2+R^2}}; \quad u_R = \frac{m}{4\pi R}\left\{\frac{x}{\sqrt{x^2+R^2}} + 1\right\}$$

(b) By noting that the radial velocity at $R = b_g$ is equal to the entrainment velocity, find an expression for the sink strength m in terms of the jet initial momentum flux M_o.

(c) Consider the two free jet discharges in Prob. 2.13 (Fig. 13). As the jet induced external flow is irrotational, the flow induced by the two jets can be determined by superposition of u. Using the results in (a), find an expression for the twin jet induced velocity along the line NM.

(d) Find the pressure along the line NM by using the Bernoulli's equation, and show that the twin jet induced flow results in a negative pressure zone that causes the jets to bend towards each other (see also Fig. 1 in Chap. 11).

Chapter 3
Turbulent Buoyant Plumes

Plumes are fluid motions that are produced by continuous sources of buoyancy. Convective currents set up by heated bodies such as a cigarette and a space heater are examples. The fluid in contact with the body attains a higher temperature than its surrounding fluid, rises as a result of its lower density, and in so doing draws ambient fluid radially inwards to mix with the warm fluid in the plume. Whereas the plume generated in this fashion may be laminar near the body, at some distance above it will break up into eddies by virtue of the momentum it derives from the force of buoyancy. Other examples of plumes include sewage effluent from sea outfalls, the localised high temperature fluid discharge from the earth's crust at the bottom of the deep ocean (hydrothermal vents), the injection of concentrated brine (from desalination plants) into sea water, fire plumes, hot gases from smokestacks and volcanic eruptions.

Figure 2 in Chap. 1 and Fig. 1 show two examples of the turbulent flow in the plume. The turbulent flow in Fig. 2 in Chap. 1 is produced by the heat released from a fire. The flow in Fig. 1 is produced in the laboratory by discharging warm water in the form of a buoyant jet from a tube into a tank of cold water. The fire and the buoyant jet are characterized in the near field by the source densimetric Froude number:

$$Fr = \frac{w_o}{\sqrt{(\Delta\rho_o/\rho_a)gD}} \tag{1}$$

where w_o = velocity at the source, $\Delta\rho_o = |\rho_o - \rho_a|$ = density difference between the source fluid and the ambient fluid, ρ_a = ambient density, and D = size of the

Supplementary Information The online version contains supplementary material available at https://doi.org/10.1007/978-3-032-01187-9_3.

J. H. W. Lee et al., *Turbulent Jets and Plumes*,
https://doi.org/10.1007/978-3-032-01187-9_3

source. The buoyant jet in the laboratory has a source densimetric Froude number $Fr = 5.6$. In the fire, the initial velocity w_o is zero, hence the densimetric Fr is also zero. Despite the very difference source characteristics, the far field of both plumes are determined by their source buoyancy flux. Any flow produced by a continuous source of buoyancy is expected to behave like a plume at large distances from the source regardless of the generating mechanism, although the plume in the transition between the near field and the far field may be affected by the source densimetric Froude number.

The pioneering work on plumes was due to Rouse et al. (1952), who studied the convection above line and point sources of heat in air. By assuming similarity of velocity, density deficiency, and shear stress distribution at all heights, they derived the variation of the characteristic plume parameters and confirmed these relationships experimentally. Subsequently, turbulent plumes in stagnant fluid have been extensively studied in the context of buoyant jets for wastewater disposal by e.g. Abraham (1963a, b), Cederwall (1968), Koh and Brooks (1975), Jirka et al. (1975) and Jirka and Harleman (1979). These studies related to wastewater disposal were primarily concerned with the mean concentration field, and relatively few measurements were made for the velocity field. More refined measurements including the turbulence properties were obtained using advanced experimental techniques by Kotsovinos and List (1977) for the plane buoyant jet in water and by Chu et al. (1981) and Papanicolaou and List (1988) for the round buoyant jet in water.

The results obtained from the measurements in the buoyant jets and plumes are reviewed in this chapter using the analytical framework based on both Eulerian and Lagrangian methods. The Eulerian method, due to Morton et al. (1956) and Turner (1986), is based on the entrainment hypothesis. The Lagrangian method is based on a spreading hypothesis and the concept of dominant eddy. Both the Eulerian and Lagrangian methods lead to the same asymptotic results that are consistent with experimental observation. However, as it will be demonstration in subsequent chapters, the Lagrangian method generally leads to a more consistent formulation in jet and plume problems in coflows and crossflows.

1 Buoyancy and Reduced Gravity

The plume is produced by a continuous source of buoyancy. The flux of the buoyancy in the plume is calculated by considering an element of the buoyant fluid of volume δV and density ρ in an environment of density ρ_a. The weight of the fluid in the element is $W = \rho g \delta V$, which is a downward force. The pressure force on the surface of the fluid element is equal to the weight of *ambient* fluid displaced by the element is $P = \rho_a g \delta V$, which is an upward force. The net force is the difference between the upward pressure force P and the downward force of the weight W, and that is the buoyancy force $(\rho_a - \rho) g\, \delta V$. The buoyancy force per unit volume of fluid is

$$\text{buoyancy} = (\rho_a - \rho) g$$

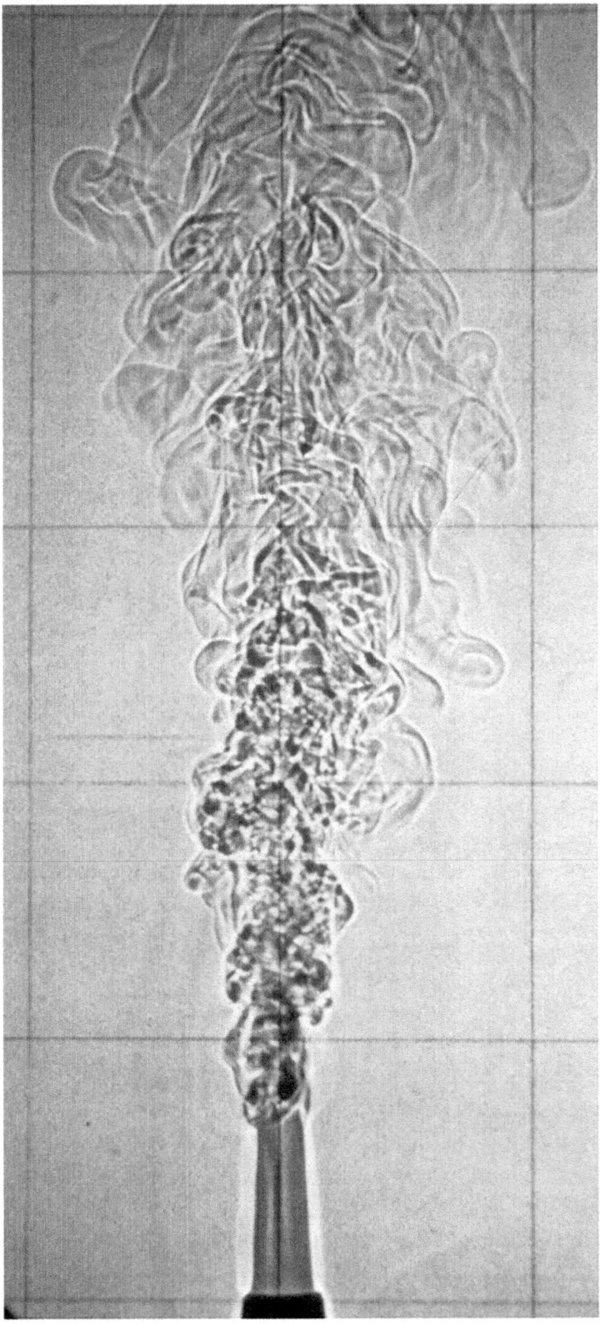

Fig. 1 Shadow graph of a vertical turbulent buoyant jet in nearly stagnant ambient fluid ($Fr = 5.6$)

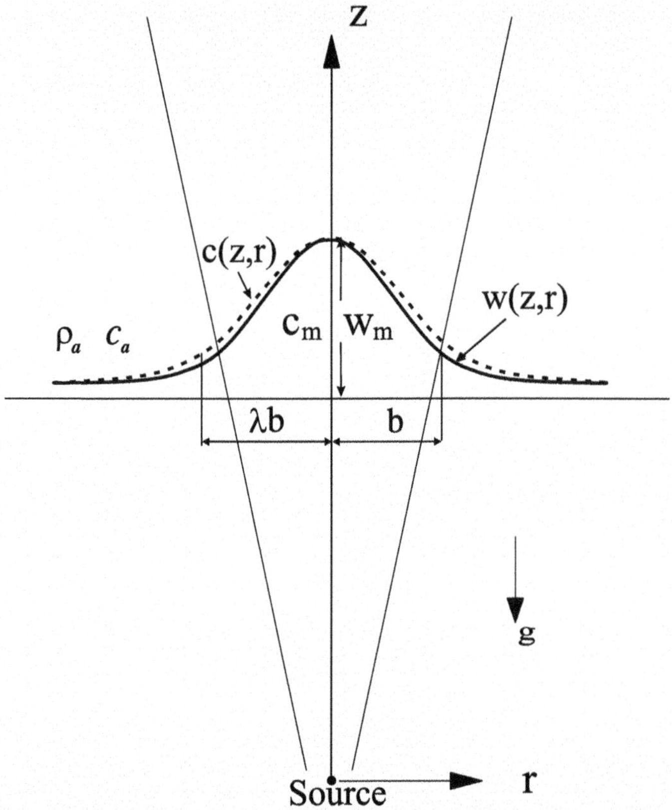

Fig. 2 Sketch of the mean velocity and concentration profiles in a round plume. The width of the concentration profile, λb, is slightly wider by a factor λ compared with the width of the velocity profile, b

The buoyancy force per unit mass of the fluid is

$$\text{reduced gravity} = \frac{(\rho_a - \rho)}{\rho} g$$

The *reduced gravity* is a term introduced to describe the distribution of the force per unit mass of the fluid while the *velocity* is to specify the distribution of momentum per unit mass of the fluid. The momentum in a fluid of non-uniform density is produced continuously by the force of buoyancy. Many physical processes can lead to changes in fluid density and hence give rise to sources of buoyancy; e.g. heating by radiation and conduction, diffusion of solute, or simply the discharging of fluids and mixtures into a receiving environment of different density. In the presence of buoyancy, momentum would be generated at a rate equal to the buoyancy force (Newton's second law of motion).

If the buoyancy in the plume is produced by discharging fluid of density ρ_o at a rate Q_o (volume flux) into an ambient fluid of density ρ_a, the buoyancy flux associated with the discharge would be

$$F_o = (\rho_a - \rho_o) g Q_o. \tag{2}$$

If the buoyancy in the plume is produced by a heat source, such as in the case of a fire, the buoyancy flux F_o is related to the heat flux H by the following relation:

$$F_o = g \frac{\gamma_T H}{c_p} \tag{3}$$

where

$$\gamma_T = -\frac{1}{\rho}\left(\frac{\partial \rho}{\partial T}\right)$$

is the volume coefficient of thermal expansion (percentage of volumetric change per degree of temperature change) and c_p is the specific heat at constant pressure (change in energy per unit mass of fluid and per degree of temperature change). The thermal expansion coefficient is determined by the equation of state, which for water is given in Appendix A. For ideal gases, the equation of state is $p = \rho R T$, where p = pressure, R = gas constant, and T = absolute temperature. For a given pressure, the fractional changes in density of an ideal gas is negatively proportional to the fractional changes in temperature, i.e., $\delta\rho/\rho = -\delta T/T$. The thermal expansion coefficient for ideal gas is $\gamma_T = 1/T$.

The flux of the buoyancy force in a plume may change due to the variation of the volume expansion coefficient, or due to the variation of the density in a stratified ambient fluid. These effects on the variations are examined in Chap. 5 and Appendix A, respectively. However, the buoyancy flux is assumed to be a constant in the consideration of the plume in this chapter.

Example 3.1 *The effluent from a desalination plant has a saline concentration of 10% and a flow rate of 10 m³/s. Find the buoyancy flux of this effluent if the salinity in the surrounding sea is 3%.*

Solution: In this case the density of the effluent is heavier than its surrounding sea water. The buoyancy force is downward. Assuming density differences are proportional to salt concentration (see Appendix A), the reduced gravity associated with the source fluid is

$$g'_o = \frac{\text{buoyancy force}}{\text{unit mass}} = \frac{\rho_a - \rho_o}{\rho_o} g = (.03 - .10) \times 9.81 = -0.687\,\text{m/s}^2.$$

The buoyancy flux is

$$F_o = \frac{\text{buoyancy force}}{\text{unit time}} = \frac{\text{buoyancy force}}{\text{unit mass}} \times \frac{\text{mass}}{\text{unit time}} = g_o' \rho_o Q_o.$$

The specific buoyancy flux (buoyancy flux per unit density) is

$$\frac{F_o}{\rho_o} \simeq \frac{F_o}{\rho_a} = g_o' Q_o = \underline{-6.87}\ \mathrm{m^4/s^3}$$

In this case, the buoyancy flux is negative because the discharge fluid is heavier than the ambient fluid.

Example 3.2 *A fire plume is set up by burning a 0.6 m diameter pool of fuel (heptane) in a large room. Find the buoyancy flux associated with the fire assuming that the heat flux is $H = 500$ kW.*

Solution: The convective heat flux integrated across the plume of cross-sectional area A is

$$H = \int_A \rho c_p (T - T_a) w dA$$

Temperature change is related to density change through the thermal expansion coefficient γ_T by

$$\delta\rho = \frac{\partial \rho}{\partial T}\delta T = -\rho\gamma_T \delta T.$$

Local linearization of the relation between temperature T and T_a gives

$$\frac{\rho - \rho_a}{\rho} \simeq -\gamma_T (T - T_a).$$

For ideal gas, $\gamma_T = 1/T$. Hence, the heat flux is related to the buoyancy flux as follows:

$$H = \int_A \rho c_p T_a \frac{(T - T_a)}{T_a} w\, dA \simeq \frac{\rho c_p T_a}{g} \int_A \frac{\rho_a - \rho}{\rho_a} g\, w\, dA \simeq \frac{c_p T_a}{g} F_o$$

where $F_o = \int_A (\rho_a - \rho) g\, w\, dA$ is the buoyancy flux. (See Eq. 3, Example 3.3 and also Prob. 3.6.) For air, the specific heat at constant pressure is $c_p = 1.006$ kJ/kg °K. Assuming standard atmosphere with $T = 293$°K and $\rho_a = 1.2\ \mathrm{kg/m^3}$, the *specific* buoyancy flux is

$$\frac{F_o}{\rho_a} = \frac{gH}{\rho_a c_p T} = \frac{9.81 \times 500}{1.2 \times 1.006 \times 293} = \underline{13.9}\ \mathrm{m^4/s^3}$$

The buoyancy flux is this specific flux times the density of the air.

2 Turbulent Round Plume

The round (axisymmetric) plume is produced by a steady and continuous source of buoyancy. The source of the buoyancy could be the result of the buoyant fluid discharging from an orifice as the case of the buoyant jet shown in Fig. 1, or could be the result of the heat produced by the fire as shown in Fig. 2 in Chap. 1. The problem of the plume is sketched in Fig. 2. As in the case of a jet, the objective of the problem is to find the velocity $w(z, r)$ and the tracer concentration $c(z, r)$ field of the plume as a function of the vertical co-ordinate (positive upwards) z and the radial co-ordinate r. Experiments have shown that the plume is a boundary-layer type of flow. The velocity and the concentration profiles in the fully established flow are similar in shape at all heights, and well-described by Gaussian profiles. The problem then reduces to the prediction of the maximum velocity, w_m, the width, b, and the maximum concentration, c_m, as a function of the lengitudinal distance z from the source.

2.1 Dimensional Considerations

A number of plume properties is derived from dimensional considerations. Neglecting the effects of the initial momentum flux and the size of the source, the width of the plume b is a function of its buoyancy flux at the source, F_o, the fluid density ρ_a, and the elevation from the source, z, as follows:

$$b = \text{fn}(F_o, \rho_a, z). \tag{4}$$

There are 4 independent variables with 3 basic dimensions and the only dimensionless variable in this functional relation is

$$\Pi = \frac{b}{z}. \tag{5}$$

which must be constant. Hence,

$$b = \beta z. \tag{6}$$

The width of the plume increase linearly with the distance from the source according this dimensional consideration. The constant β in this linear relation is the spreading rate of the plume.

The method of dimensional analysis can also be applied to find the velocity of the plume. The vertical velocity induced by the plume's buoyancy, w, is a function of its buoyancy flux at the source, F_o, the fluid density ρ_a, the elevation above the source, z, and the radial distance from the plume centerline, r, as follows:

$$w = \text{fn}(F_o, \rho_a, r, z). \tag{7}$$

There are 5 independent variables involving 3 basic dimensions in this functional relation. The dimensionless parameters are

$$\Pi_1 = \frac{w}{[F_o/(\rho_a z)]^{\frac{1}{3}}} \quad \text{and} \quad \Pi_2 = \frac{r}{z}. \tag{8}$$

These then lead to the following functional form for the velocity profile

$$\frac{w}{w_m} = \text{fn}\left(\frac{r}{z}\right), \tag{9}$$

where the maximum velocity at the centerline is proportional to the plume characteristic velocity $[F_o/(\rho_a z)]^{\frac{1}{3}}$, i.e.,

$$w_m \sim \left[\frac{F_o}{\rho_a z}\right]^{\frac{1}{3}} \tag{10}$$

The dimensionless form of the concentration profile

$$\frac{c}{c_m} = \text{fn}\left(\frac{r}{z}\right) \tag{11}$$

is obtained in a similar manner. The maximum concentration at the centerline (c_m) times the characteristic velocity and flow area, $c_m[F_o/(\rho z)]^{\frac{1}{3}} b^2$ is proportional to the mass flux at the source, Γ_o. Hence,

$$c_m \sim \frac{\Gamma_o}{[F_o/(\rho z)]^{\frac{1}{3}}}. \tag{12}$$

Figure 3 shows the time-mean velocity and concentration data obtained from the experiment by Papanicolaou and List (1988). A close approximation of these data are the Gaussian profiles

$$\frac{w}{w_m} = \exp\left[-\frac{r^2}{b^2}\right] \tag{13}$$

$$\frac{c}{c_m} = \exp\left[-\frac{r^2}{(\lambda b)^2}\right] \tag{14}$$

in which b and λb are the width of the time-mean velocity and concentration profiles, respectively. The reduced gravity, that is the buoyancy force per unit mass of the plume fluid $g' = g(\rho_a - \rho)/\rho_a$, have the same profile as the concentration of any tracer. For a heated jet, the temperature excess, $\Delta T = T - T_a$, is also expected to be Gaussian across the plume. Hence,

$$\frac{g'}{g'_m} = \exp\left[-\frac{r^2}{(\lambda b)^2}\right] \tag{15}$$

$$\frac{\Delta T}{\Delta T_m} = \exp\left[-\frac{r^2}{(\lambda b)^2}\right] \tag{16}$$

The data of Papanicolaou and List (1988) in Fig. 3 show the width of the plume increases with distance from the source following the linear relations $b_v = b = 0.105\,z$ and $b_c = \lambda b = 0.112\,z$ with a value of $\lambda = 1.06$ (see also Fig. 5). The spreading rate for the velocity profile is different from the rate for the concentration profile; $\lambda = b_c/b_v$ is the ratio of the concentration to velocity width. Some variation in the spreading rate has been observed in different experimental investigations. The average of the experiments by Rouse et al. (1953), Morton et al. (1956) and Abraham (1963a) gives a spread rate $\beta = db_v/dz = 0.102$ and $\lambda = 1.16$. The plume spread rate of around 0.11 has also been observed in related buoyant plume studies (Lee and Cheung 1991). It appears that the spread rate of the plume is approximately the same as the jet. Over the jet-plume range, the spreading rate varies from $\beta = 0.102 \sim 0.114$. For most practical purposes, a constant spreading rate of $\beta = db_v/dz \simeq 0.108$ may be adopted for both jets and plumes.

Dimensional consistency requires that the width of the round plume increases linearly with distance from the source while the velocity decreases with distance as $(F_o/\rho_a)^{\frac{1}{3}} z^{-\frac{1}{3}}$. The volume flux of a round plume is $Q = \int_0^\infty w 2\pi r\, dr$. Since $w \sim (F_o/\rho_a)^{\frac{1}{3}} z^{-\frac{1}{3}}$ and $b_v \sim z$, the volume flux increase with elevation follows a five-third power law: $Q \sim (F_o/\rho_a)^{\frac{1}{3}} z^{\frac{5}{3}}$. The tracer concentration follows the minus-five-third law: $c \sim \Gamma_o/Q \sim \Gamma_o (F_o/\rho_a)^{-\frac{1}{3}} z^{-\frac{5}{3}}$. If the plume is created by a heat source, the temperature variation will be given by $\Delta T \sim H/(\rho_a c_p)\,(F_o/\rho_a)^{-\frac{1}{3}} z^{-\frac{5}{3}}$.

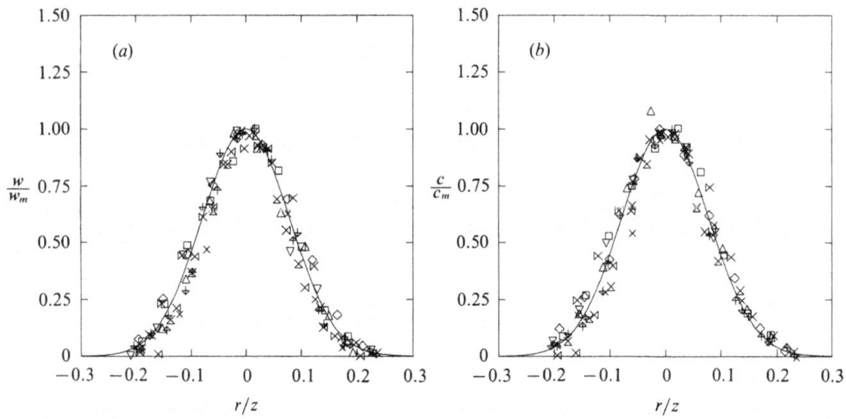

Fig. 3 Measured mean velocity and concentration profile across a round plume (from Papanicolaou and List 1988, $z/l_s = 8.6 - 62$); w_m, c_m = centerline velocity and concentration

The above dimensional analysis also shows that the key parameter in the solution is F_o/ρ_a, i.e. the specific buoyancy flux. In most applications we will be concerned with small density differences $\Delta\rho/\rho_a \ll 1$ and working with the specific buoyancy flux $F_o/\rho \simeq F_o/\rho_a$; henceforth unless otherwise stated $F_o = Q_o \frac{\Delta\rho_o}{\rho_a} g \approx Q_o \frac{\Delta\rho_o}{\rho_o} g$ denotes the specific buoyancy flux.

2.2 *Eulerian Integral Model*

Governing Equations

Consider the vertical axisymmetric plume in Fig. 2, with the following velocity and scalar concentration excess profiles in the established flow.

$$w(z,r) = w_m(z) e^{-(\frac{r}{b})^2}$$

$$c(z,r) = c_m(z) e^{-(\frac{r}{\lambda b})^2} \tag{17}$$

where w_m and c_m are the centerline maximum velocity and concentration (excess), and b is the plume radius defined by location with velocity $e^{-1} w_m$. In the axisymmetric co-ordinate system (z, r), the governing equations for turbulent incompressible flow are then:

$$\text{Continuity}: \quad \frac{\partial w}{\partial z} + \frac{1}{r}\frac{\partial}{\partial r}(rv) = 0 \tag{18}$$

$$\text{Momentum}: \quad \rho\left(w\frac{\partial w}{\partial z} + v\frac{\partial w}{\partial r}\right) = -\rho g - \frac{\partial p}{\partial z} - \rho\frac{1}{r}\frac{\partial}{\partial r}(r\overline{w'v'}) \tag{19}$$

$$\text{Mass or heat conservation}: \quad w\frac{\partial c}{\partial z} + v\frac{\partial c}{\partial r} = -\frac{1}{r}\frac{\partial}{\partial r}(r\overline{v'c'}) \tag{20}$$

where w, v are the turbulent-mean velocities in the vertical and radial (z, r) directions respectively, and w', v', c' are the velocity and concentration fluctuations. Assuming a hydrostatic pressure distribution, $\frac{\partial p}{\partial z} = -\rho_a g$, we invoke the Boussinesq approximation - that for small density differences, $\frac{\Delta\rho}{\rho} \ll 1$, density differences can be neglected in the governing equations ($\rho \approx \rho_a$), except in terms multiplied by g. This amounts to assuming *constant mass* but *variable weight*. Equation 19 then becomes

$$w\frac{\partial w}{\partial z} + v\frac{\partial w}{\partial r} = -\frac{(\rho - \rho_a)}{\rho_a} g - \frac{1}{r}\frac{\partial}{\partial r}(r\overline{w'v'}) \tag{21}$$

Integral Model Equations

By applying the integral method (as outlined in Chap. 2) to the continuity equation, Eq. 18, and integrating across the plume from $r = 0$ to $r = \infty$, we obtain

$$\frac{d}{dz}\int_0^\infty w\, 2\pi r\, dr = -2\pi r v|_0^\infty = Q_e \tag{22}$$

As $r \to \infty$, $v \to 0$, but rv remains finite. The entrainment flow Q_e can be written as $Q_e = 2\pi b v_e$, where v_e is the entrainment velocity at $r = b$. The change in volume flux is due to lateral entrainment. Similarly, the integrated form of the momentum equation, Eq. 21, can be shown to be:

$$\frac{d}{dz}\int_0^\infty w^2\, 2\pi r\, dr = \int_0^\infty \frac{\rho_a - \rho}{\rho_a}\, g\, 2\pi r\, dr \tag{23}$$

where the right hand side represents the buoyancy force per unit z-length. The buoyant force on a fluid parcel is the gravitational force resulting from non-uniform density. For the infinitesimal plume element, the change in momentum flux is due to the buoyant force associated with a density deficit distribution across the plume.

Assuming $\Delta\rho \propto \Delta c$, the conservation of tracer mass, Eq. 20, leads to the conservation of specific buoyancy flux:

$$F = \int_0^\infty w\, \frac{\Delta\rho}{\rho_a} g\, 2\pi r\, dr = \text{constant} = Q_o \frac{\Delta\rho_o}{\rho_a} g \tag{24}$$

Substituting the assumed profiles into Eqs. 22, 23, 24, and using 28, we obtain the governing equations for the integral model in terms of the three unknowns w_m, b, and $\Delta\rho_m$ (or alternatively Q, M, F):

$$\frac{dQ}{dz} = \frac{d}{dz}(\pi w_m b^2) = 2\pi b v_e \tag{25}$$

$$\frac{dM}{dz} = \frac{d}{dz}\left(\frac{\pi}{2} w_m^2 b^2\right) = \pi \frac{\Delta\rho_m}{\rho_a} g \lambda^2 b^2 \tag{26}$$

$$F = \left(\pi \frac{\lambda^2}{1+\lambda^2} w_m \frac{\Delta\rho_m}{\rho_a} g b^2\right) = F_o \tag{27}$$

Entrainment Hypothesis

A supplementary condition for the entrainment velocity v_e is required for a unique solution of the above equations (i.e. for closure of the turbulent flow equations). The classical method is to introduce the entrainment hypothesis (Morton et al. 1956). It is assumed that the eddies (resulting from a velocity gradient) are characterized by velocities proportional to a characteristic relative velocity. Thus the entrainment flow per unit z-length, Q_e, is given by:

$$Q_e = 2\pi(\alpha w_m)b \tag{28}$$

where the entrainment velocity at the nominal boundary of the plume, $r = b$, is assumed proportional to the local centerline velocity, with $v_e = \alpha w_m$, where α = entrainment coefficient, a constant for a round plume.

Defining $Q = \pi w_m b^2$, $M = \frac{\pi}{2} w_m^2 b^2$, F_o = specific buoyancy flux = $\int \frac{\Delta\rho}{\rho} g\, wdA = \frac{\pi\lambda^2}{1+\lambda^2} w_m \frac{\Delta\rho_m}{\rho_a} gb^2$, we have $w_m = \frac{2M}{Q}$, $b = \frac{Q}{\sqrt{M}} \frac{1}{\sqrt{2\pi}}$. In terms of the set of variables (Q, M, F), the governing equations then reduce to

$$\frac{dQ}{dz} = 2\pi\alpha\sqrt{\frac{2M}{\pi}} \tag{29}$$

$$\frac{dM}{dz} = \frac{F_o(1+\lambda^2)}{2}\frac{Q}{M} \tag{30}$$

Equations 29 and 30 give

$$\frac{d^2}{dz^2}(M^2) = F_o(1+\lambda^2)2\sqrt{2\pi}\alpha M^{\frac{1}{2}} \tag{31}$$

Asymptotic Solution

The asymptotic solution for a pure plume with zero momentum flux at the source $(M_o = M(0) \approx 0)$ is

$$M(z) = az^n \tag{32}$$

Substituting this power law form of solution into Eq. 31, and equating coefficients of equal power of z, gives:

$$M(z) = (2\pi)^{\frac{1}{3}}\left[\frac{9\alpha(1+\lambda^2)}{20}\right]^{\frac{2}{3}} F_o^{\frac{2}{3}} z^{\frac{4}{3}} \tag{33}$$

With this solution for M, the solution for Q can be obtained from the continuity equation:

$$Q = \frac{6\alpha}{5}(2\pi)^{\frac{2}{3}}\left[\frac{9\alpha(1+\lambda^2)}{20}\right]^{\frac{1}{3}} F_o^{\frac{1}{3}} z^{\frac{5}{3}} = \frac{6\alpha}{5}(2\pi M)^{\frac{1}{2}} z \tag{34}$$

The centerline maximum velocity, reduced gravity and the width, w_m, g'_m and b, are uniquely determined by the volume flux, momentum flux and buoyany flux of the plume, Q, M and F_o, respectively. Hence,

$$b = \frac{6\alpha}{5} z \tag{35}$$

$$w_m = \frac{5}{3\alpha}\left[\frac{1}{2\pi}\frac{9\alpha(1+\lambda^2)}{20}\right]^{\frac{1}{3}} F_o^{\frac{1}{3}} z^{-\frac{1}{3}} \tag{36}$$

$$g'_m = \frac{5}{6\alpha\lambda^2}\left[\frac{20}{9\alpha}\right]^{\frac{1}{3}}\left[\frac{1+\lambda^2}{2\pi}\right]^{\frac{2}{3}} F_o^{\frac{2}{3}} z^{-\frac{5}{3}} \tag{37}$$

The solution is solely determined by the driving force F_o - the specific buoyancy flux, with $w_m \sim z^{-\frac{1}{3}}$, $b \sim z$, $\Delta\rho_m \sim z^{-\frac{5}{3}}$, and $M \sim z^{\frac{4}{3}}$, $Q \sim z^{\frac{5}{3}}$. The above solution for the velocity and concentration in a round plume may be compared with those for the round jet. In a round jet, $w_m \sim M_o^{\frac{1}{2}} z^{-1}$, $Q \sim M_o^{\frac{1}{2}} z$ and $c \sim \Gamma\, M_o^{-\frac{1}{2}} z^{-1}$. Jets are efficient devices for mixing on a small scale but the energy cost is expensive when a large volume of fluid is involved. In the atmosphere and the oceans, mixing is dependent on buoyancy. For a buoyant plume, the kinetic energy flux continually increases, with $E \sim w_m^3 b^2 \sim z$, as the potential (buoyant) energy is continually converted into kinetic energy with the ascent of the plume. If a local Reynolds number $Re(z)$ is defined as $Re = \dfrac{w_m b}{\nu}$, we see that $Re \sim z^{\frac{2}{3}}$. Thus the plume is guaranteed to be turbulent at some distance above the source, even if it is initially laminar.

Unlike the case of a jet, the momentum flux in a plume continually increases, with a much slower velocity decay. The fact that $w_m \to \infty$, $\Delta\rho_m \to \infty$ as $z \to 0$ reflects the mathematical singularity of the point plume solution. For finite sources with $Q_o, M_o \neq 0$, the flow will be described by these equations corresponding to a virtual source below the actual source. For $Q(z) \gg Q_o$, $M(z) \gg M_o$, z will be considerably larger than the difference in location between the actual and virtual source.

Using the entrainment hypothesis , a linear spread of the plume is predicted. This implies that we could have solved the plume problem by a constant spreading rate hypothesis, $db/dz = \beta$, as in Chap. 2. The plume spread rate must be related to the entrainment coefficient as:

$$\beta_p = \frac{6}{5}\alpha_p \tag{38}$$

where the subscript 'p' denotes plume values.

For a plume width defined by the Gaussian profile, the experimentally observed β value is in the range of 0.102 to 0.112. Morton et al. (1956) recommended a value $\alpha = 0.093$ based on their experimental observation of the plume in a density stratified ambient. From the experiments by Rouse et al. (1952) and Abraham (1963a), the experimental constants α, λ for a round plume can be determined to be $\alpha = 0.085$ (corresponding to a plume spread rate of $\beta = 0.102$), and $\lambda = 1.16$.

On the other hand, the experiments of Papanicolaou and List (1988) which covered a larger distance downstream of the source, give $\beta = 0.105$ (implying $\alpha = 0.088$), and $\lambda = 1.06$; a value of $\lambda = 1.19$ is however suggested for use over entire jet-plume range. The plume spread rate of around 0.11 has also been observed in related buoyant plume studies (Lee and Cheung 1991). For the pure plume, if we adopt $\alpha = 0.088$ and $\lambda = 1.19$, the plume solution as summarized in Table 1 can be obtained. The round plume solution is very useful in many environmental problems.

Given the values $\alpha \approx 0.088$ and $\lambda = 1.19$, the above asymptotic formulae for the plume become

$$w_m = 4.71\, F_o^{\frac{1}{3}} z^{-\frac{1}{3}},\ g'_m = 10.46\, F_o^{\frac{2}{3}} z^{-\frac{5}{3}},\ Q = 0.163\, F_o^{\frac{1}{3}} z^{\frac{5}{3}} \tag{39}$$

As the reduced gravity, the maximum concentration of the tracer at the centerline of the plume follows the same minus five-third power law

$$c_m = 10.46\, \Gamma_o\, F_o^{-\frac{1}{3}} z^{-\frac{5}{3}}. \tag{40}$$

where Γ_o = flux of tracer mass. If the plume is created by a heat source, the maximum temperature excess (above its ambient) will be given by

$$(\Delta T)_m = (T_m - T_a) = 10.46 \frac{H}{\rho_a c_p} F_o^{-\frac{1}{3}} z^{-\frac{5}{3}}. \tag{41}$$

where H = flux of heat. The symbol F_o in these equations denotes the *specific* buoyancy flux.

Example 3.3 *Volume of smoke produced by a fire*

A fire plume is set up by burning a 0.6 m diameter pool of fuel (heptane) in a 10 m × 10 m × 10 m large room. If the source heat release rate is $H = 500$ kW, estimate the maximum plume velocity, temperature excess, and mass flux entrained by the plume at a height of $z = 10$ m above the heat source. Estimate the time for the smoke of the fire to fill the room. Density of air at standard atmospheric condition is $\rho_a = 1.2\ kg/m^3$.

Solution: Experiments have shown that the smoky hot gas flow above the heat source of a fire can be approximated by a round plume (Heskestad 1984). For air, the specific heat at constant pressure is $c_p = 1.006$ kJ/kg°K. Assuming standard atmosphere with $T = 293$°K and $\rho_a = 1.2\ \mathrm{kg/m^3}$, the specific buoyancy flux can be evaluated using Eq. 3 as follows:

$$F_o = \frac{gH}{\rho_a c_p T} = \frac{9.81 \times 500}{1.2 \times 1.006 \times 293} = 13.87\,\mathrm{m^4/s^3}$$

Using the plume solution in Eqs. 39, 40, and 41, and the solution summarized in Table 1, the maximum velocity at $z = 10$ m is $w_m = 4.71 F_o^{1/3} z^{-1/3} = 4.71 \times$

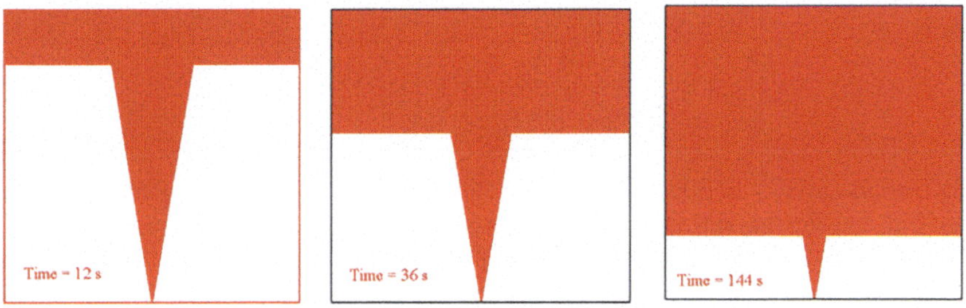

Fig. 4 Smoke produced by a 500 kW fire in a 10m×10m×10m room at time $t = 12, 36$, and 144 s. The plume boundary in this figure is defined by $\tilde{b} = 0.17z$ and the interface by Eq. 42

$13.9^{1/3}/10^{1/3} = \underline{5.3}$ m/s. The maximum excess temperature $(T_m - T_a)$ at the same level is given by

$$(\Delta T)_m = 10.46 \frac{H}{\rho_a c_p} F_o^{-1/3} z^{-5/3} = \underline{38.8}\,^\circ\mathrm{C}$$

The volume flux induced by the fire plume at this height is

$$Q = 0.163\, F_o^{1/3} z^{5/3} = 0.163 \times 13.87^{1/3} \times 10^{5/3} = \underline{18.2}\ \mathrm{m^3/s}.$$

The mass flow rate is $\rho_a Q = 1.2 \times 18.2 = \underline{21.8}$ kg/s. In well-developed fires, the convective heat flux usually accounts for a major fraction of the total heat release rate. Plume theory is useful in the design of extraction fans or sprinkler systems for shopping malls and atria. The hot and smoky gases generated by the fire rise to the top of the room and form a buoyant smoke layer below the ceiling. In the above example, the smoke of the fire is produced at a rate equal to the plume's volume flux $Q = 18.2$ m^3/s. If this volume flux is maintained, the time for the smoke to fill the room of 10^3 m^3 would be $t = 10^3/18.2 = \underline{55}$ s.

As the interface of the smoke layer moves down from the ceiling, the volume flux $Q(z)$ decreases with elevation of the interface z. (See Fig. 4 and also Fig. 15 in Chap. 5 for the downward movement of the interface.) The downward velocity of the interface, $-dz/dt$, is equal to $Q(z)/A$, i.e.,

$$-\frac{dz}{dt} = \frac{Q(z)}{A}$$

where A is area of the interface or the floor area of the room. Integrating the equation over the period of time from time $t = 0$ to t as the elevation of the interface moves down from the height of the ceiling where $z = z_c$, gives

$$\int_o^t dt = -\int_{z_c}^{z} \frac{A\,dz}{Q(z)} = -\int_{z_c}^{z} \frac{A\,dz}{0.163\, F_o^{1/3} z^{5/3}}$$

that is

$$t = \left[\frac{3A}{2 \times 0.163 F_o^{\frac{1}{3}} z_c^{\frac{2}{3}}}\right] \left[(\frac{z_c}{z})^{\frac{2}{3}} - 1\right] \tag{42}$$

The formula gives $t = 300\,\mathrm{s}$ (5 min) for a floor area of $A = 100\,\mathrm{m}^2$, interface height $z = 1\,\mathrm{m}$, ceiling height $z_c = 10\,\mathrm{m}$, and specific buoyancy flux $F_o = 13.9\,\mathrm{m}^4/\mathrm{s}^3$. According to this calculation, the smoke of the fire will fill 90% of the 10 m $\times$ 10 m $\times$ 10 m large room in a very short time, of the order of minutes! Note that the effect of ceiling boundary and extraction fans on the plume dynamics have been tacitly ignored in this worst case scenario.

Example 3.4 *Climate change and wildfire*

Climate change is a main driver of the recent increase in fire weather. Wildfire plumes can be of significant concern because the generated heat, smoke, and turbulence can severely disrupt aviation and other industries. Suppose the heat flux per unit area generated by a wildfire is measured as 90 kW/m², with an observed initial plume radius as $r_o = 100$ m, and subjected to a crosswind of $U_a = 15$ m/s. Determine rise height h when the plume is bent-over by the crosswind, assuming the plume is bent-over when the vertical velocity $w_m \ll U_a$ (see Chap. 8).

Solution: The total heat flux H can be computed as $H = 90 \times 1000 \times \pi \times 100^2 =$ 2830 MW. The specific buoyancy flux generated by the wildfire can be determined by:

$$F_o = \frac{gH}{\rho_a c_p T_a} = \frac{9.81 \times 2.8 \times 10^9}{1.2 \times 1.006 \times 293} = 77700\ \mathrm{m}^4/\mathrm{s}^3$$

The plume velocity can be determined by using Table 1 as $w_m = 4.71 F_o^{1/3} z^{-1/3} =$ $201.0 z^{-1/3}$. When $w_m \approx U_a$, we have $h^{1/3} = 201.0/15$ and $\underline{h = 2.5\ km}$. After the plume is bent-over, it will continue to rise and mix with the ambient crosswind as an 'advected line thermal', which will be described in Chap. 8.

Densimetric Froude Number

The local densimetric Froude number of the plume based on the centerline values of the velocity and reduced gravity is

$$\mathrm{Fr_L}(z) = \frac{w_m}{\sqrt{\frac{\Delta\rho_m}{\rho_a} g b}} \tag{43}$$

Based on the asymptotic solution, it can be shown that for a round plume,

$$\mathrm{Fr_L}(z) = \mathrm{Fr}_p = \left(\frac{5}{4}\right)^{\frac{1}{2}} \frac{\lambda}{\sqrt{\alpha}} = \text{constant}, \tag{44}$$

which has a numerical value of $\mathrm{Fr}_p = 4.5$ if $\alpha = 0.09$ and $\lambda = 1.19$. The constancy of $\mathrm{Fr_L}$ indicates a constant ratio of inertia to buoyancy forces throughout the plume. The rate of production of turbulent energy per unit mass of fluid from buoyant forces is $P \sim g' w_m$, while the rate of turbulent energy dissipation is $\epsilon \sim w_m^3/b$. Thus the densimetric Froude number can also be interpreted as the ratio of turbulent energy dissipation to production, with $\mathrm{Fr_L}^2 \sim \frac{w_m^2}{g'b} \sim \epsilon/P$; the constancy of the Froude number represents a fixed ratio of dissipation to production.

For any buoyant jet with both momentum and buoyancy, it can be shown that the buoyant jet will approach the asymptotic Froude number Fr_p as it rises and mixes with the surrounding fluid, regardless of whether Fr is greater or less than Fr_p (Prob. 3.8). For example, for a vertical buoyant jet, if the discharge densimetric Froude number is significantly greater than the plume value, $Fr \gg \mathrm{Fr}_p$, the effluent may initially behave like a turbulent jet, and the local densimetric Froude number would approach the asymptotic value from above (see Fig. 1). Alternatively, in the case of the plume produced by a fire, since the velocity w_m is initially zero, the densimetric Froude number $\mathrm{Fr_L}$ increases from zero at the source and asymptotically approaches the plume value of $\mathrm{Fr}_p = 4.5$ far from the source (see Fig. 2 in Chap. 1).

While the spread rate for both a jet and a plume is about the same, the entrainment coefficient can vary by as much as 60 percent over the jet-plume range, from $\alpha_j = 0.057$ for the round jet to $\alpha_p \approx 0.09$ for the round plume. A dimensional analysis of the entrainment velocity as a function of the local variables, $v_e = \mathrm{fn}(w_m, g'_m, b)$, would show the dependence of the entrainment coefficient α on the local densimetric Froude number $\mathrm{Fr_L}$. This dependence can also be deduced from the governing equations (Prob. 3.14). In buoyant jet models the variation of entrainment coefficient is typically incorporated in a form which gives the correct values in the limiting cases of a jet or plume:

$$\alpha = \alpha_j + (\alpha_p - \alpha_j)\left(\frac{\mathrm{Fr}_p}{\mathrm{Fr_L}}\right)^2 \tag{45}$$

In many practical applications, the plume solution serves a good approximation to strongly buoyant discharges (small Fr) with finite Q_o and M_o. For these cases the dilution can be inferred from Eq. 37. By definition we have ($F = F_o$)

$$F = Q\frac{\Delta\rho_m}{\rho_a} g \frac{\lambda^2}{1+\lambda^2} = Q_o \frac{\Delta\rho_o}{\rho_a} g \tag{46}$$

The relative centerline maximum concentration (inverse of centerline dilution, $S_m = c_o/c_m$) is then given by

$$\frac{c_m}{c_o} = \frac{\Delta\rho_m}{\Delta\rho_o} = \frac{5}{6\alpha\lambda^2}\left(\frac{20}{9\alpha}\right)^{1/3}\left(\frac{1+\lambda^2}{2\pi}\right)^{\frac{2}{3}} Q_o F_o^{-1/3} z^{-\frac{5}{3}} \tag{47}$$

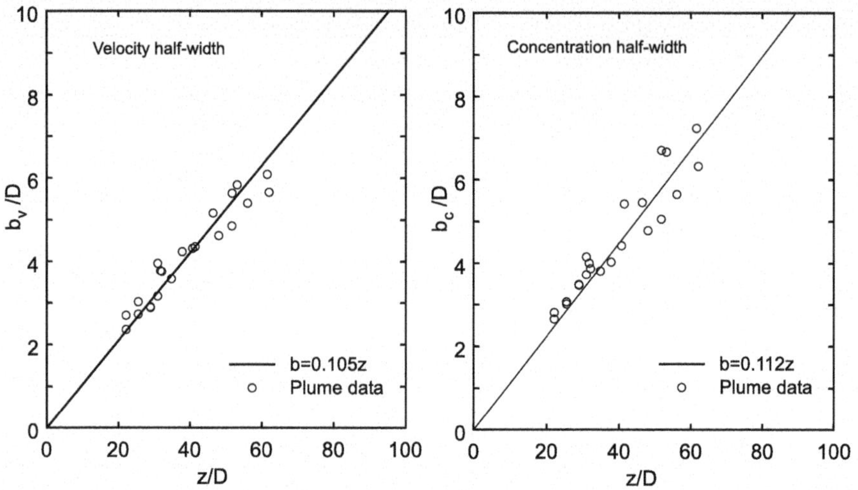

Fig. 5 Variation of plume half-width defined by velocity (b_v) and concentration (b_c) with vertical distance from the source (data from Papanicolaou and List 1988)

Experiments

Figure 3 shows measured radial profiles of velocity and scalar concentration at different elevations above a round plume; it is seen that both the velocity and concentration are self-similar and can be well-approximated by Gaussian distributions. The velocity half-width (and the concentration half-width) increases linearly with distance (Fig. 5). The data include the plume experiments of Papanicolaou and List (1988).

The plume solution (Eq. 39 or Table 1) obtained by the Eulerian integral method may be compared with the following formulae obtained from the laboratory experiments:

Papanicolau and List (1987)—tracer concentration and velocity measurements in water

$$w_m = 3.85\, F_o^{\frac{1}{3}} z^{-\frac{1}{3}},\, c_m = 11.0\, \Gamma_o\, F_o^{-\frac{1}{3}} z^{-\frac{5}{3}},\, Q = 0.14\, F_o^{\frac{1}{3}} z^{\frac{5}{3}} \tag{48}$$

where Γ_o = tracer mass flux at the source, and c_m = tracer mass concentration at the centerline.

Rouse, Yih and Humphreys (1952)—temperature and velocity measurements in air

$$w_m = 4.70\, F_o^{\frac{1}{3}} z^{-\frac{1}{3}},\, \Delta T_m = 11.0\, \frac{H_o}{\rho_a c_p} F_o^{-\frac{1}{3}} z^{-\frac{5}{3}} \tag{49}$$

where H_o = heat flux at the source, c_p = specific heat, and ΔT_m = maximum temperature excess at the plume centerline.

George, Alpert and Tamanini (1977)—temperature measurements in air

$$\Delta T_m = 9.1 \frac{H_o}{\rho_a c_p} F_o^{-\frac{1}{3}} z^{-\frac{5}{3}} \tag{50}$$

Lee and Cheung (1991)—temperature measurements in water

$$\Delta T_m = 10.0 \frac{H_o}{\rho_a c_p} F_o^{-\frac{1}{3}} z^{-\frac{5}{3}} \tag{51}$$

Chu, Senior and List (1981)—temperature measurements in water

$$\Delta T_m = 10.9 \frac{H_o}{\rho_a c_p} F_o^{-\frac{1}{3}} z^{-\frac{5}{3}} \tag{52}$$

Baines (1983)—direct measurement of the volume flux of a salty plume through a stratified interface

$$Q = 0.123\, F_o^{\frac{1}{3}} z^{\frac{5}{3}} \tag{53}$$

(see also Baines and Chu (1996) for a review of the available experimental data).

The plume theory predicts a velocity decay coefficient of 4.71; this is in excellent agreement with the measured value of 4.7 by Rouse et al. (1952), but a bit higher than the observed 3.85 by Papanicolaou and List (1988). On the other hand, the predicted centerline concentration of $c_m = 10.46\ Q_o c_o F_o^{-1.3} z^{-1/3}$ compares favorably with measured coefficients of 11.0 (Rouse et al. 1952), 10.9 (Chu et al. 1981), 9.1 (George et al. 1977), and 10.0 (Lee and Cheung 1991). The coefficient of 0.163 for the predicted volume flux also compare favorably with the experimental value of 0.14 by Papanicolaou and List (1988).

Chu et al. (1981) have carried out turbulence measurements in a plume; their data show that turbulent fluctuations in a plume are in general much more intense than in a jet. For example, the instantaneous maximum and minimum temperatures are approximately 2.7 and 0.53 times that of the turbulent mean concentration at the centerline. In Fig. 6 the measured concentration turbulent intensity across a plume is compared with those of a jet. Whereas it is customary to see a twin peak of turbulence intensity, around 0.2 to 0.25, at the edge of the jet (corresponding to the position of maximum shear), in a plume the rms fluctuation is distinctly higher, with a maximum intensity of around 0.4. The velocity turbulent intensities in a plume are however similar to a jet, with maximum axial velocity turbulent intensities of around 0.25, and similar shapes. In view of this larger plume concentration intermittency, it is perhaps not surprising that, although the spreading rate of plumes and jets are approximately the same, plumes often appear to be more 'blobby' than jets (cf Fig. 1 and Fig. 1 in Chap. 2). The larger turbulence intensity of a plume is responsible for its greater entrainment coefficient.

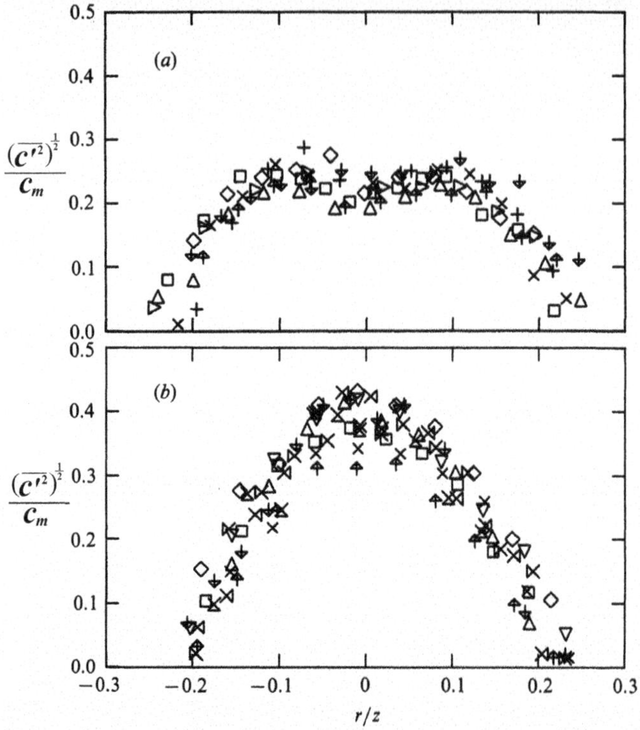

Fig. 6 Radial profile of turbulence intensity of concentration **a** across a jet; and **b** across a plume. (from Papanicolaou and List 1988)

Example 3.5 Vertical round plume
Consider the round jet in Example 2.3 again, with however the effect of buoyancy included. This is a discharge with jet diameter $D = 0.15$ m, depth $H = 14$ m, discharge velocity $w_o = 1$ m/s, and initial relative density difference $\frac{\Delta\rho_o}{\rho_a} = 0.025$. The jet discharge densimetric Froude number is $Fr = \frac{1}{\sqrt{9.81 \times 0.025 \times 0.15}} = 5.2$.

Solution: The initial volume, specific momentum and buoyancy fluxes Q_o, M_o and F_o are:

$$
\begin{aligned}
Q_o &= \frac{\pi(0.15^2)}{4}(1) = 1.767 \times 10^{-2}\ \mathrm{m^3/s} \\
M_o &= \frac{\pi(0.15^2)}{4}(1^2) = 1.767 \times 10^{-2}\ \mathrm{m^4/s^2} \\
F_o &= \frac{\pi(0.15^2)}{4}(1)(0.025)9.81 = 4.334 \times 10^{-3}\ \mathrm{m^4/s^3}
\end{aligned}
$$

Assuming Q_o and M_o are negligible, the centerline concentration can be calculated from Table 1 (Eq. 47) as:

$$\frac{c_m}{c_o} = \frac{10.46 \times \pi(0.15^2)/4}{(4.334 \times 10^{-3} \times 14^5)^{1/3}} = \underline{0.0138}$$

– the inverse of which is the centerline dilution, $S_m = \underline{73}$. Therefore, $\frac{Q}{Q_o} \gg 1$. Also Table 1 gives $\frac{M}{M_o} = 19.3 \gg 1$. Thus our starting assumption of negligible Q_o, M_o is justified.

Similarly, the maximum velocity at surface is obtained (Table 1) as:
$w_m(z = H) \; = \; 4.71 \; (\frac{4.334 \times 10^{-3}}{14})^{1/3} \; = \; \underline{0.32}$ m/s

Comparing the numerical results for the round plume and the round jet (with and without buoyancy) for this same design, we see that in the presence of buoyancy, the dilution is enhanced considerably. This can also be seen analytically by contrasting the solution for $Q(z)$. From Table 1 and Table 3 in Chap. 2, we have:

Round plume	Round jet
$Q = 0.26M^{1/2}z$	$Q = 0.28M^{1/2}z$

Whereas the momentum flux $M(z) =$ constant in a jet, it increases continuously with z in a plume, resulting in a greater total induced flow.

2.3 *Effect of Initial Momentum: Vertical Buoyant Jet*

Practical discharges possess both initial momentum and buoyancy. We have seen that jet behaviour is governed by the momentum flux, while plume behaviour is governed by the buoyancy flux. For a buoyant jet with initial momentum flux M_o and buoyancy flux F_o, we would expect that initially the jet momentum may be important. Since the kinetic energy of the jet-induced motion decreases with distance, we would expect buoyancy will eventually dominate the mixing. Considering the effect of jet momentum and buoyancy in turn, the jet momentum-induced velocity at a distance z can be estimated by $w_j \approx 7M_o^{1/2}z^{-1}$ (Table 3), while the buoyancy-induced velocity is around $w_p \approx 4.7F_o^{1/3}z^{-1/3}$. Since w_p decays much slower than w_j, the two effects will be about the same when $w_j \approx w_p$:

$$z \sim \frac{M_o^{3/4}}{F_o^{1/2}}$$

This is exactly the momentum length scale $l_s = \frac{M_o^{3/4}}{F_o^{1/2}}$ we previously introduced (Chap. 1) via pure dimensional reasoning. For $z/l_s \gg 1$, we expect the buoyant jet behaves like a plume, while for $z/l_s \ll 1$, we expect the discharge to behave like a momentum jet. For a vertical buoyant jet, the precise transition from a jet to a plume is offered by the experiments of Papanicolaou and List (1988). Figure 7 shows a plot of normalized measured volume flux as a function of dimensionless distance z/l_s from the source. Using the jet and plume solution for the volume flux Q, it can be shown that for the jet region, $Q^* = \frac{Q(x)F_o^{1/2}}{M_o^{5/4}} \sim z/l_s$, while for plumes, $Q^* \sim (z/l_s)^{5/3}$ (Prob. 4.6). The data shows that the buoyant jet behavior can be very well-correlated with this length scale; the dilution (and other variables) can be well-predicted with the plume solution for $z/l_s \geq 5$, and by the jet solution for $z/l_s \leq 2$. Except for cases when the source dimension is important ($z/D \sim 1$), this length scale provides a good means of defining jet/plume behaviour.

Based on Fig. 7 and other experimental data, a single formula to describe the average dilution $\overline{S}$ throughout the jet-plume range of a vertical round buoyant jet has been suggested (Wright and Wallace 1991):

$$\frac{\overline{S} l_Q}{z} = 0.29\,(1 + 0.16(z/l_s)^2)^{1/3} \tag{54}$$

where z = vertical distance above the source and $l_Q = Q_o/M_o^{1/2}$. It can be shown that this dilution formula is consistent with the limiting cases of a pure jet and plume (Prob. 4.7). For a round buoyant jet, it can be shown that the average dilution $\overline{S} = Q/Q_o$ and centerline minimum dilution $S_m = c_o/c_m$ are related by $\overline{S}/S_m = \frac{1+\lambda^2}{\lambda^2} \approx 1.7$ (Prob. 2.1).

Example 3.6 Mixing of vertical round buoyant jet
Consider a round jet discharge with diameter $D = 0.1$ m, discharge velocity $w_o = 1$ m/s, initial relative density difference $\frac{\Delta\rho_o}{\rho_a} = 0.025$, and discharging in depth $H = 12$ m. The jet discharge densimetric Froude number is $Fr = \frac{1}{\sqrt{9.81 \times 0.025 \times 0.1}} = 6.4$. Estimate the maximum velocity and minimum dilution at $z = 5$ m and $z = 1$ m from the jet origin.

Solution: The initial volume, specific momentum and buoyancy fluxes Q_o, M_o and F_o are:

$$\begin{aligned}
Q_o &= \frac{\pi(0.1^2)}{4}(1) = 7.854 \times 10^{-3}\ \mathrm{m^3/s}\\
M_o &= \frac{\pi(0.1^2)}{4}(1^2) = 7.854 \times 10^{-3}\ \mathrm{m^4/s^2}\\
F_o &= \frac{\pi(0.1^2)}{4}(1)(0.025)9.81 = 1.9262 \times 10^{-3}\ \mathrm{m^4/s^3}
\end{aligned}$$

The momentum length scale can be computed to be l_s = 0.6 m. At $z = 5$ m, $z/l_s = 8.3$; the mixing is plume like at this location. We can then use the plume solution (Table 1)

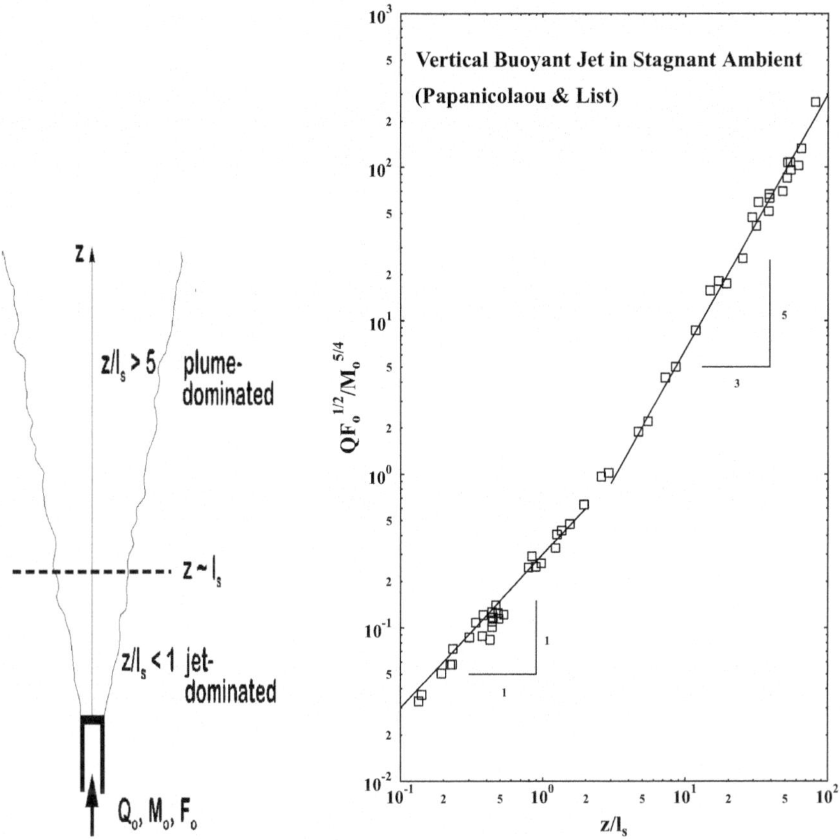

Fig. 7 Measured volume flux in a vertical turbulent buoyant jet plotted against dimensionless distance from the origin (data from Papanicolaou and List 1988)

to compute the centerline velocity as $w_m = \underline{0.35}$ m/s and the centerline dilution as $S_m = \underline{21}$. Similarly, at $z = 1$ m, $z/l_s = 1.7$, the mixing is jet like. The jet solution can then be used to give $w_m = \underline{0.62}$ m/s, $S_m = \underline{2}$. The use of Eq. 54 would have resulted in $S_m = \underline{22}$ and $S_m = \underline{2.2}$ for the two cases respectively.

Example 3.7 *Chemical waste gas discharge*
A chemical laboratory discharges a waste gas (sulphur dioxide) from a 0.2 m diameter smokestack located on the roof top of a building. The jet velocity is $w_o = 5$ m/s, and the temperature of the waste gas is $T_o = 33\,°C$. If the temperature of the outside air is $T_a = 23\,°C$, determine if the plume dilution would be adequate at a vertical height of 10 m above the roof top. The gas concentration at the source is $c_o = 1000$ ppm; the maximum allowable concentration for 8-hour exposure is 10 ppm (i.e. requiring a dilution of 100).

Solution: For small density and temperature differences, we can assume $\frac{|\rho_a - \rho_o|}{\rho_o} \approx \frac{|T_a - T_o|}{T_o}$, where T_o, T_a are the absolute temperature of the jet and ambient air respectively (Prob. 3.5). The initial relative density difference of the buoyant waste discharge is hence $\Delta\rho_o/\rho_a \approx (33 - 23)/(273 + 23) = \underline{0.0338}$. The jet densimetric Froude number is hence $Fr = 5/\sqrt{0.0338 \times 9.81 \times 0.2} = \underline{19.4}$.

The initial volume, specific momentum and buoyancy fluxes Q_o, M_o and F_o are:

$$\begin{aligned} Q_o &= \frac{\pi(0.2^2)}{4}(5) = 0.157 \text{ m}^3/\text{s} \\ M_o &= \frac{\pi(0.2^2)}{4}(5^2) = 0.785 \text{ m}^3/\text{s} \\ F_o &= \frac{\pi(0.2^2)}{4}(5)(0.0338)9.81 = 0.052 \text{ m}^4/\text{s}^3 \end{aligned}$$

Compute the length scales as $l_Q = \sqrt{\pi/4} \times 0.2 = 0.177$ m, $l_s = 0.785^{3/4}/0.052^{1/2} = 3.66$ m. At a height of 10 m, $z/l_s = 2.73$, the buoyant jet is in the transition region from a jet to a plume. Equation 54 can be used to compute the average dilution:

$$\frac{\overline{S} \times 0.177}{10} = 0.29\,(1 + 0.16 \times 2.73^2)^{1/3} = 0.377$$

with $\overline{S} = 0.377 \times 10/0.177 = \underline{21.3}$. The centerline dilution can be obtained by $S_m = \overline{S}/1.7 = \underline{12.5}$. Thus the plume dilution due to buoyancy is about 12, much less than the required dilution of 100. In the real situation, the effect of wind (crossflow) would act to bring down the concentration further (see Chaps. 7–10).

2.4 *Buoyancy Reduction Due to Density-Temperature Nonlinearity*

In the foregoing development, we have assumed a linear density-concentration dependence. In general, density differences are linearly proportional to salinity differences (see Appendix A); e.g. for freshwater discharges into the ocean. It is then legitimate to write $\frac{(\rho - \rho_a)}{\rho} \approx \gamma_S(S - S_a)$ where $\gamma_S = \frac{1}{\rho}\frac{\partial\rho}{\partial S}|_{S_a,T_a}$. The conservation of buoyancy flux is a direct consequence of this assumed linearity. However the variation of water density with temperature is quite nonlinear (Fig. 8); a linear approximation, $\Delta\rho/\rho \approx \gamma_T(T - T_a)$, is good only for small ΔT of a few degrees. In laboratory experiments, it is often convenient to use heat as a source of buoyancy. Due to the nonlinear equation of state, conservation of the heat flux does not guarantee buoyancy flux conservation. For large initial temperature excess, there can be a substantial loss of buoyancy within a short distance from the source—a point that can be noted in many previous experimental studies.

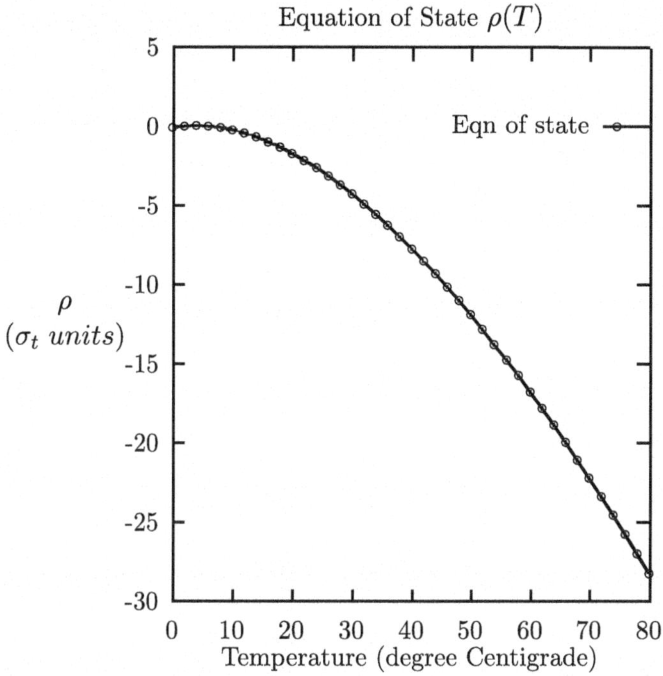

Fig. 8 Density of water as a function of temperature

Example 3.8 *Heated water jet: buoyancy conservation*
A thermal plume in a laboratory experiment is produced by using a heated discharge with an initial temperature difference of 40° Centigrade. The discharge is designed to simulate a plume with initial relative density difference of around 0.015. The plume has the following characteristics:

Initial velocity	$w_o = 0.1075$ m/s
Source temperature	$T_o = 61.0\,°$C
Ambient temperature	$T_a = 21.0\,°$C
Jet diameter	$D = 0.75$ cm
Initial density difference	$\frac{\Delta\rho_o}{\rho_a} = 0.01527$

where the initial density difference has been determined with the use of equation of state (Appendix A). This is a plume like discharge with $Fr = 3.2$.

At a distance of 0.1 m above the source, compute the centerline temperature and density deficit, and comment on the conservation of buoyancy flux.

Solution: The initial volume and specific buoyancy fluxes Q_o, and F_o can be computed to be:

$$Q_o = \frac{\pi(0.0075^2)}{4}(0.1075) = 4.7492 \times 10^{-6}\ \mathrm{m^3/s}$$
$$F_o = \frac{\pi(0.0075^2)}{4}(0.1075)(0.01537)9.81 = 0.7161 \times 10^{-6}\ \mathrm{m^4/s^3}$$

The centerline dilution, $S_m = \frac{\Delta T_o}{\Delta T_m}$, can then be computed to be (from Table 1):

$$S_m = 0.0956\,\frac{(0.7161 \times 10^{-6})^{1/3}(0.1)^{5/3}}{4.7492 \times 10^{-6}} = \underline{3.9} \tag{55}$$

Hence the centerline (maximum) temperature at $z = 0.1\ m$ is $T_m = T_a + \Delta T_o / S_m = 21.0 + 40.0/3.9 = 31.3\,°\mathrm{C}$. If the $\rho(T)$ relation were linear, the buoyancy corresponding to this dilution would be simply $\Delta\rho_m/\rho(linear) = \Delta\rho_o/\rho/S_m = 0.0153/3.9 = 0.00396$. However, from the equation of state, the centerline temperature of 31.3 °C results in an actual centerline density deficit of $\Delta\rho_m/\rho(actual) = \underline{0.00275}$. Thus at only 0.1 m above the source, where the discharge can be shown to be behaving like a pure plume, only about 0.69 of the original buoyancy flux remains (see also Appendix A)! This point can be of some importance in designing experiments and interpreting experimental data. Fortunately, in the prediction of dilution using the plume formula, $S \sim F_o^{1/3}$, such that the dilution is at most over-predicted by $0.69^{-1/3} = 1.13$, or around 13 percent. For non-vertical heated jets, this loss of buoyancy results in a *lower* observed jet trajectory compared to the case if the buoyancy is created using salinity difference. In some cases, the jet trajectory can also be sensitive to ambient temperature (Prob. 4.8).

3 Lagrangian Approach for Plume Modelling

The Eulerian integral model in the previous section is obtained from the basic Eulerian equations of fluid motion. At this junction we take a detour and introduce an alternative Lagrangian approach to model buoyant jets. Consider a round plume discharge as shown in Fig. 9, with initial velocity $w_o = U_o$. The marked material volume issuing from the source over a time interval Δt has a streamwise length of $U_o\Delta t$, cross-section area $A_o = \pi D^2/4$, and velocity U_o. This material volume then has initial mass, momentum, and buoyancy as follows:

Mass Q	$= \rho A_o(U_o\Delta t)$	[kg]
Momentum M	$= \rho A_o(U_o\Delta t)U_o$	[kg-m/s]
Buoyancy force F	$= \Delta\rho_o g A_o(U_o\Delta t)$	[N]

Instead of using the Eulerian equations, the motion of this material volume is followed with time. This plume element mixes with its surrounding by turbulent

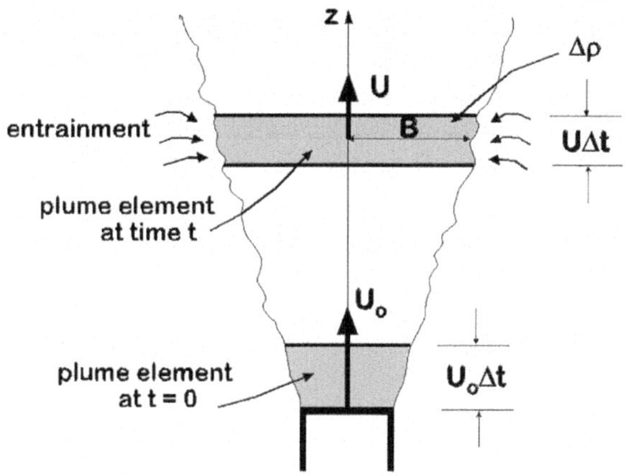

Fig. 9 Lagrangian formulation for the motion of a plume element

entrainment, and in so doing, its width, mass, momentum, and species concentration (buoyancy) change correspondingly. In general, at any time t (or elevation z) in the established flow, the plume element is characterized by

$$\begin{aligned}
&\text{Mass } Q &&= \rho A(U\Delta t)\\
&\text{Momentum } M &&= \rho A(U\Delta t)U\\
&\text{Buoyancy force } F_B &&= \Delta\rho g A(U\Delta t)
\end{aligned}$$

where B, $U\Delta t$ = half-width and thickness of plume element, $A = \pi B^2$, U, $\Delta\rho =$ average velocity and density deficit respectively (this is the 'top-hat' profile introduced in Chap. 2). Viewed in this manner, the mass, momentum, and buoyancy fluxes at time t are proportional to the mass, momentum, and buoyancy of the corresponding plume element. Applying Newton's Law to this material volume, the rate of change of the vertical momentum of the plume element is equal to the buoyancy force acting on it:

$$\frac{dM}{dt} = F_B$$

$$\frac{d}{dt}(\rho A(U\Delta t)U) = \Delta\rho g A(U\Delta t) = \rho\frac{F_o}{U}(U\Delta t) \tag{56}$$

or in a more general form,

$$\frac{d}{dt}(AU^2) = \int u\frac{\Delta\rho}{\rho_a} g\, dA = F_o \tag{57}$$

Integration yields an exceedingly simple solution for the kinematic momentum flux $M = AU^2$:

$$M(t) = F_o t + M_o \tag{58}$$

where M_o is the initial momentum flux and F_o the specific buoyancy flux. In the presence of buoyancy, the momentum of the plume element increases linearly with *time*; significant fluid motions can hence be generated by apparently small density differences. Coupled with a plume spreading hypothesis, we show how the plume solution can be easily obtained.

First, the spreading hypothesis, $db/dz = \beta$, can be written in a more general Lagrangian form (multiplying throughout by U):

$$U\frac{dB}{dz} = U\beta \tag{59}$$

where $\beta = \frac{dB}{dz} = \sqrt{2}\frac{db}{dz}$. Since $U\frac{dB}{dz} = \frac{dB}{dt}$ is the rate of change of plume radius with time (following the material volume), the changes in properties of the plume element are then governed by the following two equations (for $M_o = 0$):

$$\frac{dB}{dt} = \beta U \tag{60}$$

$$\pi U^2 B^2 = F_o t \tag{61}$$

Noting that $U = (1/\beta)dB/dt$, and $z = (1/\beta)B$, the solution can then be obtained by simple integration:

$$B = \left(\frac{4\beta}{3}\right)^{1/2}\left(\frac{F_o}{\pi}\right)^{1/4} t^{3/4} \tag{62}$$

$$U = \left(\frac{3}{4\beta}\right)^{1/2}\left(\frac{F_o}{\pi}\right)^{1/4} t^{-1/4} \tag{63}$$

$$z = \left(\frac{4}{3\beta}\right)^{1/2}\left(\frac{F_o}{\pi}\right)^{1/4} t^{3/4}$$

$$t = \left(\frac{3\beta}{4}\right)^{2/3}\left(\frac{\pi}{F_o}\right)^{1/3} z^{4/3} \tag{64}$$

This leads to the following expressions for the volume and momentum of the plume element (corresponding to the Eulerian volume and momentum fluxes):

$$Q = \pi U B^2 = \sqrt{\pi}\beta\left(\frac{3\sqrt{\pi}\beta}{4}\right)^{1/3} F_o^{1/3} z^{5/3} \tag{65}$$

$$M = F_o t = \left(\frac{3\sqrt{\pi}\beta}{4}\right)^{2/3} F_o^{2/3} z^{4/3} \tag{66}$$

Noting the correspondence between the top-hat and Gaussian profiles, $\beta = \sqrt{2}\beta_G = \sqrt{2}\,\frac{6\alpha}{5}$, Eqs. 65 and 66 are exactly the same as the plume solution developed in the previous section for $\lambda = 1$.

The underlying assumption of the Lagrangian approach is that there is negligible streamwise mixing between the plume elements at consecutive times. In the above development, we have tacitly assumed $\lambda = 1$, i.e. neglecting differences in velocity and concentration width. Nevertheless, the model gives very satisfactory results. Adopting $\beta = \sqrt{2} \times 0.105 = 0.149$, the predicted coefficients for the volume and momentum fluxes (Eqs. 65 and 66) are 0.153 and 0.34 respectively, and that of centerline velocity (Eq. 64, with $w_m = 2U$) is 4.4. Knowing the volume flux, the centerline concentration can be obtained by mass conservation, with a predicted coefficient of 11.1. Compared to the measured values of Q and c_m (0.14 and 11.0), the Lagrangian model gives the same degree of agreement with data as the Eulerian formulation. The treatment of $\lambda > 1$ and related details can be found in Chu (1994).

The Lagrangian approach is physically appealing, and the solution can be much simpler to obtain and interpret. Unlike the Eulerian framework, the Boussinesq approximation is not a requirement in the Lagrangian formulation (see also Prob. 5.5 and Prob. 10.7). More important, the extension to the complex and practically important case of a buoyant jet in crossflow is relatively straightforward. The use of the Lagrangian approach for complex buoyant jet problems was first proposed by Chu (1977; 1985) in a line impulse model. A fundamental formulation and its interpretation in terms of 'dominant eddies' (Chu 1994) is developed in the later chapters. A Lagrangian model (JETLAG) based on the heuristic approach outlined above (Lee and Cheung 1990) will be described in Chap. 10.

Example 3.9 *Plume versus jet mixing*

Buoyant fluid is discharged at a rate of 0.3 m³/s through a 0.6 m diameter opening on the sea floor of the deep ocean. Find the specific momentum flux, M, the volume flux, Q, the average dilution, $\overline{S} = Q/Q_o$*, and the reduced gravity, g', at elevation 30, 60, and 120 m above the source. The density of the sea water is 2.5% heavier than the effluent at the source.*

Solution: Calculations are carried out first for the momentum flux and buoyancy flux at the source. The cross section area of the source is $A_o = \pi D^2/4 = 0.283$ m^2. The velocity at the source is $V_o = Q_o/A_o = 1.06$ m/s. The specific momentum flux is

$$M_o = Q_o V_o = 0.318 \text{ m}^4/\text{s}^2$$

The reduced gravity at the source is $g'_o = 9.81 \times 0.025 = 0.245$ m/s^2. The specific buoyancy flux is

$$F_o = g'_o Q_o = 0.0736 \text{ m}^4/\text{s}^3$$

The length scale defined by the buoyancy and momentum fluxes is

$$l_s = \frac{M_o^{\frac{3}{4}}}{F_o^{\frac{1}{2}}} = 1.56 \text{ m}$$

Within a short distance away from the source (when $z \gg l_s$), momentum would become negligible and buoyancy would be the dominant effect. Neglecting the initial momentum at the source, the time for the material volume to move a distance $z = 30$ m, according to Eq. 64, is

$$t = (\frac{3\beta}{4})^{2/3}(\frac{\pi}{F_o})^{1/3} z^{4/3} = \underline{75.6} \text{ s}.$$

where $\beta = 0.149$. Over this period of time, the momentum is increased by the buoyancy force to

$$M = F_o t = 0.0736 \times 82.5 = \underline{5.6} \text{ m}^4/\text{s}^2,$$

that is 18 times greater than the initial momentum flux at the source, $M_o = 0.318$ m^4/s^2. The radius of the plume at this position ($z = 30$ m) is

$$B = \beta z = \underline{4.8} \text{ m}.$$

The corresponding upward velocity of the plume element is

$$W = \sqrt{\frac{M}{\pi B^2}} = 0.298 \text{ m/s}.$$

The volume flux at this time is

$$Q = \pi W B^2 = \underline{18.8} \text{ m}^3/\text{s}$$

At $z = 30$ m above the source, the dilution of the plume is $\overline{S} = Q/Q_o = \underline{62}$. The table below summarizes the calculations.

	Plume Model				Jet Model		
z (m)	t (s)	M (m^4/s^2)	Q/Q_o	g' (10^{-2}m/s^2)	t (s)	M (m^4/s^2)	Q/Q_o
30	76	5.6	62	0.394	240	0.318	17
60	191	14.0	198	0.124	960	0.318	34
90	327	24.1	389	0.063	2160	0.318	51
120	480	35.3	628	0.039	3940	0.318	68

It is clear from these calculations that buoyancy is the dominant effect. The dilution would be only $Q/Q_o = 68$ at $z = 120$ m if mixing were due to the jet momentum alone. With buoyancy, the momentum flux at 120 m is $M = 35.3\ \mathrm{m^4/s^2}$ that is 111 times the initial momentum flux at the source. Enormous momentum is generated by the minute difference in density between the buoyant fluid and its surrounding sea.

In the absence of buoyancy, the jet momentum would have to be dramatically increased to obtain the same dilution. Since $Q \sim \sqrt{M_o}$ in a jet, to acheive a dilution of 628 by the jet momentum alone, the momentum should be increased by $(628/68)^2 = 9.2^2$ times. For the same Q_o, this would produce a source velocity $V_o = 1.06 \times 9.2^2 = 89.7$ m/s; the diameter of the source opening would have to be $d_o = 6.5$ cm (a 9.2 times reduction in the size of the opening). The power consumption to produce such a high-velocity jet would be:

$$\frac{1}{2}\rho V_o^2 Q_o = \frac{1}{2}(1000) \times 89.7^2 \times 0.3\ \mathrm{Nm/s} = 1207\ \mathrm{kW}$$

The pumping cost to maintain such a jet would be prohibitively costly (in addition to possible bed erosion and other problems).

4 Negatively Buoyant Jets

In many situations, negatively buoyant fluids are discharged upwards into surrounding fluid of lighter density. An example is the discharge of concentrated brine from desalination plants. Waste gas exhaust from air-conditioned buildings in the hot summer is another example. Figure 10 shows the rise of the negatively buoyant jet to a maximum height and then the falling back of the curtain of the negatively buoyant fluid. The initial momentum flux is not negligible in the negatively buoyant jet. The negative buoyancy acts to oppose the initial jet momentum until the height of the maximum rise, z_m, after which the dense fluid will fall back due to its greater density. This case provides an example to demonstrate the versatility of the Lagrangian method.

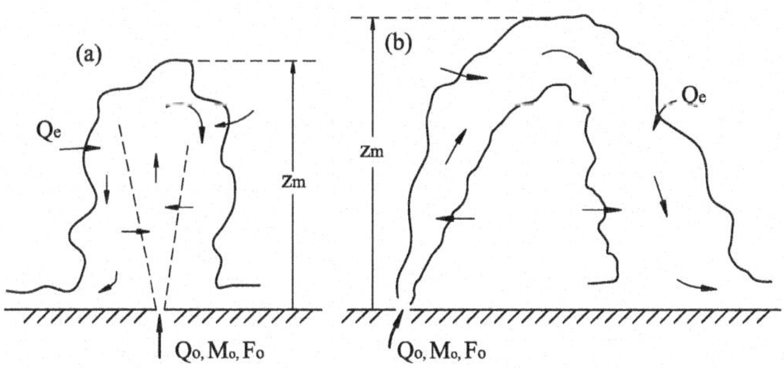

Fig. 10 Negatively buoyant plume produced by upward momentum flux

Applying the Lagrangian method, the momentum equation gives:

$$\frac{dM}{dt} = -F_o \tag{67}$$

Since the density deficiency flux is conserved, we have $M(t) = M_o - F_o t$, and when the vertical momentum vanishes at $z = z_m$, $t = M_o/F_o$. Applying the spreading hypothesis $dB/dt = \beta U$ again, we have ($M = \pi U^2 B^2$):

$$B\frac{dB}{dt} = \beta U B = \frac{\beta}{\sqrt{\pi}}(M_o - F_o t)^{1/2} \tag{68}$$

from which the solution can be easily found to be:

$$B(t^*) = \left(\frac{4\beta}{3\sqrt{\pi}}\right)^{1/2} l_s\,(1 - (1 - t*)^{3/2})^{1/2} \tag{69}$$

where $t^* = F_o t/M_o$ varies between 0 and 1 from the source to the maximum rise height, and $l_s = \frac{M_o^{3/4}}{F_o^{1/2}} = (\frac{\pi}{4})^{1/4} Fr D$ is the jet momentum length scale. Thus the maximum height of rise is given by:

$$z_m = \frac{1}{\beta}B(t* = 1) = \sqrt{\frac{2}{3\beta}}FrD = 1.98\ FrD \tag{70}$$

where $Fr = u_o/\sqrt{|g_o'|D}$, and a value of $\beta = 0.17$ is used as suggested by Baines and Chu (1996). The predicted rise coefficient of 1.98 is somewhat greater than the experimental coefficient of 1.74 according to Turner (1966). In the experiment, the falling negatively buoyant fluid is re-entrained into the rising jet. The re-entrainment, however, can be avoided by tilting the dense plume at a small angle (say 7°) from the vertical as shown in Fig. 10. Without the interaction of the upward flow with the falling curtain, as would be expected, the rise height is larger. Measurements of the slightly inclined fountains give $z_m = 2.04 FrD$ (Baines et al. 1990), in remarkable agreement with the prediction of the simple Lagrangian model (Eq. 70).

It is instructive to compute the jet volume flux for this case, $Q = \pi U B^2$. Substituting $U = \frac{1}{\beta}\frac{dB}{dt}$, and noting that $B = \beta z$, it can be shown that the volume flux is:

$$Q(t^*) = \sqrt{\pi}\beta\, M_o^{1/2} z(1 - t^*)^{1/2} = 0.30\, M_o^{1/2} z(1 - t^*)^{1/2} \tag{71}$$

Close to the source, $z \ll z_m$ or $t^* \ll 1$, the jet volume flux varies linearly with elevation as for the pure jet (cf $Q = 0.28 M_o^{1/2} z$ for round jet). However, the volume flux actually decreases towards zero at the maximum rise height, $z = z_m$ or $t^* = 1$. Denoting $\tau = 1 - t^*$, we have $Q \sim z\tau^{1/2} \sim [(1 - \tau^{3/2})\tau]^{1/2}$. It follows then that the positive entrainment is a maximum at τ_c, where $\tau_c^{3/2} = 2/5$, i.e. $\tau_c = 0.543$ or

$t_c^* = 0.457$. At this critical time, the maximum amount of 'clean' fluid flow entrained by the dense plume is given by:

$$Q_c = \sqrt{\pi}(\frac{4\beta}{3\sqrt{\pi}})^{1/2}[(1 - \tau_c^{3/2})\tau_c]^{1/2} \, M_o{}^{1/2} l_s \quad (72)$$

The maximum average dilution $\overline{S} = Q_c/Q_o$ can be shown to depend only on the jet densimetric Froude number:

$$\overline{S} = (8\beta/3)^{1/2}(3/5 \times 0.543)^{1/2} Fr = 0.384 \, Fr \quad (73)$$

The centerline minimum dilution is then given by ($\overline{S}/S_m \approx 1.7$)

$$S_m = 0.23 \, Fr \quad (74)$$

Since entrainment is the incorporation of the ambient fluid into the turbulent source fluid, by definition Q should always increase. The fact that Q decreases suggests that there is "detrainment"—i.e. the turbulent fluid is peeling off from the jet to form the falling part of the dense plume.

Example 3.10 *Dense gas plume*
Consider the waste gas discharge from the chemical laboratory in Example 3.7 again, with $D = 0.2$ m, and the jet velocity is $w_o = 5$ m/s. In the hot summer, however, the exhaust from an air-conditioned building has a temperature lower than the ambient. Assume now the temperature of the waste gas is $T_o = 23\,°C$. If the temperature of the outside air is $T_a = 33\,°C$, estimate the maximum height of rise of the waste gas in a calm atmosphere and the minimum dilution achieved.

Solution: This is a negatively buoyant jet with the effluent denser than the surrounding environment; the other parameters are similar to Example 3.7 with the buoyancy reversed. The effluent discharged upwards has a relative density difference of $\Delta\rho_o/\rho_a \approx (23 - 33)/(273 + 23) = \underline{-0.0338}$. The jet densimetric Froude number is hence $Fr = 5/\sqrt{|0.0338| \times 9.81 \times 0.2} = \underline{19.4}$. The maximum height of rise is $z_m = 1.74 Fr D = 1.74 \times 19.4 \times 0.2 = \underline{6.8}$ m. The minimum dilution can be estimated as that achieved at the level of maximum positive entrainment, with $S_m \approx 0.23 Fr = 0.23 \times 19.4 = \underline{4.5}$.

The dense plume will eventually sink back to the discharge level and spread over on the roof or ground. From the air pollution point of view, this is important as the pollutant may contaminate the area near the discharge point. Since the most concentrated part of the plume mixes further beyond t_c and on its descent, the dilution is expected to be somewhat greater than estimated. More details on dense plumes can be found in Baines and Chu (1996).

5 Turbulent Line Plume

The flow generated by a line source of buoyancy is called a line plume or plane plume. The flow in this line plume is governed by the buoyancy flux per unit length of the source. There have been only a few studies of the line plumes. Our present knowledge is mainly based on the studies of Rouse et al. (1952) who studied convection over line fires, and more recently by Kotsovinos and List (1977) who carried out detailed measurements with a plane heated water jet. Chu and Baines (1989) measured the entrainment flow rate into a plane plume using a direct method. The axial velocity and tracer concentration profiles in a plane plume are found to be self-similar and can be represented by Gaussian profiles as in the plane jet. The formulation for the two dimensional case is similar in almost every respect to the axisymmetric plume.

Consider the flow generated by a line source of buoyancy (located at $z = 0$) in an otherwise stagnant ambient fluid of constant density ρ_a. The motion is sustained by a steady constant buoyancy flux $F_o = \int_{-\infty}^{\infty} \frac{\Delta\rho}{\rho} g w \, dy$ at the origin, $z = 0$. The continuity, momentum, and tracer mass conservation equations for the plume are (in the $y - z$ plane, where $z =$ upward vertical co-ordinate from source) respectively:

$$\frac{\partial w}{\partial z} + \frac{\partial v}{\partial y} = 0 \tag{75}$$

$$w\frac{\partial w}{\partial z} + v\frac{\partial w}{\partial y} = -\frac{\partial \overline{w'v'}}{\partial y} - \frac{(\rho - \rho_a)}{\rho_a} g \tag{76}$$

$$w\frac{\partial c}{\partial z} + v\frac{\partial c}{\partial y} = -\frac{\partial \overline{v'c'}}{\partial y} \tag{77}$$

An integral model can be obtained from the above equations by suitable turbulent closure. By assuming a linear equation of state, an analytical solution can be obtained. In particular, using the Lagrangian model with a spreading hypothesis, $dB/dt = \beta W$, the following solution can be easily obtained in terms of the 'top-hat' velocity and half-width W, B, and the spreading rate, $\beta = \frac{dB}{dz}$ (Prob. 3.13).

$$B = \left(\frac{\beta}{\sqrt{2}}\right)^{2/3} F_o^{1/3} t$$

$$W = (2\beta)^{-1/3} F_o^{1/3} \tag{78}$$

For the line plume, it is interesting to note the plume velocity is a constant, implying that $z = Wt$. The volume and momentum fluxes (per unit length of source), and the centerline concentration can then be obtained as (see Prob. 3.12 for relation between average and centerline variables for plane buoyant jets):

$$Q(z) = (2\beta)^{2/3} F_o^{1/3} z$$

$$M(z) = (2\beta)^{1/3} F_o^{2/3} z$$

$$c_m(z) = (2\beta)^{-2/3} \sqrt{\frac{1+\lambda^2}{\lambda^2}} Q_o F_o^{-1/3} z^{-1} \tag{79}$$

An analysis of the earlier experiments of Rouse et al. (1952) gives $\beta_G = 0.147$ and $\lambda = 1.24$ (Jirka and Harleman 1979). More detailed velocity and temperature measurements by Kotsovinos and List (1977) give $\beta_G = 0.116$ and $\lambda = 1.35$. This suggests a spread rate of $\beta = \sqrt{\frac{\pi}{2}}\beta_G = \sqrt{\frac{\pi}{2}} \times 0.116 = 0.145$ and $\lambda = 1.35$. A recent review of data by Chu (1994) has suggested a spreading rate $\beta = dB/dz = 0.17$. Adopting $\beta = 0.145$, the solution in terms of the fluxes and centerline variables can be obtained; it is summarized in Table 1.

For the line plume, the volume flux or dilution varies linearly with distance. The predicted momentum coefficient (0.74) and centerline concentration (2.68) are in excellent agreement with the measured values of Rouse et al. (1952), which are 0.72 and 2.6 respectively. It is also consistent with Kotsovinos and List's (1977) measured concentration coefficient of 2.38. The predicted coefficient for the volume flux is 0.47. Based on direct measurements of the entrainment flow rate into a plane buoyant jet, Chu and Baines (1989) have obtained the following asymptotic formulae for the plane plume:

$$Q = 0.48\, F_o^{1/3} z \tag{80}$$

The formula proposed by Kotsovinos and List (1977) to fit their data is

$$Q = 0.34\, F_o^{1/3} z \tag{81}$$

The classical experiment of the plane plume by Rouse et al. (1952) gives the formula

$$Q = 0.57\, F_o^{1/3} z \tag{82}$$

There is considerable variation in the results obtained by the different investigations. The measured centerline dilution of $S_m = 0.42 F_o^{1/3} z$ given by Kotsovinos and List (1977) is also not consistent with their measured average dilution coefficient of 0.34. The nature of the plane plume is sensitive to the ambient condition (see Baines and Chu, 1996). The measured spreading characteristics may depend on this condition. Clearly, further experimental data are needed.

Plane Buoyant Jet —For a two-dimensional slot buoyant jet with initial unit discharge Q_o m^2/s, specific momentum flux $M_o = Q_o w_o \text{m}^3/\text{s}^2$, and specific buoyancy flux $F_o = Q_o g'_o$ m^3/s^3, a momentum length scale similar to that of the round buoyant jet can be defined as:

$$l_s(2D) = \frac{M_o}{F_o^{2/3}} \tag{83}$$

The momentum length scale is a measure of the distance over which the initial jet momentum is important. The dilution data of Kotsovinos and List (1977) shows that the buoyant jet flow is plume-like for $z/l_s \geq 4$.

For practical application an important equation for dilution prediction is that of a line plume in a stagnant unstratified ocean. The experimental result of Rouse (1952) is often used for the prediction of centerline dilution:

$$S_m = \frac{c_o}{c_m} = 0.38\frac{F_o{}^{1/3}z}{Q_o} \tag{84}$$

The above results are for purely two-dimensional plumes. Roberts (1979; 1980) studied the mixing of a line pume of finite length in a current. For near stagnant conditions, the minimum surface dilution measured in his experiments give:

$$S_m = 0.27\frac{F_o{}^{1/3}H}{Q_o} \tag{85}$$

for a depth of H. Compared with Eq. 84, the smaller dilution coefficient reflects the effect of blocking of the surface layer. Dilution charts to account for the effect of an unstratified current can be found in Roberts (1980).

Example 3.11 *Thermal discharge from slot diffuser*
An oil-fueled steam electric generating station is located on one bank of a tidal river (Fig. 11). Condenser cooling water is withdrawn from the river (at a flowrate of $Q_o = 20$ m^3/s*) into an upstream intake channel, and discharged back into the river at an excess temperature of 20 °C. It is proposed to discharge the thermal effluent vertically through a submerged slotted pipe (a 2D slot diffuser) running along the canal bank, at* $H = 10$ m *below the free surface. The regulatory environmental standard stiputates that the temperature in the upper 5 m layer cannot exceed 28 °C. In addition, to minimize any offshore momentum that may be hazardous to navigation, the maximum allowable transverse velocity (perpendicular to river axis) is 0.5 m/s.*

If the maximum ambient temperature that can occur in the year is 23 °C, determine the required dilution of the diffuser discharge so that the temperature standard will be met all the year around. Determine the diffuser length and discharge velocity that will satisfy all the environmental constraints. Assume a worst case scenario of zero river velocity. The ambient water has a typical salinity of 32 ppt. Typical channel depths at the site are about 12–15 m. To minimize head losses, the initial jet velocity should not exceed 2 m/s.

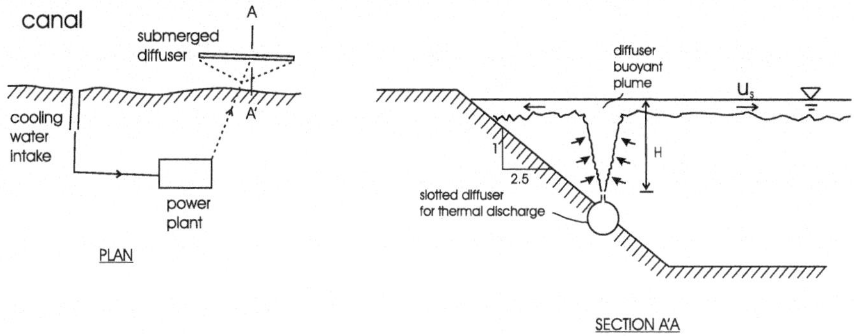

Fig. 11 Slot plume thermal discharge

Solution: The thermal discharge from the slotted diffuser can be modelled as a two-dimensional vertical buoyant plume. For the highest ambient temperature $T_a = 23\,°\text{C}$ and and a maximum allowable surface temperature of 28 °C, the required minimum dilution is $S_m = \Delta T_o/\Delta T_m = 20/(28 - 23) = \underline{4}$. For a plane buoyant jet, the ratio of minimum to average dilution is given by $\sqrt{\pi\lambda^2/(1+\lambda^2)} = 0.8$. The required average dilution is hence $\overline{S} = 4/0.8 = \underline{5.0}$.

Under stagnant ambient conditions, the vertical plume rises to the surface and spreads into two symmetrical layers. For a 2D plume it can be shown the upper layer thickness is approximately 1/6 H (Jirka and Harleman 1979). For a salinity of $S = 32$ ppt and the temperatures given, the initial density difference can be determined to be (Appendix A) $\Delta\rho_o/\rho = 0.0071$. The average dilution achieved by the time the thermal effluent enters the surface layer at $z = 5/6\ H$ is then (Table 1):

$$\overline{S} = 5 = 0.465 F_o^{\frac{1}{3}} \left(\frac{5}{6} H\right) / Q_o = 0.465(0.0071 \times 9.81)^{\frac{1}{3}} \left(\frac{5}{6} \times 10\right) / Q_o^{\frac{2}{3}}$$

The required unit discharge is then $Q_o = 20.0/L_d = 0.18\ \text{m}^2/\text{s}$, corresponding to a diffuser length of $L_d = 20/0.18 = \underline{111}$ m. Adopting a vertical discharge velocity of $U_o = 2.0$ m/s, the slot width is then $d_o = 20/111/2. = \underline{0.09}$ m. When the mixed thermal effluent reaches the surface, it flows away symmetrically in two directions. The transverse velocity of the mixed flow can be estimated to be $U_s - 0.5\overline{S}Q_o/(H/6) = (5/2) \times 0.18/1.67 = \underline{0.27}$ m/s. Thus the navigation constraint is also met. For this case, the unit momentum flux is $M_o = 0.18 \times 2 = 0.36\ \text{m}^3/\text{s}^2$, and the buoyancy flux is $F_o = 0.18 \times 0.0071 \times 9.81 = 0.125\ \text{m}^3/\text{s}^3$, giving a momentum length scale of $l_s = M_o/F_o^{2/3} = 1.44$ m. Thus $z/l_s = 5.8$, and the use of the 2D plume solution is well-justified.

An alternative to the use of a slot diffuser would be to discharge in the form of a large number of round jets from a multiport diffuser (Prob. 3.3). If located in sufficient depth, wastewater discharges from closely spaced jets on a submerged multiport diffuser tend to form line plumes before they surface or form a trapped spreading layer, and the dilution in stagnant water can often be estimated by the line plume formula (e.g. Fischer et al. 1979, Roberts et al. 1989a, b). It can be shown that for a given diffuser length, optimal dilution can be achieved using a line plume (Prob. 3.10).

6 Summary

The development of plane and round plumes, and the positive and negative buoyant jets, are analysed in this chapter using both the Eulerian method and the Lagrangian method. The entrainment coefficient of a plume differs greatly from that of a jet, but the spreading rate for the plume is practically the same as the rate for the jet. The Lagrangian method differs from the Eulerian method in the manner in which the velocity and concentration profiles are specified. In the Eulerian method, the plume is typically described in terms of the centerline variables, and the solution obtained by specifying the Gaussian velocity and concentration profiles. Such an assumption is not required in the Lagrangian formulation which adopts a 'top-hat' profile for the velocity of the dominant eddies. The most important distinction between the Lagrangian and Eulerian methods lies in the formulation of the turbulent entrainment hypothesis. The entrainment coefficient defined from a Eulerian point of view such as the one given in Eq. 45 is flow dependent. In the positively buoyant jet this entrainment coefficient α_G varies from a jet value of 0.057 to a plume value of 0.09. The entrainment coefficient may become negative in the negatively buoyant jet and is equal to zero in the far field region of the coflowing jet (see Chap. 6). However, the spreading coefficient β (defined in a Lagrangian reference frame following the dominant eddies) is relatively constant and not dependent on the flow. The Lagrangian method is simple and convenient in its implementation because the important results can be obtained without actually specifying the velocity and concentration profiles, which are not generally known for more complex turbulent flows. In the subsequent chapters, the Lagrangian method will be adopted to study the inclined buoyant jet, jets and plumes in stratified fluid, coflowing jets, puffs and thermals, and for arbitrary-inclined buoyant jet in non-uniform stratified crossflow.

Table 1 Summary of plume properties (Based on Gaussian profiles)

Parameter	Round plume	Plane plume
Plume volume flux $Q = \int w dA$	Dimensions L^3T^{-1} $Q = \pi w_m b^2$	L^2T^{-1} $Q = \sqrt{\pi} w_m b$
Specific momentum flux $M = \int w^2 dA$	L^4T^{-2} $M = 0.38\ F_o^{\frac{2}{3}} z^{\frac{4}{3}}$	L^3T^{-2} $M = 0.743\ F_o^{\frac{2}{3}}$
Specific buoyancy flux $F = \int w \frac{\Delta\rho}{\rho} g dA$	L^4T^{-3} $F = \frac{\pi\lambda^2}{1+\lambda^2} w_m g_m' b^2$	L^3T^{-3} $F = \sqrt{\frac{\pi\lambda^2}{1+\lambda^2}} w_m g_m' b$
Maximum time-averaged velocity w_m	$w_m = 4.71\ F_o^{\frac{1}{3}} z^{-\frac{1}{3}}$	$w_m = 2.26\ F_o^{\frac{1}{3}}$
Plume width b	$b = 0.105\ z$	$b = 0.116\ z$
Maximum time-averaged tracer concentration c_m	$c_m = 10.46 Q_o C_o F_o^{-\frac{1}{3}} z^{-\frac{5}{3}}$	$c_m = 2.68 Q_o C_o F_o^{-\frac{1}{3}} z^{-1}$
Average dilution $\overline{S} = Q/Q_o$	$\overline{S} = 0.163\ F_o^{\frac{1}{3}} z^{\frac{5}{3}} Q_o^{-1}$	$\overline{S} = 0.465\ F_o^{\frac{1}{3}} z Q_o^{-1}$
Plume spreading angle β	$\beta_G = 0.105$ $(\beta = \frac{6}{5}\alpha)$	$\beta_G = 0.116$ $(\beta = \frac{2}{\sqrt{\pi}}\alpha)$
Entrainment coefficient α	$\alpha_G = 0.088$	$\alpha_G = 0.103$
Ratio of concentration to velocity width, λ	$\lambda = 1.19$	$\lambda = 1.35$
Plume densimetric Froude number	$\text{Fr}_p = (\frac{5}{4})^{1/2} \frac{\lambda}{\sqrt{\alpha}} = 4.50$	$\text{Fr}_p = (\frac{\pi}{2})^{1/4} \sqrt{\frac{\lambda}{\alpha}} = 4.06$

Problems

3.1 The vertical centerline velocity of a round plume in near stagnant fluid is given by $w_m = C F_o^{1/3} z^{-1/3}$, where F_o and z are the specific buoyancy flux and elevation above source respectively, and $C = 4.71$. Show that the average travel time t_H from the source to the free surface can be estimated by $t_H = 3/(2C) F_o^{-1/3} H^{4/3}$, where H is the depth of the plume. Calculate the travel time for the discharge in Prob. 4.1 for $U_a = 0$. Comment on your results.

How will the result change for the same discharge in an ambient current of $U_a = 1$ m/s (see Chap. 8). Justify any assumptions made in your estimate.

3.2 Consider the discharge of fresh water from a round nozzle into ambient salt water in the laboratory, with $D = 0.015$ m, $w_o = 0.07$ m/s, $\frac{\Delta\rho_o}{\rho_a} = 0.03$. Assuming that the plume will be turbulent if the local Reynolds number defined by $Re = \frac{w_m b}{\nu}$ exceeds 2000, estimate the distance over which the plume may be laminar. This is a consideration in the design of laboratory experiments of buoyant discharges.

3.3 Merging plumes

Wastewater plumes are often discharged into rivers or coastal waters through a multiple port outfall laid along the bottom (a multiport diffuser). Wastewater is discharged as a number of equally-spaced vertical buoyant jets into water of depth H. Field observations showed that as long as the depth of discharge below the surface was less than three times the distance between adjacent outlets, the measured dilution at the surface is the same as that of a single jet (Rawn and Palmer (1930), *Trans. ASCE*, Vol. 94, 1036).

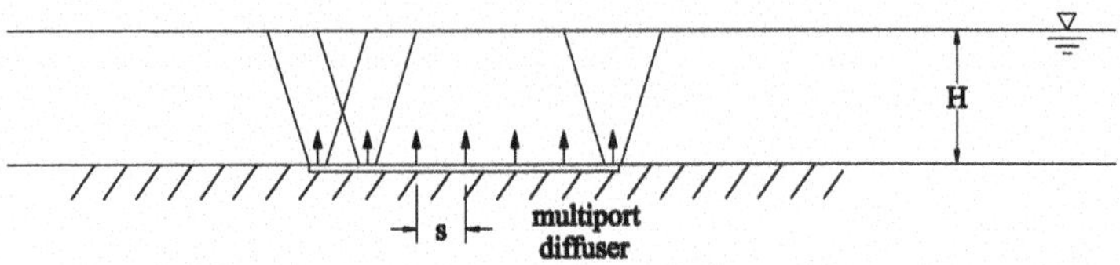

Use jet and plume theory to establish a criterion for the optimal spacing (s) between adjacent buoyant jets. Compare your result with the observations. Assume that the optimal dilution is given by the case when the jets do not interfere.

3.4 For a heated water jet discharge into a receiving water of constant ambient temperature T_a, show that the source heat flux is given by

$$H = c_p \int \rho w \Delta T dA = c_p \rho_a Q_o \Delta T_o$$

where $\Delta T = T - T_a$, and c_p = specific heat of water. Show that the predicted centerline temperature excess ΔT_m in a round thermal plume (Eq. 51) can be written as:

$$S_m = \frac{\Delta T_o}{\Delta T_m} = 0.1 \frac{F_o^{1/3} z^{5/3}}{Q_o}$$

For a discharge of $Q_o = 0.1\ \mathrm{m^3/s}$ and $\Delta T_o = 10\,°\mathrm{C}$, calculate the heat flux H (kW) if $c_p = 4.18$ kJ/kg − °C (at 20 °C).

3.5 Consider an atmospheric plume that satisfies perfect gas law, $p = \rho R T$, where p = pressure, R = gas constant, and T = absolute temperature. Show that at constant pressure we have:

$$\frac{|\Delta T|}{T} \approx \frac{|\Delta \rho|}{\rho}$$

Show also the volume coefficient of thermal expansion (percentage change of volume per degree Kelvin) is given by $\gamma_T = 1/T$.

3.6 For a buoyant gas plume arising from a heat source, the convective heat flux is given by $H = c_p \int \rho w \Delta T dA$. Show that for small temperature excess $\Delta T/T \ll 1$ and assuming constant pressure, the heat flux is related to the specific buoyancy flux F_o as:

$$H = \frac{c_p \rho_a T}{g} \int w g' dA = \frac{c_p \rho_a T}{g} F_o$$

For the standard atmosphere, with $R = 287$ J/kg°K, $p = 101.3$ kPa, $c_p = 1.006$ kJ/kg°K, show that

$$H\ (kW) \approx 36.2\ F_o\ (\mathrm{m}^4/\mathrm{s}^3)$$

3.7 A desalination plant discharges concentrated brine (specific gravity 1.032) into the adjacent sea ($\rho_a = 1025\ \mathrm{kg/m^3}$) of depth 15 m. The dense effluent contains copper which can be toxic to marine phytoplankton. It is proposed to design a multiport diffuser to mix the effluent (flow $Q_o = 0.6\ \mathrm{m^3/s}$) sufficiently to reach a safe copper concentration by dilution. Physical model tests show that maximum trajectory and hence dilution are obtained by having the jet nozzles discharging at 60° to the horizontal .

The experiments show that for $\phi_o = 60°$, the maximum height of rise z_m and the minimum dilution at the top of the jet are correlated with the jet discharge parameters as $z_m = 2.04\ Fr\,D$, and $S_m = 0.5\ Fr$ respectively.

If the required dilution of the copper in the effluent is $S = 15$, and the jet nozzle diameter is $D = 0.1$ m, determine the jet velocity and the number of jets required. Assume non-interfering jets. Determine the maximum height of rise; will the dense effluent impact on the surface?

3.8 *Asymptotic densimetric Froude number of vertical round buoyant jet*
Consider a vertical round buoyant jet with initial velocity w_o, relative density difference $\Delta\rho_o/\rho_a$, and jet diameter D, discharging in the z-direction in ambient fluid of constant density ρ_a. The initial jet densimetric Froude number is $Fr = w_o/\sqrt{g'_o D}$. The velocity and concentration (density-deficiency) distribution can be assumed to be self-similar and Gaussian, with velocity $w(z,r) = w_m e^{(-r^2/b^2)}$, and density deficiency, $\Delta\rho(z,r) = \Delta\rho_m e^{-r^2/(\lambda b)^2}$, where $\Delta\rho = \rho_a - \rho$.

An integral model based on the entrainment hypothesis results in the following governing equations for the centerline maximum velocity and density deficit, u_m and $\Delta\rho_m$, and jet-width b:

$$\frac{d}{dz}(w_m b^2) = 2\alpha w_m b$$

$$\frac{d}{dz}\left(\frac{w_m^2 b^2}{2}\right) = g'\lambda^2 b^2$$

$$\frac{d}{dz}(w_m g' b^2) = 0$$

where $g' = \Delta\rho_m/\rho_a$ and α = entrainment coefficient.

It can be shown from the governing equations of the integral model that the local densimetric Froude number defined by
$\mathrm{Fr_L} = \frac{w_m}{\sqrt{\Delta\rho_m/\rho_a g b}} = \frac{w_m}{\sqrt{g'b}}$ satisfies

$$\frac{d\mathrm{Fr_L}}{dz} = \frac{2\alpha}{\mathrm{Fr_L} b}\left(\frac{5\lambda^2}{4\alpha} - \mathrm{Fr_L}^2\right)$$

(a) Show that the continuity equation and the momentum equation results in the following expressions:

$$2w_m^2 b\frac{db}{dz} + w_m b^2 \frac{dw_m}{dz} = 2\alpha w_m^2 b$$
$$w_m^2 b\frac{db}{dz} + w_m b^2 \frac{dw_m}{dz} = g'\lambda^2 b^2$$

Hence show that

$$\frac{db}{dz} = 2\alpha - \frac{\lambda^2}{\mathrm{Fr_L}^2}$$

(b) From the momentum equation, and using the result in (a), deduce that:

$$w_m\frac{dw_m}{dz} = \frac{g'\lambda^2 b^2 - w_m^2 b\, db/dz}{b^2}$$
$$= 2\,g'(\lambda^2 - \alpha\mathrm{Fr_L}^2)$$

(c) From the buoyancy conservation equation, show that $d(g'b)/dz = -\frac{g'}{w_m}$ $d(w_m b)/dz$. Since from the momentum equation, we have $w_m b\frac{d}{dz}(w_m b) = g'\lambda^2 b^2$, show that:

$$\frac{d}{dz}(g'b) = -\frac{g'\lambda^2}{\mathrm{Fr_L}^2}$$

(d) The local densimetric Froude number is a measure of the ratio of the inertia to buoyancy forces, with $\mathrm{Fr_L}^2 = w_m^2/(g'b)$. We then have:

$$\frac{d\mathrm{Fr_L}}{dz} = \frac{(g'b)\frac{dw_m^2}{dz} - w_m^2\frac{d}{dz}(g'b)}{(g'b)^2}$$
$$= \frac{2\alpha}{\mathrm{Fr_L} b}\left(\frac{5\lambda^2}{4\alpha} - \mathrm{Fr_L}^2\right)$$

The above expression shows that the vertical buoyant jet attains an asymptotic densimetric Froude number of $\mathrm{Fr}_p = \sqrt{\frac{5\lambda^2}{4\alpha}}$. If $Fr \le \mathrm{Fr}_p$, $d\mathrm{Fr_L}/dz \ge 0$, and the local

Froude number will increase. Conversely, if $Fr \geq \mathrm{Fr}_p$, $d\mathrm{Fr_L}/dz \leq 0$, and the local Froude number will decrease. The asymptotic value of $\mathrm{Fr_L}$ for a vertical buoyant jet with any Fr is Fr_p.

Determine the value of Fr_p using the values of $\alpha = 0.088$ and $\lambda = 1.2$ for a round plume.

3.9 (a) Under what conditions will wastewater plumes from a multiport diffuser resemble a "line plume ".

(b) Consider a turbulent line plume (freshwater) in otherwise still sea water of constant density (and at the same temperature). The initial buoyancy flux is $F_o = Q_o g'_o$, where Q_o is the initial volume flux per unit diffuser length. Deduce from first principles the conservation of specific buoyancy flux at any elevation above the source (cf Eq. 24).

(c) Using dimensional reasoning, arrive at a prediction (to within a constant) of the centerline velocity $w_m(z)$ and the average dilution $Q(z)/Q_o$ of a line plume in terms of the important physical parameters. Determine also the plume width and the local densimetric Froude number. Comment on your results.

For a fixed wastewater flow and depth, how will the dilution depend on the diffuser length?

3.10 Pure plume solutions are very useful for estimating the initial dilution of buoyant discharges. The average dilution for a round and line plume at elevation z above the source are given by:

$$S_{3D} = 0.156\,\frac{F_o^{1/3} z^{5/3}}{Q_o}$$

$$S_{2D} = 0.535\,\frac{F_o^{1/3} z}{Q_o}$$

Consider a submerged multiport diffuser of given length L in depth H. It is desired to handle a design flow Q_d and two opposing port arrangements are proposed:

- very closely spaced ports to simulate a line plume
- N ports spaced to avoid jet interference before the plumes arrive at the surface

(a) Show that the ratio of the average dilution (total flow entrained) at any z for the two cases is

$$R = \frac{S_{row\ of\ N\ ports}}{S_{line\ plume}} = 0.29\,(zN/L)^{2/3}$$

(b) If $w = 2b(z)$ is taken as the nominal half-width of each round buoyant jet (what percentage of $Q(z)$ is accounted for by this definition?), calculate R at the surface, $z = H$. What about R for $z < H$? For a round plume, $db/dz = \beta_G = 0.105$.

3.11 *Turbulent line plume*
(a) The velocity and tracer concentration in a *two-dimensional plume* are confirmed to be self-similar and well-approximated by Gaussian profiles. Employing the entrainment hypothesis and a control volume analysis, deduce the governing equations of a line plume in terms of the centerline velocity $w_m(z)$, plume width $b(z)$, and centerline density deficit $\Delta\rho_m(z)$. Neglect initial volume and momentum fluxes.

(b) Adopting the values of the entrainment coefficient $\alpha = 0.103$ and the ratio of concentration/velocity width $\lambda = 1.35$, deduce the relations for the variation of the characteristic properties of a line plume.

3.12 For a plane buoyant jet, the volume, momentum, and tracer mass fluxes can be obtained in terms of either the average properties (U, B, C) or the centerline variables of the Gaussian profile (u_m, b, C_m), where the terms have their usual meaning. Show by mass, momentum, and tracer mass conservation that:

$$Q = 2UB = \sqrt{\pi} u_m b_g$$

$$M = 2U^2 B = \sqrt{\frac{\pi}{2}} u_m^2 b_g$$

$$QC = \sqrt{\frac{\pi\lambda^2}{1+\lambda^2}} u_m C_m b_g$$

Hence show that the average and centerline properties are related by: $U = u_m/\sqrt{2}$, $B = \sqrt{\pi/2} b_g$, and $C = \sqrt{\lambda^2/(1+\lambda^2)} C_m$.

3.13 *Lagrangian model for line plume in stagnant fluid*
(a) Assume the line plume cross-section can be represented by an average velocity U and a half-width B. Using a Lagrangian plume spreading hypothesis, $dB/dt = \beta U$, develop the governing equations for the plume element as a function of *time*.

(b) Derive the line plume equations, Eq. 78, by simple integration.

(c) Noting that, $w_m = \sqrt{2} U$, $b_g = B/(\sqrt{\pi}/2)$, obtain the prediction of centerline velocity and centerline concentration. For $\beta_G = 0.116$ and $\lambda = 1.35$ evaluate the coefficients in the expressions. Compute the local densimetric Froude number of the plume.

3.14 Dependence of entrainment coefficient on Froude number
When a discharge has both momentum and buoyancy, the entrainment coefficient α is in general not a constant but a function of the local jet densimetric Froude number $\mathrm{Fr_L}$. The dependence of α on $\mathrm{Fr_L}$ can be developed from the mechanical energy equation and is illustrated herein for an arbitrarily inclined plane buoyant jet in stagnant fluid (Jirka and Harleman 1973).

Consider a two-dimensional buoyant jet with velocity $[u, v]$ (s, n) in the streamwise (s) and normal (n) co-ordinates respectively. In this natural co-ordinate system (s, n) following the jet trajectory, the governing continuity and axial momentum equations can be shown to be (invoking the boundary layer approximations):

$$\frac{\partial u}{\partial s} + \frac{\partial v}{\partial n} = 0$$

$$u\frac{\partial u}{\partial s} + v\frac{\partial u}{\partial n} = g' \sin\phi - \frac{\partial \overline{u'v'}}{\partial n}$$

where $g' = \Delta\rho g$, and ϕ = local inclination of jet trajectory to horizontal.

(a) Show that by multiplying the continuity equation with u^2 and the momentum equation with u, the mechanical energy equation for the two-dimensional buoyant jet can be obtained as follows:

$$\frac{\partial u^3}{\partial s} + \frac{\partial u^2 v}{\partial n} = 2\,u\frac{\Delta\rho}{\rho_a} g \sin\phi - 2u\frac{\partial \overline{u'v'}}{\partial n}$$

(b) Assume Gaussian profiles for u and $\Delta\rho$, with $u(s, n) = u_m e^{-n^2/b^2}$, and density deficiency, $\Delta\rho(s, n) = \Delta\rho_m e^{-n^2/(\lambda b)^2}$. Assume that the shear stress is also self-similar (see e.g. Mih and Hoopes 1972), $\overline{u'v'} = u_m^2 f(n/b)$, where $f(n/b)$ is a universal but unknown function.

The integral model equations for continuity and axial momentum can be shown to be:

$$\frac{d}{ds}(u_m b) = \frac{2}{\sqrt{\pi}}\alpha w_m$$

$$\frac{d}{ds}(u_m^2 b) = \sqrt{2} g' \lambda b sin\phi$$

By integrating the mechanical energy equation across the buoyant jet, show that the entrainment function must be of the form

$$\alpha = \sqrt{3} I_1 + \left(\sqrt{2} - \sqrt{\frac{3}{1+\lambda^2}}\right) \sin\phi\, \frac{\sqrt{\pi}\lambda}{\mathrm{Fr_L}^2}$$

where $I_1 = \int_{-\infty}^{\infty} f(n/b) d(n/b)$. Determine the constant $\alpha_1 = \sqrt{3} I_1$ using available data on jets and plumes.
(Hint: Use the identity $\frac{d}{ds}(u_m^3 b) = 2u_m \frac{d}{ds}(u_m^2 b) - u_m^2 \frac{d}{ds}(u_m b)$)

3.15 For an unbounded 2D heated gas plume, show from the plume solution in Table 1 that the mass flux of the plume is given by:

$$M_a = C_m \rho_a \, H^{1/3} L^{2/3} z$$

where L is the lateral length of the line plume, H = heat flux, and ρ_a is the ambient density. Evaluate the dimensionless entrainment coefficient C_m.

Experiments were performed for a laboratory scale model of a line fire (length L). For $H = 8.2$ kW, $L = 1.0$m, find the mass flow rate in the fire plume at a height of 2 m above the source.

3.16 Settled sewage from a population of about 3 million is to be discharged from a 1200 m long submarine diffuser located in a harbour. The design sewage flow is 20 m^3/s, and the depth is 12 m. Assuming a line plume and $\Delta\rho_o/\rho = 0.025$, estimate the minimum surface dilution in near stagnant water. Will the initial dilution be adequate?

3.17 Plume interference

It has been shown that the velocity field of a jet group can be determined by superimposing u^2 (Prob. 2.13). The superposition approach can be extended to compute the velocity field of a plume group by superimposing u^3 (Chap. 11), which can be developed from the mechanical energy equation of a round buoyant jet. Consider a vertical buoyant jet with turbulent-mean velocity $[u, v](z, r)$ in the vertical (z) and radial (r) direction respectively. The governing continuity and axial momentum equations can be shown to be:

$$\frac{\partial u}{\partial z} + \frac{1}{r}\frac{\partial (rv)}{\partial r} = 0$$

and

$$u\frac{\partial u}{\partial z} + v\frac{\partial u}{\partial r} = g' - \frac{1}{r}\frac{\partial (r\overline{u'v'})}{\partial r}$$

where g' is the local reduced gravity, and u' and v' are the turbulent velocity fluctuations in the $z-$ and $r-$direction respectively. By multiplying the continuity equation by u^2, and the momentum equation by $2u$, the mechanical energy equation can be obtained by summing both equations as:

$$\frac{\partial u^3}{\partial z} + \frac{1}{r}\frac{\partial r u^2 v}{\partial r} = 2ug' - \frac{2u}{r}\frac{\partial (r\overline{u'v'})}{\partial r}$$

(a) Assume Gaussian profiles for u and $\Delta\rho$, with $u = u_m e^{-(\frac{r}{b_g})^2}$ and density deficiency $\Delta\rho = \Delta\rho_m e^{-(\frac{r}{\lambda b_g})^2}$, show that the integral model equations for continuity and axial momentum are:

$$\frac{d(u_m b_g^2)}{dz} = 2\alpha u_m b_g$$

and

$$\frac{d(u_m^2 b_g^2)}{dz} = 2\lambda^2 g'_m b_g^2$$

respectively.

(b) By integrating the mechanical energy equation across the jet cross-section, and assuming the shear stress is also self-similar $\overline{u'v'} = u_m^2 f(r/b_g)$, show that the vertical variation of the kinetic energy flux can be expressed as:

$$\frac{d}{dz}\int u^3 dA = \frac{2\pi\lambda^2}{1+\lambda^2} u_m g'_m b_g^2 - 4\pi u_m^3 I_1 b_g$$

where $I_1 = \int_0^\infty e^{-(r/b_g)^2} \frac{\partial f(r/b_g)}{\partial r} dr$ is the shape integral.

(c) In Prob. 3.14 we have shown the dependence of the entrainment coefficient on the local jet densimetric Froude number for a 2D buoyant jet. Following the same approach, by using the identity $\frac{d}{dz}(u_m^3 b_g^2) = 2u_m \frac{d}{dz}(u_m^2 b_g^2) - u_m^2 \frac{d}{dz}(u_m b_g^2)$, show that the entrainment function of a round buoyant jet can be expressed as:

$$\alpha = 6I_1 + \left(2 - \frac{3}{1+\lambda^2}\right)\frac{\lambda^2}{Fr_L^2}$$

where $Fr_L = u_m/\sqrt{g'_m b_g}$ is the local densimetric Froude number. Show that the equation in (b) can be expressed as:

$$\frac{d}{dz}\int u^3 dA = \frac{2\pi\lambda^2}{1+\lambda^2} u_m g'_m b_g^2 - \frac{2}{3}\pi u_m^3 \alpha_j b_g$$

where $\alpha_j = 0.057$ is the entrainment coefficient for a pure round jet (Table 3 in Chap. 2).

(d) Consider the asymptotic case of a round pure plume. Using the pure plume solutions in Table 1, show that:

$$\frac{2\pi\lambda^2}{1+\lambda^2} u_m g'_m b_g^2 \approx 2.0\,\mathrm{F}$$

and

$$\frac{2}{3}\pi u_m^3 \alpha_j b_g \approx 1.3\,\mathrm{F}$$

where $F = \int u^3 g' dA$ is the specific buoyancy flux.

(e) Show that the plume kinetic energy flux can be related to the buoyancy flux as:

$$\frac{dE}{dz} \sim \frac{d}{dz}\int u^3 dA \approx 0.7\,\mathrm{F}$$

and thus the kinetic energy flux increases linearly with elevation:

$$\int u^3 dA \approx kFz$$

with $k = 0.7$.

(f) Show by conservation of buoyancy flux that:

$$\int u_{N_J}^3(x, y, z) dA = \int u_1^3 dA + \int u_2^3 dA + \cdots + \int u_N^3 dA$$

A solution to this equation is:

$$u_{N_J}^3(x, y, z) = u_1^3 + u_2^3 + \cdots + u_N^3$$

The above shows heuristically that the local kinetic energy at each point of a given level z of individual buoyant jets can be superimposed to give the kinetic energy and thus velocity field) of the plume group.

Chapter 4
Inclined Buoyant Jet in Stagnant Environment

The analysis of straight jets and plumes in the previous chapters has provided us with a basic understanding of turbulence and mixing in jets and plumes, and some tools to obtain reliable estimates in asymptotic cases. However, for many engineering problems it is far more convenient and often necesssary to use a general integral jet model. In this chapter we develop such a model for the case of inclined buoyant jets. The formulation for the inclined buoyant jets is an extension to the problems of the straight jets and plumes, and is a prelude to the general formulation of the jet-and-plume problems to include the effects of density stratification and crossflow in the later chapters.

Figure 1 shows the inclined buoyant jet in an otherwise still fluid. Momentum and buoyancy are discharged at a rate equal to M_o and F_o, respectively and at an angle ϕ_o to the horizontal. Figure 2 shows a buoyant jet in a model test for a wastewater discharge (Wah Fu Outfall, Hong Kong) at an inclination of 20 degrees to the horizontal. Close to the source, jet momentum dominates. The jet spreads linearly as it entrains the ambient fluid. The buoyant force acts in the vertical direction of the gravity. It changes the vertical component of the momentum at a rate equal to the buoyant force. The horizontal component of the momentum flux is constant. The initial specific momentum and buoyancy fluxes, M_o and F_o, are the two parameters characterising the source. The length scale

$$l_s = \frac{(M_o)^{\frac{3}{4}}}{(F_o)^{\frac{1}{2}}} \tag{1}$$

defines the near field and the far field. Since $M_o = \pi D^2 V_o^2/4$ and $F_o = \pi D^2 V_o g'_o/4$,

$$l_s = \left(\frac{\pi}{4}\right)^{\frac{1}{4}} D Fr_o \tag{2}$$

Supplementary Information The online version contains supplementary material available at https://doi.org/10.1007/978-3-032-01187-9_4.

J. H. W. Lee et al., *Turbulent Jets and Plumes*,
https://doi.org/10.1007/978-3-032-01187-9_4

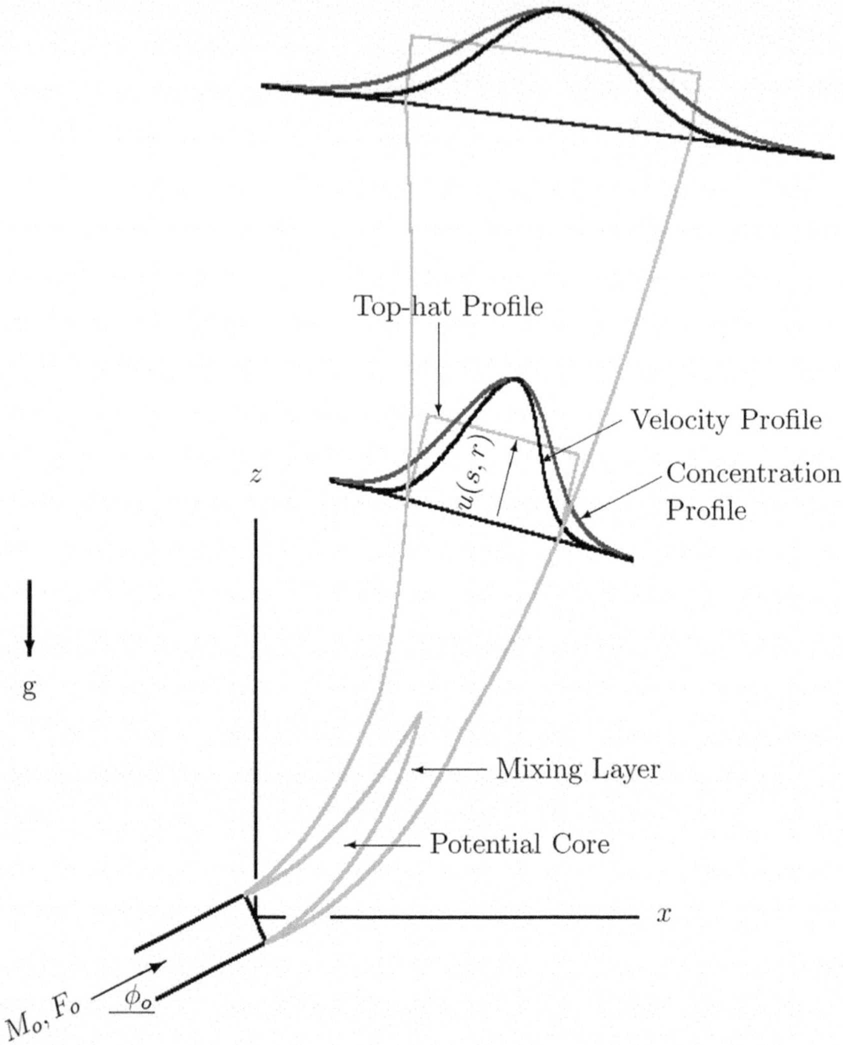

Fig. 1 Schematic diagram of an inclined buoyant jet in stagnant ambient fluid. Gaussian velocity and concentration profiles, and the top-hat profile, are shown at two cross sections along the jet trajectory

where D is the diameter of the source and $Fr = V_o/\sqrt{g_o' D}$ is the densimetric Froude number. The buoyancy effect becomes comparable to that due to the initial momentum at distance $z = l_s$ from the source. Eventually at large distances from the source, $z \gg l_s$, the buoyant jet behaves like a plume. The strength of the jet is directly related to the value of its densimetric Froude number, Fr. The greater the value of the densimetric number the stronger the flow is determined by the jet's initial momentum. The jet turns rapidly into a plume when Fr is small.

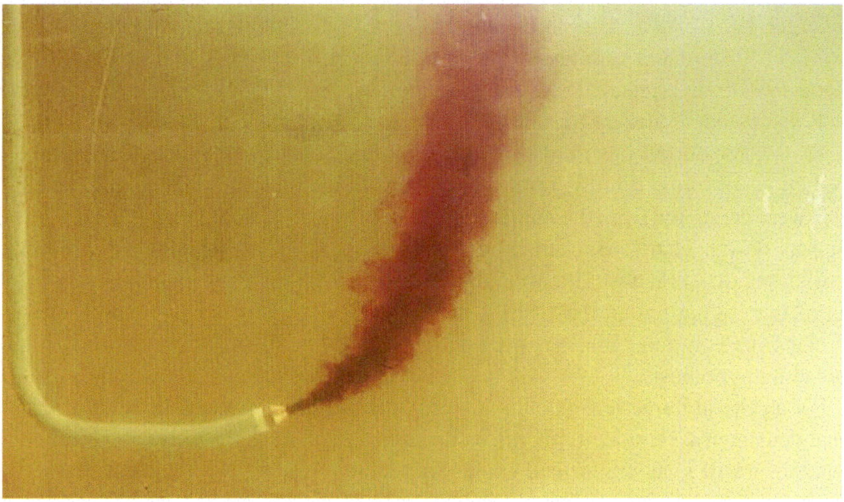

Fig. 2 Observed buoyant jet trajectory in a model test: $Fr = 6.8$, $\phi_o = 20°$

The jet path is in general unknown a priori and it is governed by the initial discharge angle and the balance of inertia and buoyancy. The flow and concentration distributions (and hence the buoyancy distribution) are coupled and have to be solved simultaneously by numerical integration of the governing equations.

The inclined buoyant jet in stagnant fluid has been extensively studied by the integral method in connection with environmental applications (e.g. Brooks 1972; Koh and Brooks 1975; Jirka et al. 1975; Fischer et al. 1979). In a model of this type, the governing boundary-layer equations for continuity, momentum, and tracer mass conservation can be written in a local, 'natural' co-ordinate system (s, r), where s is a streamwise co-ordinate along the jet centerline (defined by e.g. point of maximum concentration), and r the radial co-ordinate away from the centerline. As the shear flow is induced by jet momentum and buoyancy, locally the mixing is analogous to straight jets and plumes. The same boundary-layer approximation and assumptions of self-similar Gaussian profiles for velocity and concentration in the jet cross-section (Fig. 1) can be made. The governing equations can be integrated across the buoyant jet to yield a set of differential equations in terms of the centerline variables, or the mass, momentum, and tracer mass (buoyancy) fluxes. Given the initial conditions, the numerical integration of the resulting system of equations gives a prediction of the jet trajectory, width, velocity, and tracer concentration (dilution).

In most of the models reported in the literature, turbulent closure is formulated using the entrainment hypothesis (e.g. Fan 1967; Fan and Brooks 1969; Ayoub 1971; Hirst 1972; Schatzmann 1979). As noted in the previous chapter, the entrainment coefficient α is not a constant but varies as the turbulent flow changes from a jet to a plume. The dependence of α on the local densimetric Froude number was recognised and a relation formulated for vertical buoyant jets (Fox 1970; List and

Imberger 1973). As an alternative, models based on the spreading hypothesis have also been formulated (Abraham 1963a, b; Jirka et al. 1975; Lee and Jirka 1981; Wood 1993); the equivalence of the two approaches for the limiting cases has also been discussed. Different variants of these models have been extensively applied to wastewater outfall design and related environmental applications. For example, predictions of initial dilution contours as a function of Fr and dimensionless depth y/D were generated from the numerical model of Fan and Brooks (1969), and widely used as design charts. Backed by extensive laboratory experiments, the dilution predictions of Abraham (1963a) and Cederwall (1968) were also extensively used in practice. In particular, the USEPA model UPLUME (Muellenhoff et al. 1985) is an elaborate computer implementation of Abraham's (1963b) theory based on the spreading hypothesis.

For a general numerical calculation, provided the correct dependence of α on the local densimetric Froude number is incorporated, both the spreading or entrainment hypothesis will give similar and equally acceptable results. Nevertheless, the great majority of the models either neglect the potential core, or assume a straight jet of fixed core length for coupling source conditions and those at the start of the zone of established flow (ZEF) where Gaussian profiles can be assumed. In many applications, this may be a reasonable approximation, as the distance from source to the point of interest (e.g. free surface or trapping level) may be much larger than the source dimension (jet diameter), $z \gg D$. Since far away from the source, all discharges behave like plumes, neglecting the potential core will not introduce significant error. Nevertheless, there are many situations where the source dimension or initial core length cannot be conveniently ignored. In the general case, it would be highly desirable to be able to predict the path and length of the potential core.

In this chapter we present a Lagrangian model for an inclined buoyant jet in stagnant fluid, using a spreading hypothesis following the motion of the dominant eddy in the buoyant jet. The constancy of spread rate over the jet-plume range is employed to advantage. The formulation is made first for the zone of established flow (ZEF). The treatment of the zone of flow establishment (ZFE) in the potential core is then described along with its coupling with the flow in the ZEF. The connection between the model and Eulerian formulations is discussed. The model predictions are illustrated and compared with experimental data.

1 Lagrangian Model for Buoyant Jet in Stagnant Fluid

The formulation is illustrated for the round buoyant jet in a stagnant fluid of uniform density (Fig. 3). At sufficiently high Reynolds number, the fluid emerging from the circular nozzle of a round jet would immediately begin to mix with the ambient fluid. A turbulent mixing layer is formed surrounding a core of essentially irrotational fluid from the nozzle. As the mixing layer grows by entrainment of the surrounding fluid, the area of the core decreases correspondingly. The velocity and concentration in the ZFE depend on the potential core and mixing layer surrounding the core. In

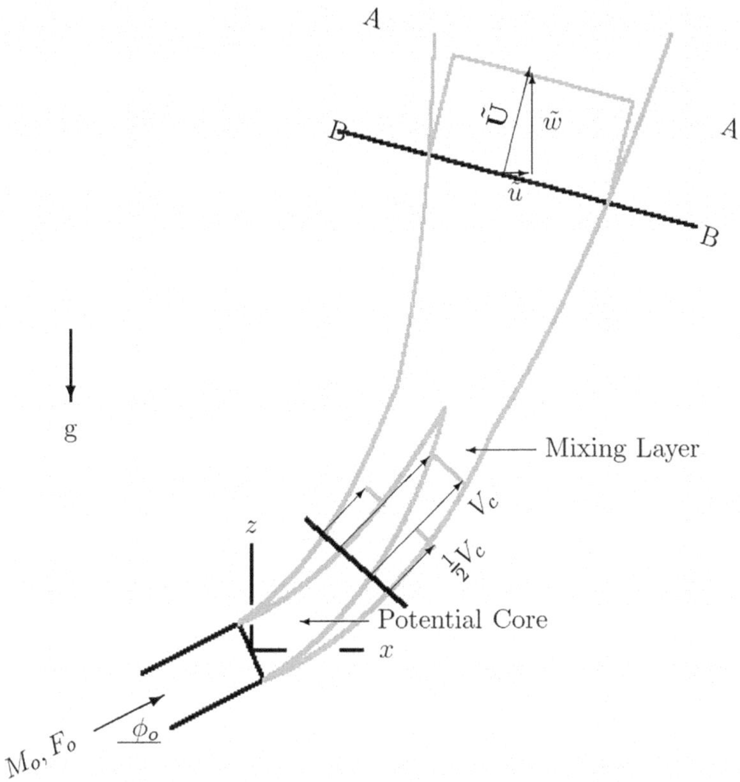

Fig. 3 Schematic diagram of an inclined buoyant jet in stagnant ambient fluid. The top-hat profiles for the velocity of the dominant eddies are shown at a cross section in the ZFE and a cross section in the ZEF

the potential core, the velocity and concentration, V_c and C_c, are unaffected by the mixing process. In the mixing layer, the velocity and concentration are $V_c/2$ and $C_c/2$. These are the average velocity and concentration of the dominant eddies in the mixing layer as envisaged in the dominant-eddy hypothesis proposed by Chu (1994). Entraining fluid from the potential core, and fluid from the stagnant environment, the dominant eddies in the mixing layer move forward with a velocity equal to $\frac{1}{2}V_c$ that is the average of the velocities of the fluid on either side of the mixing layer. Beyond the ZFE, the velocity and concentration of the dominant eddies are $\tilde{U}$ and $\tilde{C}$, respectively. The governing equations are derived first for the ZEF. This is followed by the treatment of the ZFE in a subsequent section.

1.1 Zone of Established Flow (ZEF)

We adopt a Lagrangian approach following the motion of the dominant eddies (Chu 1994). The formulation is carried out using a material volume of fluid that is defined by the release of momentum and buoyancy from the source over a period of one time unit. The initial momentum and buoyancy force associated with this material volume is equal to the momentum flux M_o and buoyancy flux F_o. At a subsequent time, this material volume as schematised in Fig. 3 is bounded between section AA and section BB. The fluid associated with section AA left the source one unit of time earlier than those associated with section BB. If the average velocity of the plume element in the direction of the flow (s-direction) is $\tilde{U} = \sqrt{\tilde{u}^2 + \tilde{w}^2}$, the length of this material volume would be equal to the velocity $\tilde{U}$ times one time unit and that is $\tilde{U}$. If the radius of the cylinderical volume is $\tilde{b}$, the cross section area will be $\pi\tilde{b}^2$ and the volume of the element AA-BB would be $\pi\tilde{b}^2|\tilde{U}|$. Momentum and buoyancy associated with the material volume are $\mathbf{M} = (\pi\tilde{b}^2|\tilde{U}|)\tilde{\mathbf{U}}$ and $F = (\pi\tilde{b}^2|\tilde{U}|)\tilde{g}'$, respectively. In an ambient fluid of uniform density, the buoyancy force associated with the material volume is a constant and is equal to the buoyancy flux at the source as follows:

$$F = (\pi\tilde{b}^2|\tilde{U}|)\tilde{g}' = F_o \tag{3}$$

(assuming without loss of generality a linear equation of state). The momentum flux (that is the momentum associated with the material element AA-BB) has two components. The horizontal component is constant:

$$M_x = (\pi\tilde{b}^2|\tilde{U}|)\tilde{u} = M_o \sin\phi_o \tag{4}$$

since the buoyancy force acts in the vertical direction. The vertical component is changed at a rate equal to the buoyancy force; i.e.,

$$\frac{dM_z}{dt} = \frac{d}{dt}[(\pi\tilde{b}^2|\tilde{U}|)\tilde{u}] = F_o \tag{5}$$

where d/dt is the material derivative for the rate with time in a Lagrangian co-ordinate system moving with the dominant eddies.

A Lagrangian spreading hypothesis is introduced for turbulent closure. The dominant eddies in the buoyant jet are assumed to spread at a rate proportional to the relative velocity between the dominant eddies and its surroundings:

$$\frac{d\tilde{b}}{dt} = \beta|\tilde{U}| \tag{6}$$

where β is the speading coefficient which is assumed to be a constant equal to the value adopted for straight jets and plumes. The path of the jet defined by the co-ordinates $\tilde{x} = x(t)$ and $\tilde{z} = z(t)$ are related to the velocity components of the dominant eddies as follows:

$$\frac{d\tilde{x}}{dt} = \tilde{u} \tag{7}$$

$$\frac{d\tilde{z}}{dt} = \tilde{w} \tag{8}$$

The six unknowns, $\tilde{b}, \tilde{u}, \tilde{g}', \tilde{w}, \tilde{x}$ and $\tilde{z}$, are determined by numerical integration of the sets of the four ordinary differential equations, Eqs. 5, 6, 7, 8, and the two algebraic relations, Eqs. 3 and 4, for constant F and constant M_x.

In a steady flow, the Lagrangian operator is related to the Eulerian advection operator by

$$\frac{d}{dt} = \tilde{U}\frac{d}{ds} \tag{9}$$

where s= distance along the jet trajectory. With this relation, the above system of ordinary differential equations can be cast into Eulerian form as follows:

$$\frac{d\tilde{b}}{ds} = \beta \tag{10}$$

$$\frac{dM_z}{ds} = \frac{F_o}{U} \tag{11}$$

$$\frac{dx}{ds} = \cos\phi \tag{12}$$

$$\frac{dz}{ds} = \sin\phi \tag{13}$$

where $\phi = \tan^{-1}(M_z/M_x)$. This system of ordinary differential equations are in standard (canonical) form and can be solved using standard numerical techniques, e.g. a 5th order Runge-Kutta scheme. The above equations can also be obtained by integrating the equations of motion in natural co-ordinates across the buoyant jet, and can be extended to account for ambient density stratification (see Chap. 5 and Prob. 5.1 in Chap. 5). The Lagrangian formulation however allows a straightforward extension of the procedure to include effects such as unsteadiness, density stratification and crossflow. The general derivation for some of these effects using the Lagrangian method are considered in the later chapters.

1.2 The Potential Core (ZFE)

Most of the existing integral models ignore the presence of the potential core by assigning a virtual origin for the jet which is assumed to be located at a distance $s_e \simeq 6.2D$ from the source. In many cases a straight jet behaviour is postulated close

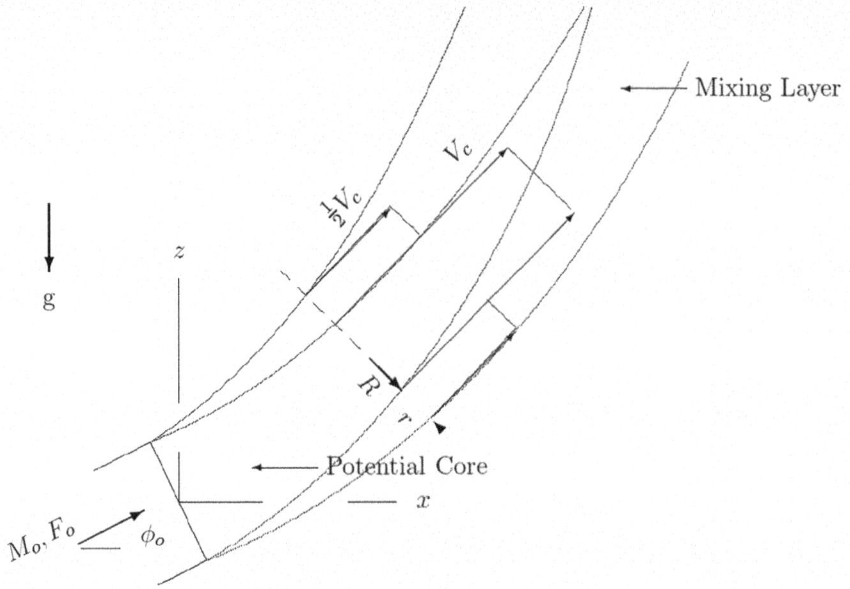

Fig. 4 Development of the mixing layer around the potential core in the ZEF. The radius of the core is R and the size of the dominant eddies in the mixing layer is r

to the source; a potential core of fixed length 6.2 D is assumed. The initial conditions for the governing equations in the zone of established flow (ZEF) are then obtained at $s = s_e$ by momentum conservation (Prob. 4.2). In many practical applications, particularly for buoyancy-dominated jets, the initial mixing of the flow around the core is significant and the mixing process cannot be correctly represented by a point source model. We present here a model of the potential core, and the mixing layer surrounding the core, as shown in Fig. 4. The radius of the core is R. The width of the mixing layer surrounding the core is r. The core velocity is V_c and is uniform across the core. The dominant eddies in the mixing layer moves with a velocity equal to $\frac{1}{2}V_c$ which is the average of the velocity in the core, V_c, and the velocity in the stagnant environment. Hence,

$$\frac{ds}{dt} = \frac{1}{2}V_c. \tag{14}$$

where s is advection distance of the dominant eddies. In a co-ordinate moving with the dominant eddies, the size of the eddies (that is the thickness of the mixing layer), r, increases at a rate proportional to the velocity difference across the mixing layer; i.e.,

$$\frac{dr}{dt} = \beta\frac{1}{2}V_c. \tag{15}$$

The ratio of the above relations gives the spreading rate of the mixing layer

$$\frac{dr}{ds} = \beta. \tag{16}$$

This rate for the mixing layer is known to be the same as the jets and the plume; that is $\beta = \sqrt{2}\beta_G \approx 0.17$ (cf Eq. 61 in Chap. 6 and Fig. 11 in Chap. 2). With this spreading rate and an overall conservation of momentum relation, the length of the potential core can be computed. For a pure momentum jet, the core length ($s_e \simeq 6.2D$) is fixed. For a buoyant jet, the core length depends on buoyancy force and also on the direction of the source momentum relative to this buoyancy force. In general, the velocity in the core, V_c, is not constant but varies with the elevation from the source, z, according to the Bernoulli equation

$$\frac{1}{2}V_c^2 - g_o' z = \frac{1}{2}V_o^2. \tag{17}$$

where g_o' is the reduced gravity and V_o the jet discharge velocity. The gravity term is negative in this equation since the buoyancy force is in the upward direction. The average concentration and buoyancy across the mixing layer are $\frac{1}{2}c_o$ and $\frac{1}{2}g_o'$ respectively. The specific momentum flux in the core is $\pi R^2 V_c^2$. The specific momentum flux associated with the mixing layer is $\pi[(R+r)^2 - R^2](\frac{1}{2}V_c)^2$. The total momentum flux is

$$\begin{aligned} M &= \pi R^2 V_c^2 + \pi[(R+r)^2 - R^2]\left(\frac{V_c}{2}\right)^2 \\ &= \pi V_c^2\left[R^2 + \frac{1}{2}Rr + \frac{1}{4}r^2\right] \end{aligned} \tag{18}$$

Similar to the analysis of the ZEF, the horizontal momentum flux is constant, i.e.,

$$M_x = M_o \cos\phi_o, \tag{19}$$

while the vertical momentum flux increases with the height at a rate equal to the buoyant force:

$$\begin{aligned} \frac{dM_z}{ds} &= \pi R^2 g_o' + \pi[(R+r)^2 - R^2]\frac{g_o'}{2} \\ &= \pi g_o'\left[R^2 + Rr + \frac{1}{2}r^2\right] \end{aligned} \tag{20}$$

The path of the potential core is defined by

$$\frac{dx}{ds} = \cos\phi \tag{21}$$

$$\frac{dz}{ds} = \sin\phi \tag{22}$$

We thus have a system of 4 differential equations (Eqs. 16, 20, 21 and 22) and two algebric relation (Eqs. 17 and 19) for the 6 unknowns in terms of $r(s)$, $R(s)$, $M_x(s)$, $M_z(s)$, $x(s)$, $z(s)$. At any longitudinal location s, we have $M = \sqrt{M_x^2 + M_z^2}$, and $\phi = \sin^{-1}(M_z/M)$. Once $M(s)$ and $r(s)$ are calculated, the radius of the core $R(s)$ can be obtained. To facilitate calculations, we non-dimensionalise all lengths with the source diameter D and momentum fluxes by the initial momentum flux $M_o = V_o^2 \pi D^2/4$. The dimensionless governing equations are then:

$$\frac{d\hat{r}}{d\hat{s}} = \beta \tag{23}$$

$$\frac{d\hat{M}_x}{d\hat{s}} = 0 \tag{24}$$

$$\frac{d\hat{M}_z}{d\hat{s}} = \frac{2}{Fr^2}(2\hat{R}^2 + 2\hat{R}\hat{r} + \hat{r}^2) \tag{25}$$

$$\frac{d\hat{x}}{d\hat{s}} = \cos\phi \tag{26}$$

$$\frac{d\hat{z}}{d\hat{s}} = \sin\phi \tag{27}$$

The dimensionless momentum flux along the jet, after eliminating V_c, is

$$\hat{M} = (1 + 2Fr^{-2}\hat{z})(4\hat{R}^2 + 2\hat{R}\hat{r} + \hat{r}^2) \tag{28}$$

The development of the potential core can then be determined by numerical integration of the system of non-linear relations, Eqs. 23 to 27. The initial conditions at the source are: $(\hat{M}_x, \hat{M}_z)_o = (\cos\phi_o, \sin\phi_o)$, $\hat{R} = 0.5$, $\hat{r} = 0$ and $(\hat{x}, \hat{z}) = (\hat{x}_o, \hat{z}_o)$. At any $\hat{s}$, given the magnitude of $\hat{M}$, the radius of the core $\hat{R}$ is obtained from the solution of the quadratic equation, Eq. 28; i.e.,

$$\hat{R} = \frac{-\hat{r}}{4} + \frac{1}{4}\sqrt{4\frac{\hat{M}}{(1 + 2Fr^{-2}\hat{z})} - 3\hat{r}^2} \tag{29}$$

The core radius R reduces with distance from the nozzle. The core reaches its end when the radius of the core $\hat{R}$ is reduced to zero. At the end of the ZFE, the computed location and orientation, velocity, width, and concentration then become the initial conditions for the calculations in the ZEF. The half-width or radius of the jet at the end of the potential core $\tilde{b}_e$ is the same as the width of the mixing layer, r_e. The initial streamwise velocity and angle at $s = s_e$ are $V_c/2$ and ϕ_e respectively, from which $(M_x, M_z)_e$ can be determined.

2 Numerical Solution

2.1 Jet Trajectory and Potential Core Development

Figure 5 and 6 show typical examples of the computed jet trajectory including the potential core development for a selected range of jet densimetric Froude numbers Fr and discharge angles. For comparison with available experimental data, the results are normalized using the length scale $l_s = \frac{M_o^{\frac{3}{4}}}{F_o^{\frac{1}{2}}} = (\frac{\pi}{4})^{\frac{1}{4}} Fr\,D$. Similarly we can define time and velocity scales based on M_o and F_o. The time scale is

$$t_s = \frac{M_o}{F_o} = \frac{\frac{\pi}{4} D^2 V_o^2}{\frac{\pi}{4} D^2 V_o g'_o} = \frac{V_o^2}{g'_o D} \frac{D}{V_o} = Fr^2 \frac{D}{V_o} \tag{30}$$

and the velocity scale is

$$u_s = \frac{l_s}{t_s} = \frac{F_o^{\frac{1}{2}}}{M_o^{\frac{1}{4}}} = \left(\frac{\pi}{4}\right)^{\frac{1}{4}} Fr^{-1} V_o \tag{31}$$

Figure 5 shows an inclined buoyant jet with $Fr = 4$ and $Fr = 1$ discharged at $\phi_o = -20$ to $45°$. It can be seen that the potential core can occupy a significant portion of the jet path before it turns vertical by virtue of buoyancy; it is interesting that the trajectory of a horizontal plume with $Fr = 1$ is quite different from a vertical plume. The resemblance of the computed jet path for $Fr = 4$, $\phi_o = 20$ to that of the experiment for a similar Fr should also be noted (Fig. 2). The generality of the present model is demonstrated in Fig. 6 which shows the predicted initial mixing of a vertical buoyant jet for $Fr = 20 - 0.1$. It can be seen that for a momentum-dominated discharge, $Fr = 20$, the potential core length is about $z_e/D \approx 6$, but the core length decreases to less than 5 for $Fr = 2$, and smaller than 3 for $Fr = 0.1$. The examples show that depending on the value of the densimetric Froude number and the initial orientation of the jet, the development of the core can be significantly altered by the buoyancy forces.

For a discharge from a nozzle, the limiting condition for which a full flow from the nozzle can be defined is $Fr \approx 1$ (Wilkinson 1988); smaller Fr (e.g. lower jet velocity) will cause intrusion of the ambient fluid into the nozzle. Thus for practical aquatic discharges only $Fr \geq 1$ need to be considered. On the other hand, the case of a pure thermal source of buoyancy (i.e. entirely temperature-induced) has a $Fr \approx 0$ ($u_o \approx 0$). For these extreme cases, the acceleration of the source fluid (or fluid near the source) can lead to contraction of the core such as the example shown in Fig. 6 ($Fr = 0.1$). The densimetric Froude number is zero in the case of plume produced by a fire; the core of the fire is contracting and highly unstable. A swirl is often observed in such strongly buoyant flows as the surrounding fluid is drawn into the low pressure region of the core.

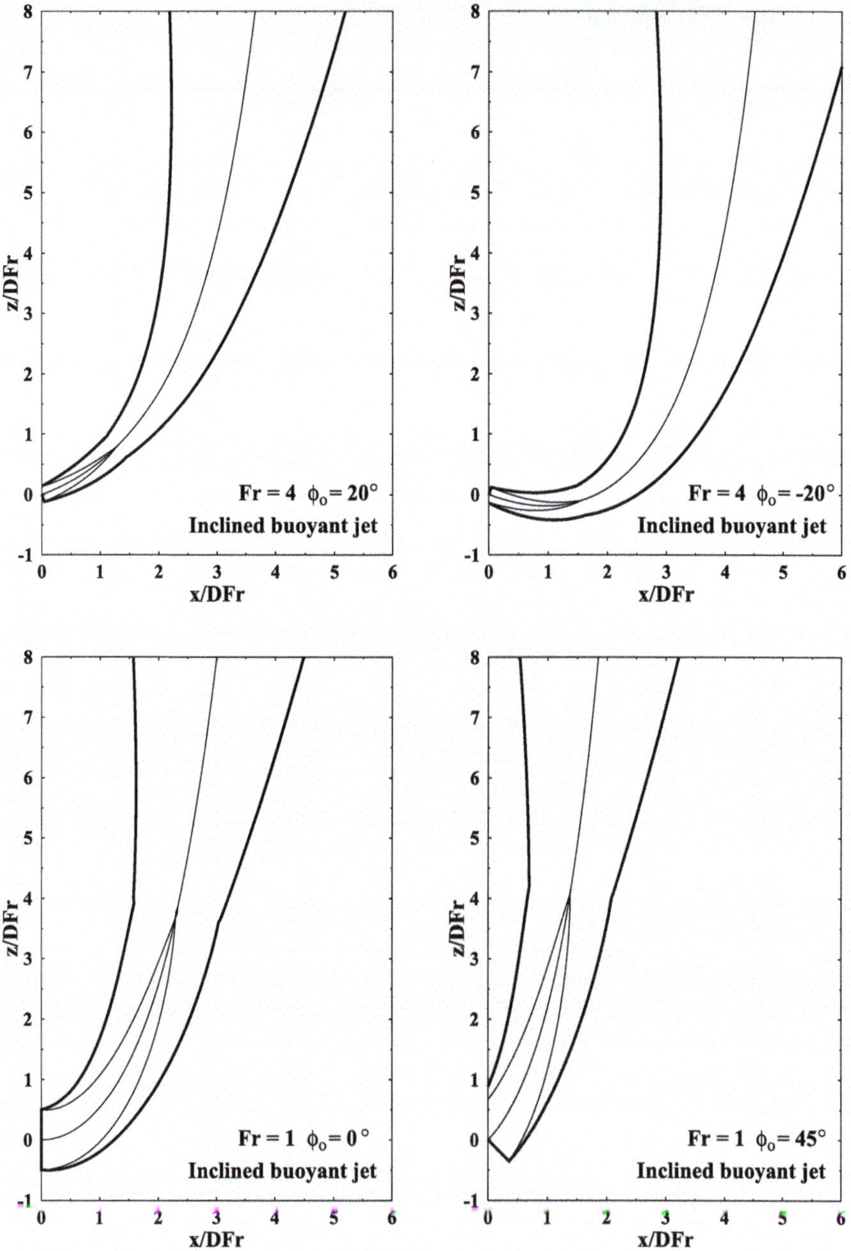

Fig. 5 Computed inclined buoyant jet trajectory showing development of potential core, $Fr =$ 4, 1 and different discharge angles $\phi_o = -20° - 45°$. The jet path is shown in dimensionless coordinates $x/(FrD)$ and $z/(FrD)$

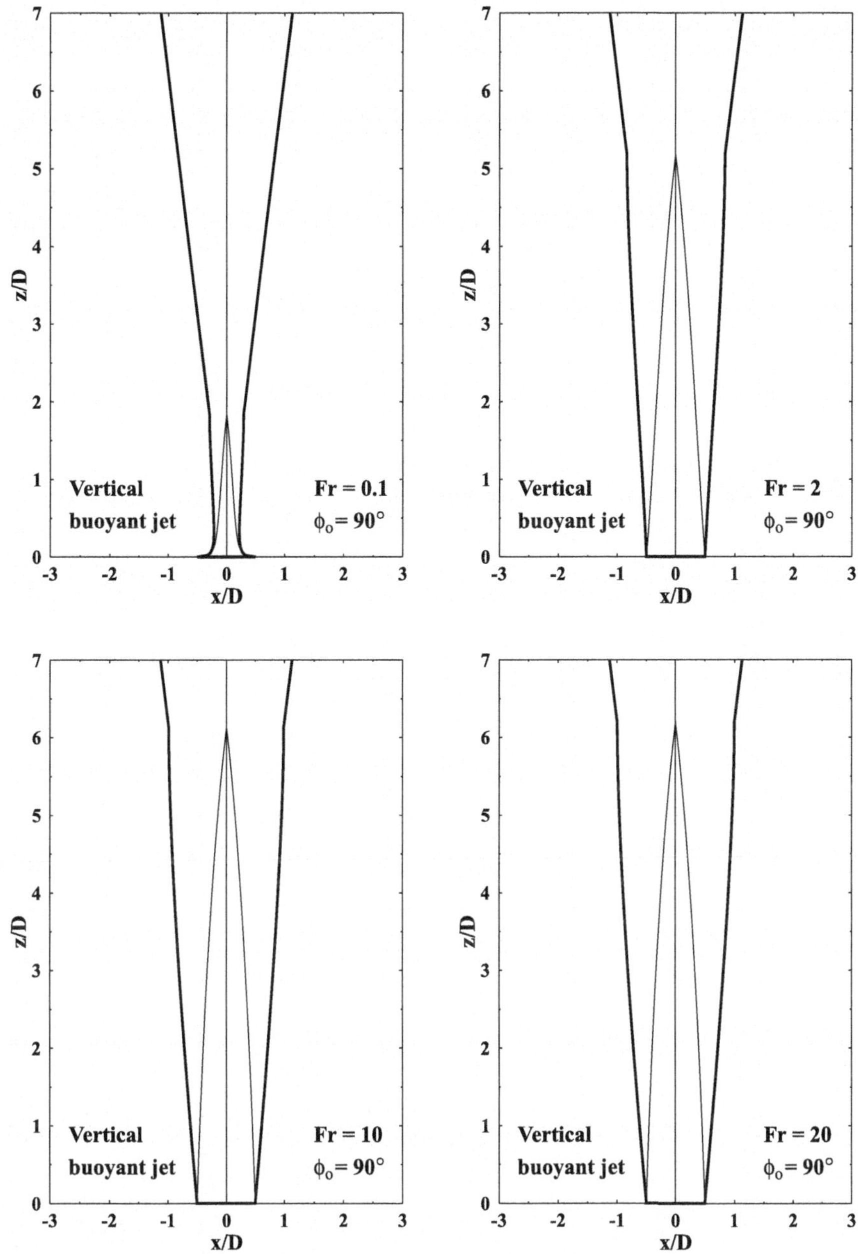

Fig. 6 Potential core development of a vertical buoyant jet for different densimetric Froude number Fr; $Fr = 0.1 - 20$

The results of the numerical integration for the jet trajectory obtained for the horizontal buoyant jet ($\phi_o = 0$) are given in Fig. 7. The dimensionless vertical coordinate of the jet path, $z/(FrD) = (\pi/4)^{\frac{1}{4}}(z/l_s)$, is plotted against the horizontal coordinate, $x/(FrD) = (\pi/4)^{\frac{1}{4}}(x/l_s)$. Model predictions obtained for three densimetric Froude numbers, $Fr = 1{,}4$ and $Fr = 20$, are also shown in the figure as lines in comparison with corresponding experimental data obtained by Anwar (1969), Syrbin and Lyakhovskiy (1936), and Cederwall (1968). Plotted on this scale, it is seen that, supported by the data, the dimensionless trajectories for $Fr \geq 4$ differ very little; the trajectories for $Fr \geq 20$ collapse to the same curve.

Figure 8 shows the observed trajectory and potential core of a turbulent horizontal round buoyant jet in stagnant ambient. These experiments were performed with a negatively buoyant jet, with the initial relative density difference created by salt, $\frac{\Delta\rho_o}{\rho} = 0.025$; at a jet densimetric Froude number of $Fr = 1, 2, 4, 10$. The jet diameter is $D = 33$ mm for $Fr = 1$, $D = 15$ mm for $Fr = 2, 4$, and $D = 5$ mm for $Fr = 10$. A fluorescent dye tracer, Rhodamine 6B, is added to the discharge and the tracer concentration field in the centerline plane made visible by the laser-induced fluorescence (LIF) technique (see Chap. 6). It can be seen the predicted potential core length and shape are well-supported by the experiments, with the core length $s_e/D \approx 3$ for $Fr = 1$ and $s_e/D \approx 4 - 5$ for $Fr = 2 - 4$ respectively.

2.2 Dilution

The mass flux of a conservative tracer is constant. If the average concentration (mass per unit volume) in the material volume AB is $\tilde{c}$, the mass flux would be

$$\Gamma = \tilde{U}\pi\tilde{b}^2\tilde{c} = \Gamma_o. \tag{32}$$

where $\Gamma_o = V_o\frac{\pi}{4}D^2c_o$ is the mass flux at the source. The ratio of the concentration at a particular section of the jet to the source concentration is

$$\frac{\tilde{c}}{c_o} = \frac{\frac{\pi}{4}D^2V_o}{\pi\tilde{b}^2\tilde{U}} = \frac{Q_o}{Q} = \frac{1}{\tilde{S}} \tag{33}$$

where $\tilde{S} = c_o/\tilde{c} = Q/Q_o$ is the average dilution, defined as the ratio of the jet volume flux Q and the source volume flux Q_o. In dimensionless form, we have

$$\tilde{S} = \frac{c_o}{\tilde{c}} = \left(\frac{4}{\pi}\right)^{\frac{1}{4}} Fr\pi\hat{b}^2\hat{U} \tag{34}$$

where $\hat{b} = \tilde{b}/l_s$, and $\hat{U} = \tilde{U}/u_s$. For a Gaussian velocity and concentration profile, defined by velocity half-width (based on the $1/e$ centerline velocity point) b_v and concentration half-width $b_c = \lambda b_v$, where $\lambda \approx 1.2$, application of a mass balance

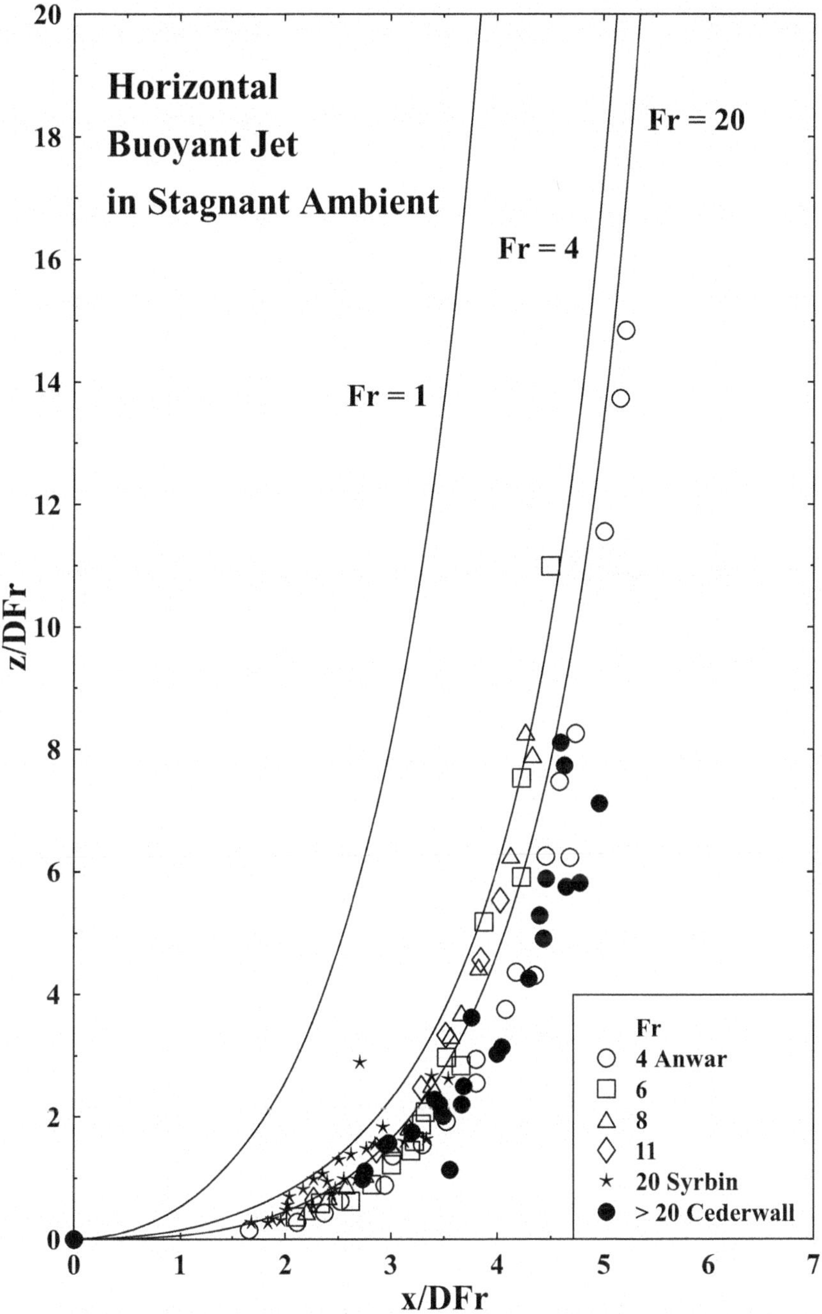

Fig. 7 Predicted normalized trajectory of the horizontal buoyant jet

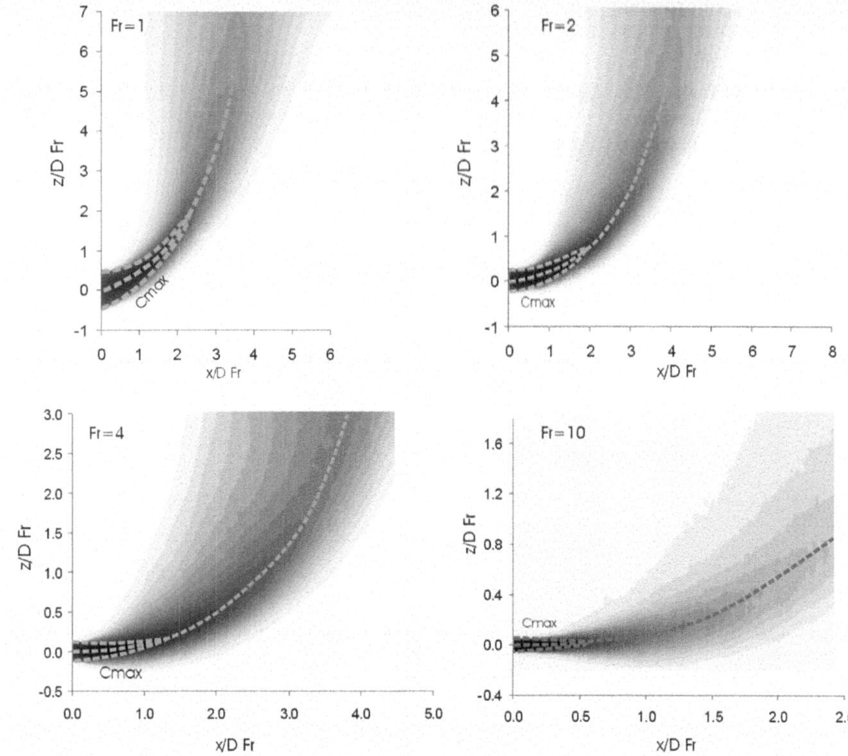

Fig. 8 Observed concentration field and potential core of horizontal round buoyant jet in stagnant fluid

(Prob. 2.1 in Chap. 2) gives the relation between the centerline dilution (minimum dilution based on the maximum concentration in the jet cross-section) and the average dilution:

$$S_m = \frac{c_o}{c_m} = \frac{\lambda^2}{1+\lambda^2}\frac{Q}{Q_o} = \frac{\lambda^2}{1+\lambda^2}\tilde{S} \tag{35}$$

Hence,

$$S_m = \left(\frac{4}{\pi}\right)^{\frac{1}{4}} Fr\, \pi\hat{b}^2\hat{U}\frac{\lambda^2}{1+\lambda^2} \tag{36}$$

The model predictions of this minimum dilution ratio, S_m, divided by the densimetric Froude number Fr is plotted as a function of the dimensionless elevation $z/(FrD)$, and compared with the experimental data of the same range of densimetric Froude number in Fig. 9. In the potential core, the concentration c_m is a constant equal to c_o and the minimum dilution $S_m = c_o/c_m = 1$. The horizontal lines in the figure, $S_m/Fr = 1/4 = 0.25$ and $S_m/Fr = 1/10 = 0.1$, are the values associated with the potential core for $Fr = 4$ and 10, respectively. On the whole the model predictions

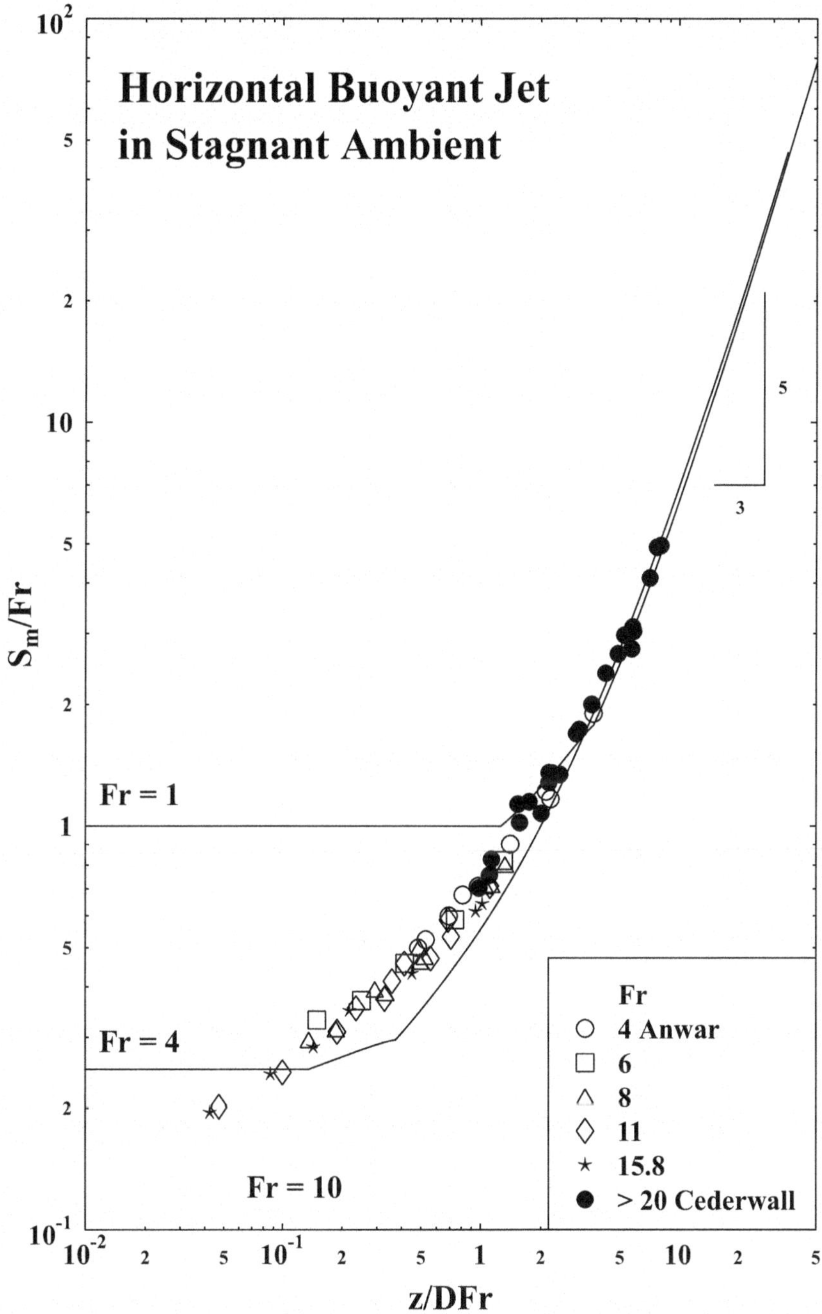

Fig. 9 Predicted centerline (minimum) dilution of the horizontal buoyant jet, S_m/Fr, as a function of $z/(FrD)$

of the dilution follow quite closely the trend of the experimental data but appears to be slightly below the values observed in the experiment.

It should be noted that a secondary circulation is produced on the plane perpendicular to the path of the jet due to buoyancy. This circulation, in the form of a vortex pair, promotes turbulent entrainment and may cause a further dilution of the jet that has not been accounted for by the present model (see Chap. 8 on advected line thermal).

For the same source densimetric Froude number Fr, a horizontal buoyant jet will result in a longer jet trajectory to rise to a given depth; hence the dilution is greater for a horizontal jet than a corresponding vertical jet. Figure 10 compares the predicted dilution obtained for a horizontal jet and a vertical jet. The dilution of the horizontal jet is clearly the higher up to a dimensionless depth of $z/DFr \approx 10$.

Based on experiments and numerical model calculations, Cederwall (1968) has given a widely used equation for predicting centerline dilution for a horizontal round buoyant jet in still fluid.

$$S_m = 0.54\ Fr \left(0.38\frac{z}{FrD} + 0.66\right)^{5/3} \quad \text{for} \quad \frac{z}{FrD} > 0.5 \tag{37}$$

$$S_m = 0.54\ Fr \left(\frac{z}{FrD}\right)^{7/16} \quad \text{for} \quad \frac{z}{FrD} \leq 0.5 \tag{38}$$

It can be shown the above equation reduces to the point plume equation as $z/(FrD) \to \infty$. For $z/D \geq 10$, the present model predictions agree closely with the experimentally derived Cederwall equation (not shown).

2.3 *Boundary Effects*

The theoretical model generally works well for free buoyant jets in stagnant ambient. However, if the jet nozzle is located very close to the bed, the pressure reduction due to the confinement of the entrainment flow demand may cause the plume to cling to the bottom boundary (Coanda effect). For nearly horizontal buoyant jets inclined at angle ϕ_o to vertical, a simple criterion for bottom attachment is given by Jirka and Doneker (1991) as $\tan\phi_o \leq 0.2 - z_o/l_s$, where z_o = height of nozzle centerline above the bed (Prob. 4.9). Experiments by Sharp and Vyas (1977) and Sobey et al. (1988) have shown the horizontal buoyant jet will be attached if $z_o/l_s \geq 0.1 - 0.15$. Similarly, for high momentum discharges in shallow water, the stratified flow established at the free surface may be hydrodynamically unstable. Recirculating flow cells may be established adjacent to the jet; partially diluted effluent is hence re-entrained into the jet, leading to reduced dilution. A simple criterion for jet stability has been given as $l_s/H \leq 4.3$ or $H/D \geq 0.22Fr$ with apparently little sensitivity to the discharge angle (Lee and Jirka 1981; Jirka 1982). For jet discharges with strong bottom or surface interaction, alternative numerical solution methods may be used (see Chaps. 7 and 8).

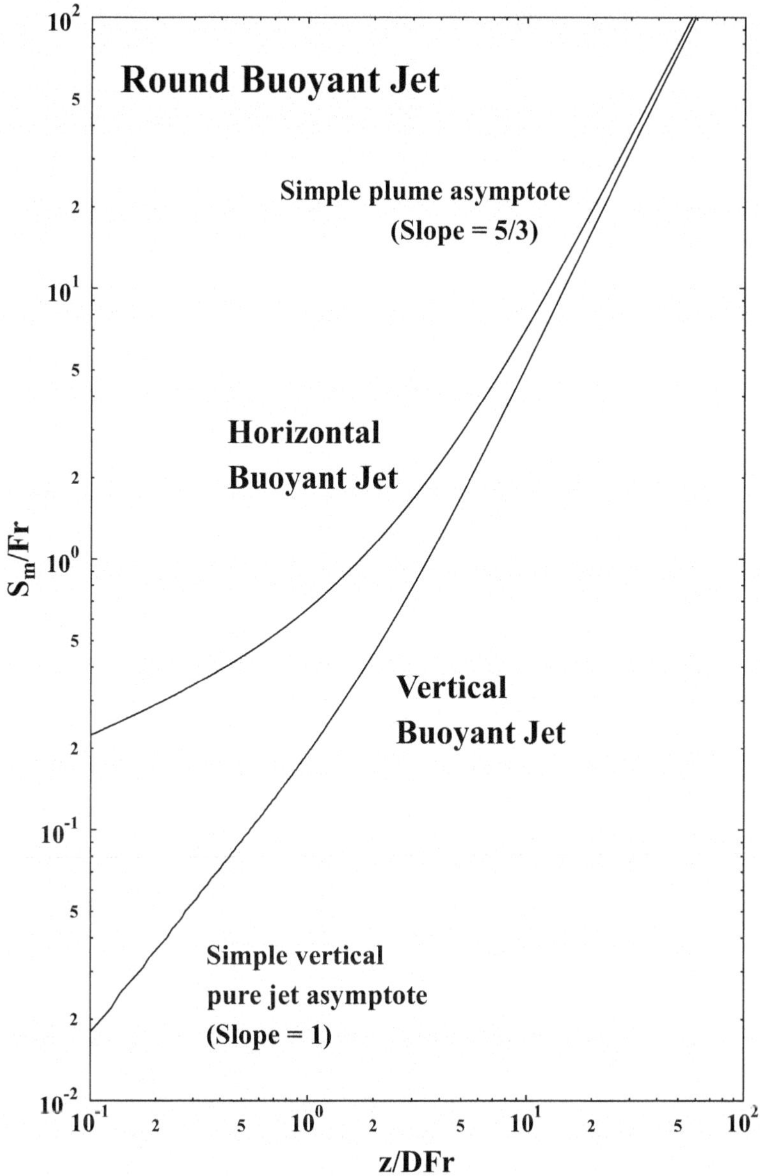

Fig. 10 Predicted normalized centerline dilution of horizontal and vertical buoyant jet

3 Application Examples

The numerical model along with the general numerical solution (Fig. 9) can be used to illustrate many principles underlying the design of environmental mixing devices. They are illustrated in the following worked examples.

Example 4.1 *Consider a single submerged round buoyant jet with* $V_o = 0.5$ *m/s*, $D = 0.25\ m$, *and* $\frac{\Delta\rho_o}{\rho_a} = 0.025$. *Estimate the minimum surface dilution in a depth of 25 m for* $\phi_o = 0°$ *and* $\phi_o = 90°$.

Solution: The source densimetric Froude number for this inclined buoyant jet is

$$Fr = \frac{0.5}{\sqrt{0.025 \times 9.81 \times 0.25}} = 2.0.$$

Compute $z/FrD = 25/(2 \times 0.25) = 50$. The data in Fig. 10 shows that the dimensionless dilution is $S_m/Fr \approx 80$, that is $S_m = 160$ for both the horizontal jet (Case a) and vertical jet (Case b). Alternatively, the Cederwall equation can be used to compute S_m and that gives almost an identical value $S_m = 0.54(2.0)(0.38(50) + 0.66)^{5/3} = \underline{155}$.

For this low value of $Fr = 2.0$, the forced plume trajectory is near vertical over a large portion of its trajectory; thus it is not surprising the dilution is virtually independent of the initial discharge angle.

Example 4.2 *Consider now a submerged jet discharging at high velocity in depth of 10 m,* $V_o = 3.1\ m/s$, $D = 0.1\ m$ *and* $\frac{\Delta\rho_o}{\rho_a} = 0.025$. *Estimate the minimum surface dilutions for both vertical and horizontal discharge cases. Consider also the case of a jet discharging downwards at an angle of 30 degrees to the horizontal.*

Solution: The source densimetric Froude namber of this discharge at high velocity is $Fr = 3.1/\sqrt{0.1 \times 0.025 \times 9.81} = 19.8$. Compute $z/FrD = 10/(19.8 \times 0.1) = 5.1$. From Fig. 9, we obtain $S_m/Fr \approx 1.8$, that gives a minimum dilution $S_m = \underline{34}$ for the vertical jet. For the horizontal jet, we obtain a minimum dilution of $S_m = \underline{55}$, or a similar result of $S_m = 53$ if the Cederwall Equation is used to do the calculation.

For a relatively high momentum discharge, a horizontal discharge is more effective than a vertical one because of its longer trajectory, and hence larger surface area for entrainment. With a downward discharging jet, the momentum flux opposes the buoyancy initially in very much the same way as a dense plume (Chap. 3); however, the much longer trajectory results in higher dilution. Figure 11 shows the computed trajectory and average dilution at z = 10 m and z = 17 m for the three cases. It can be seen the effect of discharge angle is relatively more important at smaller depths. In some cases it may be advantageous to issue the jet downwards to increase initial dilution; however a jet discharging in proximity to a solid boundary may become attached to the bottom due to the *Coanda Effect* (Prob. 4.9). The dilution of such a wall jet can be significantly increased than a free jet, with however potential impacts on the benthos on the sea bed.

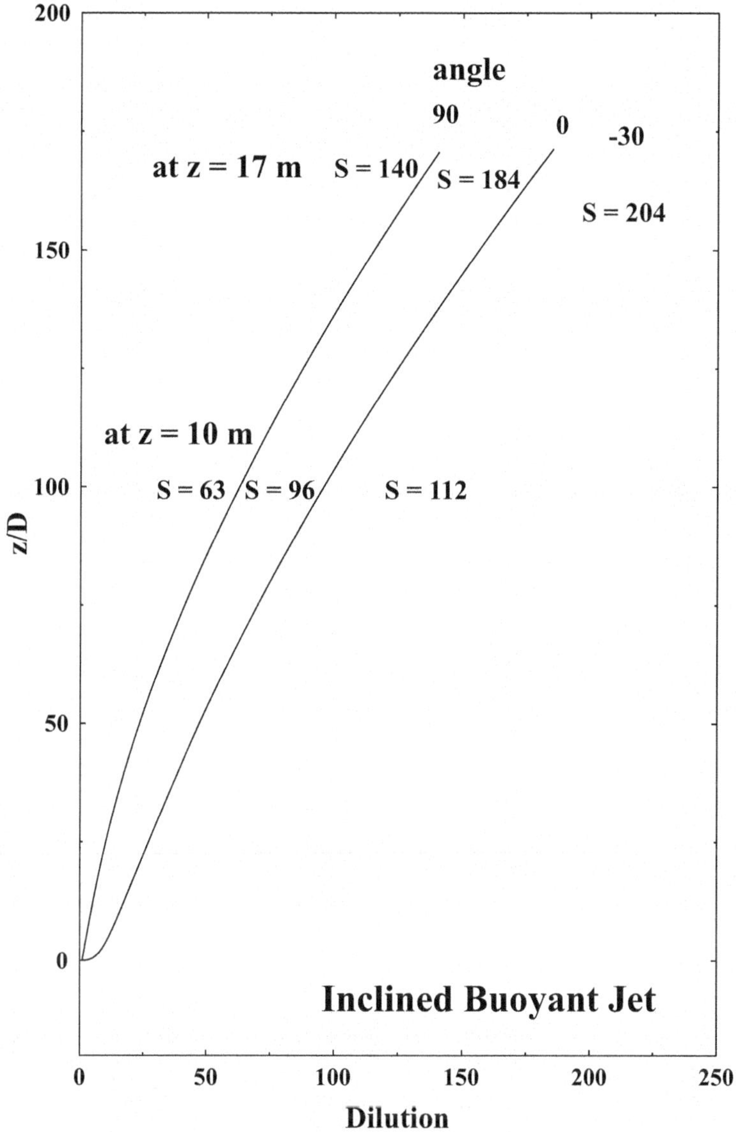

Fig. 11 Predicted trajectory and dilution for three discharge angles

Example 4.3 *A sewage treatment plant discharges* 0.088 m^3/s *of screened domestic wastewater in the form of horizontal jets into a harbour of mean depth* 12.2 *m. As some brackish water is used for toilet flushing, the relative density difference is about* 0.014. *Evaluate the two methods of disposal: (a) a single submerged horizontal jet with diameter* 0.37 *m; and (b) discharge in the form of* $N = 14$ *identical jets (such* *a design is called a multiport diffuser). The velocity in each case is kept the same. The horizontal jets are adequately spaced apart such that they do not merge before surfacing.*

Solution: For case (a), the discharge velocity can be computed to be 0.8 m/s, with $Fr = 3.55$. For case (b), the discharge velocity is kept the same at 0.8 m/s, and the corresponding jet diameter for the same total flow is 0.1 m.

	Case a $N = 1$			Case b $N = 14$	
V_o	=	0.8 m/s	V_o	=	0.8 m/s
D	=	0.37 m	D	=	0.1 m
Fr	=	3.55	Fr	=	6.8
$\frac{z}{FrD}$	=	9.3	$\frac{z}{FrD}$	=	17.9

Using Cederwall equation or the dilution nomogram (Fig. 10), the computed dilution for z = 12.2 m is approximately 21 for the single jet (case a), and 105 for the multiple jets with same velocity (case b). By dividing the given discharge flow into a number of buoyant jets, the total surface area for jet entrainment is increased. A multiple port outfall is therefore a much more powerful mixing device than a single port outfall—in this example a gain in surface dilution by over a factor of six and at no extra pumping cost!

Example 4.4 *In a Scandinavian country, it is proposed to make use of the freshwater used for hydropower to enrich the primary productivity of fjords; the surface layers of these waters are often poor in nutrients. In a fjord of weak currents and relatively stable stratification the design is to discharge freshwater as a submerged buoyant jet. The objective is to lift by jet entrainment the nutrient rich water from depths below 30 m to a depth of about 10 m throughout the long summer, thus extending the growing season of marine organisms for aquaculture (McClimans and Eidnes 2000).*

Figure 12 shows the ambient salinity profile for two cases. The jet is discharged at a downwards angle at 30 m depth. Two cases tested in the laboratory are as follows: (a) $Q_o = 6\ m^3/s$, $D = 1\ m$, $\phi_o = -35°$ and $Fr = 15.3$; the ambient salinity at source level is $S_a = 32.1\ ppt$. (b) $Q_o = 6\ m^3/s$, $D = 2\ m$, $\phi_o = -45°$ and $Fr = 2.65$; the ambient salinity at source level is $S_a = 33.6\ ppt$. Compute the unknown jet trajectory and average dilution at the surface or submerged level.

Solution: The ambient density can be computed as a function of salinity and temperature via the equation of state (Appendix A). Figure 12 shows there is a density difference of $5.4\sigma_t$ units between the surface and bottom waters in the fjord. Natural stratification is typically nonlinear and calculation of initial mixing necessitates the use of a general mathematical model. The source diameter is also not negligible compared with the scale of interest, of the order of 10–20 m. The general integral model (accounting for potential core) is used to compute the trajectory and dilution using the source conditions and ambient density profile (from the equation of state,

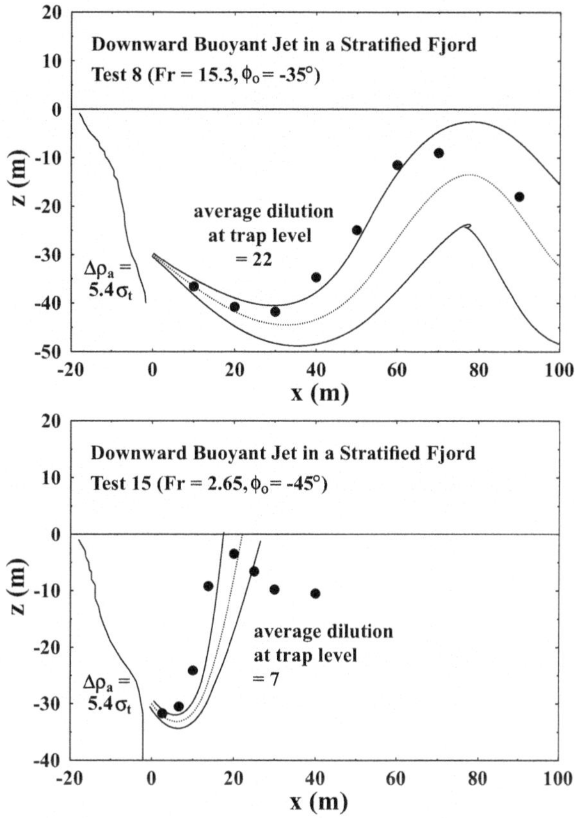

Fig. 12 Artificial upwelling by buoyant plume (solid circles: observed trajectory)

Appendix A). For the first case the downward jet discharges at a jet densimetric Froude number of $Fr = 15.3$; the jet momentum results in an extended trajectory. The jet mixes with sufficient amounts of the ambient lower layer fluid so that the density of the mixed fluid is less than that of the surface layer; the plume thus stays submerged at an equilibrium trap level of depth around 20 m. The predicted average dilution is about 22. This means that the buoyant plume continually brings a flow of $22 \times 6 = 132\ \text{m}^3/\text{s}$ to the euphotic zone. Based on measured nutrient profiles, the supply of nutrient-rich flow from the deep layers can then be computed; the primary production (photosynthesis) of the surface layer is enhanced. On the other hand, case (b) is a relatively buoyant discharge, with $Fr = 2.65$. The buoyant plume does not penetrate deep enough into the lower layer, and the plume impinges onto the water surface with a dilution of only 7. It can be seen that the predictions compare well with the laboratory data for both cases. Calculations have shown for this situation that a downward directed jet would give a good mixing in the deeper water below the euphotic zone, and would be less sensitive to changes in stratification. On the other hand, the discharge must be large enough so that the mixed flow has sufficient buoyancy to force it high enough to promote primary production (McClimans and Eidnes 2000).

Example 4.5 *Deposition of a sediment-laden buoyant jet*
Preliminarily treated domestic sewage discharges often contain polluted suspended solid or sediment that are harmful to the environment, and the prediction of the deposition of the sediment is essential to ensure the discharge will not cause unacceptable environmental impact. Consider an inclined buoyant jet discharge from an outfall with a daily averaged flow of $0.088\,m^3/s$ *into a stagnant and unstratified ambient. The ambient sea water and sewage densities are* $\rho_a = 1018\,kg/m^3$ *and* $\rho_o = 1004\,kg/m^3$ *respectively. About 40 tonnes of sewage solids with diameter 300* μm, *density* $\rho_s = 2650\,kg/m^3$, *and settling velocity* $w_s = 4.3\,cm/s$ *are discharged to the marine environment through the outfall each year. Determine the sediment deposition profile. The jet discharge parameters are given as follows:*
Jet discharge $Q_o = 0.00626\,m^3/s$
Jet diameter $D = 0.1\,m$
Jet velocity $u_o = 0.8\,m/s$
Water depth $H = 7\,m$
Discharge angle $\phi_o = 20°$

Solution: This is an example of a two-phase buoyant jet flow problem—wastewater as the fluid phase, and sediment as the solid phase. The initial concentration of the sediment can be estimated as:

$$C_o = \frac{mass}{volume} = \frac{40 \times 1000\ \text{kg}}{0.00626 \times 3600 \times 24 \times 365\ \text{m}^3} = 0.2\ \text{kg/m}^3 = 200\ \text{mg/L}$$

Volume fraction of sediment can be estimated as $\alpha_o = C_o/\rho_s = 0.2/2650 = 0.008\%$. At such a low volume fraction, it is expected that the buoyant jet flow will not be affected by the sediment (at high sediment concentration, a negative buoyancy will be imparted to the jet flow), and its mixing characteristics can be predicted using the general integral buoyant jet model developed in this chapter.

The model predicted jet trajectory and width of the sediment laden jet is shown in Fig. 13. The densimetric Froude number is $Fr = 6.9$, and is almost horizontal, hence the predicted trajectory is close to that of Fig. 7. Assuming a Gaussian jet velocity profile and a distributed point sink model to determine the jet-induced external flow (Chap. 11), the complete buoyant jet induced flow field can be predicted.

The sediment will gradually fall out from the jet due to its settling velocity. The motion of the sediment due to the buoyant jet flow and its settling can be predicted using a particle-tracking model that is coupled with the buoyant jet induced flow field (Chan 2013). The model has been validated extensively with experimental data, and the predicted sediment jet trajectory and deposition profile are shown in Figs. 13 and 14, respectively. The deposition profile consists of a narrow concentrated zone corresponding to fall out from the jet margin, and a spread-out circular region corresponding to fall out from the buoyant surface current (see Chan 2013 for more

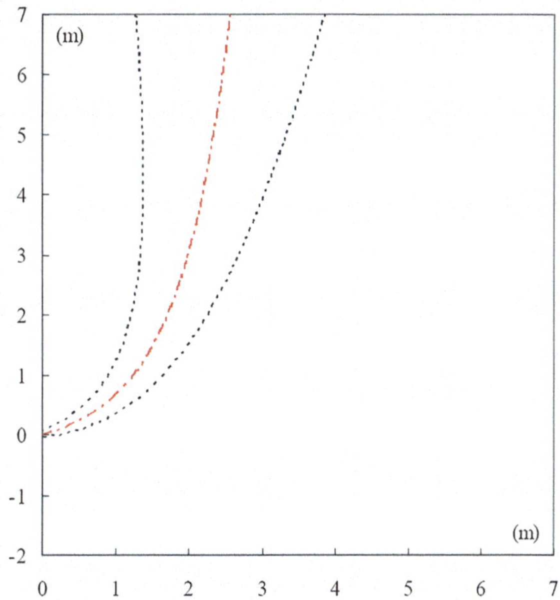

Fig. 13 Sediment laden jet trajectory predicted by an integral model in Example 4.5 (Chan 2013)

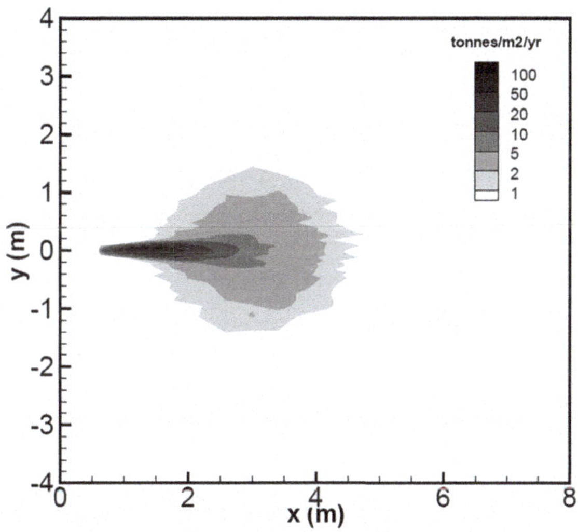

Fig. 14 Sediment deposition profile predicted by coupling a particle tracking model with a jet integral model in Example 4.5 (Chan 2013)

details). This gives a general idea on how the sediment deposits close to the outfall and its potential impact on marine water quality. In reality, the ambient tidal current would also affect the sediment transport and erosion. Particle size distribution and flocculation are also complicating factors.

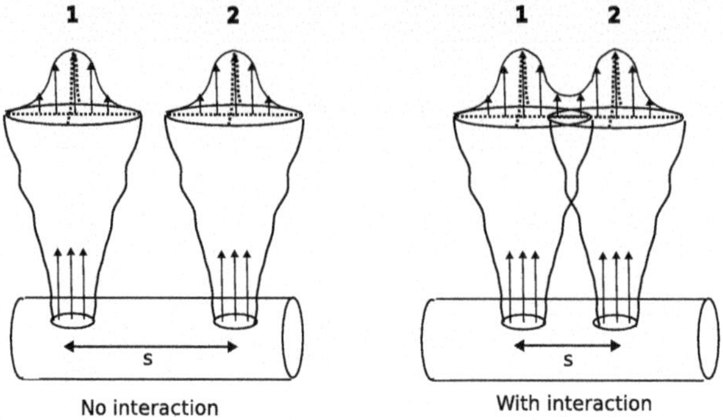

Fig. 15 The two outfall openings damaged by the dredger

Example 4.6 *Buoyant jet interference—outfall damaged by dredger*
Problem 4.11 describes an accident in which a dredger damaged an outfall pipe, resulting in two large openings that discharged most of the sewage. It is assumed that the openings are sufficiently wide apart that the two plume discharges can be regarded as independent and analyzed using single buoyant jet theory (Fig. 15). In this example, we consider the more general case of two openings close to each other such that the plumes overlap and interact. Using the discharge parameters in Problem 4.11, suppose the separation of the two holes $s = 2$ m, denoting u as the streamwise jet velocity, determine the centerline velocity of each plume at $H = 20$ m assuming (i) u; (ii) u^2; and (iii) u^3 is additive.

Solution: After the outfall was damaged by the dredger, the discharge from each of the two holes has the characteristics:

- Flow rate $Q_o = 0.25$ m^3/s
- Diameter $D = 0.35$ m
- Density difference $\Delta\rho/\rho_a = 0.025$
- Initial jet velocity $u_o = Q_o/(\pi D^2/4) = 2.60$ m/s
- Buoyancy flux $F_o = Q_o g' = 0.0613$ m^4/s^3
- Momentum flux $M_o = Q_o u_o = 0.65$ m^4/s^2

Then $l_s = M_o^{3/4}/F_o^{1/2} = 2.92$ m and $H/l_s = 6.8 \gg 1$, hence the discharge is plume-like and the plume solution can be applied.

The velocity $u(z, r)$ of each independent plume at centerline is $u_1(20, 0) = u_2(20, 0) = 4.71 F_o^{1/3} H^{-1/3} = 4.71 \times 0.0613^{1/3} \times 20^{-1/3} = 0.684$ m/s. Noting that the plume velocity profile is Gaussian, the plume velocity at a radial distance of $s = 2$ m is $u_1(20, 2) = u_2(20, 2) = u_m e^{-(r/b)^2} = 0.684 e^{-[2/(0.105\times 20)]^2} = 0.276$ m/s, where $b = 0.105H$ using the spreading rate of a round plume.

(i) If u is additive, the interacting plume centerline velocity $u_{1_{\text{int}}}(20, 0) = u_1(20, 0) + u_2(20, 2) = 0.684 + 0.276 = \underline{0.96 \text{ m/s}}$

(ii) If u^2 is additive, then $u_{1_{\text{int}}}(20,0)^2 = u_1(20,0)^2 + u_2(20,2)^2 = 0.684^2 + 0.276^2 = 0.544\ \text{m}^2/\text{s}^2$, hence $u_{1_{\text{int}}}(20,0) = \sqrt{0.544} = \underline{0.74\ \text{m/s}}$

(iii) If u^3 is additive, then $u_{1_{\text{int}}}(20,0)^3 = u_1(20,0)^3 + u_2(20,2)^3 = 0.684^3 + 0.276^3 = 0.342\ \text{m}^3/\text{s}^3$, hence $u_{1_{\text{int}}}(20,0) = (0.342)^{1/3} = \underline{0.69\ \text{m/s}}$

By symmetry, $u_{2_{\text{int}}}(20,0) = u_{1_{\text{int}}}(20,0)$. Apparently there are three different approaches to determine the interacting plume velocity and can differ by 40%, which is correct? The topic of jet and plume interaction will further be discussed in Chap. 11.

4 Summary

A general mathematical model is needed for solving environmental mixing problems encountered in practice. In this chapter we have presented a general integral model for an arbitrarily inclined round buoyant jet in stagnant ambient fluid. The model adopts a Lagrangian spreading hypothesis along with the use of top-hat profiles; it predicts the jet characteristics in both the potential core and the region of established flow. Model predictions of dilution and trajectory are well supported by experimental data. The model can readily be extended to treat ambient density stratification. The subject of density stratification is treated in detail in the next chapter.

Problems

4.1 A discharge of $0.15\ \text{m}^3/\text{s}$ is released at depth into a fluid of uniform density, with a relative density difference of 0.025 at the point of release. Calculate the initial dilution for a depth of 30 m above the discharge site assuming the release is:

(a) a round buoyant plume

(b) a round buoyant jet (initial velocity 4 m/s) discharged (i) vertically; and (ii) horizontally. Compute the momentum length scale l_s, and estimate the jet width at the free surface for case (ii).

4.2 The numerical solution of a typical Eulerian integral jet model assumes that the velocity and tracer concentration profiles are Gaussian at the origin (i.e. the flow is fully developed right at source). In reality, it takes a certain distance before turbulent entrainment penetrates into the jet centerline. In dilution prediction, the effect of the zone of flow establishment (ZFE) is sometimes accounted for by a correction factor:

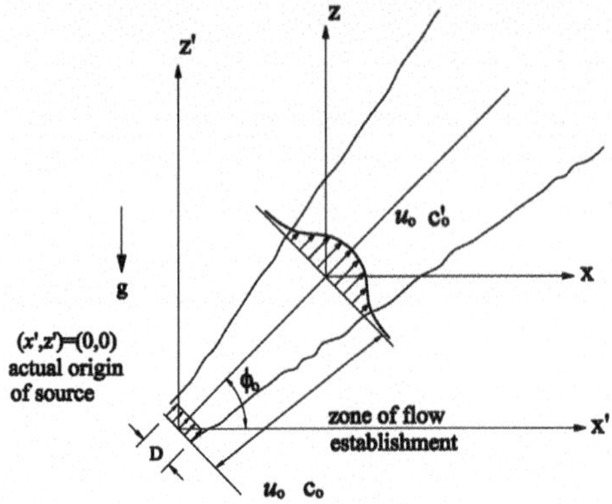

Assuming that buoyancy effects are negligible close to the discharge, apply momentum and tracer mass conservation between the discharge and the end of ZFE at $(x, z) = (6.2D\cos\phi_o, 6.2D\sin\phi_o)$ to establish the following relations for a round buoyant jet:

$$b_o = D/\sqrt{2}$$

$$c_o' = c_o \frac{1+\lambda^2}{2\lambda^2}$$

$$S = \frac{2\lambda^2}{1+\lambda^2} S_o$$

where S_o = centerline dilution computed without correction for ZFE.

4.3 Consider a vertical round buoyant jet with the following characteristics:

Jet velocity	w_o	=	1 m/s
Jet diameter	D	=	0.15 m
Density difference	$\Delta\rho_o/\rho_a$	=	0.025
Water depth	H	=	10 m

Compute the length scale l_s and the initial dilution in still water. How would the dilution and length scale change if the relative density difference is 0.001 instead? (what situation does this correspond to?) Calculate the power of this jet in kW.

4.4 (a) A screening plant discharges a sewage flow of $0.25\ m^3/s$ horizontally through a submerged round outfall of diameter 0.4 m in coastal waters of 15 m depth. Estimate the initial dilution for (i) this outfall, and (ii) if the outfall is moved to 100 m depth. The mean salinity and temperature of the receiving water are 32 ppt and 20 °C respectively. For case (i), plot the predicted plume geometry.

(b) By keeping the discharge velocity the same as in (a), produce a non-interfering multiport diffuser design that will perform better than the single round outfall in (a) as a water quality control device (for $H = 15$ m). Comment on your results and discuss briefly other advantages, if any, of your design. Practical considerations dictate that the minimum port diameter is 0.1 m.

4.5 A 500 MW nuclear power plant is located on an open coastline. A single port submerged outfall is used to discharge the condenser cooling water into the ocean. The following design data apply:

- Condenser cooling water flow = 10 m^3/s
- Condenser temperature rise $\Delta T_o = 11$ °C
- Discharge angle $\phi_o = 0°$
- Discharge velocity $u_o = 5$ m/s
- Ambient water is stagnant and non-stratified; $T_a = 10$ °C Salinity = 25 ppt
- Discharge depth $H = 60$ m

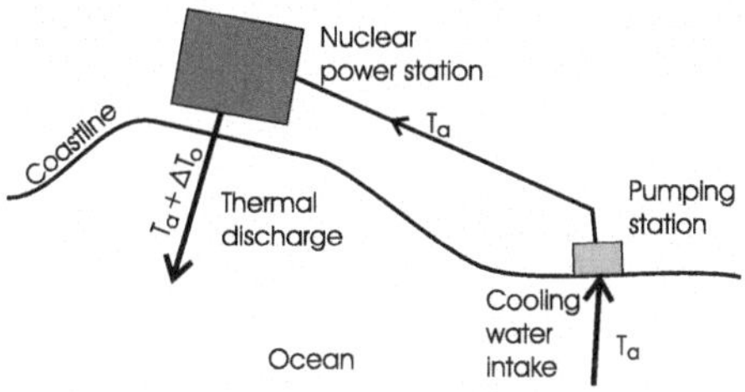

Determine

(a) The plume location where the centerline temperature is 1.5 °C above the ambient.

(b) The centerline temperature and width of the plume when it reaches the surface.

(c) The stability of this thermal discharge.

4.6 Consider a vertical buoyant jet discharge in otherwise stagnant ambient fluid, with initial volume flux Q_o, kinematic momentum flux M_o, and specific buoyancy flux F_o; the jet densimetric Froude number is Fr. Explain the significance of the momentum length scale $l_s = M_o^{3/4}/F_o^{1/2}$. By using the functional form of the jet and plume solutions in Table 1 in Chap. 3 and Table 3 in Chap. 2, show that the centerlinevelocity w_m, centerlinedilution S_m, and average dilution $\overline{S}$ can be expressed in the following dimensionless forms as a unique function of dimensionless vertical distance above source, z/l_s, over the entire jet-plume range.

$$\frac{M_o^{1/2}}{z w_c} \sim f\left(\frac{z}{l_s}\right)$$

$$\frac{S_m Q_o}{z M_o^{1/2}} \sim f\left(\frac{z}{l_s}\right)$$

$$\frac{\overline{S} Q_o F_o^{1/2}}{M_o^{5/4}} \sim f\left(\frac{z}{l_s}\right)$$

$$\frac{S_m}{Fr} \sim f\left(\frac{z}{l_s}\right)$$

4.7 Analysis of experimental data of the mixing of a vertical buoyant jet suggest that the *average dilution*, $\overline{S}$, can be expressed in terms of the length scales $l_Q = Q_o/M_o^{1/2}$ and $l_s = M_o^{3/4}/F_o^{1/2}$ by a single formula throughout the jet-plume range:

$$\frac{\overline{S} l_Q}{z} = \frac{\overline{S} Q_o}{z M_o^{1/2}} = C_1\,(1 + C_2(z/l_s)^2)^{1/3}$$

where $z =$ vertical distance above the source, and $C_1 = 0.29$ and $C_2 = 0.16$ are empirical constants. Show that this dilution formula is consistent with the limiting cases of a pure jet and plume.

Estimate the initial dilution for a vertical single port sewage discharge with jet velocity $u_o = 1$ m/s, diameter 0.15 m, in depth of 15 m into (i) a freshwater lake; and (ii) the ocean with relative density difference of 0.025. Assume density of sewage = 1000 kg/m^3.

4.8 A horizontal turbulent heated water jet in the laboratory discharges with jet velocity 1 m/s, jet diameter 0.01 m, and the design jet densimetric Froude number is $Fr = 20$. The jet with the same Froude number can be simulated for two situations, with different ambient temperatures.

For an ambient water temperature of (i) $T_a = 15\,°\text{C}$, and (ii) $T_a = 25\,°\text{C}$, find the corresponding discharge temperature T_o which will give the required Froude number (use the approximate equation of state in Appendix A).

Will the two discharge with the same Fr and submergence result in the same jet trajectory? Please explain.

4.9 *Coanda attachment of buoyant jet to sea bed*
For buoyant jets located close to the bed, the pressure reduction due to the confinement of the entrainment flow demand may cause the plume to cling to the bottom (the so called *Coanda* effect). It is desired to determine the minimum height of the nozzle from the bottom boundary to avoid plume attachment.

(a) Consider a horizontal buoyant jet (jet diameter D) located at a height of z_o above the bottom, discharging with velocity U_o in stagnant fluid. The jet densimetric Froude number is Fr. For a given z_o, the vertical pressure differential increases as the jet velocity (and hence the momentum flux and entrainment velocity). At some point the buoyant force on the plume cannot balance the downward pressure force, and the plume attaches to the bottom. By dimensional reasoning, the limiting height depends on the jet momentum and buoyancy fluxes:

$$z_o = f(M_o, F_o)$$

Hence show that the criterion for bottom attachment can be written as $z_o = Cl_s$ where $C =$ experimental constant. The experiments of Sharp and Vyas (1977) show that $C = 0.15$. Figure 16 shows the data of heated jet experiments by Wong (C.F. Wong, unpublished 1987) who obtained $C = 0.14$. In contrast, Sobey et al. (1988) obtained $C \approx 0.1$. Sharp's attachment criterion can alternatively be written as:

$$\frac{z_o}{D} = 0.14\ Fr$$

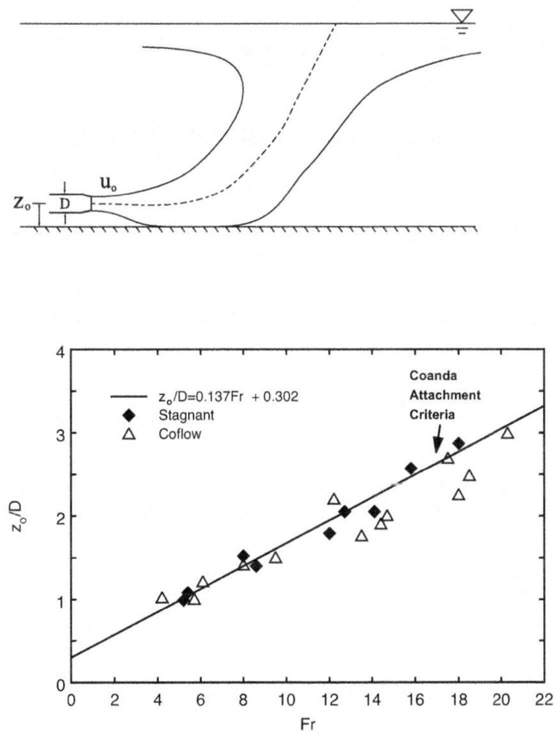

Fig. 16 Bottom attachment criterion of a horizontal buoyant jet (Wong 1987)

This means the nozzle must be very close to the bed for the Coanda effect to be noticeable. For $Fr = 10$, the nozzle needs to be located only $1.4D$ from the bottom to avoid plume attachment. In the case of a coflow, the limiting height required is even less.

(b) The attachment criterion can also be deduced from kinematic considerations. By assuming a straight horizontal jet over a distance of l_s with a spreading rate of $dB/dx = \beta$, show that the plume will attach if:

$$\beta l_s \geq z_o$$

For a nearly horizontal jet inclined at vertical angle ϕ_o, show that the general attachment criterion is:

$$\tan\phi_o \leq \beta - \frac{z_o}{l_s}$$

where $\beta \approx 0.2$ (Jirka and Doneker 1991).

4.10 *Thermal diffuser discharge in shallow water*:
Thermal power stations located in coastal waters often employ once-through cooling systems. Large flows of condenser cooling water are sometimes discharged from submerged diffusers in the form of high velocity jets. The temperature of the heated effluent can be brought down to acceptable levels within a small mixing zone. After initial mixing, the high momentum flow in the near field is typically vertically mixed, and flows away from the discharge in the form of a 2D jet with bottom friction dissipation (Fig. 17).

(a) Using boundary layer approximation, the 2D jet beyond the vicinity of the discharge can be described by the following x-momentum equation:

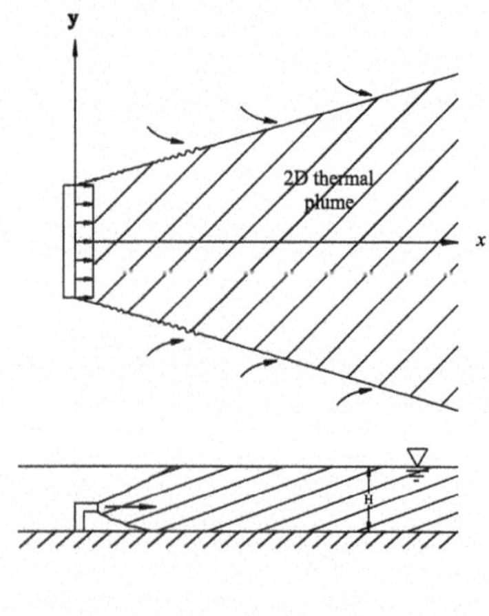

Fig. 17 Thermal diffuser plume in shallow water

$$x\text{-Momentum:} \quad u\frac{\partial u}{\partial x} + v\frac{\partial u}{\partial y} = -\frac{\partial \overline{u'v'}}{\partial y} - \frac{f_o}{8H}u^2$$

where H = water depth, f_o = bottom friction coefficient, and the other terms have their usual meaning. Using the integral method, show that the diffuser jet momentum, $M = \rho \int u^2 dy$, is dissipated exponentially downstream.

(b) Assuming top-hat profiles, obtain an analytical expression for the variation of jet velocity and half-width with distance downstream. Does the jet spread linearly?

4.11 Pollution from sewage outfall damaged by dredger

In a densely populated city, wastewater effluent (after screening to remove the large solids) is discharged into heavily navigated coastal water through a 1.5 km long submarine outfall of diameter 1.5 m. In normal operation the screened flow of $Q_o = 0.75\ \text{m}^3/\text{s}$ is discharged through a multiport diffuser section fitted with 70 round ports (0.15 m diameter). The jets discharge vertically in a depth of 20 m. On a windy night, a dredger operating in the area accidentally damaged the outfall. In subsequent days the bacteria concentration on a nearby beach 3 km away rose sharply. This caused a public outcry and it was thought that the dredger may be held responsible. You are asked to make an evaluation of this pollution incident.

The concentrations of the key water quality parameters in the screened effluent are: Biochemical Oxygen Demand (BOD) = 300 mg/L; Bacteria concentration = $10^7/100$ mL.

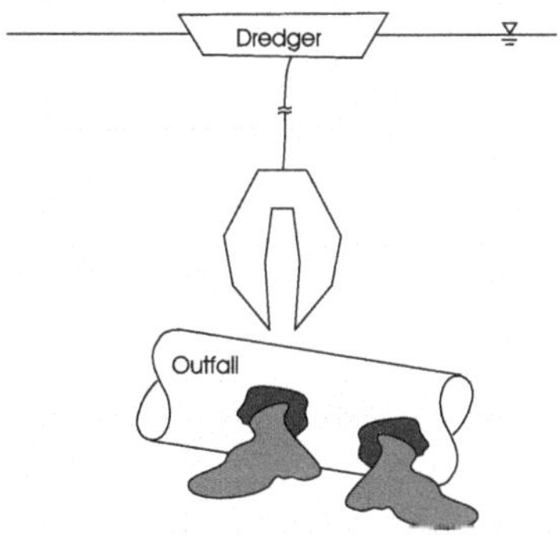

(a) Under normal operating conditions, determine the initial dilution in still water achieved by the outfall at the water surface. Assume an initial relative density difference of 0.025 and uniform discharge from all the ports. The jets can be regarded as independent of each other. Estimate the BOD and bacteria concentration at the water surface above the outfall.

From field observations, it is known that if the initial dilution is above 50, sewage slicks are not likely to form; black 'sewage boils' will be absent. In addition, typically it takes 8 h for the mixed sewage to travel to the nearby beach. If the bacteria concentration further decreases (due to sunlight, sedimentation etc.) according to e^{-kt} where k= decay rate, and t=travel time, determine the bacteria concentration on the beach. $k = 0.6$ per hour. The bacteria concentration in the beach water should not exceed 1000/mL.

(b) After the outfall was damaged by the dredger, diver inspection revealed two large openings (equivalent diameter 0.35 m each) were punctured on the outfall. An internal hydraulics calculation shows that the sewage flow 'short circuits' and two-thirds of the flow will be discharged through these two holes ($Q = 0.25$ m^3/s per hole) instead. Under these conditions, determine the initial dilution in still water. Estimate the BOD and bacteria concentration at the water surface above the outfall and the bacteria concentration on the beach.

Chapter 5
Density Stratification

Natural and man-made water bodies are generally density-stratified. The variations in density can be due to temperature, salinity, and particulate matter in suspension. The thermal stratification of lakes and reservoirs is an example (e.g. Harleman (1982)). In the presence of the earth's gravity, the fluid of non-uniform density would move towards a state of stable stratification—the heavier fluid tends to stay at the bottom while fluid of smaller density is generally found at a higher elevation. The stable stratification is disturbed when momentum and buoyancy sources, in the form of jets, plumes, puffs and thermals, are present to produce the vertical motions. The rise of buoyant fluid draws fluid from a lower elevation and discharges it into higher elevation. The fall of dense fluid would produce the opposite effect. Small density differences can have a marked effect on the vertical exchange. A variety of significant mixing processes observed in the atmosphere, rivers and lakes, estuaries and the ocean can be modeled as the interaction of jets/plumes/puffs/thermals (JPPT) with a stratified fluid.

Figure 1 shows an example of thermal stratification. At the bottom of a stratified lake, the temperature of the colder water tends to stay constant independent of the seasonal fluctuations but the surface temperature is more variable due to solar heating and inflows such as rivers and industrial and municipal discharges. The stable thermal stratification in the summer is often associated with oxygen depletion in the hypolimnion of eutrophic lakes (Imberger and Patterson 1990). The water quality of these lakes can however be modified by thermal de-stratification (Zic and Stefan 1988; Schladow and Fisher 1995; Lawrence et al. 1997). The temperature stratification of air in the interior space of buildings is another example. Cold dense air tends to stay on the floor as warm air flows towards the ceiling. The placement of heating and cooling equipment and the openings for natural ventilation are important design considerations that can modify the temperature stratification and hence the indoor

Supplementary Information The online version contains supplementary material available at https://doi.org/10.1007/978-3-032-01187-9_5.

J. H. W. Lee et al., *Turbulent Jets and Plumes*,
https://doi.org/10.1007/978-3-032-01187-9_5

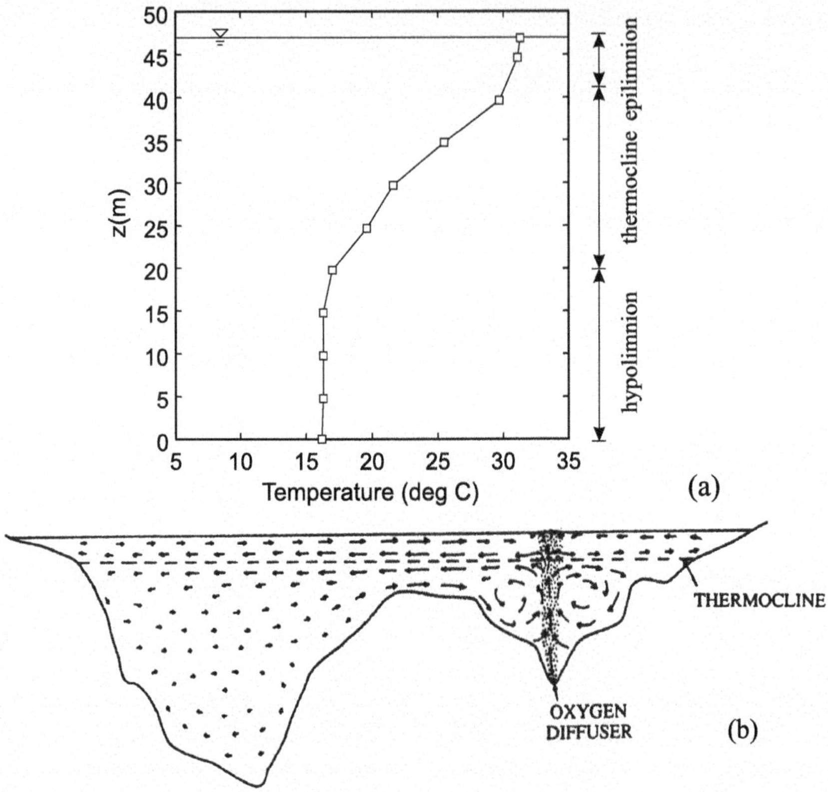

Fig. 1 **a** Water temperature measured in High Island reservoir, Hong Kong, in July 1990; **b** Thermal de-stratification scheme at Amisk Lake (Lawrence et al. 1997)

air quality in modern buildings (Baines and Murphy 1986; Linden 1999). In addition to temperature effect, the density of water is affected by turbidity and salinity. The density of the air in the atmosphere is also determined by thermodynamics effects such as compressibility and moisture content (see, e.g., Iribarne and Godson (1981)).

1 Buoyancy Variation

The first step in quantifying the effect of density stratification is to evaluate the variation of the buoyancy force with elevation. According to Archimedes Principle, the force of buoyancy in a fluid is the difference in weight between the fluid and it's displaced volume, which per unit *volume* of the fluid, is

$$\mathbf{B} = B\hat{\mathbf{k}} = g(\rho_a - \rho)\hat{\mathbf{k}} \tag{1}$$

where $B = g(\rho_a - \rho)$ is the buoyancy, g is the gravitational acceleration, $\hat{\mathbf{k}}$ is the base vector in the $+z$ direction (opposite direction to the force of gravity), and $(\rho - \rho_a)$ is the density of the fluid above that of the ambient. The buoyancy B varies with the elevation z as the fluid moves through and mixes with its surroundings, due to the changes in both ρ and ρ_a. The density of air may vary depending on temperature and moisture content. The density of water is a function of temperature and salinity (see Appendix A). Figure 2 shows the mixing of the fluid of density $\tilde{\rho}$ in an elemental volume with its surrounding fluid of density ρ_a. The elemental volume is not a material volume. As puffs and thermals or jets and plumes advance through its surrounding fluid, the mass and associated constituents in the elemental volume increase continuously due to the entrainment of the ambient fluid.

1.1 Salinity Equation

If the density variation is due to dissolved salts, the tracer concentration is the salinity–the mass of dissolved salts per unit mass of the carrying fluid. The salinity is $\tilde{S}$ in the elemental volume and S_a in the ambient. The carrying mass is $\tilde{\rho}\tilde{V}$ in the elemental volume and $\delta(\tilde{\rho}\tilde{V})$ in the small volume from the ambient. Therefore, the mass of the salt is $\tilde{S}\,(\tilde{\rho}\tilde{V})$ in the elemental volume and is $S_a\,\delta(\tilde{\rho}\tilde{V})$ in the small volume from the ambient. After the mixing, the concentration of salt becomes $\tilde{S} + \delta\tilde{S}$ and the mass in the elemental volume becomes $\tilde{\rho}\tilde{V} + \delta(\tilde{\rho}\tilde{V})$. The carrying mass is unchanged. Equating the tracer mass of the system before and after the entrainment (Fig. 2),

$$S_a\delta(\tilde{\rho}\tilde{V}) + \tilde{S}\tilde{\rho}\tilde{V} = (\tilde{S} + \delta\tilde{S})[\tilde{\rho}\tilde{V} + \delta(\tilde{\rho}\tilde{V})], \tag{2}$$

which gives the salinity equation

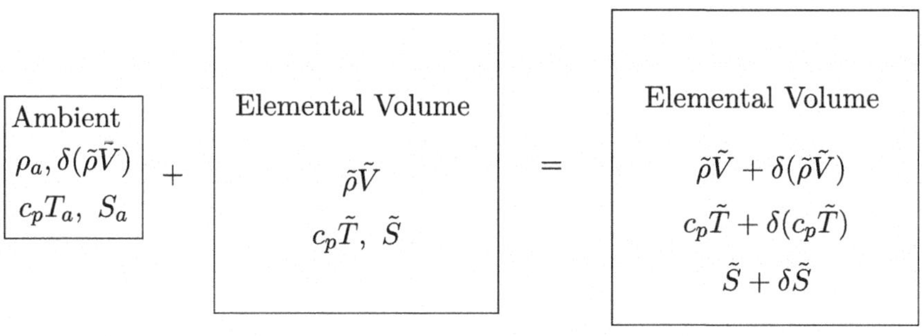

Fig. 2 Mixing of fluid of density $\tilde{\rho}$ in an elemental volume with a small volume of ambient fluid of density ρ_a. The carrying mass, salt mass and heat in the elemental volume are $\tilde{\rho}\tilde{V}$, $\tilde{S}\tilde{\rho}\tilde{V}$, and $c_p\tilde{T}\tilde{\rho}\tilde{V}$, respectively

$$\delta[(\tilde{S} - S_a)\tilde{\rho}\tilde{V}] = -\tilde{\rho}\tilde{V}(\delta S_a). \tag{3}$$

Once the salinity is calculated using this equation, the density of the water can be computed using the equation of the state. If density is a function of only salinity, we have:

$$\frac{d\rho}{\rho} = \frac{1}{\rho}(\frac{\partial\rho}{\partial S})dS = \gamma_s dS, \tag{4}$$

where

$$\gamma_s = \frac{1}{\rho}(\frac{\partial\rho}{\partial S})_T \tag{5}$$

is the salinity expansion coefficient. For small salinity differences between the elemental volume and the ambient,

$$(\tilde{\rho} - \rho_a) = \overline{\rho}\,\overline{\gamma}_s(\tilde{S} - S_a), \qquad \delta\rho_a = \rho_a\gamma_{Sa}\delta S_a. \tag{6}$$

Through these relations, the salinity in Eq. 3 is eliminated to give the density equation

$$\delta[(\tilde{\rho} - \rho_a)\tilde{\rho}\tilde{V}] \simeq -\tilde{\rho}\tilde{V}(\delta\rho_a), \tag{7}$$

since $\overline{\rho}\,\overline{\gamma}_s \simeq \rho_a\gamma_{Sa}$. The density and salinity is linearly related; the proportionality constant $\rho\gamma_s \simeq 0.8$ is practically independent of the salinity and temperature (see Fig. A.3 in Appendix A). Multiplying Eq. 7 by g and dividing by δz, the result is the buoyancy equation

$$\frac{dB}{dz} \simeq \tilde{\rho}\tilde{V}\frac{g}{\tilde{\rho}}\frac{d\rho_a}{dz} \tag{8}$$

where $B = g(\rho_a - \tilde{\rho})\tilde{V}$ is the buoyant force acting on the elemental volume. The buoyancy force B is positive if $\rho < \rho_a$. The gradient of this force dB/dz is positive if $d\rho_a/dz > 0$.

If density variation is due to suspended solid, such as sediments, or other dissovled solids, the density change in the fluid is expected to be proportional to tracer mass concentration, and Eq. 8 would be equally applicable under those conditions. See method of excess in Chap. 9.

1.2 Temperature Equation

If density variation is due to temperature, the tracer concentration is heat per unit mass of carrying fluid. The heat in the elemental volume is $c_p\tilde{T}\,(\tilde{\rho}\tilde{V})$. The heat in the small volume from the ambient is $c_pT_a\delta(\tilde{\rho}\tilde{V})$, where c_p = specific heat at constant pressure. Equating the heat before and after the entrainment (Fig. 2),

$$c_p\tilde{T}\,(\tilde{\rho}\tilde{V}) + c_p T_a\,\delta(\tilde{\rho}\tilde{V}) = [\tilde{\rho}\tilde{V} + \delta(\tilde{\rho}\tilde{V})][c_p\tilde{T} + \delta(c_p\tilde{T})] \tag{9}$$

which gives the heat equation

$$\delta[c_p(\tilde{T} - T_a)\tilde{\rho}\tilde{V}] = -\tilde{\rho}\tilde{V}\,\delta(c_p T_a). \tag{10}$$

The temperature variation in this equation is related to the density variation:

$$(\tilde{\rho} - \rho_a) = -\overline{\rho}\,\overline{\gamma}_{_T}(\tilde{T} - T_a), \qquad \delta\rho_a = -\rho_a\,\gamma_{_{Ta}}\delta T_a \tag{11}$$

where

$$\gamma_{_T} = -\frac{1}{\rho}\left(\frac{\partial\rho}{\partial T}\right)_s \tag{12}$$

is a thermal expansion coefficient; $\overline{\gamma}_{_T}$ is the average of the coefficient over the temperature range $(\tilde{T}, T_a)$ and $\gamma_{_{Ta}}$ is the coefficient at temperature T_a. The thermal expansion coefficient is temperature dependent. For ideal gas, $\gamma = T^{-1}$ (where T is the absolute temperature). For water, the coefficient can be determined by the equation of the state given in Appendix A. Eliminating the temperature in Eq. 10 using Eq. 11, gives

$$\delta\left[\left(\frac{c_p}{\overline{\gamma}_{_T}\overline{\rho}}\right)(\rho_a - \tilde{\rho})\tilde{\rho}\tilde{V}\right] = -\tilde{\rho}\tilde{V}\left(\frac{c_p}{\gamma_{_{Ta}}\rho_a}\right)\delta\rho_a. \tag{13}$$

Multiplying both sides by g and let $\tilde{B} = (\rho_a - \tilde{\rho})g\tilde{V}$, gives the buoyancy equation

$$\delta\left[\left(\frac{c_p}{\overline{\gamma}_{_T}\overline{\rho}}\right)\tilde{\rho}\tilde{B}\right] = -\tilde{\rho}\tilde{V}\left(\frac{c_p}{\gamma_{Ta}\rho_a}\right)g\delta\rho_a. \tag{14}$$

According to this equation, the buoyancy in the elemental volume $\tilde{B}$ is changed on the one hand due to the variation of the thermal expansion coefficient and on the other hand by the variation of the density in the ambient with elevation. If the thermal expansion coefficient and specific heat are assumed to be a constant and if $\overline{\rho}/\tilde{\rho} \simeq 1$ over a small range of temperature, Eq. 14 becomes

$$\frac{dB}{dz} \simeq \tilde{\rho}\tilde{V}\,\frac{g}{\rho_a}\frac{d\rho_a}{dz}. \tag{15}$$

Either Eqs. 8 or 15 can be written as:

$$\frac{dB}{dz} \simeq -\tilde{\rho}\tilde{V}\,N^2 \tag{16}$$

in terms of the stratification frequency

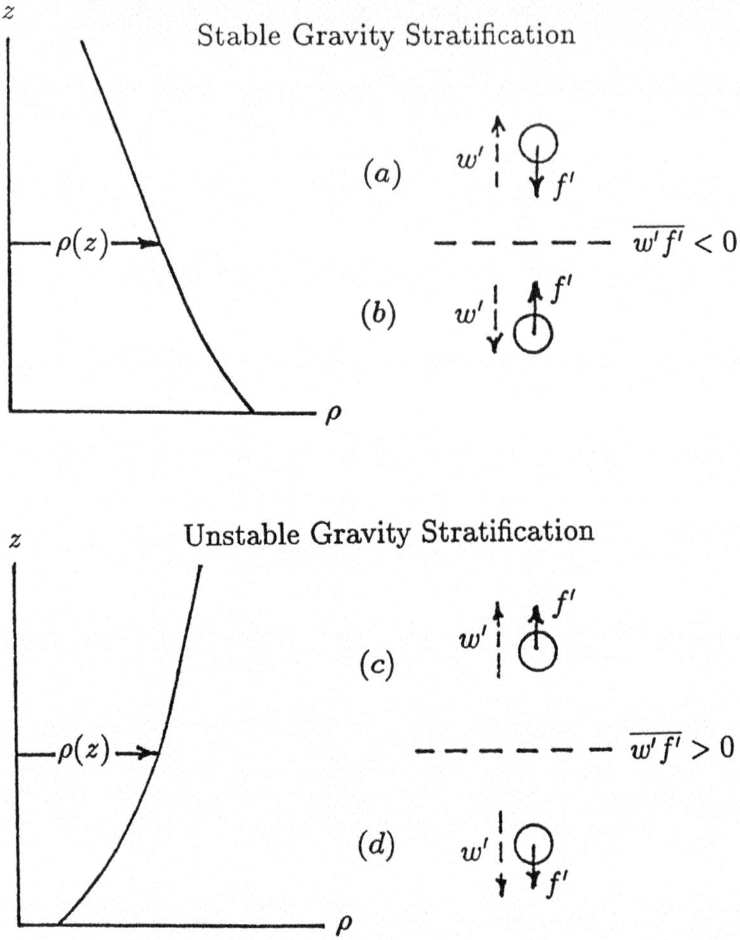

Fig. 3 Displacement of a fluid parcel in a stratified fluid: **a** and **b** show oscillation about the static equilibrium level in stable stratification, while **c** and **d** show the tendency for the displaced fluid to move further from the equilibrium in unstable stratification; w' and f' are the velocity and buoyancy fluctuations associated with the motion

$$N = \sqrt{-\frac{g}{\rho_a}\frac{d\rho_a}{dz}}. \tag{17}$$

Displacement of a parcel of fluid in a stably stratified environment is expected to oscillate about its equilibrium position close to this frequency (see Fig. 3).

Example 5.1 The water temperature in High Island reservoir (see Fig. 1a) can be represented by a three-layer profile. The temperature near the surface of the reservoir, in the epilimnion from $z = 42$ m to 47 m, is essentially constant at 31 ^{o}C. The bottom layer, in the hypolimnion from $z = 0$ to 20 m, has a lower temperature of 17 ^{o}C. Find the stratification frequency N in the thermocline assuming that the density variation in the middle layer is linear from $z = 20$ m to 42 m.

Solution: The density of fresh water can be calculated by the formula in Appendix A. In the surface layer (epilimnion), the temperature is $T_e \simeq 31\ ^o$C and the density is $\rho_{31} = 995.35$ kg/m^3. The density in the bottom layer (hypolimnion), where the temperature $T_h \simeq 17\ ^o$C, is $\rho_{17} = 998.78$ kg/m^3. The density changes rapidly over the 22 m depth of the thermocline. According to Eq. 17, the stratification frequency in the thermocline is

$$N = \sqrt{-\frac{g}{\rho_a}\frac{d\rho_a}{dz}} = \sqrt{\frac{9.81(998.78 - 995.35)}{\frac{1}{2}(998.78 + 995.35) \times 22}} = \underline{0.0392}\ \text{rad/s}.$$

The corresponding period of oscillation is

$$T = \frac{2\pi}{N} = \underline{160}\ \text{s}.$$

The water mass in this upper layer of the lake is expected to oscillate with a dominant internal wave period of 160 s (Fig. 4).

1.3 Stratification Frequency

Density stratification, or gravity stratification, is said to be stable if $d(\rho_a g)/dz < 0$ and unstable if $d(\rho_a g)/dz > 0$. Fluid in a stable stratification tends to approach the static equilibrium. As shown in Fig. 3, the displacement of a fluid in stable stratification upward would make the fluid heavier than its surrounding and hence tends to fall back to its equilibrium position. Conversely, a fluid element moved downward would become lighter than its surrounding and tends to return to its equilibrium position. The oscillation of a displaced fluid about its equilibrium level is defined by the stratification frequency, or the so-called Brunt-Väisälä frequency

$$N = \sqrt{-\frac{g}{\rho_a}\frac{d\rho_a}{dz}}. \tag{18}$$

In the atmosphere, the density gradient can be determined by detecting the oscillation frequency of the weather balloon, according to the above equation. In surface waters and the sea, the density structure is measured by a conductivity-temperature-depth (CTD) probe, from which the salinity and density can be determined.

The stratification frequency, N, is a parameter for the relative effect between gravity stratification and the inertia. We speak of gravity stratification because the effect comes from the variation of the fluid's specific weight, $\delta(\rho_a g)$, not the variation in

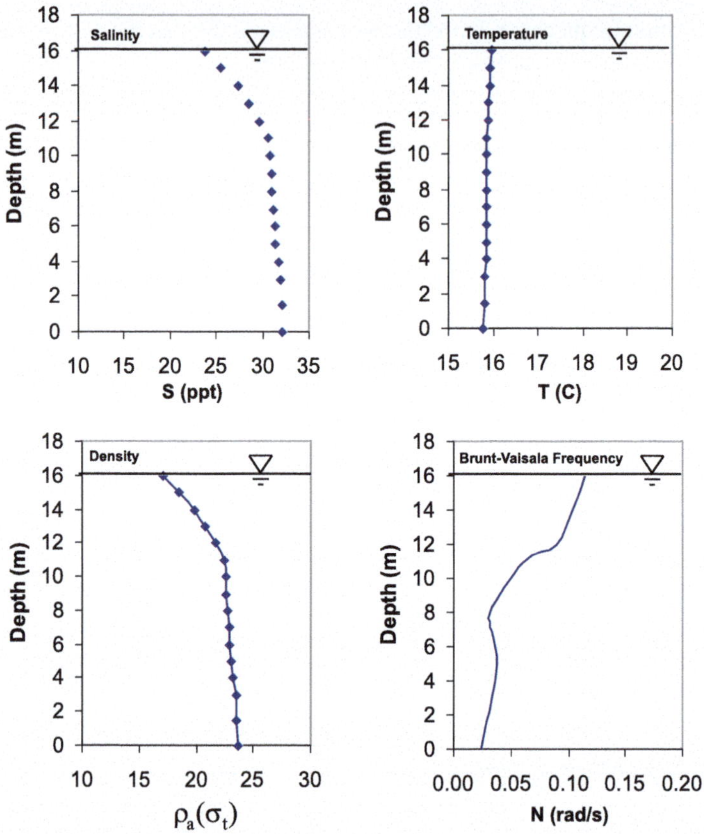

Fig. 4 Profile of salinity, temperature, density and Brunt-Väisälä frequency in the Pearl River Estuary, China, obtained in March 1995. Measurements were made in the northwestern waters of Hong Kong

the fluid's density. The distinction between specific weight and density is significant in dimensional analysis. However, 'density stratification' is the common usage to describe the gravity effect.

In marine and coastal waters buoyancy variation is often a combination of the temperature and salinity changes over the depth. The so-call sigma unit ($\sigma_t = \rho$ (kg/m^3) $- 1000$) is defined as the density excess relative to the density of pure water at 4°C. The density of sea water can be determined from the equation of state (Appendix A), from which the Brunt-Väisälä frequency can be obtained. Figure 4 shows a vertical profile of salinity and temperature measured in the Pearl River estuary in the Spring season. The measurements were made in the northwestern waters of Hong Kong as part of a post-operation monitoring study at the Urmston Road Sewage Outfall (see Chap. 10). The salinity is 24 ppt near the free surface and 32 ppt at 16 m depth. The surface and bottom temperatures are 16°C and 15.8^{o}C respectively.

As the temperature is practically constant over the depth, the density stratification is primarily due to variation of salinity. It has essentially a two-layer structure: the sigma value varies rapidly from $\sigma_t = 17$ to $\sigma_t = 22$ over the top 5 m surface layer and then changes from $\sigma_t = 22$ to $\sigma_t = 23.5$ over the 11 m of bottom layer. The Brunt-Väisälä frequency in the upper 5-m layer is

$$N = \sqrt{-\frac{g}{\rho_a}\frac{d\rho_a}{dz}} \simeq \sqrt{\frac{9.81 \times (22 - 17)}{1000 \times 5}} = 0.099 \text{ rad/s}. \tag{19}$$

The 11-m of water below the surface layer has an average Brunt-Väisälä frequency approximately equal to

$$N = \sqrt{-\frac{g}{\rho_a}\frac{d\rho_a}{dz}} \simeq \sqrt{\frac{9.81 \times (23.5 - 22)}{1000 \times 11}} \simeq 0.036 \text{ rad/s}. \tag{20}$$

The two-layer density stratification in this case is primarily influenced by the fresh water outflow from the Pearl River, which changes with season. The order of magnitude of buoyancy frequency in this coastal water is about 10^{-1} rad/s to 10^{-2} rad/s. These values obtained in the river estuary are large compared with data obtained in the open sea. As will be seen later, the existence of even much weaker density variations can have a significant impact on environmental mixing and transport (Fig. 11 and Example 5.6).

2 Thermals in Stratified Fluid

Thermals are isolated volumes of turbulent fluid produced by instantaneous sources of buoyancy. These are classified as round thermals if the source of buoyancy is derived essentially from an instantaneous point source and as line thermals if the buoyancy is from a line source. As the thermal moves through a stratified environment, the buoyancy force associated with the thermal varies with the changing densities of both the thermal and its surrounding fluid following Eq. 8.

The rise and fall of the thermals in stratified fluid are determined by two basic equations: the buoyancy equation and momentum equation. Multiplying Eq. 8 by dz/dt gives the buoyancy equation

$$\frac{d\tilde{B}}{dt} = -(-\frac{g}{\rho_a}\frac{d\rho_a}{dz})\rho_a \tilde{V}\frac{dz}{dt}, \tag{21}$$

which in compact form is

$$\frac{d\tilde{B}}{dt} = -N^2 \frac{\rho_a}{\tilde{\rho}} \tilde{M} \tag{22}$$

where $\tilde{M} = \tilde{\mathcal{M}}\tilde{w}$ is the momentum, $\tilde{\mathcal{M}} = \tilde{\rho}\tilde{V}$ the mass, $\tilde{w} = dz/dt$ the velocity of the thermal and z elevation of the thermal above its source. The rate of momentum change is equal to the buoyancy force acting on the thermal (Newton's second law). The rate of the change is determined by the momentum equation

$$(1 + k)\frac{d\tilde{M}}{dt} = \tilde{B} \tag{23}$$

where k is an added mass coefficient for the fraction of the momentum in the non-turbulent (irrotational) fluid surrounding the thermal (see also Prob. 8.3 for an explanation of added mass). Given a distribution of stratification profile, $N(z)$, and initial values for the momentum and buoyancy at the source, M_o and B_o, respectively, the values of $\tilde{B}(t)$ and $\tilde{M}(t)$ at subsequent time are determined by numerical integration of the buoyancy and momentum equations, Eqs. 22 and 23.

The space-time relation is obtained by integration of the equation $\tilde{M}(t) = \tilde{\mathcal{M}}\tilde{w} = \tilde{\mathcal{M}}dz/dt$ where $\tilde{\mathcal{M}} = 4\rho\pi R^3/3$ for a round thermal and $\tilde{\mathcal{M}} = \rho\pi R^2$ for a line thermal, where R is the radius of the thermal. Hence,

$$\tilde{M}(t) = \rho\frac{4\pi R^3}{3}\frac{dz}{dt} \quad \text{for round thermals}$$

$$\tilde{M}(t) = \rho\pi R^2\frac{dz}{dt} \quad \text{for line thermals} \tag{24}$$

Using a linear spreading hypothesis, the radius R is related to the height of the thermal:

$$dR = \beta|dz|. \tag{25}$$

Given the spreading coefficient β and the added-mass coefficient k, Eqs. 22, 23, 24, and 25 form a complete set of equations to find $R(t)$ and $z(t)$ by numerical integration.

2.1 *Round Thermals*

The round thermals in a linear density-stratified ambient are first examined as an example. For a linear density profile, $d\rho_a/dz$ = constant. The density structure of the environment is defined by a single parameter–the Brunt-Väisälä or stratification frequency. Under this condition, the asymptotic solutions of the thermals can be obtained by dimensional considerations to within a constant.

The maximum rise, z_m, of the round thermal is a function of fluid density, ρ, the initial buoyancy, B_o, and the buoyancy frequency, N. The dimensions of these variables are

$$[\frac{B_o}{\rho}] = [g_o'][V] = \frac{L^4}{T^2}, [N] = \frac{1}{T} \tag{26}$$

The dimensionless parameter of this problem is

$$\frac{z_m N^{\frac{1}{2}}}{(B_o/\rho)^{\frac{1}{4}}}. \tag{27}$$

Since this is the only dimensionless parameter, it must be equal to a constant; that means

$$z_m = \text{constant } (\frac{B_o}{\rho})^{\frac{1}{4}} N^{-\frac{1}{2}} \tag{28}$$

For the case of linear ambient stratification, an analytical solution for an instantaneous point source of buoyancy can be obtained. Differentiating the momentum equation (Eq. 23) once with time, and then making use of the buoyancy equation (Eq. 22) to eliminate $\tilde{B}$, gives the equation of the simple harmonic oscillator

$$\frac{d^2\tilde{M}}{dt^2} + N'^2\tilde{M} = 0 \tag{29}$$

where

$$N' = N\sqrt{\frac{\rho_a}{\rho(1+k)}} \tag{30}$$

is a modified frequency. A similar equation for B,

$$\frac{d^2\tilde{B}}{dt^2} + N'^2\tilde{B} = 0, \tag{31}$$

can be obtained. The solution for both the momentum $\tilde{M}$ and the buoyancy force $\tilde{B}$ associated with the thermal are simple harmonic with an oscillation frequency equal to N'.

In the stable stratification, the buoyancy is not a constant. If momentum is initially zero, M and B would vary sinusoidally as:

$$\tilde{B} = B_o \cos(N't), \tag{32}$$

$$\tilde{M} = -\frac{\rho}{\rho_a N^2}\frac{d\tilde{B}}{dt} = \frac{B_o}{N'(1+k)}\sin(N't) \tag{33}$$

In the presence of a stable density stratification, the thermal buoyancy reduces with the depth and becomes zero at a depth when $N't = \pi/2$. At this depth the momentum of the thermal is maximum as $\tilde{M}$ is 90 degree out of phase with $\tilde{B}$. Due to inertia, the thermal continues to move downward until $N't = \pi$. The momentum becomes zero when $N't = \pi$ as the maximum depth of the thermal penetration is attained.

For a point source, since $R = \beta z$ and $\tilde{M} = 4\pi\rho R^3\tilde{w}/3$, Eq. 33 becomes

$$\frac{4\pi}{3}\rho\beta^3 z^3\frac{dz}{dt} = \frac{B_o}{N'(1+k)}\sin(N't). \tag{34}$$

Integrating once with time, gives

$$\frac{\pi}{3}\rho\beta^3 z^4 = \frac{B_o}{N'^2(1+k)}[1-\cos(N't)]; \tag{35}$$

that is

$$z = [\frac{3B_o(1-\cos(N't))}{\rho\pi\beta^3 N'^2(1+k)}]^{\frac{1}{4}} = [\frac{3B_o(1-\cos(N't))}{\rho_a\pi\beta^3 N^2}]^{\frac{1}{4}} \tag{36}$$

where the constant of integration has been selected so that $z = 0$ at $t = 0$. The maximum depth of penetration, z_m, where $\tilde{M} = 0$, is obtained by setting $N't = \pi$:

$$z_m = [\frac{6B_o}{\rho_a\pi\beta^3 N^2}]^{\frac{1}{4}} = 2.64\,(\frac{B_o}{\rho_a})^{\frac{1}{4}}N^{-\frac{1}{2}} \tag{37}$$

The equilibrium level, z_e, where $\tilde{B} = 0$, is obtained by setting $N't = \frac{1}{2}\pi$:

$$z_e = [\frac{6B_o}{\rho_a\pi\beta^3 N^2}]^{\frac{1}{4}} = 2.22\,(\frac{B_o}{\rho_a})^{\frac{1}{4}}N^{-\frac{1}{2}} \tag{38}$$

where $\beta = 0.34$ is adopted (see also Eq. 20 in Chap. 9). Compared to a spreading angle of 0.17 for a round plume, round thermals have been observed to spread at a much wider angle, and thus they dilute more with height (cf Prob. 1.5). The added mass coefficient is assumed to be the same as that for a sphere, with $k = 0.5$. Note that the added mass coefficient affects the frequency but not the maximum level of the plume. Experiments on discrete buoyant clouds in a stably stratified fluid (Morton et al. 1956) have resulted in a maximum rise coefficient of 2.66 for z_m, in close agreement with the prediction. It should be noted that the Lagrangian approach outlined above is not limited to small density differences (Escudier and Maxworthy (1973); see Prob. 5.5).

Example 5.2 *Thermal in Linear Density Stratification* A 1.0 m^3 of dredge spoil contaminated with 10.0 kg of cyanide is accidentally released into a stably-stratifed lake with a linear temperature stratification. The temperature varies linearly with the depth–20 ^oC at the surface and 10 ^oC at a depth of 20 m. The Brunt-Väisälä period for such a temperature profile is 232 seconds. Find the depth of penetration of the toxic spill and estimate the cyanide concentration as the cloud begins to spread out into a horizontal layer in the stratified ambient. Assume that the cloud spreads at a rate $\beta = d\tilde{R}/dz = 0.34$. The density of the dredge spoil is 1100 kg/m^3 (relative density difference 0.1).

Solution: The density of the dredge spoil $\rho_o = 1100$ kg/m^3. The density of the surrounding fluid is $\rho_a = \rho_w \simeq 1000$ kg/m^3. The initial negatively buoyancy force associated with the contaminated waste cloud is $B_o = g(\rho_o - \rho_w)V_o = 9.81 \times (1100 - 1000) \times 1.0 = 981$ N. The buoyancy frequency is $N = 2\pi/232 = 0.0271$ rad/s. The maximum depth of penetration is then given by Eq. 37.

$$z_m = 2.64 \times [\frac{981}{1000}]^{\frac{1}{4}}/\sqrt{0.0271} = \underline{16.0}\text{ m}. \quad (39)$$

As the cloud entrains fluid from its surrounding, the radius of the cloud is assumed to spread at a rate $\beta = dR/dz = 0.34$. At this maximum depth the radius of the thermal is $R = 0.34 \times 16.0 = 5.44$ m. The volume of the thermal is $V = 4\pi R^3/3 = 674$ m^3. The average cyanide concentration in the thermal is $\tilde{c} = 10.0/674 = \underline{0.0148}$ kg/m^3 (i.e. 14.8 mg/L).

Numerical integration of Eqs. 22, 23, 24, and 25 is conducted for an initial momentum $M_o = 0$, an initial buoyancy force $B_o = 9.81 \times (1100 - 1000)$ N, and a uniform Brunt-Väisälä frequency $N = 0.0271$ rad/s. Figure 5 a–d shows the results of the numerical integration: the stratification frequency profile $N(z)$, thermal momentum

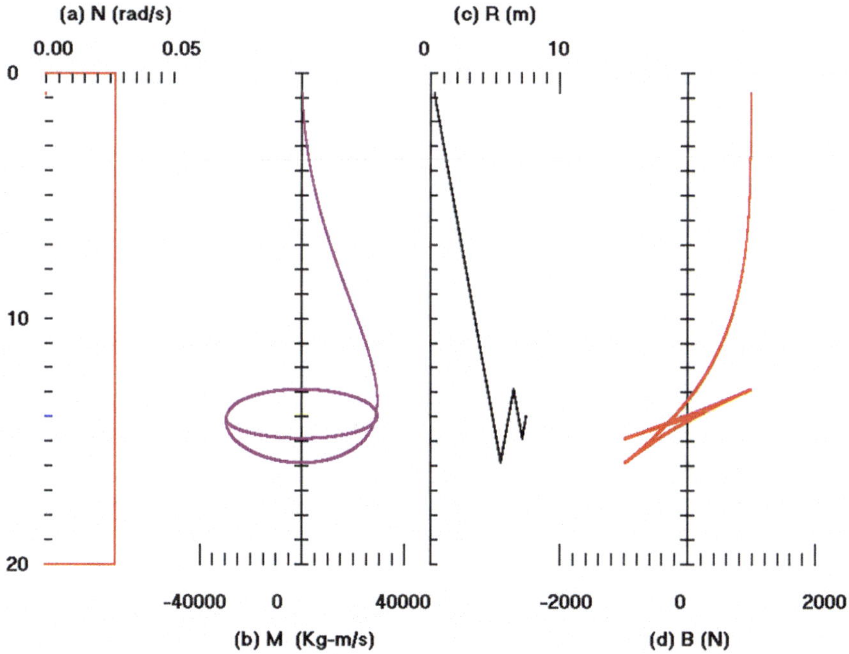

Fig. 5 Round thermal in a uniform density stratification; $N = 0.0271$ rad/s. Momentum (M), radius (R), and buoyancy (B) of the thermal vs depth of advance

$\tilde{M}(z)$, radius R, and buoyancy force $\tilde{B}(z)$, respectively. The cyanide cloud penetrates the thermocline to a maximum depth of $z = 16$ m and then starts to oscillate with frequency equal to N' at an equilibrium level of $z = 14$ m. The oscillation is manifested in the momentum and buoyancy of the cloud (Fig. 5b, d) which varies respectively from $\tilde{M}$ =+3000 Kg-m/s to -3000 Kg-m/s and $\tilde{B}$ = -1000 N to +1000 N; $\tilde{M}$ and $\tilde{B}$ are 90 degrees out of phase. The buoyancy force at maximum depth is upward. Subsequently, after reaching the maximum depth, the thermal reverses its direction and rises back to its equilibrium level. Further mixing will take place before the intrusion of the thermal fluid into the stratified ambient to form a horizontal layer but the concentration would probably not be further reduced by more than a factor of 2. Internal waves are radiated from the oscillating cloud. The amplitude of the oscillation is quickly diminished due to radiation damping, which is however ignored in the present analysis.

Example 5.3 *Thermal in Nonlinear Density Stratification* Consider the accidental release of 1.0 m^3 of toxic dredge spoil into the 20 m deep lake, in this example for a more realistic temperature profile. The temperature is constant at T_s=20 ^{o}C in the surface layer, and T_b=10 ^{o}C in the bottom layer. The temperature varies linearly from T_s to T_b over a 5 m thick thermocline. Study the thermal behaviour if the surface layer thickness is (a) 5 m and (b) 4.5 m. The density of the dredge spoil is 1100 kg/m^3 as in Example 5.2.

Solution: Numerical integration of Eqs. 22, 23, 24, and 25 is conducted for an initial momentum $M_o = 0$, an initial buoyancy force $B_o = 9.81 \times (1100 - 1000)$ N, and a uniform Brunt-Väisälä frequency $N = 0.0271$ rad/s are conducted for two cases with nonlinear density stratification. The results for two case of slightly different thermocline thickness, to be referred to as case A and case B, are shown in Figs. 6 and 7, respectively.

Case A: Thermal Trapped by the Thermocline

For the first case, the upper edge of the thermocline is located at a depth of 5 m below the surface–i.e. $T(z = 5) = 20^o$C and $T(z = 10) = 10^o$C. Figure 6 shows the numerical solution of the thermal behaviour in the non-linearly stratified lake. The change of temperature from 20°C to 10°C over a 5 m thick thermocline gives a stratification frequency $N = 0.0542$ rad/s within the thermocline–twice that in the previous example due to a four-fold increase in ambient density gradient. As the thermal advances through the surrounding fluid, the radius R is assumed to increase linearly with depth of advance z following a linear relation as shown in Fig. 6c.

In this case, the buoyancy force acting on the thermal stays constant in the upper layer but begins to reduce in the thermocline and becomes negative as the thermal pass through the lower interface at a depth of 10 m. Below the 10 m depth, the buoyancy force is upward. Despite the upward force of buoyancy, due to the inertia the thermal continues the downward movement until the maximum depth of penetration is reached at z =13.4 m. The upward force finally brings the thermal back to the thermocline. After a few oscillations, the thermal finally settles at an equilibrium depth of 9.8 m. In this case, the cyanide cloud is trapped in an area near the lower interface but within the thermocline.

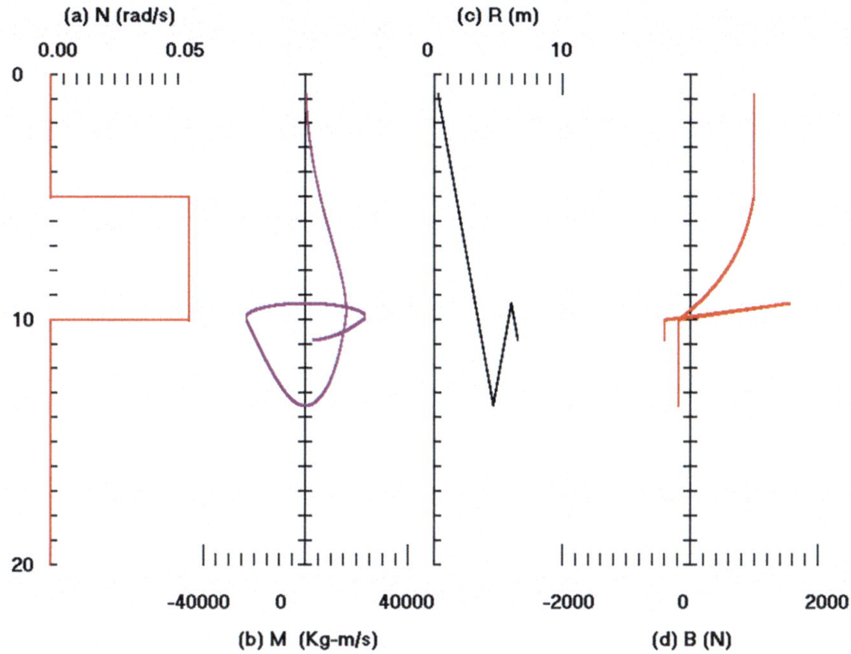

Fig. 6 Round thermal in non-uniform stratification, Case A. Thermal advance arrested by a thermocline with $N = 0.0542$ rad/s for $z = 5 - 10$ m. The thermal continues its advance to maximum depth before it reverses and oscillates about an equilibrium depth

Case B: Thermal Penetration Through the Thermocline

The situation is quite different if the thermocline is located at a slightly shallower depth with slightly smaller thickness (4.5 m to 9 m). The buoyancy frequency is slightly increased to $N = 0.0542 \times \sqrt{(10-5)/(9-4.5)} = 0.0571$ rad/s. In this case, as shown in Fig. 7, the buoyancy force is still downward after passing through the lower interface of the thermocline at 9 m depth. With the downward force the cyanide cloud is not trapped as in case A but continues its advance to the hypolimnion of the lake.

Whether a thermal penetrates the thermocline is dependent on the volume of the thermal $\tilde{V}$ and the change in density across the thermocline $\delta\rho_a$, according to the basic formula given by Eq. 8. The thermal volume, $\tilde{V}$, increases with the distance from its source. The reduction in buoyancy is directly proportional to the density difference and volume. The deeper the depth of the thermocline, the greater would be the volume of the thermal and consequently harder for the thermal to penetrate the thermocline. Large volume leads to high rate of reduction in buoyancy force. The rebound of the thermal by the deep thermocline is the consequence of the high reduction rate that reverses the direction of the buoyancy force. In other words, by the time the thermal hits the thermocline, it would have mixed sufficiently with the top layer fluid to have a density smaller than that of its surroundings.

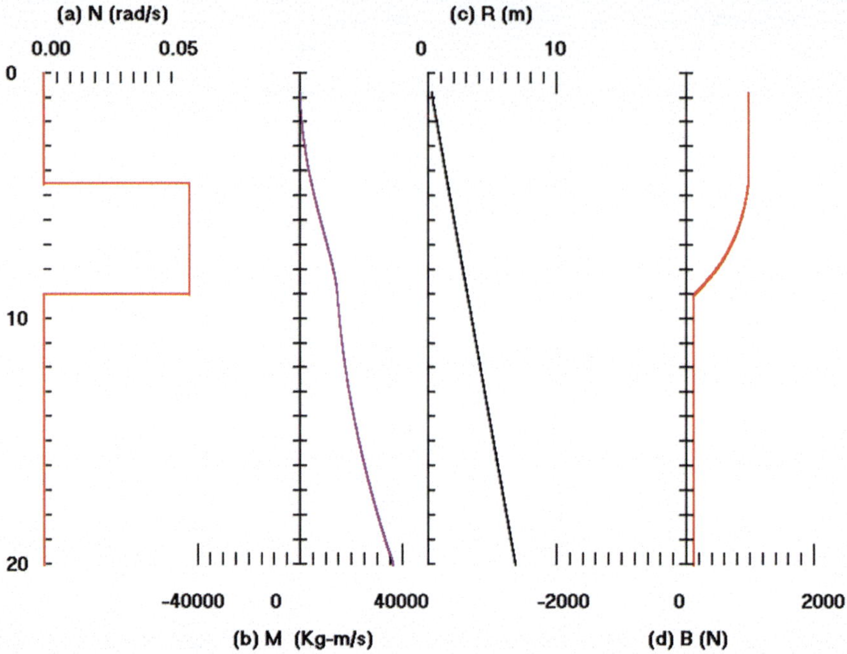

Fig. 7 Round thermal in non-uniform stratification, Case B. Thermal penetrates a thermocline and continues into the hypolimnion; $N = 0.0571$ rad/s for $z = 4.5$ m to 9.0 m

2.2 *Line Thermal*

As for the round thermal, a simple dimensional analysis can be made for the line thermal. The dimensions of the variables for the line thermal are

$$[\frac{B_o}{\rho}] = [g_o'][\mathcal{V}] = \frac{L^3}{T^2}, [N] = \frac{1}{T} \tag{40}$$

In this case, B_o is the buoyancy of the instantaneous source per unit length of the thermal. The dimensionless parameter of this problem is

$$\frac{z_m(B_o/\rho)^{-\frac{1}{3}}}{N^{\frac{2}{3}}}. \tag{41}$$

This is the only dimensionless parameter of the problem and it must be equal to a constant. Hence,

$$z_m = \text{constant } (\frac{B_o}{\rho})^{\frac{1}{3}} N^{-\frac{2}{3}} \tag{42}$$

The constant was found from the experimental observation by Wright (1984) to be 2.30.

The momentum of the line thermal is

$$\pi \rho \beta^2 z^2 \frac{dz}{dt} = \frac{B_o}{N'(1+k)} \sin(N't). \tag{43}$$

Integrating once with respect to the time, gives

$$\frac{\pi}{3} \rho \beta^2 z^3 = \frac{B_o}{N'^2(1+k)} [1 - \cos(N't)]; \tag{44}$$

that is

$$z = [\frac{3B_o(1 - \cos(N't))}{\rho \pi \beta^2 N'^2 (1+k)}]^{\frac{1}{3}} = [\frac{3B_o(1 - \cos(N't))}{\rho_a \beta^2 \pi N^2}]^{\frac{1}{3}} \tag{45}$$

the constant of integration has been selected so that $z = 0$ at $t = 0$. The maximum rise occurs when $N't = \pi$; i.e.,

$$z_m = [\frac{6B_o}{\rho_a \beta^2 \pi N^2}]^{\frac{1}{3}} \tag{46}$$

Assuming $\beta = 0.34$ and the added mass coefficient of a line thermal is similar to that of a cylinder, $k = 1$ (see numerical determination of k in Chap. 8), we have

$$z_m = 2.54 \, (\frac{B_o}{\rho})^{\frac{1}{3}} N^{-\frac{2}{3}} \tag{47}$$

The equilibrium level is determined by setting $N't = \pi/2$:

$$z_e = [\frac{3B_o}{\rho_a \beta^2 \pi N^2}]^{\frac{1}{3}} = 2.02 \, (\frac{B_o}{\rho})^{\frac{1}{3}} N^{-\frac{2}{3}} \tag{48}$$

These levels may be compared with formula for the advected line thermal (buoyancy-dominated far field , BDFF, of a plume in crossflow, see Chap. 8):

$$z_m = 2.30 \, (\frac{B_o}{\rho})^{\frac{1}{3}} N^{-\frac{2}{3}} \tag{49}$$

$$z_e = 1.85 \, (\frac{B_o}{\rho})^{\frac{1}{3}} N^{-\frac{2}{3}} \tag{50}$$

obtained from the experimental observation by Wright (1984). The formula derived from atmospheric plume data in stratified cross wind by Briggs (1965),

$$z_e = 1.78 \, (\frac{B_o}{\rho})^{\frac{1}{3}} N^{-\frac{2}{3}}$$

is also in agreement with the Eq. (48) obtained by Lagrangian method. The buoyancy per unit length of the line thermal is equal to the buoyancy flux F_o divided by the crossflow velocity U_a, i.e., $B_o = F_o/U_a$ (see Chap. 8). The source size in the laboratory experiment is finite, but the formulae in Eqs. 49 and 48 are derived for the line source of infinitesimally small size. As pointed out by Wright (1984), the coefficients for the maximum rise and the equilibrium level generally should be lower for the laboratory sources of finite size.

3 Plumes in Stratified Fluid

The continuous release of buoyancy produces a plume. In stable density stratification, the buoyancy flux of the plume is not a constant but reduces with height. The sketch in Fig. 8 shows the rise and leveling off of the plume at the equilibrium level. Initially, the plume fluid at the source is lighter than its surrounding environment. The reduction of the density in the environment also causes the buoyancy flux of the plume to decrease with elevation. The buoyancy flux becomes zero at the neutral buoyancy level as the density difference between the plume and its surrounding is reduced to zero. Due to inertia the plume overshoots the neutral buoyancy level to form a dome. The fluid in the dome is heavier than its surrounding. Further mixing is expected as the plume

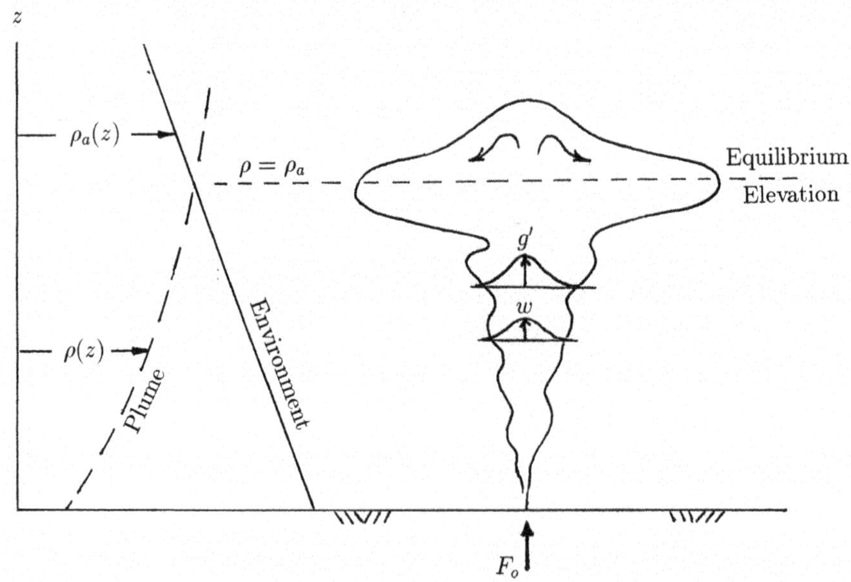

Fig. 8 Plume rise in a stratified fluid. The plume entrains fluid from the lower levels. Buoyancy flux becomes negative as the plume overshoots the neutral buoyancy level. The mixed fluid settles back and spreads at the equilibrium level

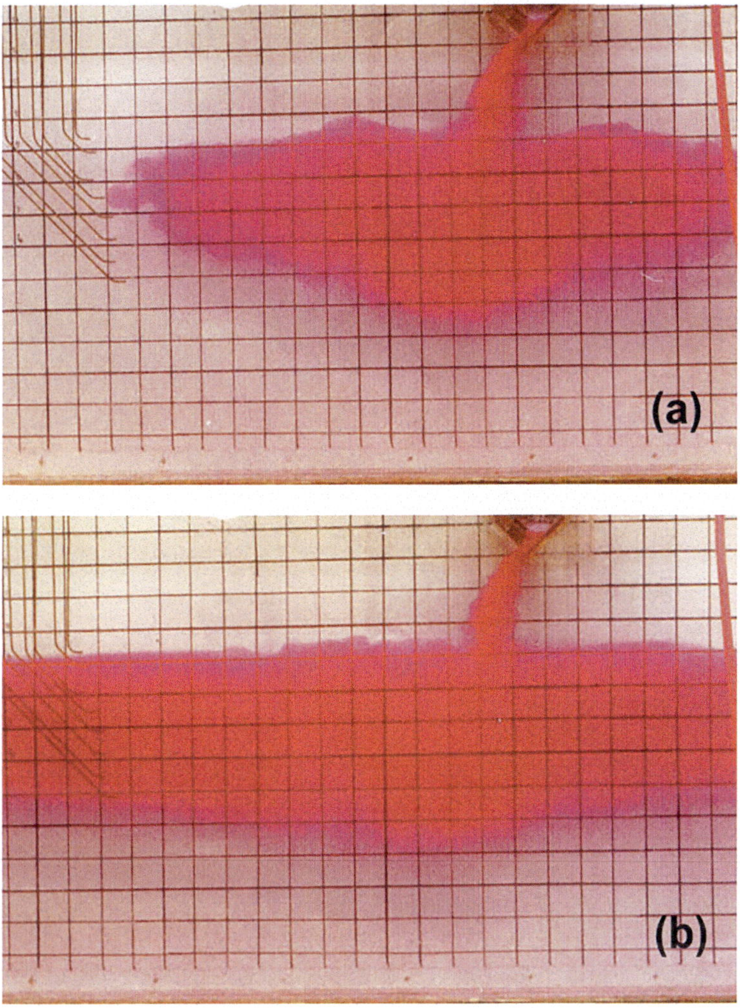

Fig. 9 Inclined plane negatively buoyant jet in a linear density stratified fluid: **a** shortly after discharge; **b** when spreading layer is established ($\phi_o = 45°$, $Q = 5.84\ \mathrm{cm}^2/\mathrm{s}$, $g'_o = 20.31\ \mathrm{cm/s}^2$, $N = 0.485\ \mathrm{s}^{-1}$)

fluid from the dome falls back to spread out laterally to form a horizontal layer (equilibrium level). Figure 9 shows a negatively buoyant two-dimensional plume (inclined at 45° to vertical) in a linearly stratified fluid in a laboratory experiment. The spreading layer in this case is quite thick compared with the depth of penetration. The maximum depth of the dome, z_m, and the depth of the horizontal layer, h_e, can be determined from such a laboratory experiment.

3.1 Round Plume in Linearly Stratified Environment

The Brunt-Väisälä frequency is constant in a linearly stratified fluid. Asymptotic formulae of plumes in uniform density stratification can be derived using dimensional analysis. The dimensions of the *specific* buoyancy flux of the round plume F_o is,

$$[F_o] = [g'_o][Q] = \frac{L^4}{T^3} \tag{51}$$

It can then be shown that the height of rise is proportional to a length scale based on this buoyancy flux, F_o, and the buoyancy frequency N (Prob. 5.3):

$$z_e \sim F_o^{\frac{1}{4}} N^{-\frac{3}{4}} \tag{52}$$

Briggs (1969) has correlated field and laboratory data of plume rise and proposed the following formula for the rise of the plume:

$$z_e = 3.76\, F_o^{\frac{1}{4}} N^{-\frac{3}{4}}. \tag{53}$$

Figure 10 shows Briggs' correlation of the data with this formula. The laboratory experiments of Wong and Wright (1988) gave a similar formula for the maximum height of the plume

$$z_m = 4.5\, F_o^{\frac{1}{4}} N^{-\frac{3}{4}}. \tag{54}$$

Wong and Wright (1988) also provided a formula for the thickness of the horizontal spreading layer at the equilibrium level as follows:

$$h_e = 1.7\, F_o^{\frac{1}{4}} N^{-\frac{3}{4}}. \tag{55}$$

This thickness of the spreading layer is quite large. It accounts for 38% of the maximum rise of the plume.

Lagrangian method may be generally employed in the formulation for linear and nonlinear stratification effects. The elemental volume is defined as the volume due to release of buoyancy at the source of the plume over a period of one time unit. Substituting volume flux, Q, for $\tilde{V}$, and the specific buoyancy flux, $F = g'Q$, for $\tilde{B}$, the mass balance equation (Eq. 8) becomes

$$\frac{dF}{dz} = Qg\frac{d\rho_a}{dz}, \tag{56}$$

which can be re-written, in terms of the buoyancy frequency as

$$\frac{dF}{dt} = -N^2 M \tag{57}$$

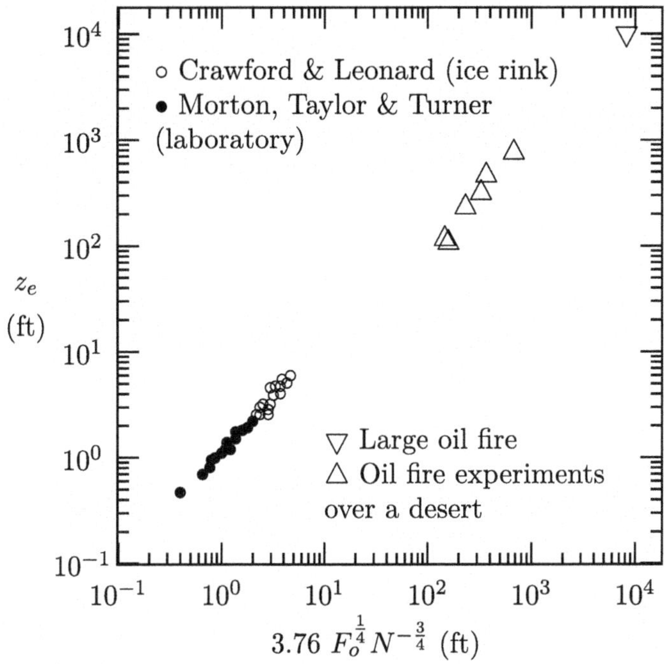

Fig. 10 Field and laboratory experimental data for the rise of plumes in linearly stratified environment; data reproduced from Briggs (1969) and Turner (1973)

where $M = \tilde{w}Q$ is the *specific* momentum flux. This momentum in the material volume, $\tilde{\rho}M$, increases with time at a rate equal to the buoyancy force, ρF:

$$\frac{dM}{dt} = F. \tag{58}$$

Equations 57 and 58 combine to give

$$\frac{d^2M}{dt^2} + N^2M = 0. \tag{59}$$

If the frequency N is a constant, the equation is simple harmonic and the solution for a source with an initial momentum M_o is

$$M = M_o \cos(Nt) + \frac{F_o}{N} \sin(Nt), \tag{60}$$

An asymptotic solution is obtained by ignoring the initial momentum flux at the source. Equation 60 with $M_o = 0$ is

$$\pi \tilde{b}^2 \tilde{w}^2 = \frac{F_o}{N} \sin Nt \tag{61}$$

The square root of the above is

$$\sqrt{\pi} \tilde{b} \tilde{w} = \sqrt{\pi} \beta z \frac{dz}{dt} = \sqrt{\frac{F_o}{N} \sin Nt}. \tag{62}$$

Integrating with time once, gives

$$\frac{1}{2} \beta \sqrt{\pi} z^2 = \sqrt{\frac{F_o}{N}} \int_0^t \sqrt{\sin Nt} dt \tag{63}$$

This leads to the equation for the penetration distance $z(t)$

$$z = \sqrt{\frac{2}{\beta \pi^{\frac{1}{2}}}} I(\tau)^{1/2} F_o^{\frac{1}{4}} N^{-\frac{3}{4}} \tag{64}$$

where $\tau = Nt$ and $I(\tau) = \int_o^\tau \sqrt{\sin \tau} d\tau$. The maximum height of the plume rise occurs when $M = 0$. Equation 63, integrating from $t = 0$ to $t = \pi/N$, gives the maximum height

$$z_m = \sqrt{\frac{2}{\beta}} \pi^{-\frac{1}{4}} [I(\pi)]^{\frac{1}{2}} F_o^{\frac{1}{4}} N^{-\frac{3}{4}}. \tag{65}$$

where the integral $I(\pi) = \int_0^\pi \sqrt{\sin \tau} d\tau = 2.396$. For $\beta = 0.17$ we have

$$z_m = 4.0 \, F_o^{\frac{1}{4}} N^{-\frac{3}{4}}. \tag{66}$$

Strictly speaking, the integral formulation is invalid shortly after the plume passes the neutral buoyancy level, when the plume starts to re-entrain mixed fluid from the spreading layer. Nevertheless, the linear spreading hypothesis gives a reasonable solution, which is essentially the same as the following solution obtained by Morton et al. (1956) using the entrainment hypothesis:

$$z_m = 1.15 \alpha_G^{-\frac{1}{2}} \, F_o^{\frac{1}{4}} N^{-\frac{3}{4}} = 4.02 \, F_o^{\frac{1}{4}} N^{-\frac{3}{4}} \tag{67}$$

where the plume entrainment coefficient of $\alpha_G = 0.082$ is adopted. The predicted plume rise compares also favorably with the observed rise coefficient of 4.5 by Wong and Wright (1988) (Eq. 54).

The volume flux, $Q = \pi \tilde{b}^2 \tilde{w}$, in the plume increases with distance from the source. The formula for the round plume in stratified fluid is derived using Eqs. 62 and 64 as follows:

$$Q = \pi \tilde{b}^2 \tilde{w} = \beta z \, \pi^{\frac{1}{2}} F_o^{\frac{1}{2}} N^{-\frac{1}{2}} \sqrt{\sin Nt} \tag{68}$$

Hence,

$$Q = \sqrt{2\beta\pi^{\frac{1}{2}}}\,[I'(\tau)]^{\frac{1}{2}}\,F_o^{\frac{3}{4}}N^{-\frac{5}{4}} \tag{69}$$

where $\tau = Nt$ and $I'(\tau) = \sin\tau \int_o^\tau \sqrt{\sin\tau'}d\tau'$. The maximum volume flux (for $\beta = 0.17$) is

$$Q_m = 0.94\,F_o^{\frac{3}{4}}N^{-\frac{5}{4}}, \tag{70}$$

which occurs at $\tau = Nt = 0.66\,\pi$ with $I' = 1.481$. According to this analysis, the average dilution of the round plume as it impinges onto the equilibrium layer is

$$S = \frac{Q_m}{Q_o} = 0.94\,Q_o^{-1}F_o^{\frac{3}{4}}N^{-\frac{5}{4}}. \tag{71}$$

The above average dilution coefficient is consistent with the laboratory observations. Wong and Wright (1988) obtained the formula $S_m = 0.80\,Q_o^{-1}F_o^{\frac{3}{4}}N^{-\frac{5}{4}}$ based on their laboratory data; S_m is the *minimum* dilution obtained in the horizontal spreading layer, which is expected to be less than the average dilution. Similar to previous investigations (e.g. Fan (1967)), their experiments show that the entrainment of ambient fluid up to the maximum height of rise is not significantly affected by the presence of a stable ambient density gradient.

3.2 *Plane Plume in Linearly Stratified Environment*

The plane plume is also called the line plume because it is produced by a line source of buoyancy. Similar derivation can be carried out to find the asymptotic solution of the plane plume in an environment of uniform density stratification. The *specific* buoyancy flux per unit length of the line source has a dimension

$$[F_o] = [g_o'][Q] = \frac{L^3}{T^3} \tag{72}$$

By dimensional analysis the maximum height of rise can be shown to correlate with a length scale based on F_o and the buoyancy frequency

$$z_m \sim F_o^{\frac{1}{3}}N^{-1} \tag{73}$$

The Lagrangian method can similarly be applied for the plane plume. The momentum per unit length of line plume is

$$M = \frac{F_o}{N}\sin(Nt); \tag{74}$$

which is,

$$2\tilde{b}\tilde{w}^2 = \frac{F_o}{N} \sin Nt. \tag{75}$$

The square root of the above is

$$\sqrt{2\beta} z^{\frac{1}{2}} \frac{dz}{dt} = \sqrt{\frac{F_o}{N} \sin Nt} \tag{76}$$

Integrating with time once gives

$$\frac{2}{3}\sqrt{2\beta} z^{\frac{3}{2}} = \sqrt{\frac{F_o}{N}} \int_0^t \sqrt{\sin Nt}\, dt \tag{77}$$

that leads to the depth of the plume penetration

$$z = (\frac{9}{8\beta})^{\frac{1}{3}} F_o{}^{\frac{1}{3}} N^{-1} [I(\tau)]^{\frac{2}{3}} \tag{78}$$

where $\tau = Nt$ and $I(\tau) = \int_o^{\tau} \sqrt{\sin \tau'} d\tau'$. The maximum height of the plume rise occurs when $M = 0$. The integration of the above, from $Nt = 0$ to $Nt = \pi$, gives $I(\pi) = 2.396$ and the formula for the maximum height

$$z_m = 3.36\ F_o{}^{\frac{1}{3}} N^{-1}. \tag{79}$$

Wright and Wallace (1979) and Wallace and Wright (1984), based on their laboratory data, have proposed a coefficient of 3.6 for the maximum rise and a coefficient 1.84 for the horizontal spreading layer thickness. The thickness of the layer is 51% of the maximum height (Fig. 9). The laboratory data give the height of the horizontal spreading layer

$$z_{et} = 2.99\ F_o{}^{\frac{1}{3}} N^{-1} \tag{80}$$

at the top of the layer, and

$$z_{eb} = 1.15\ F_o{}^{\frac{1}{3}} N^{-1}. \tag{81}$$

at the bottom of the horizontal spreading layer. Similar values for a plume have been obtained for an inclined buoyant jet (Lee and Cheung 1986).

The volume flux per unit length of the line plume is equal to the width of the top-hat profile, $2\tilde{b}$, times the velocity of the top-hat profile, $\tilde{w}$; i.e.,

$$Q = 2\tilde{b}\tilde{w} = 2\beta z \frac{dz}{dt}. \tag{82}$$

Given Eq. 78 for the variation of z as a function of time t,

$$Q = (3\beta)^{\frac{1}{3}} F_o{}^{\frac{1}{3}} N^{-1} [I(\tau)]^{\frac{1}{3}} \sqrt{\sin \tau} \tag{83}$$

If $\beta = 0.17$, the *average* dilution at the plume impingement of the equilibrium layer would be

$$S = \frac{Q_{\max}}{Q_o} = 0.76\, Q_o^{-1} F_o^{\frac{2}{3}} N^{-1}. \tag{84}$$

The maximum volume flux $Q_{\max}$ occurs shortly after the plume pass the equilibrium level at time $Nt = 0.62\pi$. The *minimum* dilution formula proposed by Wallace and Wright (1984) based on laboratory data obtained in the horizontal spreading layer is

$$S_m = 0.88\, Q_o^{-1} F_o^{\frac{2}{3}} N^{-1} = 0.24 \frac{F_o^{\frac{1}{3}} z_m}{Q_o} \tag{85}$$

After impingement the plume undergoes further mixing; the dilution in the spreading layer is expected to be somewhat higher than the dilution at the point of plume impingement.

In addition to the above two-dimensional plume experiments, Roberts *et al.*(1989) studied the dilution of the waste field formed with a line plume of finite length in a stratified crossflow. Three-dimensional effects are considered in his experiments which generally revealed higher dilutions and hence lower trap levels. For near stagnant conditions, the rise height and dilution coefficients are 2.6 and 0.37 respectively (cf 3.36 and 0.24 in Eqs. 79 and 85).

Example 5.4 The diffuser pipe of the Orange County wastewater outfall is 1829 m (6000 ft) long and is located at 61 m (200 ft) below the sea surface. The monthly average density profiles in the vicinity of the outfall in the coastal waters of Southern California are shown in Fig. 11 (data given in Koh and Brooks (1975)). The total flow from the outfall is 13.2 m^3/s. Neglecting the effect of coastal current, estimate the dilution of effluent plume in February when the density of the coastal water varies from $\sigma_t = 25.5$ to 24.5 over the 60 m depth of water. Assume $\sigma_t = 0$ for the effluent.

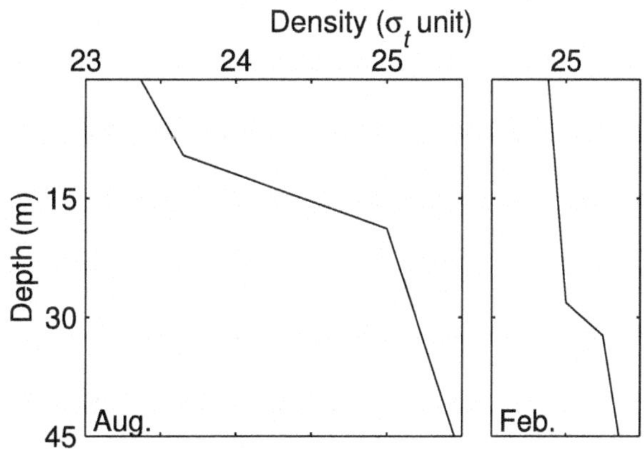

Fig. 11 Monthly-averaged density profiles (in σ_t units) used for design of Orange County Outfall, California. Reproduced from Fischer *et al.* (1979)

Solution: The water in February is least stratified. Over the entire depth of 60 m, σ_t varies from 25.5 to 24.9. Assuming that the variation $\Delta\sigma_t = 0.6$ is linear over the 60 m depth, the Brunt-Väisälä frequency is

$$N = \sqrt{-\frac{g}{\rho_a}\frac{d\rho_a}{dz}} \simeq \sqrt{\frac{9.81 \times (25.5 - 24.9)}{1000 \times 60}} = 0.0099 \text{ rad/s}.$$

The buoyant effluent from the diffuser rises essentially as a line plume. The discharge per unit length of the 1829 m diffuser diffuser is $Q_o = 13.15/1829 = 7.19 \times 10^{-3}$ m^2/s. The specific buoyancy flux is

$$F_o = g'_o Q_o = 9.81 \times \frac{25.5}{1000} \times \frac{13.15}{1829} = 1.799 \times 10^{-3} \text{ m}^3/\text{s}^3$$

The length scale associated with the rise of this line plume is $l_s = F_o^{\frac{1}{3}} N^{-1} = 12.28$ m. The maximum rise of the effluent plume is (Eq. 79)

$$z_m = 3.36\, F_o^{\frac{1}{3}} N^{-1} = \underline{41.3} \text{ m}.$$

The top and bottom of the horizontal spreading layer are $z_t = 2.99\, l_s = 38.2$ m and $z_b = 1.15\, l_s = 14.1$ m above the source at the diffuser, respectively (Eqs. 80 and 81). The average dilution at the plume impingement is $S = 0.76\, Q_o^{-1} F_o^{\frac{2}{3}} N^{-1} = \underline{157}$ (Eq. 84). The minimum dilution ratio in the horizontal spreading layer is estimated to be $S_m = 0.88\, Q_o^{-1} F_o^{\frac{2}{3}} N^{-1} = \underline{182}$ using Eq. 85.

It is remarkable how such a small density stratification of the ocean in February can arrest the rise of the plume and cause the spreading of the plume along a horizontal layer. The dilution of the plume is significantly reduced by the presence of the density stratification. In an un-stratified ocean, the same line plume at the base of the spreading layer would have a volume flux per unit length of the diffuser equal to

$$Q = 2\beta(1 - \beta)z\left(\frac{F_o}{2\beta}\right)^{\frac{1}{3}}$$

$$= 2 \times 0.17 \times (1 - 0.17) \times 60 \left(\frac{1.799 \times 10^{-3}}{2 \times 0.17}\right)^{\frac{1}{3}} = 2.95 \text{ m}^3/\text{s}$$

and an average dilution equal to $S = Q/Q_o = 410$, which is 2.6 times greater than the same plume trapped at depth 21.8 m to 45.9 m (i.e., from $z = 38.2$ m to 14.1 m) by the density stratification. In the above, the plume is assumed to reach the ocean surface when its elevation is within one half-width from the surface, that is when the elevation $z = 60(1 - \beta)$ m.

The above calculations of the plume in linear density stratification using the one-dimensional model has produced results that are generally in agreement with formulae derived from laboratory experimental data. The results are also consistent with the numerical solution of the 2D Navier-Stokes equations using the Large Eddy Simulation (LES) method (Altai and Chu 2001; Chu and Altai 2001a, b). The LES simulation results in Fig. 12 shows how the turbulent line plume develops in the stratified environment. The dynamics of plume penetration and the spreading of the horizontal layer in stratification are shown as a movie attached with this monograph.

Example 5.5 The Harbour Area Treatment Scheme (HATS) is a major environmental infrastructure in Hong Kong. Sewage is collected from a densely populated urban area through a tunnelled system to a centralized sewage treatment works for Chemically Enhanced Primary Treatment (CEPT). An average of 1.4 million m^3/d of this partially treated wastewater is discharged into a stratified ambient water with depth $H = 12$ m via a 1.1 km long submarine outfall. It is given that in the wet season, $Q_o = 29.5$ m^3/s, $\Delta\rho/\rho_o = 0.025$, $N = 0.08$ rad/s, and discharge *E. coli* concentration $C_o = 10^5$ counts/100 mL. The sewage plume can be trapped below the surface under stratified conditions; the trapped plume undergoes turbulent dispersion by an ambient current of $u_a = 0.5$ m/s, with the *E. coli* decay process governed by first order kinetics, i.e. $C = C_i e^{-kt}$, where C_i is the initial concentration, $k(d^{-1})$ is the decay rate, and t is time (d). A modified form of the Mancini equation Mancini (1978) has been developed to determine the depth dependent *E. coli* decay rate in subtropical coastal water as (Chan et al. 2015):

$$k(z) = (0.68 + 0.017S) \times 1.07^{(T-20)} + 1.1 I_A e^{-e_t z}$$

where z = depth from the water surface, S = salinity (ppt) and T = ambient temperature (^{o}C). I_A is the average intensity of surface solar radiation (ly/h), e_t is the light extinction coefficient (m^{-1}).

Given that $I_A = 1.5$ MJ/m^2/h $= 35.8$ly/h, $e_t = 0.7$ m^{-1}, S = 31.5 ppt, and T = 28°C. Estimate the *E. coli* concentration at a beach 6 km away from the outfall (a) without and (b) with density stratification. Assume the the outfall discharge as a line plume in stagnant ambient. The secondary dilution of the plume surfaced and trapped layer can be estimated by the Brooks' '4/3' dispersion law (Fischer et al. 1979).

Solution: This example illustrates the water quality concern for beaches located close to an wastewater discharge in a densely populated city. The *E. coli* concentration is governed by (i) the initial dilution achieved by the plume mixing, (ii) the secondary dilution due to the turbulent dispersion of the surfaced or trapped plume layer, and (iii) *E. coli* decays primarily due to solar radiation.

The specific buoyancy flux per unit length of the line plume is computed as $F_o = Q_o g'_o = 29.5/1100 \times 0.025 \times 9.81 = 0.00658$ m^3/s^3 (per unit length of diffuser).

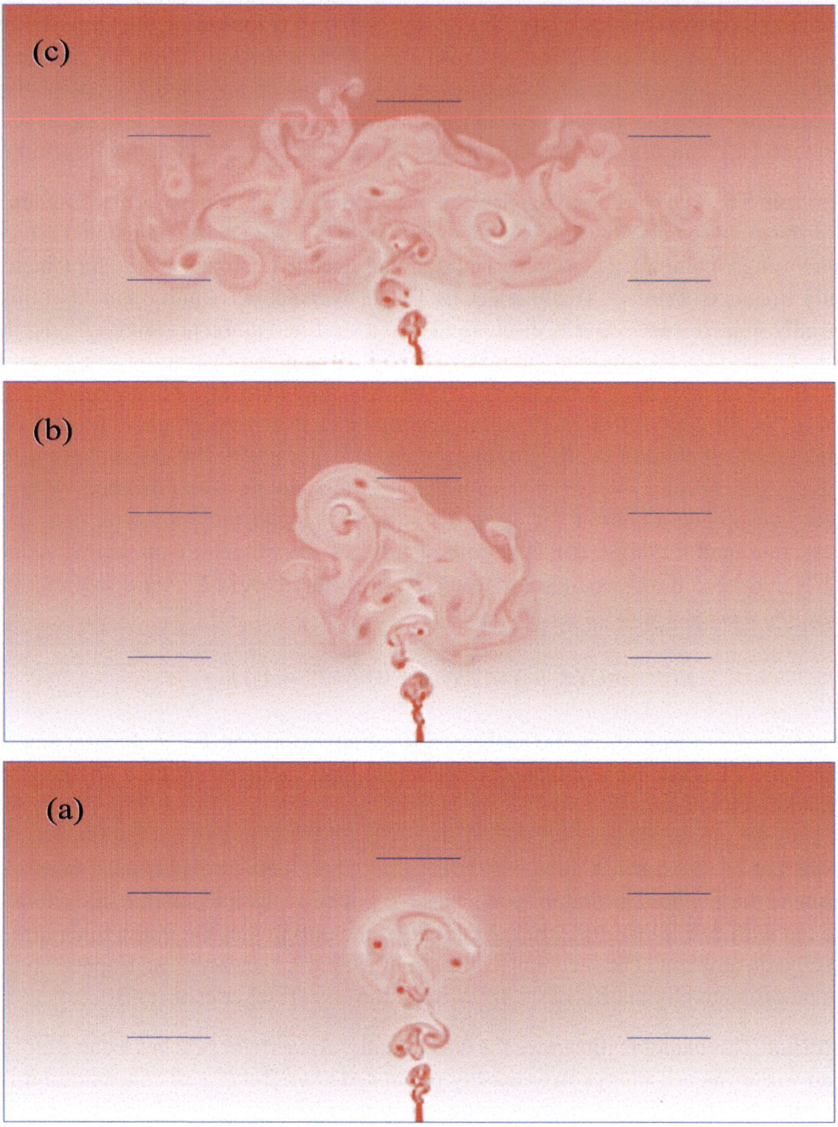

Fig. 12 LES (Large Eddy Simulation) of plane plume in stagnant linearly stratified fluid. LES based on the data of Orange County California Outfall in February; σ_t varies linearly from 24.9 to 25.5 over the 60m depth. Computed plume at: **a** $t = 352$ s, **b** $t = 701$ s, **c** $t = 1430$ s. The horizontal lines are the predicted maximum rise ($z_m = 44$ m) at the dome, the top ($z_{et} = 38.2$ m) and bottom ($z_{eb} = 14.1$ m) elevation of the horizontal spreading layer from Eqs. 79, 80 and 81, respectively

(a) Without density stratification:

- Initial dilution S_1: Using Table 1 of Chap. 3, the average plume dilution is $S_1 = 0.465 F_o^{1/3} z/Q_o = 0.465 \times 0.00658^{1/3} \times 12/(29.5/1100) = \underline{39}$.

- Secondary dilution S_2: The centerline concentration of the spreading layer due to passive turbulent dispersion is determined by Brooks (1960) law as:

$$\frac{C_m}{C_i} = erf(\sqrt{\frac{1.5}{(1+\frac{2}{3}P)^3 - 1}})$$

where C_i is the concentration after the initial dilution, $P = \beta(x/b)$, and $\beta = (12\epsilon_o)/(u_a b)$, ϵ_o is diffusivity at the location of diffuser ($x = 0$), and b plume width after initial dilution $\approx$ diffuser length. Assume $\epsilon_o \approx 10^1$ m^2/s for the coastal water, $\beta = (12 \times 10^1)/(0.5 \times 1100) = 0.22$, $P = \beta(x/b) = 0.22 \times (6000/1100) = 1.2$, and $S_2 = C_i/C_m = 1/0.56 = \underline{1.79}$.
- Bacteria decay S_3: The decay rate at the water surface is determined as $k = (0.68 + 0.017 \times 31.5) \times 1.07^{(28-20)} + 1.1 \times 35.8 \times 1 = 41.5$ /d. It takes the plume 3 hours to reach the beach. The 'dilution' that can be achieved from bacteria decay is then $S_3 = 1/e^{[-41.5\times(3/24)]} = \underline{179}$.

E.coli concentration at the beach is $C = C_o/(S_1 S_2 S_3) = \underline{8}$ counts/100 mL.

(b) With density stratification:

- Initial dilution S_1: The plume will be trapped at the equilibrium level z_e with density stratification. Using Table 1, $z_m = 3.60 F_o^{1/3} N^{-1} = 3.60 \times 0.00658^{1/3} \times 0.08^{-1} = 8.4$ m. Using Eq. 84, the corresponding average dilution $S_1 = 0.76\ Q_o^{-1} F_o^{2/3} N^{-1} = \underline{12.4}$
- Secondary dilution due to horizontal turbulence and spreading is assumed to be similar to that of a surface plume and hence $S_2 = \underline{1.79}$.
- Bacteria decay S_3: The decay rate at the equilibrium level is determined as $k = (0.68 + 0.017 \times 31.5) \times 1.07^{(28-20)} + 1.1 \times 35.8 \times e^{-[0.7\times(12-8.4)]} = 5.3$/d, and $S_3 = 1/e^{[-5.3\times(3/24)]} = \underline{2}$.

E.coli concentration at the beach is $C = C_o/(S_1 S_2 S_3) = \underline{2250}$ counts/100 mL, significantly more than that without density stratification ! The main reasons for the much less dilution achieved in this case are (i) less water depth for plume mixing; (ii) less sunlight for bacteria decay. Using Hong Kong's beach ranking system with *E.coli* concentration less than 24 counts/100 mL as good, and more than 610 as very poor (Thoe 2010), a contradictory conclusion can be reached if density stratification is not accounted for.

3.3 Plumes in Arbitrary Density Stratification

The Lagrangian method has been illustrated for the case of linear density profile, for which basic laboratory data is available to validate the theory. In practice, the ambient density profile is typically nonlinear. Many practical problems cannot be described by the asymptotic formulae. Nevertheless, they can be readily analyzed by the Lagrangian method which is generally applicable for arbitrary density stratification profile. Discharge below a thermocline may or may not be trapped by the thermocline. The dilution characteristics of the un-trapped plume above the thermocline are very different from the trapped plume below the thermocline.

Example 5.6 The diffuser pipe of the Orange County wastewater outfall is 1829 m (6000 ft) long and is located at 61 m (200 ft) below the sea surface. The total flow from the outfall is 13.2 m^3/s. Neglecting the effect of coastal current, estimate the dilution of effluent plume in August when the density structure of the coastal water is essentially two-layer. Ambient density is approximately constant in the upper and lower layer. The density changes rapidly across the thermocline from $\sigma_t = 23.5$ to 25.5 from 10 m to 20 m depth (see Fig. 11).

Solution: Fig. 11 shows the ambient density profile near the Orange County outfall in southern California. The density stratification is greatest in August. Across the thermocline in August, σ_t varies from 25.5 to 23.5 over a depth of 10 m with a Brunt-Väisälä frequency at the thermocline equal to

$$N = \sqrt{-\frac{g}{\rho_a}\frac{d\rho_a}{dz}} \simeq \sqrt{\frac{9.81 \times (25.5 - 23.5)}{1000 \times 10}} = 0.044 \text{ rad/s.} \tag{86}$$

A simple analysis of the line plume in August may be made by assuming the density to be constant and neglecting the effect of the stratification for the first 40 m rise of the line plume. The dilution of the plume below the thermocline is calculated by the plume formula in un-stratified fluid

$$Q = 2\beta z(\frac{F_o}{2\beta})^{\frac{1}{3}}$$

The plume impinges on the thermocline as the plume rises a distance of $z = 40(1 - \beta)$ m. The volume flux at the plume impingement is

$$Q = 2 \times 0.17 \times (1 - 0.17) \times 40\,(\frac{1.799 \times 10^{-3}}{2 \times 0.17})^{\frac{1}{3}} = 1.97 \text{ m}^2/\text{s}$$

The corresponding dilution ratio is $S = Q/Q_o = \underline{294}$. The effluent density from the diffuser is practically the same as the fresh water, which has a sigma unit $\sigma_t \simeq 0$. The sigma unit of the surrounding sea is 25.5. The sigma unit of the effluent is initially below its surrounding by a deficit of $\Delta\sigma_t = 25.5$. Dilution of the deficit gives $\Delta\sigma_t =$

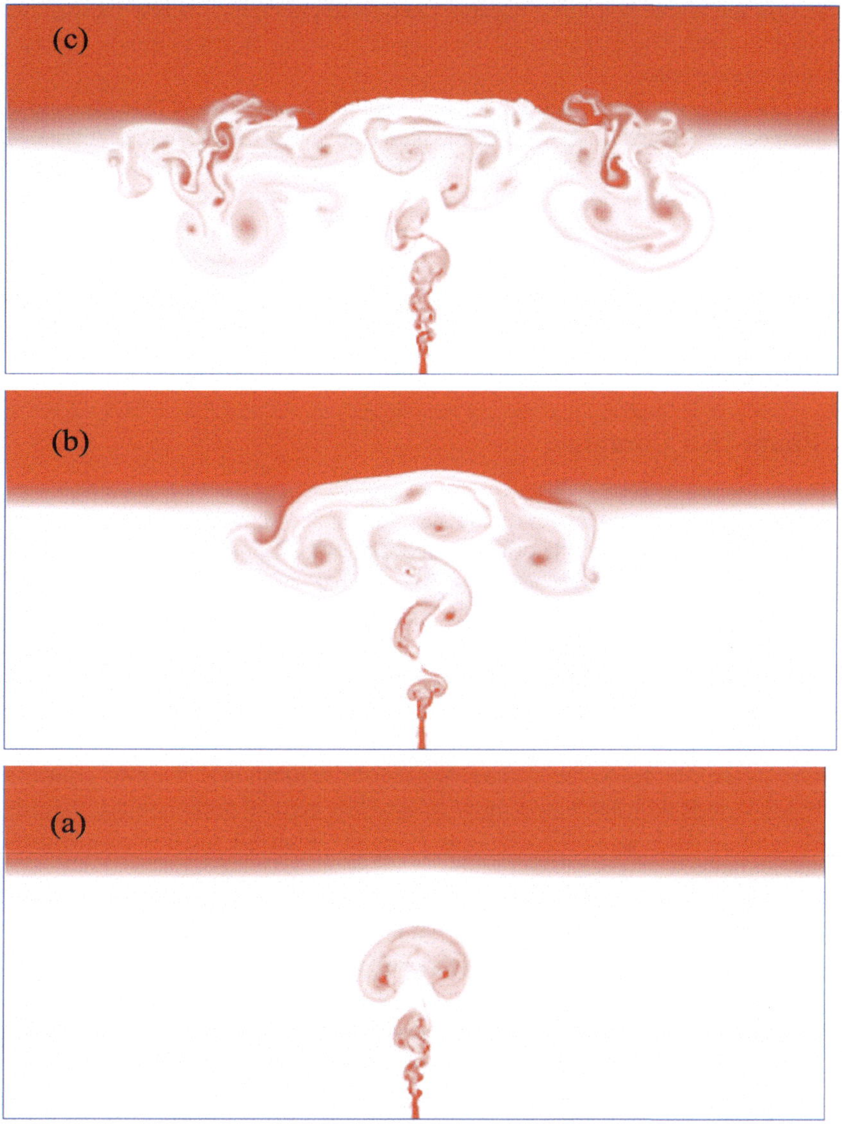

Fig. 13 LES of plane plume impingement at a thermocline. LES using the data of the Orange County California Outfall and August stratification; σ_t varies from 23.5 to 25.5 from $z = 10$ to $z = 20$ m. Predicted concentration field at (bottom to top) time **a** $t = 251$ s, **b** $t = 499$ s, **c** $t = 752$ s

$25.5/294 = 0.0867$ at the plume impingement to the thermocline. Therefore, $\sigma_t = 25.5 - 0.0867 \simeq 25.4$ at the impingement. The plume is so diluted at the thermocline by the surrounding sea, such that its density is now heavier than the water above the thermocline. The LES (Large Eddy Simulation) of the line plume in Fig. 13

shows how the plume is trapped because the diluted plume fluid is heavier than the surrounding fluid. The dynamics of the trapped plume is shown as movie attached with this monograph.

The plume may not be trapped if the diluted plume fluid is lighter than the fluid at the thermocline. Figure 14 shows how the plume advances through a thermocline of weak density stratification. The dynamics of plume penetration through a thermocline is shown as a movie attached with this monograph. For this 2D situation, as the plume fluid is added to the upper layer above the thermocline, the interface between the layers moves down at a rate equal to the rate the volume of the plume fluid enters the upper layer. The following section decribes how the stratification in the ambient is developed as buoyant fluid is fed to the environment by the plume.

4 Plume in a Container

The supply of buoyant fluid by a plume into a container introduces a stratified environment, which affects the rise of the plume and, at the same time, depends on the rate of buoyancy supplied by the plume. The mutual interaction of the plume with its environment can lead to very complex phenomena. The heating and cooling of a room is one example. In the absence of a fan to mix the air, a stove is known to be able to produce strong temperature stratification. As the plume produced by the stove impinges on the ceiling, it spreads out laterally to form a warm and stratified layer. Heat waves move down from the ceiling as plume fluid accumulates in the stratified layer below the ceiling. A similar development of stratification is observed as lateral inflows of cold and turbid waters enter a lake. The vertical progression of density stratification in the lake, and in the room, is closely related to the fluid entrainment into the plume.

The development of density stratification in a container of constant cross-sectional area by a round plume was examined by Baines and Turner (1969). Figure 15 shows the problem examined. The density stratification is produced by a steady inflow of buoyant fluid located at the bottom of the container. This sketch shows how the density profile in the environment develops from density fronts which originate at the top where the plume spreads laterally. The location of the initial front from starting the flow is shown in the sketch on the right. The density profiles at these times are shown on the left. The downward velocity of the stratified interfaces in the environment is induced by the upward flow of the plume. The volume flux of the plume, Q, is equal to the downward flow of the environment:

$$Q = -AW \tag{87}$$

in which A is the horizontal cross section of the container and W the vertical velocity in the stratified environment. Neglecting mixing across the stratified interfaces, the density is constant following the fluid motion in the environment; i.e.,

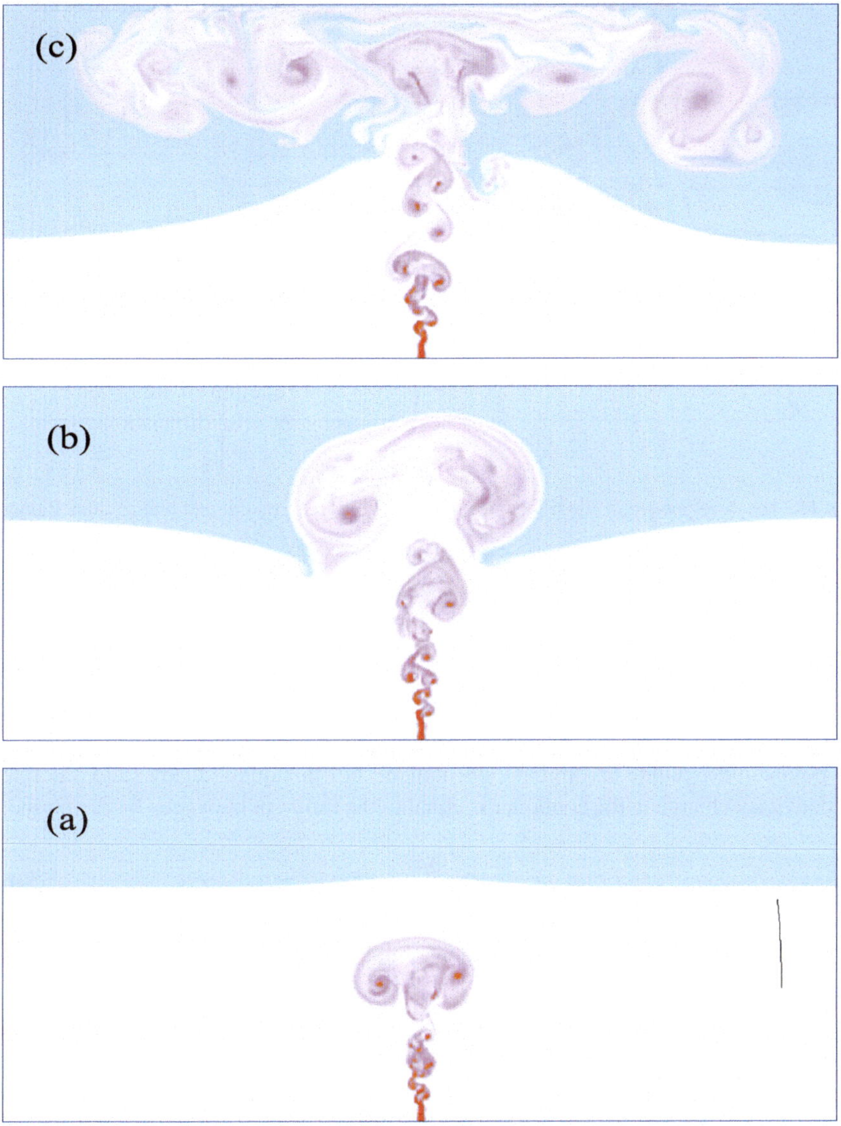

Fig. 14 LES of plane plume penetration at a weak thermocline. In this example, σ_t varies only slightly from 25.4 to 25.5 across the thermocline. The depth of water and the buoyancy flux of the plume is the same as the case shown in Fig. 13. Sequence of images are predicted plume concentration fields at (from bottom to top) t = 251 s, 503 s, and 1004 s

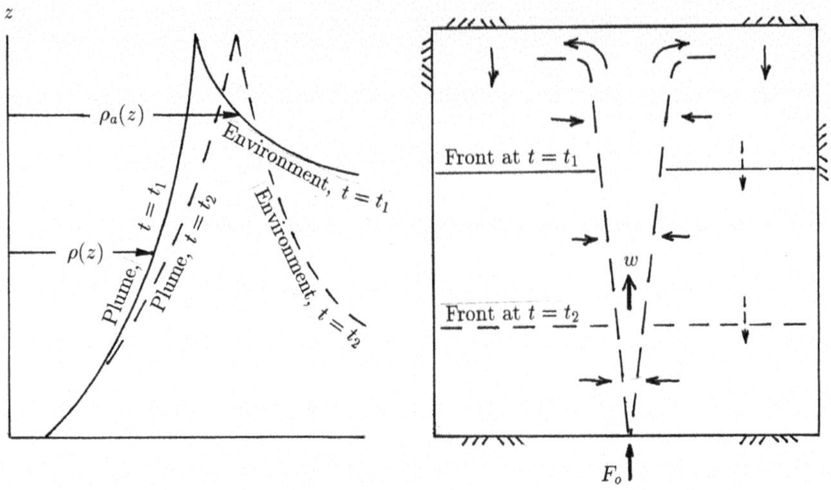

Fig. 15 The development of density stratification in a container by a round plume, after Baines and Chu (1996)

$$\frac{\partial \rho_a}{\partial t} + W \frac{\partial \rho_a}{\partial z} = 0. \tag{88}$$

The solution by Baines and Turner (1969) is based on the quasi-steady assumption by which the density in the environment was assumed to increase at the same rate at all points in the plume and in the environment. For this quasi-steady state, the rate of buoyancy change in the container is equal to the influx of buoyancy by the plume:

$$-V_R g \frac{\partial \rho_a}{\partial t} = F_o \tag{89}$$

where V_R is the volume of the container. First, eliminating $\partial \rho_a / \partial t$ from Eqs. 88 and 89,

$$V_R W g \frac{\partial \rho_a}{\partial z} = F_o. \tag{90}$$

Then, eliminating W from Eqs. 87 and 90,

$$Qg \frac{\partial \rho_a}{\partial z} = -\frac{A}{V_R} F_o = -\frac{F_o}{H} \tag{91}$$

where $H = A/V_R$ is the depth of the container. With this density stratification, and applying to Eq. 56,

$$\frac{dF}{dz} = -\frac{F_o}{H}. \tag{92}$$

Hence, the buoyancy flux in the plume decreases linearly with height above the source in the quasi-steady state. The buoyancy flux is F_o at the bottom of the container and it decreases to zero at the top where the plume impinges the ceiling of the container; i.e.,

$$F = F_o(1 - \frac{z}{H}). \tag{93}$$

The momentum equation for the motion of the plume is

$$\frac{dM}{dt} = F_o(1 - \frac{z}{H}) \tag{94}$$

(see Eq. 58). The momentum flux of the round plume is

$$M = \pi \tilde{b}^2 \tilde{w}^2. \tag{95}$$

If $\tilde{b} = \beta z$,

$$M = \pi \beta^2 z^2 (\frac{dz}{dt})^2 \tag{96}$$

which can be re-written as

$$\frac{dz^2}{dt} = \frac{2}{\beta}\sqrt{\frac{M}{\pi}}. \tag{97}$$

The pair of equations, Eqs. 94 and 97, is integrated numerically to give M and z^2. The volume flux of the plume is

$$Q = \beta z \sqrt{\pi M}. \tag{98}$$

The ambient density stratification, according to Eq. 91, is

$$\frac{d\rho_a}{dz} = -\frac{F_o}{gHQ}. \tag{99}$$

Figure 16 shows the two density profiles produced by the plume with curve 2 identifying a later time. The initial density front approaches the floor asymptotically with time while the density in the ceiling of the container increases steadily toward the source density.

The mutual interaction between plumes and its stratified environment has many applications. Baines and Murphy (1986) analyzed and verified the density distribution produced by the failure of a steam line in the interior of a power plant. Germeles (1975) extended it to the unsteady case and applied it to the filling of tanks of liquid natural gas. Baines et al. (1993) examines the effect of fountain and its application to clean room technology. The method has been applied to line plumes, periodic thermals, unsteady source discharge (Killworth and Turner 1982) and laminar plumes (Worster and Leith 1985). The equations differ in each case but are derived from the

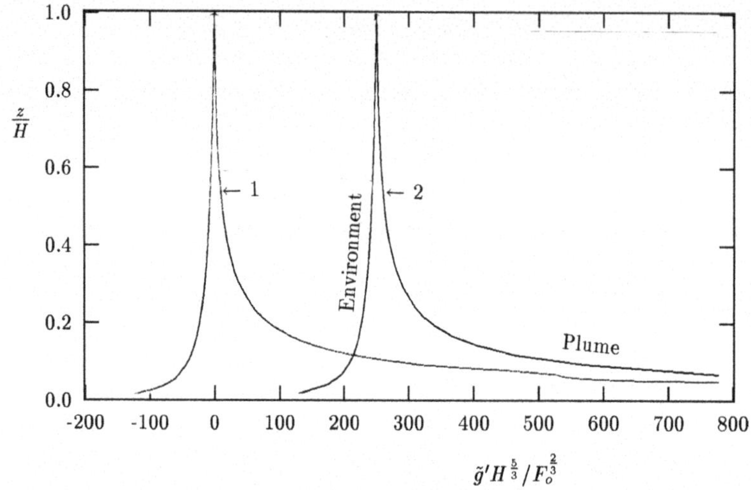

Fig. 16 Buoyancy profiles in plume and environment, after Baines and Chu (1996). Curve 1 at time t. Curve 2 at time $(t + 200\, A / F_o^{\frac{1}{3}} H^{\frac{4}{3}})$

same method of separating the flow into plume and stratified regions and then connecting the regions by entrainment relation. The quasi-steady solution is presented here as an example. The general solution for the unsteady flow involves partial differential equations (Eq. 88) and can be obtained numerically, for example, by method of characteristics.

5 Summary

Thermals and plumes in linear stratification are analyzed in this chapter using the Lagrangian method. The predicted rise heights are in very close agreement with experimental and field observations in all cases. Table 1 summarizes the asymptotic formulae obtained using the Lagrangian method and the formulae obtained from experimental and field data.

In practice, the stratification is often not linear, the initial momentum may not be small, and the size of source is not generally negligible. Solution to these more complex problems of finite momentum and buoyancy sources in non-unform stratification are readily obtained by numerical integration of the governing equations. We have carried out the numerical calculations in this chapter of the thermal penetration problem in a nonlinear stratification using the equation obtained by the Lagrangian method and compared successfully with the results of LES (Large Eddy Simulation). Further development of the Lagrangian method for jets and plumes in coflows and crossflows will be considered in subsequent chapters.

Table 1 Asymptotic formulae for buoyancy sources in linear density stratification obtained using the Lagrangian method and the formulae derived from experimental and field data

Flows	Lagrangian model	Experimental & field data
Round thermal	$z_m = 2.64\,(\frac{B_o}{\rho})^{\frac{1}{4}} N^{-\frac{1}{2}}$ $z_e = 2.22\,(\frac{B_o}{\rho})^{\frac{1}{4}} N^{-\frac{1}{2}}$ $(\beta = 0.34, k = 0.5)$	$z_m = 2.66(\frac{B_o}{\rho})^{\frac{1}{4}} N^{-\frac{1}{2}}$ (Morton et al. 1956)
Line thermal	$z_m = 2.54\,(\frac{B_o}{\rho})^{\frac{1}{3}} N^{-\frac{2}{3}}$ $z_e = 2.02\,(\frac{B_o}{\rho})^{\frac{1}{3}} N^{-\frac{2}{3}}$ $(\beta = 0.34, k = 1.0)$	$z_m = 2.30\,(\frac{B_o}{\rho})^{\frac{1}{3}} N^{-\frac{2}{3}}$ (Wright 1984) $z_e = 1.85\,(\frac{B_o}{\rho})^{\frac{1}{3}} N^{-\frac{2}{3}}$ (Wright 1984) $z_e = 1.78\,(\frac{B_o}{\rho})^{\frac{1}{3}} N^{-\frac{2}{3}}$ (Briggs 1965)
Round plume	$z_m = 4.00\,F_o{}^{\frac{1}{4}} N^{-\frac{3}{4}} (\beta = 0.17)$	$z_m = 4.50 F_o{}^{\frac{1}{4}} N^{-\frac{3}{4}}$ (Wong and Wright 1988) $z_e = 3.76\,F_o{}^{\frac{1}{4}} N^{-\frac{3}{4}}$ (Briggs 1969)
Plane plume	$z_m = 3.36\,F_o{}^{\frac{1}{3}} N^{-1} (\beta = 0.17)$	$z_m = 3.60\,F_o{}^{\frac{1}{3}} N^{-1}$ $z_{et} = 2.99\,F_o{}^{\frac{1}{3}} N^{-1}$ $z_{eb} = 1.15\,F_o{}^{\frac{1}{3}} N^{-1}$ (Wright and Wallace 1979)

B_o = buoyancy force; F_o = specific buoyancy flux

The significance of density stratification in engineering applications, such as the environmental impact to recreational beaches from an wastewater outfall, has been illustrated in a worked example. A numerical model is normally needed for environmental impact assessment, since the ambient conditions in practice is much more complex. A model that couples a near-field general integral jet model and a far-field recirculation model for such purpose, will be introduced in Chap. 11.

Problems

5.1 The integral jet model formulation outlined in Chaps. 3 and 4 (for the case of constant ambient density) can be extended to the case of a stratified ambient fluid. Consider a stable ambient density distribution $\rho_a(z)$, with the density decreasing vertically upwards. For this case, it is also reasonable to assume that the velocity profile and the concentration excess or density deficit with respect to the **local** ambient value are also Gaussian:

$$c(s,r) - c_a[z(s)] = (c_m - c_a)e^{-\frac{r^2}{\lambda^2 b^2}}$$

$$\rho_a[z(s)] - \rho(s,r) = (\rho_a - \rho_m)e^{-\frac{r^2}{\lambda^2 b^2}}$$

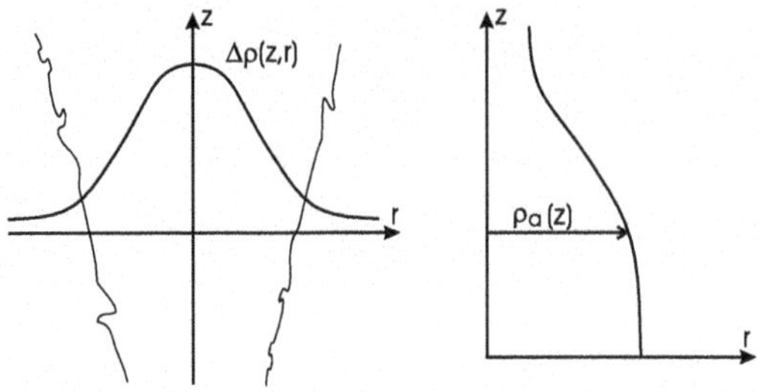

Show that the governing equations remain the same as that for the case of $\rho_a =$ constant, except with the equations for conservation of tracer mass and density deficit flux replaced by:

$$\frac{d}{ds}\int u\Delta c\, dA = -Q\frac{dc_a}{dz}\sin\phi$$

$$\frac{d}{ds}\int u\Delta\rho\, dA = Q\frac{d\rho_a}{dz}\sin\phi$$

where c_a(z) is the ambient tracer concentration, $\Delta c = c - c_a$, $\Delta\rho = \rho_a - \rho, dz/ds = \sin\phi$, Q is the volume flux, and s=natural streamwise co-ordinate. (Hint: use both the conservation of tracer mass and the continuity equations).

5.2 Consider a total discharge flow of 2.2 m^3/s from a multiport diffuser at 50 m depth when there is ambient stratification in the ocean. The 50 ports are spaced apart to avoid interference between adjacent jets, and the density of the discharge is $\rho_o = 1000\ kg/m^3$.

Fig. 17 Ambient density profile in summer

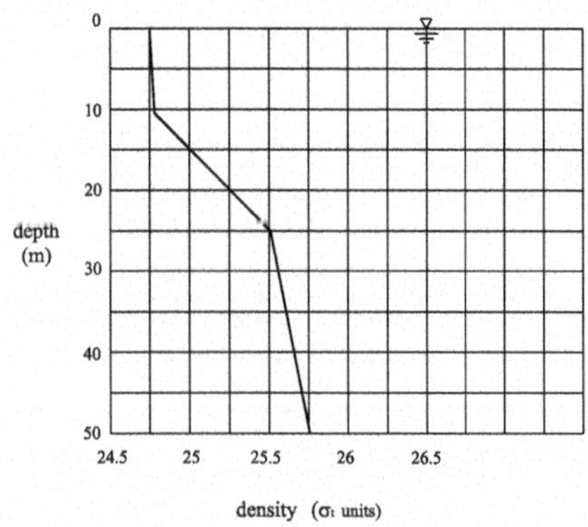

A typical vertical ambient density profile in the summer is shown in the accompanying graph (Fig. 17). Determine if the sewage field will stay submerged beneath the surface under these conditions.

Use VISJET or any reasonable approximate method (e.g. assume linear density stratification over height of rise).

5.3 A solid sphere of diameter d (volume V) is released from the free surface of a linearly stratified liquid with density variation $\rho_a(z) = \rho_o + (d\rho_a/dz)\, z$, where $d\rho_a/dz$ = constant, and z = vertical distance below free surface. The density of the sphere is $\rho_d > \rho_o$. The sphere sinks to an equilibrium position z_t within the fluid.

(a) At the equilibrium position, show using hydrostatics that regardless of the size of the sphere, the buoyant force acting on the sphere (which balances the weight of the sphere $W = \rho_d g V$) depends only on the density at z_t (= vertical distance from free surface to center of sphere):

$$F_b = \rho_a(z_t) g V = (\rho_o + \frac{d\rho_a}{dz} z_t) g V$$

Show that the equilibrium level of the sphere is

$$z_t = g'_o \left(\frac{g}{\rho_d} \frac{d\rho_a}{dz}\right)^{-1}$$

where z = positive downwards (same direction as gravity), and $g'_o = (\rho_d - \rho_o)g/\rho_d$.

(b) The above experiment corresponds to a negatively buoyant release in linearly stratified fluid. Thus for a general positively buoyant plume released at $z = 0$ in a linearly stratified environment, it is reasonable to postulate that the maximum height of rise depends on:

$$z_m = fn(Q_o, g'_o, -\frac{g}{\rho_d} \frac{d\rho_a}{dz})$$

where z=vertical co-ordinate (positive upwards) above source. Since for small density differences, Q_o and g'_o are both represented in the specific buoyant flux, $F_o = Q_o g'_o$, we can write:

$$z_m = fn(F_o, -\frac{g}{\rho_d} \frac{d\rho_a}{dz})$$

Hence z_m can be obtained by dimensional reasoning (cf Eq. 52 and Prob. 5.6).

5.4 *Generation of linear stratification*
In laboratory studies it is often desired to generate a linear vertical stratification, i.e. $d\rho_a/dz$ = constant. This can be efficiently done using the following method (Oster, G 1965, *Scientific American*, 213(2), 70-79).

Assume a linear stratification is to be prepared using a salt solution in a rectangular flume of plan cross-section area A_f and volume V_f. Prepare in a separate container a salt solution with density equal to the desired ambient density at the bottom layer. The container has an initial volume of salt solution equal to about $V_o = V_f/2$. The salt solution (constantly well-stirred by mixing) is then pumped into the flume at a certain flow rate Q. Simultaneously freshwater (with zero salt concentration) is fed into the container at a flow rate of $Q/2$. It can be shown the salt concentration of the solution in the container (which is pumped meticulously into the flume with minimal disturbance) decreases linearly with time.

(a) The time required to fill the flume is $t_f = V_f/Q = 2V_o/Q$; this is the same as the time required to empty the container, $t_d = V_o/(Q/2)$.
If $V(t)$ and $c(t)$ are the volume and concentration of the salt solution in the container, By performing a mass and salt balance (assuming full mixing in the container) show that:

$$\frac{dV}{dt} = -\frac{Q}{2}$$
$$\frac{d(Vc)}{dt} = -Qc$$

(b) Show that the concentration of the salt solution pumped into the flume will be linearly decreaing with time:

$$\frac{c(t)}{c_o} = 1 - \frac{t}{t_d}$$

where c_o = initial salt concentration (corresponding to the lowest layer in the flume), and $t_d = V_o/(Q/2)$ is the "detention time" in the container.

As the flume is filling up with the salt solution, $A_f dh/dt = Q$, and $t = (A_f/Q)h$, show that the salt concentration (hence density) in the flume varies linearly with depth as:

$$\frac{c(h)}{c_o} = 1 - \frac{h}{h_f}$$

where $h(t)$ = depth of salt solution in the flume at time t, and h_f = final depth of linear stratification in the flume.

5.5 *Model for turbulent round thermal*
Consider an isolated mass of buoyant fluid (initial density ρ_o, volume V_o) released at time $t = 0$ in ambient fluid of uniform density ρ_a, with $\rho_o < \rho_a$. The flow is driven by the initial buoyant force, $B_o = (\rho_a - \rho_o)gV_o$, which stays constant by virtue of conservation of tracer mass. The thermal mixes with the ambient fluid by turbulent entrainment. Both experiments and numerical calculations show that the round

thermal is a vortex-ring like flow, which is self-similar and can be described by its characteristic radius $R(t)$, average velocity $W(t)$, and density $\rho(t)$.

(a) Applying the equation of motion for the round thermal (assumed to be a sphere), show that the momentum equation is:

$$\frac{d}{dt}[\frac{4}{3}\pi R^3(\rho + k\rho_a)W] = \frac{4}{3}\pi R^3(\rho_a - \rho)g$$

where k accounts for the virtual mass of the unsteadily moving thermal. The virtual mass coefficient is assumed to be the same as a solid sphere, $k = \frac{1}{2}$.

Show that the thermal velocity is given by:

$$W = \frac{\rho_a - \rho}{\rho + k\rho_a} gt$$

Obtain limiting expressions for (i) $\Delta\rho/\rho_a \ll 1$ and (ii) $\rho_a \gg \rho$ $(\Delta\rho/\rho \approx 1)$.

(b) Entrainment hypothesis–If it is assumed that the ambient fluid is entrained into the thermal at a rate proportional to its surface area and velocity:

$$\frac{d}{dt}(\frac{4}{3}\pi R^3) = (4\pi R^2)\alpha W$$

where α is the entrainment coefficient (assumed constant). Noting that $W = dz/dt$ show that a linear spread is predicted, with $R = \beta z$, and $\beta = \alpha$ for this case.

5.6 (a) Consider a round buoyancy-dominated wastewater plume in a linearly stratified ocean. Deduce to within a constant the maximum height of rise of the mixed effluent, z_m, in terms of the specific buoyancy flux F_o and the stratification parameter $\epsilon = -g/\rho_o(d\rho_a/dz) = N^2$, where N = buoyancy frequency, and the other terms have their usual meaning. Explain briefly your physical reasoning.

(b) In the deep ocean, hot sulfide ore-bearing solutions (~ 350°C) are sometimes released from local spots on the sea bed. The fine sulfide particles (typical settling velocity = 1/500 cm/s) are carried with the hot, grey-black water from these hydrothermal 'submarine vents'. When the buoyant plume ('black smoker') reaches a certain level, it spreads sideways and the suspended sulfide particles may settle back to the sea floor to form sulfide deposits. A typical mean horizontal velocity and temperature at 1-2 km depth are 1 cm/s and 10°C respectively.

(i) Assuming the flowrate and temperature of the source to be 10^4 cm^3/s and 350°C ($\Delta\rho/\rho = 0.01$) respectively, estimate the maximum height of rise of the black smoker. The stratification frequency can be taken as $N = 3 \times 10^{-4}$ s^{-1}. Neglect the effect of ambient current.

(ii) Massive sulfide deposits of horizontal dimensions 100 to 1000 m are sometimes found on the sea floor. Comment on the possibility that these confined ore deposits are due to the settling out of particles from the rising effluent or from the spreading layer of the black smoker plume.

5.7 (a) Consider a vertical round jet in a linearly stratified ocean. Deduce to within a constant the maximum height of rise of the mixed effluent, z_m, in terms of the specific momentum flux M_o and the stratification parameter $\epsilon = -g/\rho_o(d\rho_a/dz) = N^2$, where $N =$ stratification frequency. Show that:

$$z_m = C\,(\frac{M_o}{\epsilon})^{1/4} = C M_o{}^{1/4} N^{-1/2}$$

Experiments show that $C = 3.8$.

(b) Wastewater from a small town next to a popular beach is discharged in the form of a single round jet in coastal water of 12 m depth. The jet diameter is $D = 0.036$ m and the discharge flow is $Q_o = 0.02\ \mathrm{m^3/s}$. In the Spring month of April the receiving water is weakly stratified with a salinity gradient of 0.1 ppt/m. The bottom salinity and temperature are 32.5 ppt and 26°C respectively, and the effluent density is $\rho_o = 998\ \mathrm{kg/m^3}$.

Compute the jet densimetric Froude number Fr, the jet momentum and buoyancy fluxes, and the momentum length scale L_s. If the jet is discharged vertically, assuming a momentum jet, estimate the maximum height of rise z_m and the dilution. Will the beach be protected?

Determine z_m for a corresponding plume.

5.8 For the wastewater discharge described in Prob. 5.7, suppose the jet is discharged horizontally in the offshore direction with the same velocity and other parameters unchanged. Assess if the wastewater jet will be trapped beneath the free surface. (a) If $\overline{S}$ is the average dilution up to the trapping level z_t, then the density of mixed sewage at the trapping level ρ_m can be estimated by assuming the entrained ambient flow has an average density between source level and z_t. Show that:

$$\rho_m = \frac{\rho_o Q_o + \frac{1}{2}(\rho_a(0) + \rho_a(z_t))(\overline{S} - 1)Q_o}{\overline{S} Q_o}$$

Since $\rho_m = \rho_a(z_t)$, show that:

$$z_t = \frac{2(\rho_a(0) - \rho_o)}{(-d\rho_a/dz)\overline{S}}$$

(b) The above equation shows that if $\overline{S}$ is sufficient, trapping can occur, with $z_t \leq H$. Experiments show that the mixing up to the trapping level is not much affected by the stratification. Use the Cederwall equation to estimate the dilution of the horizontal buoyant jet and assess the likelihood of trapping. (Hint: first assume $z_t = H$, determine horizontal jet dilution $\overline{S}$, then compute z_t and iterate).

5.9 For a plane buoyant jet in a linearly stratified fluid, with source momentum flux Q_o, specific buoyancy flux F_o, and stratification frequency $N = \sqrt{-g/\rho_a(d\rho_a/dz)}$, show that the following length scales can be obtained:

$$l_s = M_o/F_o^{2/3}$$
$$l_b' = F_o^{1/3}/N$$
$$l_m' = M_o^{1/3}N^{-2/3}$$

l_s/l_b' measures the relative importance of the initial momentum, initial buoyancy, and the ambient density stratification. For a buoyant jet that behaves like plumes, $l_s/l_b' \ll 1$, show by dimensional analysis that maximum height of rise is $z_m \sim l_b' = C_1 F_o^{1/3} N^{-1}$ (Eq. 79). For a buoyant jet that behaves like momentum jets, $l_s/l_b' \gg 1$, show that $z_m = C_2 M_o^{1/3} N^{-2/3}$. Experiments show that $C_1 = 3.6$ and $C_2 = 2.3$ for a vertical plane buoyant jet, and $C_1 = 3.6$ and $C_2 = 1.9$ for a jet inclined at 45° to horizontal.

5.10 A 20 port submerged diffuser for a marine terminal discharges treated ship ballast water at a flow of $Q_s = 1.06\ \text{m}^3/\text{s}$ at 45° to the horizontal. The ambient water (70 m depth) is near stagnant and is weakly stratified, with a variation of 0.58 σ_t units over the bottom 50 m. The waste effluent density is $\rho_o = 1020.7\ \text{kg/m}^3$, while the bottom ambient density is $\rho_a(0) = 1024.1\ \text{kg/m}^3$. If the diffuser length is $L = 60$ m, and the jet velocity is $U_o = 4.2$ m/s, determine the volume flux $Q_o = Q_s/L$ of the equivalent slot (2D) buoyant jet, and the momentum and buoyancy fluxes (M_o and F_o). Estimate the maximum height of rise and the initial dilution.

For a 45° discharge, the minimum diluton in the spreading layer is given by $S_m \sim F_o^{1/3} l_b'/Q_o$ for plume like conditions (dilution coefficient = 0.81), and $S_m \sim (M_o l_m')^{1/2}/Q_o$ for jet like conditions (dilution coefficient = 0.6).

Chapter 6
Turbulent Round Jet in Coflow

The jet in coflowing current is examined in this chapter as a prelude to the general Lagrangian formulation of a jet in non-uniform ambient current, with a three-dimensional trajectory. Like simple jets and plumes, the axisymmetric coflowing jet is an important limiting case. It is a shear flow which has been extensively studied (e.g. Maczynski 1962; Antonia and Bilger 1973, 1974; Shirazi et al. 1974; Knudsen 1988; Wood 1993; Nickels and Perry 1996). The detailed velocity field has been measured by many investigators and more recently the tracer concentration has been studied using Laser-induced fluorescence (LIF) techniques (Chu et al. 1999; Wang and Davidson 2001). It has a number of turbulent spreading characteristics that are critical to understanding the Lagrangian method. In the following, the key observed features of a coflowing jet will be first summarized. The subtleties and difficulties of an integral formulation or turbulent closure for a jet in a current are explained and discussed. Using a general Lagrangian integral model based on a spreading hypothesis, the coflowing jet properties are then calculated. The predictions of jet width, velocity excess, and centerline dilution are compared with experimental data and the implications on jet modelling are discussed.

1 Summary of Experimental Observations

Figure 1 shows a schematic diagram of the round jet in a coflow. The jet is issued from a circular nozzle of diameter D at a velocity of U_o. The velocity of the coflow is U_a. The jet velocity is the coflow velocity, U_a, plus the excess velocity, ΔU_e, as shown in Fig. 1. For a coflowing jet, using a similar derivation as in Chap. 2, it can be readily shown that the specific *jet excess momentum*, $M_e = \int uu_e dA = \int u(u - u_a)dA$, is conserved (Prob. 6.1). Two parameters are significant for the coflowing jet. These

Supplementary Information The online version contains supplementary material available at https://doi.org/10.1007/978-3-032-01187-9_6.

J. H. W. Lee et al., *Turbulent Jets and Plumes*,
https://doi.org/10.1007/978-3-032-01187-9_6

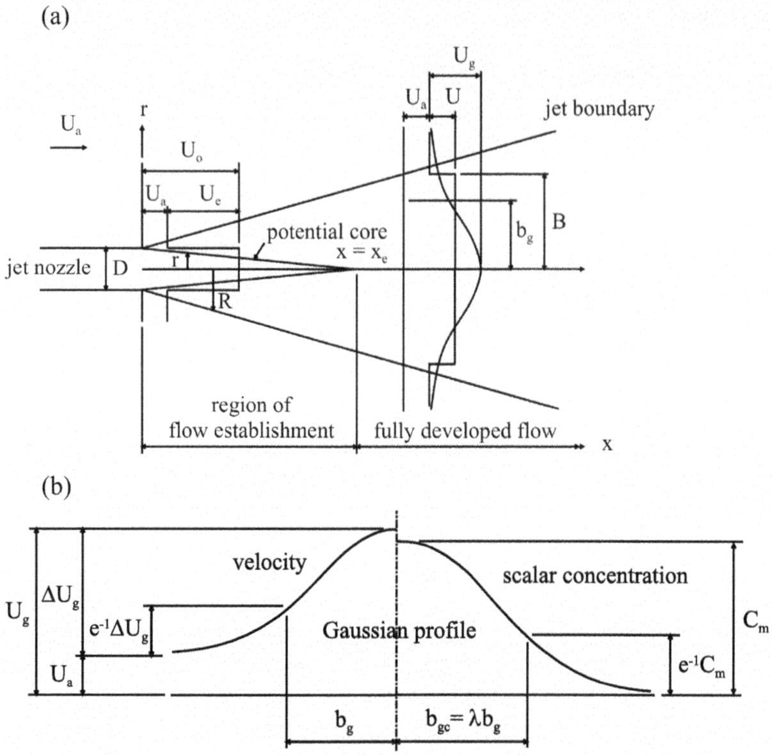

Fig. 1 **a** Schematic diagram of round jet in coflow; **b** representation of characteristic jet properties

are the coflow velocity, U_a, and the excess momentum flux, M_{eo}. Similar to previous chapters, we will use M_{eo} to denote the specific (kinematic) excess momentum flux unless otherwise stated. Since the dimension of $[M_{eo}]$ is L^4/T^2 and the dimension of $[U_a]$ is L/T, the excess momentum length scale

$$l_m^* = \frac{M_{eo}^{1/2}}{U_a} \tag{1}$$

and the coflow velocity U_a are the length scale and velocity scale of the problem, respectively.

1.1 Cross-sectional Images

The jet mixing can be studied by measuring the concentration of a tracer at different x-sections from the source using a technique known as Laser-induced fluorescence

(LIF). The source fluid is a fluorescent dye solution of known concentration. A sheet of laser light can be made to illuminate a given section against a completely dark background; the resulting laser-induced fluorescent image of the jet cross-section can be video-recorded. The fluorescence intensity at each point is proportional to the tracer concentration. With careful calibration, the dye concentration at each point can be determined from the measured light intensity. This technique provides a powerful means to visualize the phenomenon, and to measure the instantaneous tracer concentration field not achievable with probe-based techniques.

Figure 2 shows the instantaneous images and the time-averaged concentration of the jet cross-section at $x = 40D$ downstream of the nozzle in an experiment with $U_o = 0.796$ m/s, $U_a = 0.05$ m/s, and $D = 1.02$ cm (Chu 1996; Chu et al. 1999). The time-averaged image is obtained from a series of 250 video frames obtained over a one minute period of time. The cross-sectional shape of the jet is constantly varying with time as a core of turbulent eddies that moves around the jet centerline. The time-averaged profile partly reflects the variation of concentration within this dominant turbulent eddy structure, and perhaps also the movement of the dominant eddy around the centerline. This behaviour is not due to the coflow; very similar turbulent structure was observed in the cross-sectional images of the jet in stagnant environment.

1.2 Gaussian Profiles

Figure 3 shows time-averaged concentration (C) contours at two sections of the coflowing jets. The dimension of the jet is expressed in pixel coordinates (~23 pixels/cm). Despite the irregular shape of the instantaneous images, the time-averaged contours are circular in shape with very good axisymmetry. It is not surprising then that the radial profile of time-averaged concentration $C(x, r)$ is self-similar and Gaussian. In Fig. 4, the measured concentration C normalized by the centerline maximum $C_m = C(x, 0)$ is plotted against the dimensionless radial distance r/b_{gc}, where b_{gc} is the concentration half-width defined by the radial location at which $C = e^{-1}C_m$. It can be seen the radial concentration variation is self-similar; the data at different sections collapse nicely onto one curve and can be well-approximated by the Gaussian distribution:

$$\frac{C}{C_m} = \exp\left(-\frac{r^2}{b_{gc}^2}\right) \tag{2}$$

(See definitions of C_m and b_{gc} in Fig. 1.)

For a coflowing jet, it is the velocity excess above the ambient that determines the entrainment. Figure 5 shows the excess velocity measured across a coflowing jet by Laser Doppler Anemometry (LDA). The excess velocity $u_e(x, r) = u(x, r) - U_a$ normalized by the maximum ($u_{em} = \Delta U_g$) is plotted against r/b_g, where b_g is a similarly defined velocity half-width. The self-similar velocity excess can also be

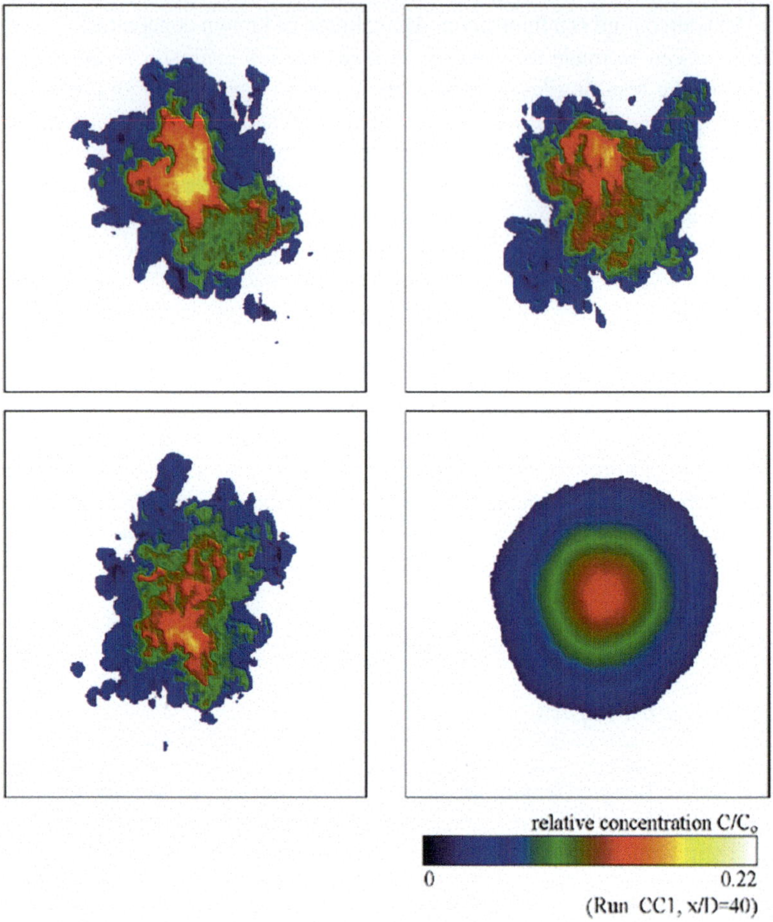

Fig. 2 Instantaneous images and time-averaged concentration field (lower right corner) in the cross-section of coflowing jet ($U_o = 0.796$ m/s, $U_a = 0.05$ m/s, and $D = 1.02$ cm). Concentration lower than the visual threshold not shown

described by:

$$\frac{u_e}{\Delta U_g} = \exp\left(-\frac{r^2}{b_g^2}\right) \tag{3}$$

(See definitions of ΔU_g and b_g in Fig. 1.)

The width of the Gaussian concentration profile, b_{gc}, is generally wider than that of the velocity profile, b_g. Unlike the case of a jet in stagnant ambient, however, the width of a coflowing jet varies nonlinearly with the distance from the source.

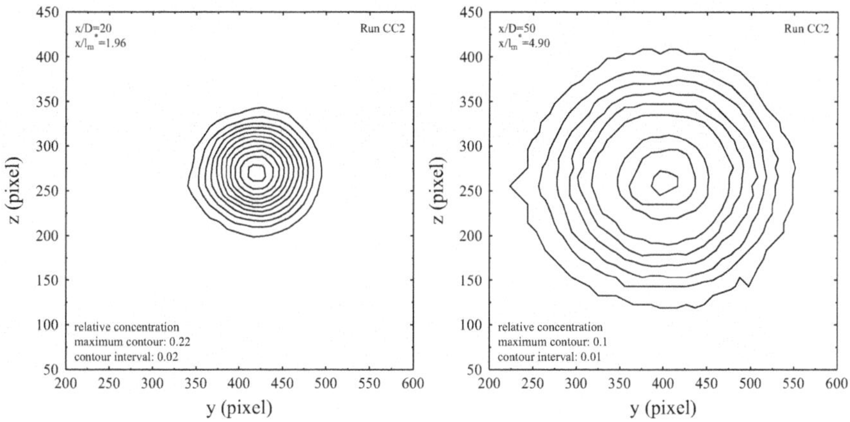

Fig. 3 Cross-sectional time-averaged concentration contours of a coflow jet

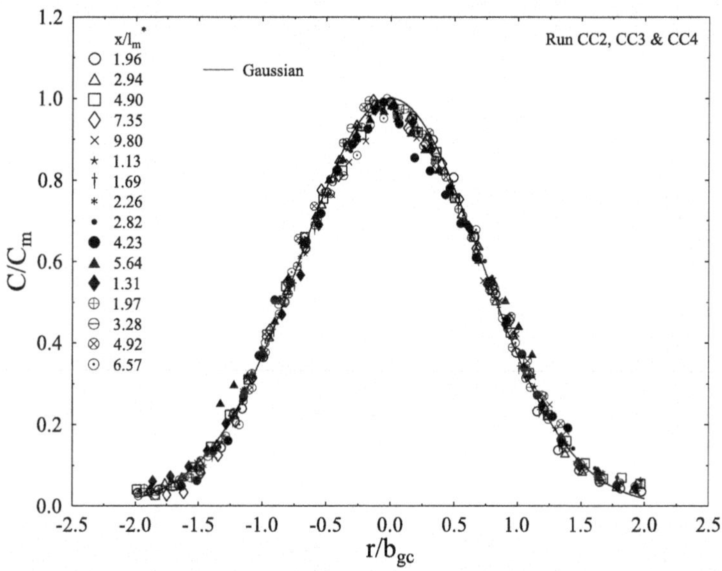

Fig. 4 Radial profile of time-averaged concentration in a coflow jet

2 Integral Model

Figure 1 shows the schematic definition of an axisymmetric turbulent jet with source diameter D and exit velocity U_o in a coflow U_a. Near the source, a jet mixing layer develops within the zone of flow establishment (ZFE) which extends from the jet nozzle up to the end of the potential core at $x = x_e$; the velocity within the potential core is constant, with an excess velocity equal to $U_o - U_a$. In the zone of established

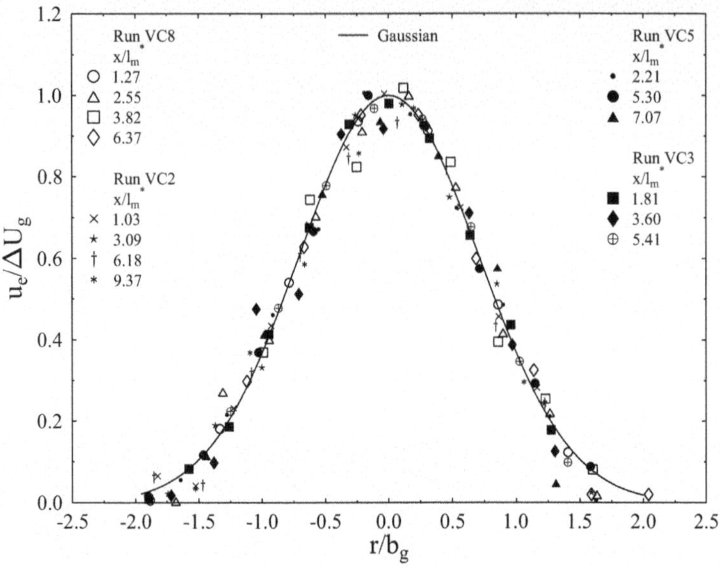

Fig. 5 Measured radial profile of time-averaged excess velocity in a coflow jet

flow (ZEF), $x \geq x_e$, the turbulent-mean velocity profile of the fully turbulent flow is approximately Gaussian, with a centerline excess velocity $\Delta U_g = U_g - U_a$ and nominal half-width b_g. Our objective is to develop an integral model that predicts the variation of the jet width b_g and the excess velocity ΔU_g, so that the dilution or the tracer concentration in the coflowing jet can be determined.

2.1 *The Natural but Incorrect Formulation*

The modelling of a jet in a current is considerably more complicated than the stagnant case. The presence of the ambient current introduces a number of subtleties and difficulties that can be illustrated using the coflowing jet. Building on the theory of jets and plumes, one can formulate the mass and momentum balances for a control volume bounded between x and $x + dx$, as shown in Fig. 6. It seems reasonable to write:

$$\frac{dQ}{dx} = \frac{d}{dx}\int u dA = Q_e \tag{4}$$

$$\frac{dM}{dx} = \frac{d}{dx}\int u^2 dA = Q_e U_a \tag{5}$$

where Q, M are the jet volume and momentum flux, and Q_e is the entrainment flow. The above simply relates the increase in jet volume flux due to turbulent entrainment,

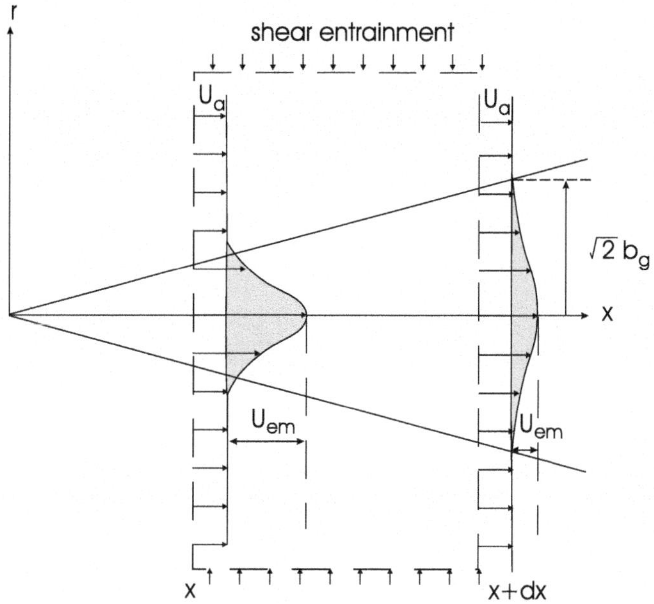

Fig. 6 Control volume formulation of a coflow jet

and the change in jet momentum flux resulting from the addition of momentum due to the entrained ambient flow. However, we note that the jet volume flux Q is not a bounded integral; hence some definition of jet edge is required. It is customary to define $r = \sqrt{2}b_g = B$ (where B is the top-hat half-width) as the edge of the jet. It is also reasonable to assume in accordance with the entrainment hypothesis that the turbulent entrainment is proportional to the centerline jet excess velocity $u_{em} = \Delta U_g$. After substituting the volume and momentum flux in terms of centerline variables of the Gaussian distribution, we then have (Prob. 6.2):

$$\frac{d}{dx}\int_0^{\sqrt{2}b_g} u dA = \frac{d}{dx}\left(\pi b_g^2\left(u_{em} + 2U_a\right)\right) = 2\pi\alpha b_g u_{em} \tag{6}$$

$$\frac{d}{dx}\int_0^{\sqrt{2}b_g} u^2 dA = \frac{d}{dx}\left(\frac{\pi}{2}b_g^2(u_{em} + 2U_a)^2\right) = 2\pi\alpha b_g u_{em} U_a \tag{7}$$

In a coflow of uniform velocity, it can be shown that the excess momentum flux is conserved (Prob. 6.1 or Chu 1996).

$$M_e = \int_0^{\infty} u_e(u_e + U_a)2\pi r dr = \frac{\pi}{2}b_g^2 u_{em}(u_{em} + 2U_a) \tag{8}$$

is constant and equal to the source excess momentum flux $M_{eo} = (U_o - U_a)U_oA_o$, where $u_{em} = U_g - U_a = \Delta U_g$ is the centerline excess velocity, and $A_o = \pi D^2/4$ the cross-sectional area of the nozzle.

The plausibility of the above commonly used integral model can be examined. First, the entrainment hypothesis as formulated includes the advective flux due to the ambient flow U_a as part of the turbulent entrainment, and is hence physically unreasonable. As shown in Fig. 6, it seems clear the entrainment due to velocity shear should affect only the excess volume flux $\Delta Q = \int u_e dA$. That is, the entrainment hypothesis should be modified to:

$$\frac{d}{dx}\int_0^{\sqrt{2}b_g} u_e dA \approx \frac{d}{dx}\int_0^{\infty} u_e dA = \frac{d}{dx}(\pi b_g^2 u_{em}) = 2\pi\alpha b_g u_{em} \tag{9}$$

Adopting this entrainment hypothesis then, by expanding Eq. 7 and making use of Eq. 9, it can be shown that

$$\frac{d}{dx}M_e = -4U_a^2\frac{db_g}{dx} \tag{10}$$

Since $\frac{db_g}{dx}$ is in general not zero, the above formulation clearly does not conserve jet excess momentum. What is at fault here? The terms on the RHS reflects the increase in jet momentum flux due to the fact the jet is expanding—i.e. the ambient advection into the conical surface of the differential control volume has been neglected in Eq. 7 (Schatzmann 1979). Only with the integration limits properly taken into consideration is the excess momentum conservation recovered. The correct equations for the coflow jet are then Eqs. 9 and 8 respectively.

We see that an apparently innocent formulation has resulted in many problems for the coflowing jet. Many integral models (e.g. Fan 1967; Abraham 1970; and Hirst 1972 upon which the USEPA model DKHDEN is based) were formulated heuristically using a control volume approach as in the above Eq. 6. Consequently they do not model shear entrainment correctly nor satisfy excess momentum conservation; they cannot be expected to predict the coflowing jet well. In addition, there is the key question of the entrainment coefficient. Is it a constant as in the simple jet? It turns out the entrainment assumption with a constant shear entrainment coefficient fails miserably for the coflowing jet.

2.2 *Jet Spreading Hypothesis*

We adopt a general spreading hypothesis (Chu and Lee 1996): the change in width of the shear layer, *in a Lagrangian frame of reference (moving with the eddies)*, is assumed to be proportional to the relative velocity between the jet element and its surroundings.

Let $\tilde{u}$ be the characteristic jet velocity and U_a be the ambient velocity. For the steady jet in coflow, the spreading hypothesis reduces to a particularly simple form:

$$\frac{D\tilde{b}}{Dt} = |\tilde{u}|\frac{d\tilde{b}}{dx} = \beta|\tilde{u}_e| \tag{11}$$

where t and x represent time and distance along jet trajectory, $\tilde{b}$ is the characteristic half-width, $\tilde{u_e} = \tilde{u} - U_a$ is the characteristic jet excess velocity, and β is the proportionality constant.

If β is a universal constant, this relation must hold for $U_a = 0$. This implies that $\beta = \beta_s$, where β_s is the jet spreading rate in stagnant ambient (e.g. Fischer et al. 1979; Wood et al. 1993); the exact value of β_s, however, depends on the choice of velocity profile in the integral model. At the outset, it is not clear what characteristic velocity should be adopted in Eq. 11. The centerline excess jet velocity has been used in some studies (Antonia and Bilger 1974; Knudsen 1988; Wood et al. 1993), while recent works (Wright 1994; Chu 1994) suggest that the average excess velocity may be more appropriate. In our notation, and supported by the observations, the physics of the flow, such as transport of mass and momentum, can be characterized by the ensemble average properties of a coherent dominant eddy structure consisting of a range of turbulent eddies which govern the mixing with the irrotational ambient fluid. To facilitate this concept, the 'top-hat' profile is employed to represent the flow phenomenon and the average velocity is then considered as the characteristic velocity. This choice of characteristic velocity must ultimately be tested against experiments.

The excess velocity can be expressed in term of a 'top-hat' profile:

$$u_e = \begin{cases} \Delta U & \text{if } r \leq \text{B} \\ 0 & \text{otherwise} \end{cases}$$

where ΔU, and B are the excess velocity and half-width of an equivalent jet with a sharp boundary and uniform velocity, $\Delta U + U_a$, carrying the same mass flow and excess momentum flux as the actual jet. By equivalence of excess volume flux (the part of the volume flux due to turbulent entrainment), $\int u_e dA$, and *excess* momentum flux $M_e = \int u u_e dA$, the following relations between the two profiles can be obtained (Problem 6.3):

$$\pi \Delta U B^2 = \pi \Delta U_g b_g^2 \tag{12}$$

$$\pi(\Delta U + U_a)\Delta U B^2 = \frac{\pi}{2}\Delta U_g^2 b_g^2 + \pi U_a \Delta U_g b_g^2 \tag{13}$$

Thus, we have,

$$\Delta U = \frac{\Delta U_g}{2} \tag{14}$$

$$B = \sqrt{2} b_g \tag{15}$$

For total tracer mass flux, we have

$$\Gamma = \int_0^\infty C(u_e + U_a) 2\pi r dr$$

$$= \int_0^\infty (\Delta U_g\, e^{-(r/b_g)^2} + U_a) C_g\, e^{-\left(\frac{r}{\lambda b_g}\right)^2} 2\pi r dr$$

$$= \pi \lambda^2\, b_g^2\, C_g \left(U_a + \frac{1}{1+\lambda^2} \Delta U_g \right) \tag{16}$$

By conservation of tracer mass flux, $\Gamma = \Gamma_o = C_o U_o A_o$, where C_o is concentration at the source. The centerline dilution S_c can hence be determined from Eq. 16,

$$S_c = \frac{C_o}{C_m} = \frac{\lambda^2\, \pi\, B^2\, (U_a + \frac{2}{1+\lambda^2} \Delta U)}{2\; Q_o} \tag{17}$$

where $Q_o = U_o A_o$ is the source volume flux. Note that the above equation only requires knowledge of the average flow properties (ΔU and B). In Eq. 16 the turbulent mass transport $\int u'c' dA$ (of the order of 10% of the total mass flux) has tacitly been ignored. This assumption is justified primarily by comparison of the predicted mean flow quantities with data.

2.3 *Governing Equations*

Adopting the top-hat half-width $B = \sqrt{2}\, b_g$ and excess velocity $\Delta U = U - U_a = \Delta U_g/2$ as the characteristic length and velocity scales (Fig. 2a), the jet spreading hypothesis (Eq. 11) can be rewritten as,

$$\frac{DB}{Dt} = |U_a + \Delta U| \frac{dB}{dx} = \beta_s\, |\Delta U| \tag{18}$$

With the well-known spreading rate $db_g/dx \simeq 0.109$ of pure jet in stagnant ambient, the value of β_s can be taken as $\sqrt{2} \times 0.109 \simeq 0.154$ in our definition of jet edge $B = \sqrt{2}\, b_g$.

The conservation of excess momentum flux M_e can be expressed as (for ZEF):

$$M_{eo} = M_e = \int u(u - U_a)\, dA$$

$$= \pi\; B^2 U\, \Delta U = \pi\; B^2 (\Delta U + U_a) \Delta U \tag{19}$$

Equation 19 can be rewritten as

$$\left(\frac{\Delta U}{U_a}\right)^2 + \left(\frac{\Delta U}{U_a}\right) - \frac{l_m^{*2}}{\pi B^2} = 0 \tag{20}$$

where $l_m^* = M_e^{1/2}/U_a$ is the excess momentum length scale.

Introducing the dimensionless variables $U^* = \Delta U/U_a$, $B^* = B/l_m^*$ and $x^* = x/l_m^*$, the momentum equation (Eq. 20) and the spreading hypothesis (Eq. 18) become

$$\left.\begin{aligned} U^{*2} + U^* - \frac{1}{\pi B^{*2}} &= 0 \\ \frac{dB^*}{dx^*} = \beta_s \frac{\Delta U}{\Delta U + U_a} &= \beta_S \frac{U^*}{1 + U^*} \end{aligned}\right\} \tag{21}$$

This pair of equations, subjected to initial conditions, can then be numerically integrated using a 5th order Runge-Kutta method to obtain the solution for $\Delta U(x)$ and $B(x)$ at different downstream locations from the source. The centerline dilution S_c is then determined from Eq. 17.

2.4 Approximate Initial Conditions

For strong jets, $x/l_m^* \leq O(10)$, the initial spreading rate of the jet at exit can be expressed as,

$$\frac{dB}{dx} = \beta_s \frac{\Delta U}{\Delta U + U_a} = \beta_s (1 - U_a/U_o) \simeq \frac{D}{2x_o} \tag{22}$$

where x_o is the virtual origin of the jet. As a first approximation the initial conditions at the jet nozzle exit relative to the virtual origin can then be expressed as,

$$\left.\begin{aligned} \Delta U_o^* &= \frac{\Delta U_o}{U_a} \\ B_o^* &= \frac{D}{2l_m^*} \\ x_o^* &= \frac{-D}{2\beta_s (1 - U_a/U_o)} \end{aligned}\right\} \tag{23}$$

Strictly speaking, the above governing equations are only applicable to the region after the jet flow is fully developed and the velocity and tracer concentration profiles become self-similar, i.e. $x > x_e$. Alternatively, the initial conditions of the zone of established flow can be obtained as a solution of a model for the zone of flow establishment. Within the ZFE ($x \leq x_e$), the potential core with constant velocity $U = U_o$ is bounded by the free shear layer developed between the jet and the ambient. The spreading of this mixing layer and the location where the mixing layer intersects the jet axis, x_e, can also be predicted using the same spreading hypothesis. After determining the length of potential core x_e, the integration of the governing equations for the ZEF can then be performed starting from this location.

2.5 Prediction of Potential Core Length

Referring to Fig. 1, let r and R be the half-width of the potential core and location of the edge of the shear layer in the ZFE respectively. The characteristic half-width $\tilde{b}$ of the mixing layer can then be expressed as $R - r$. In the ZFE, the average velocity $\tilde{u}$ of the mixing layer can be taken as the average of the initial jet velocity U_o and ambient velocity U_a, i.e. $\tilde{u} = (U_o + U_a)/2$, and the average excess velocity is chosen as the characteristic excess velocity of the mixing layer $\tilde{u}_e$, i.e. $\tilde{u}_e = \tilde{u} - U_a = (U_o - U_a)/2$.

Applying the spreading hypothesis (Eq. 11) and integrating from $x = 0$ to x_e, we have

$$\int_{x=0}^{x=x_e} d(R-r) = \int_0^{x_e} \beta_s \frac{U_o - U_a}{U_o + U_a}\, dx \tag{24}$$

By conservation of excess momentum, we have

$$M_e = \pi \frac{D^2}{4} U_o (U_o - U_a)$$

$$= \pi\, r^2\, U_o (U_o - U_a) + \pi\, (R^2 - r^2) \frac{U_o + U_a}{2} \frac{U_o - U_a}{2} \tag{25}$$

Imposing the boundary condition at $x = x_e$, i.e. at the end of the potential core, $r = 0$, we have,

$$\frac{R}{D} = \frac{1}{\sqrt{1 + U_a/U_o}} \tag{26}$$

Substitute into Eq. 24, we have

$$\frac{x_e}{D} = \frac{\sqrt{1 + U_a/U_o}}{\beta_s (1 - U_a/U_o)} \tag{27}$$

It can be seen that the length of the potential core increases with the ambient current to jet velocity ratio; as $U_a/U_o \to 1, x_e/D \to \infty$, i.e. spreading rate of the mixing layer is very slow. For the limiting case of stagnant ambient, $U_a/U_o = 0$, using $\beta_s = 0.154$, the length of the potential core is determined to be $x_e = (1/\beta_s)D \simeq 6.5D$—which agrees with the generally accepted value of $\simeq 6.2D$ (Fischer et al. 1979).

2.6 Alternative Definition of Characteristic Velocity

Alternatively, one is often tempted to choose the centerline excess velocity ΔU_g of the Gaussian profile as the characteristic excess velocity (Antonia and Bilger 1974; Knudsen 1988). The jet spread hypothesis, Eq. 11, can then be written as,

$$\frac{dB}{dx} = \beta_s \frac{\Delta U_g}{U_a + \Delta U_g} \tag{28}$$

Note that when the excess velocity is much less than the ambient velocity, i.e. in the weak jet regime with $\Delta U \ll U_a$, this equation implies a spreading rate,

$$\frac{dB}{dx} \simeq \beta_s \frac{\Delta U_g}{U_a} = \beta_s \frac{2\Delta U}{U_a} \tag{29}$$

that is **twice** that given by Eq. 18 or 21.

3 Asymptotic Solutions: Strong and Weak-Jet

The asymptotic properties of two limiting cases may be considered. The effect of the coflow velocity can be ignored in the near field of a strong jet where $\Delta U >> U_a$. The initial development of the strong jet is expected to be close to the case of the jet in stagnant fluid. The excess velocity reduces rapidly with the distance from the source; both the jet spread and excess velocity decay are governed by the pure jet solution if the jet momentum flux is replaced by the excess momentum flux M_{eo}. The width and velocity of the coflow jet in the near field (strong jet) region, $x/l_m^* \leq 10$, are then given by the following in terms of dimensionless scales (see Eq. 2 in Chap. 2):

$$\frac{b_g}{l_m^*} = 0.114 \left(\frac{x}{l_m^*}\right) \tag{30}$$

$$\frac{\Delta U_g}{U_a} = 7.0 \left(\frac{x}{l_m^*}\right)^{-1} \tag{31}$$

For a strong jet, the solution is assumed to be the same as those for the jet in stagnant environment.

However, far from the source (the strongly-advected region) where $\Delta U << U_a$, the effect of the excess velocity ΔU may be ignored except its contribution to the spreading process. The asymptotic properties of the weak jet are as follows. Let $(U_a + \Delta U) \simeq U_a$, the momentum equation becomes

$$U_a \Delta U \pi B^2 = M_{eo}, \tag{32}$$

The spreading equation is

$$U_a \frac{dB}{dx} = \beta \Delta U. \tag{33}$$

Multiplying both side of the spreading equation by $U_a \pi B^2$,

$$U_a^2 \pi B^2 \frac{dB}{dx} = \beta U_a \Delta U \pi B^2 = \beta M_{eo} \tag{34}$$

The equation can be integrated once with respect to x to give a one-third power law

$$B = \left(\frac{3\beta}{\pi}\right)^{\frac{1}{3}} \left[\frac{M_{eo}}{U_a^2}\right]^{\frac{1}{3}} x^{\frac{1}{3}} \tag{35}$$

for the spreading of the weak jet in the far field. The corresponding excess velocity follows the minus two-third power law

$$\frac{\Delta U}{U_a} = [9\beta^2 \pi]^{-\frac{1}{3}} \left[\frac{M_{eo}}{U_a^2}\right]^{\frac{1}{3}} x^{-\frac{2}{3}} \tag{36}$$

The use of $\beta = \sqrt{2} db_g/dx = \sqrt{2}(0.114) = 0.16$ for the spreading rate and using the relation between Gaussian and top-hat variables results in:

$$b_g = 0.38 \left[\frac{M_{eo}}{U_a^2}\right]^{\frac{1}{3}} x^{\frac{1}{3}} \tag{37}$$

$$\frac{\Delta U_g}{U_a} = 2.23 \left[\frac{M_{eo}}{U_a^2}\right]^{\frac{1}{3}} x^{-\frac{2}{3}} \tag{38}$$

Alternatively, the adoption of $\beta = 0.17$ based on velocity intermittency data (Chu 1994) would have resulted in almost the same coefficients for the above equations (0.39 and 2.14 respectively). The one-third power law for the width and the minus-two-third power law for the excess velocity are in good agreement with experimental data (see later discussion and Figs. 7 and 8). When normalized by the length and velocity scales, the coflow jet solution in the far field then becomes:

$$\frac{b_g}{l_m^*} = 0.385 \left(\frac{x}{l_m^*}\right)^{\frac{1}{3}} \tag{39}$$

$$\frac{\Delta U_g}{U_a} = 2.14 \left(\frac{x}{l_m^*}\right)^{-\frac{2}{3}} \tag{40}$$

4 Comparison of Theory with Experimental Data

4.1 *Jet Spreading Rate*

Although the shape of the concentration contour and the mixing characteristics are similar to those of a pure jet, the jet spread is, however, not linearly related to the distance x. For the same discharge, the half-width of coflowing jet is less than that of the pure jet and depends on the ambient to jet velocity ratio $R' = U_a/U_o$. Figure 7 shows the comparison of theory (Eq. 21) with the measured concentration half-width b_{gc} in the LIF experiments of Chu (1996) and previous studies. The normalized results of different R' generally collapse on the same curve suggesting that l_m^* is the correct length scale. The present results using LIF techniques are comparable with the experimental results of Knudsen (1988) and McQuivey et al. (1971) using probe-based techniques. It can also be seen that the jet in coflow can be well-predicted by the present spreading hypothesis using the top-hat velocity as the characteristic velocity (Eq. 18), especially in the strong jet region with $x/l_m^* \leq 10$. Inclusion of the potential core does not result in significant difference in the results; although the initial phase is better predicted. On the other hand, using the centerline maximum velocity as the characteristic velocity (Eq. 28) results in over-prediction of the jet half-width. The data give support to a suggestion of Wright (1994) on using the top-hat velocity ΔU as the velocity scale.

Using $\lambda = b_{gc}/b_g$, the velocity half-width b_g obtained from the detailed point velocity measurements (Sect. 5.3) can be converted into concentration half-width $b_{gc} = \lambda b_g$, with $\lambda = 1.19$. Figure 8 shows the inferred b_{gc} against the numerical prediction. The predicted b_{gc} compares well with the data, as well as those results determined directly using LIF technique (Fig. 7). This gives further support to the integral model and the use of $\lambda \simeq 1.2$.

4.2 *Centerline Dilution*

In the strong jet region, by dimensional analysis, the centerline dilution can be written as $S_c(x) = C_o/C_m(x) \sim (M_e^{1/2}/Q_o)\,x$. Or equivalently,

$$S_c(x)\,\frac{R'}{1-R'} = k\,\frac{x}{l_m^*} \tag{41}$$

where k is the proportionality constant. The normalized measured centerline dilution $S_c R'/(1-R')$ is plotted against the normalized distance x/l_m^* in Fig. 9. The results are comparable with the data of Knudsen (1988) and McQuivey et al. (1971). In the strong jet region, say $x/l_m^* \leq 5$, the normalized dilution is found to vary linearly with x/l_m^*, with $k = 0.174$; this is almost the same as the predicted value of 0.170 and the value inferred from the data of Ayoub (1971) by Lee (1989). The predicted centerline

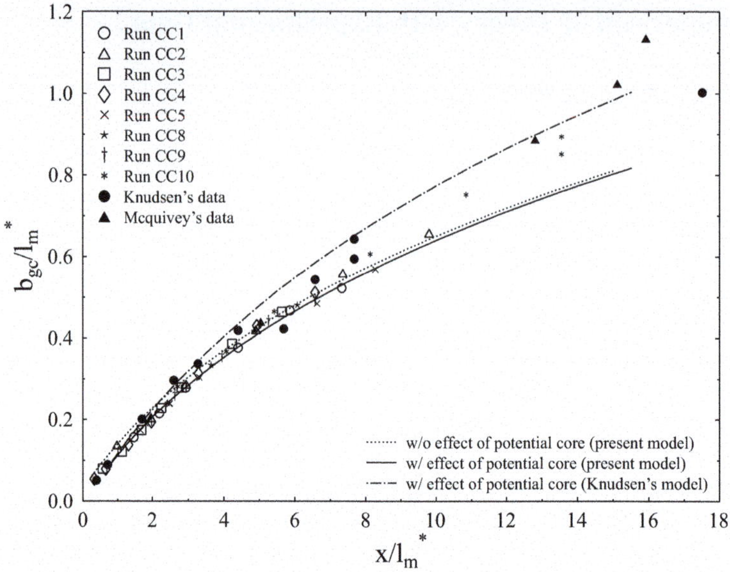

Fig. 7 Predicted concentration half-width of coflow jet versus horizontal distance (empty symbols: data of Chu 1996)

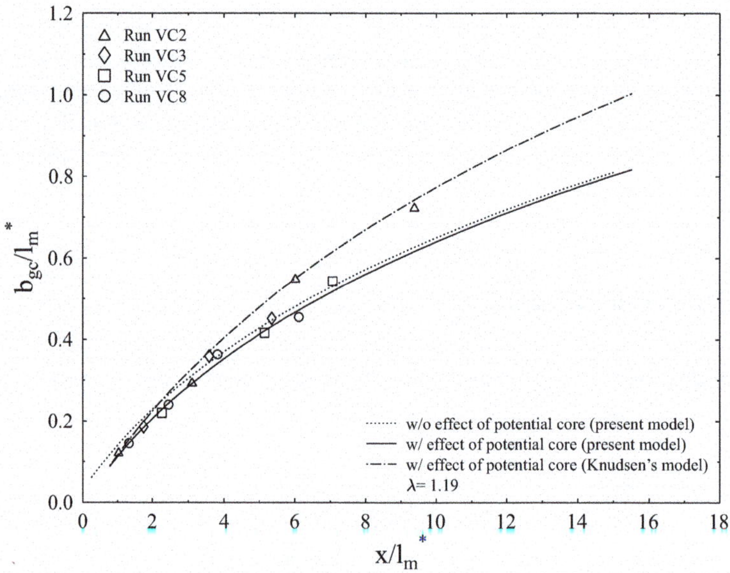

Fig. 8 Predicted concentration half-width of coflow jet (inferred from velocity) versus horizontal distance

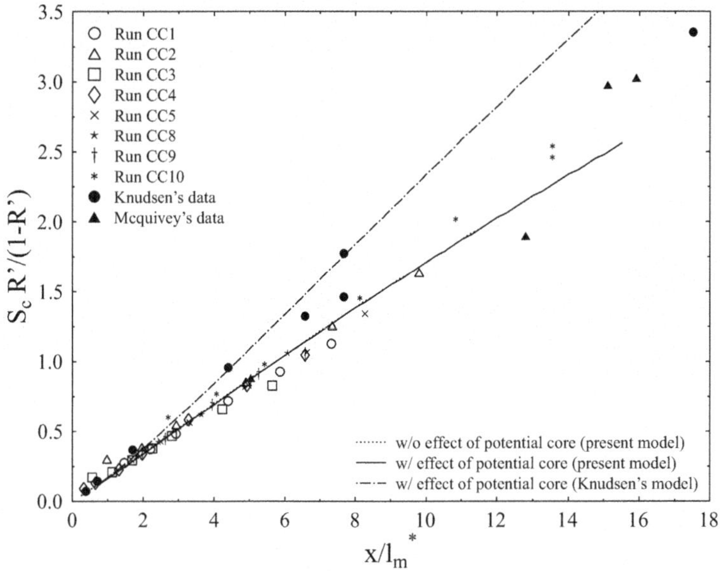

Fig. 9 Centerline dilution of coflow jet versus downstream distance

dilution using the present hypothesis (Eq. 41) is well supported by the experimental results. However, using the alternative spreading hypothesis (Eq. 28) results in over-prediction of the centerline dilution—giving $k = 0.221$. It is interesting to note that despite the narrower jet spread in a coflow, the centerline dilution is only slightly less than the linear dilution predicted by the straight jet in stagnant ambient; this is apparently related to the slower velocity decay in a coflow.

4.3 Centerline Excess Velocity Decay

The ratio of the centerline excess velocity over ambient velocity $\Delta U_g/U_a$ can be determined directly from LDA velocity measurements. The normalized excess velocity is plotted against x/l_m^* in Fig. 10 along with the data of McQuivey (1971) and the data of Nickels and Perry (1996). Alternatively, by conservation of excess momentum, $\Delta U_g/U_a$ can be determined indirectly from the concentration half-width (and using $\lambda = 1.2$). The centerline velocity excess inferred from the LIF data and previous concentration data are also shown as a comparison. It can be seen the present data are consistent with previous studies, and the theory is well-supported by the experiments. The prediction follows closely the asymptotic relation in the strong jet region (cf. Table 3 in Chap. 2), $\Delta U_g/U_a = 7.0\ M_{eo}^{1/2} x^{*-1}$, as well as the $-2/3$ law in the weak jet region; the best fit of Nickel and Perry's data (x/l_m^* up to 60), $\Delta U_g/U_a = 2.67\ M_{eo}^{1/2} x^{*-2/3}$, is in reasonable agreement with the present spread-

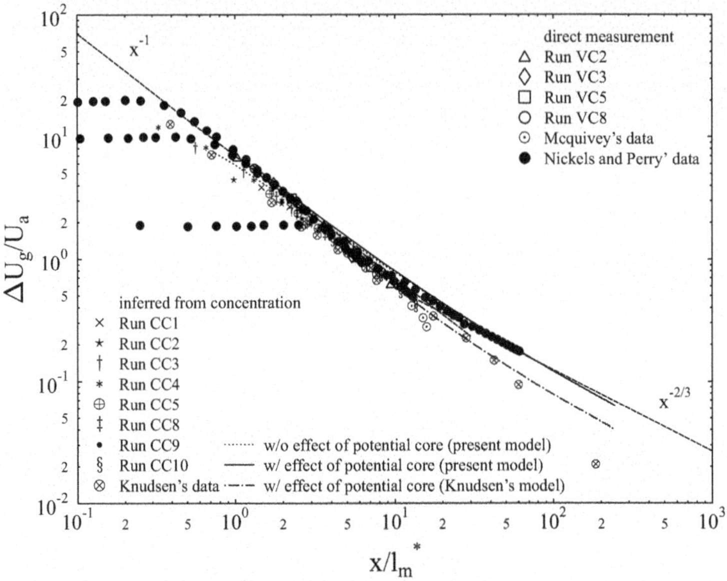

Fig. 10 Decay of centerline excess velocity of a coflow jet

ing hypothesis. Inclusion of the potential core also does not introduce any significant difference in the result. On the whole, the predicted excess velocity using $\beta \approx 0.16$–0.17 is about 20% lower than the observed values in Nickel and Perry's experiments. At present, there is no concentration data in the weak-jet region; the experiments have to be done by a towed jet, as the weak-jet behaviour is expected to be influenced by the ambient turbulence in the coflow. The present results seem to suggest a slightly lower spread rate in the weak-jet or wake region, consistent with related studies.

4.4 The Entrainment Coefficient

In many general integral models, the entrainment hypothesis is used. In view of the success of the spreading hypothesis, it is worthwhile to use the result to determine the appropriate entrainment coefficient to be used for the coflowing jet. Using the entrainment hypothesis, the problem can be formulated in terms of top-hat variables as:

$$\frac{d}{dx}(\pi B^2(U - U_a)) = 2\pi B\alpha(U - U_a) \tag{42}$$

$$\frac{d}{dx}(\pi B^2 U(U - U_a)) = 0 \tag{43}$$

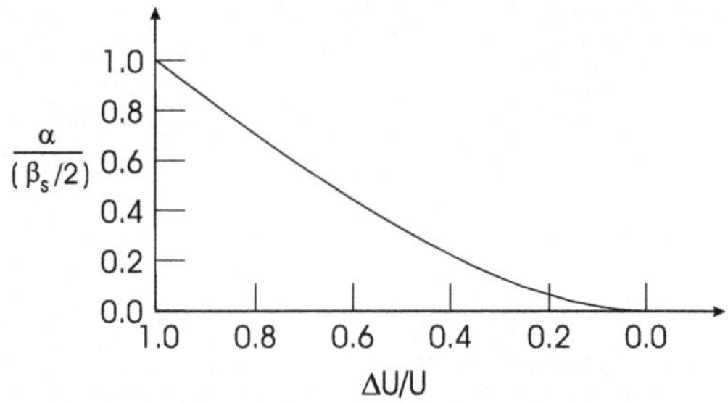

Fig. 11 Entrainment coefficient of a coflow jet

Since $U\frac{dB}{dx} = \beta_s \Delta U = \beta_s (U - U_a)$, it can be shown from the above (Problem 6.4) that:

$$\alpha = \beta_s \frac{\Delta U^2}{U(U + \Delta U)} \tag{44}$$

It is seen from Eq. 44 the entrainment coefficient for a coflowing jet exhibits even greater variability than a buoyant jet in stagnant fluid. For a strong jet, $\Delta U \to U$, $\alpha \to \beta_s/2$ (cf. Table 3 in Chap. 2). For a weak-jet, we have $\Delta U \approx 0$, and $\alpha \to 0$. Thus for this basic flow, the entrainment coefficient varies continually along the jet in accordance with the jet velocity; α changes from the stillwater value, 0.057, for a strong jet, and approaches zero in the far field weak jet region (Fig. 11). On the other hand, with the Lagrangian spreading hypothesis and the same "universal" spreading coefficient, the simple formulation predicts the jet properties very well in the strong jet region; the agreement is also satisfactory in the far field.

5 Correlation of Model Results with Experiments

In adopting a top-hat profile, the half-width and average velocity represent the characteristic properties of the dominant eddies. However, in practice, these average properties must be correlated with the experimental observations such as intermittency, visual width, centerline concentration and minimum concentration. Although some of these correlations can be theoretically established based on the observed profile distribution, model validation is necessary.

5.1 Visual Boundary

Intermittency and Top-Hat Edge

The intermittency γ at a point is the proportion of time in which that location is occupied by turbulence; it reflects the fluctuation of the turbulent/non-turbulent boundary.

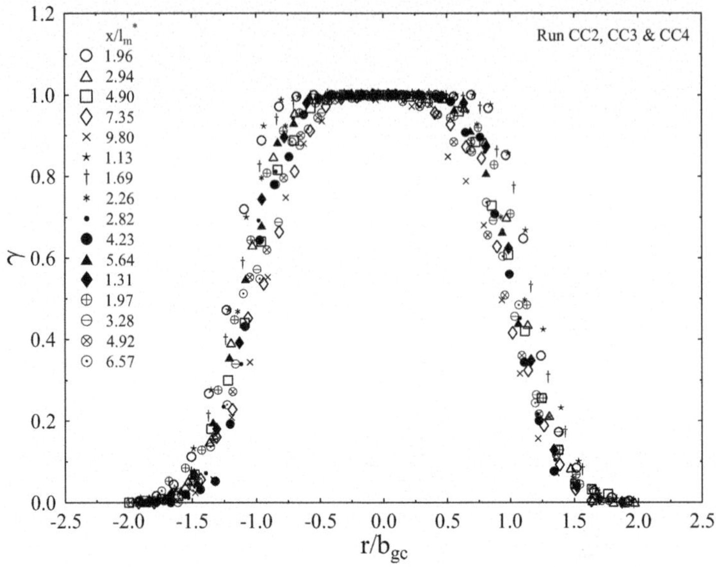

Fig. 12 Radial profile of concentration intermittency

Figure 12 shows the intermittency contours at different downstream locations x. The results show good symmetry and similar features as the pure jet case (not shown). The radial variation of γ is shown in Fig. 12. The intermittency has a peak value of 1 at centerline. It generally stays at about 1 in the central core region, $r/b_{gc} <\simeq 0.6$, and decreases to about zero at $r/b_{gc} >\simeq 1.70$. The size of this core seems to reduce as x/l_m^* increases; this means the ambient fluid is able to penetrate the core more as the turbulence grows in size downstream.

Figure 13 shows the comparison between the concentration half-width b_{gc} and the half-width b_i defined at 50% intermittency as a function of downstream distance x/l_m^* for both the coflowing jet and the pure jet (Run CC6 and CC7) cases. The ratio of b_i/b_{gc} generally remains fairly constant for $x/l_m^* < 10$, with an average of about 1.16 and 1.26 respectively; i.e. on average $b_i/b_{gc} \simeq 1.2$. Recall that our top-hat edge $B = \sqrt{2}\, b_g \simeq 1.2 b_{gc}$, (using $\lambda = 1.2$); this suggests the top-hat edge is a close representation of the 50% intermittency contour for pure jet or coflowing jet.

Visual Boundary

The LIF image without any further processing can be considered very similar to the normal flow visualization using simple dye injection technique. The visual boundary of the jet region can be identified in the LIF image and extracted by digital image processing techniques (Lee and Chu 1995); it can be defined by a threshold gray level selected based on visual judgement of thresholding an *instantaneous* image against

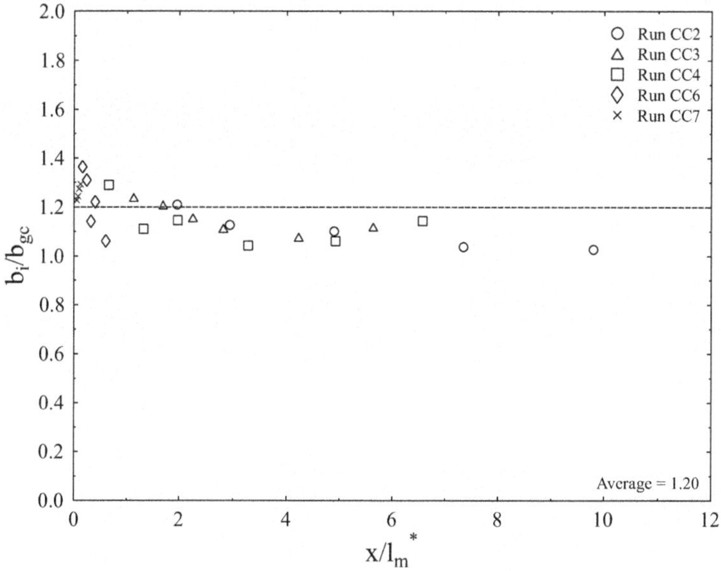

Fig. 13 Comparison of concentration half-width with half-width defined at 50% intermittency

its visual boundary. On the other hand, for a Gaussian profile, $C = C_m e^{-(\frac{r}{\lambda b_g})^2}$, our edge $B = \sqrt{2}\, b_g$ is equivalently defined at concentration $C_m e^{-1.389} \simeq 0.25C_m$, using $\lambda \simeq 1.2$. Figure 14 shows the comparison of the visual boundary and the edge defined at $0.25C_m$ contour for two typical, randomly selected instantaneous LIF images of the coflowing jet, where $C_m = C_g$ is the maximum concentration. The results show excellent agreement between the visually perceived boundary of turbulent region and the $0.25C_m$ contour (i.e. the top-hat edge). Additional examples can be found in Chu (1996).

5.2 Flow-Weighted Average Dilution

The flow-weighted average concentration $C_{avg} \simeq Q_o C_o / Q$, where $Q = \int u\, dA$ is the total volume flux, can be defined as $C_{avg} = \int uc\, dA / \int u\, dA$. The flow-weighted average dilution $S_{avg} = C_o / C_{avg}$ can be determined from the integration of the measured scalar concentration and velocity data across the section within the jet boundary. The results are plotted against distance $x/(Q_o/M_e^{1/2})$ (Fig. 15), and suggest a linear relationship between S_{avg} and $x/(Q_o/M_e^{1/2})$, with a slope of 0.290. The experimentally determined ratio of average to centerline dilution $S_{avg}/S_c = C_m/C_{avg}$ is shown in Fig. 16. The results show a constant ratio of about 1.66, roughly the same as that of the pure jet in stagnant ambient—for a Gaussian concentration and velocity distribution the theoretical value is $(1 + \lambda^2)/\lambda^2 \simeq 1.69$.

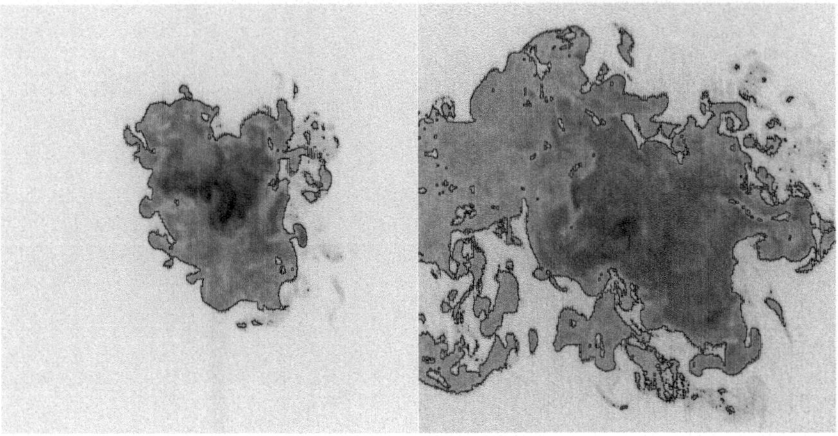

Fig. 14 Comparison of visual boundary with top-hat boundary in a coflow jet: Run CC2: x/D = 49 (left) and x/D = 98 (right)

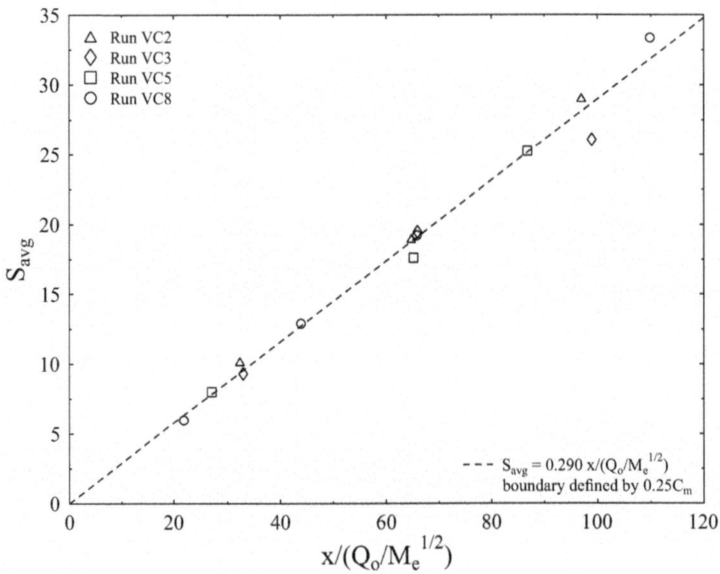

Fig. 15 Flow-weighted average dilution of a coflow jet

Example 6.1 *A circular jet discharges an industrial effluent from a pipe of 0.4 m diameter at 3 m/s into a coflowing river velocity of 0.2 m/s. The river is about 10 m deep, and the jet is located at about mid-depth. Compute the variation of jet width, excess jet velocity, and centerline dilution for the first 100 m downstream of the jet.*

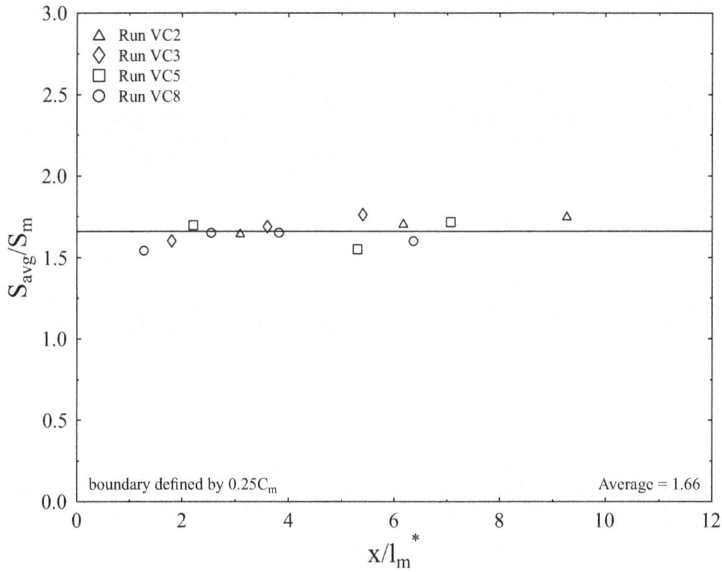

Fig. 16 Comparison of average and centerline minimum dilution of a coflow jet

Solution: For this jet, the initial volume flux is $Q_o = 3 \times \pi \times 0.4^2/4 = 0.377\,\mathrm{m^3/s}$; the excess momentum flux is $M_{eo} = Q_o(U_o - U_a) = 0.377 \times (3 - 0.2) = 1.056\,\mathrm{m^4/s^2}$. The excess momentum length scale can then be computed as $l_m^* = M_{eo}{}^{0.5}/U_a = \sqrt{1.056}/0.2 = 5.14\,\mathrm{m}$. Thus it can be expected that for distances much greater than 25 m ($\approx 5l_m^*$), the jet will be strongly advected by the coflow. Figure 17 shows the computed variation of the top-hat half-width, the centerline velocity excess, and the centerline dilution with distance downstream. It is seen that at $x = 100$ m, the half-width is quite narrow ($B = \underline{5.6}$ m) compared with the straight jet value (17 m). However, the absolute coflow jet velocity is greater. It can be seen the excess velocity decay is less rapid than the strong jet. At $x = 100$ m, the velocity excess $\Delta U_g = 0.045$ m/s is small compared with the ambient velocity, while the absolute centerline jet velocity is $u_m = 0.245$ m/s. The centerline dilution is $S_c = \underline{43.8}$, only slightly less than the corresponding strong jet value of $S_c = 47.5$. Wastewater effluents are often discharged as nearly non-buoyant coflowing jets into rivers.

5.3 *Summary of Experimental Data*

Table 1 shows a summary of time-mean as well as key turbulence properties of the scalar field of a coflow jet. Data on jet in stagnant fluid ($x/l_m^* \to 0$ limit of coflow jet), or momentum-dominated regime of a vertical buoyant jet (Papanicolaou and List 1987, 1988; Papantonious and List 1989) are also included. It can be seen the

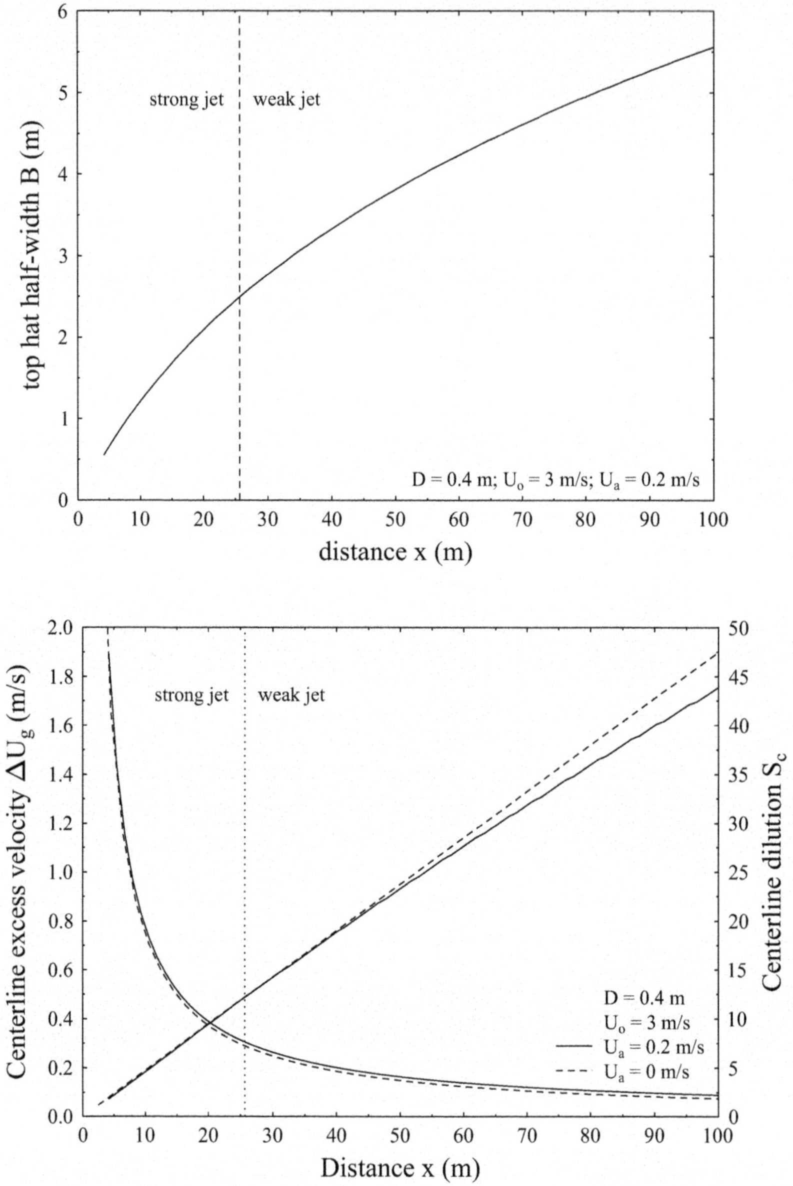

Fig. 17 Variation of coflowing jet properties with distance from source: effluent discharge in river

instantaneous maximum and minimum concentrations (C_{max}, C_{min}) can differ by more than 50% from the time averaged value; the concentration turbulent intensity C_{rms}/C_m is around 0.2, where the root-mean-square concentration $C_{rms} = \sqrt{\overline{c'^2}}$, and $c' = c - \overline{c}$ is the concentration fluctuation. The variabilities in the turbulence properties data reflect the different experimental techniques used and also the limited number of profiles measured. Further details on other statistics such as skewness skewness factor = $\overline{c'^3}/(\overline{c'^2})^{3/2}$ and flatness factor = $\overline{c'^4}/(\overline{c'^2})^2$ can be found in Chu (1996).

5.4 Ambient Turbulence

The effects of ambient turbulence on the mixing of a coflowing jet are similar to that described in Chap. 2, and is expected to be important in the field. Larger eddies in the turbulence cause jet meandering, and small eddies that are strong and comparable to the length scale of the jet can break up the jet (after which the jet will be governed by the turbulent dispersion of the coflow). The overall effects is the increased spreading of the jet, which was first observed by Wright (1994). Based on the spreading hypothesis developed in this chapter, the lumped effects of ambient turbulence can be modelled as:

$$\frac{dB}{dx} = \beta_s \frac{\Delta U_g}{U_a + \Delta U_g} + \beta_* \frac{\sigma}{U_a} \tag{45}$$

where $\beta_* = 0.4 - 1.0$ is spreading coefficient due to the ambient turbulence; and $\sigma = \sqrt{\overline{u_a'}}$ is the root mean square of the ambient velocity fluctuation above the mean, taken as a characteristic turbulent velocity (Wright 1994). The model was successfully applied by Chan et al. (2017) to predict the mixing of a coflow chlorine jet in a field scale experiment in which ambient turbulence was found to be significant. More details of this case study will be given in Chap. 8.

6 Summary

The excess momentum of a jet in coflow is conserved. Using a Lagrangian jet spreading hypothesis, an integral model of the jet in coflow has been formulated and validated against experiments. It is demonstrated that the jet spreading hypothesis using the top-hat velocity as the characteristic velocity can be effectively applied to model the time-averaged flow. Theoretical predictions of the jet spreading rate and minimum dilution are in good agreement with experiments.

The present results support the applicability of the 'top-hat' profile in the integral formulation of buoyant jet in a current. This profile appears to possess a certain physical significance in the description of the active mixing process. It is found that

Table 1 Summary of experimental results on coflowing jet; previous studies on jet in stagnant ambient are also included

Jet properties	Coflowing jet	Jet in stagnant ambient			
	Cross-sectional LIF measurement[a]	Cross-sectional LIF measurement[b]	Cross-sectional temperature measurement[c]	Point LIF measurement[d]	Line LIF measurement[e]
db_{gc}/dx	0.125	0.125	0.13	0.139	0.115
$S_m(x/l_Q)^{-1}$	0.174	0.175	0.165	0.147	0.172
C_{max}/C_m	1.4	1.4	1.5	1.65	
C_{min}/C_m	0.45	0.5	0.45	0.2	
$\gamma = 100\%$	$(r/b_{gc} < 0.6)$	$(r/b_{gc} < 0.75)$	$(r/b_{gc} < 0.89)$		$(r/b_{gc} < 0.43)$
$\gamma = 50\%$	$(r/b_{gc} \simeq 1.1)$	$(r/b_{gc} \simeq 1.2)$	$(r/b_{gc} \simeq 1.31)$		$(r/b_{gc} \simeq 1.19)$
C_{rms}/C_m	0.18	0.15	0.165	0.22	0.34
(centreline value)					
C_{rms}/C_m	0.22	0.2	0.25	0.25	0.36
(peak value)	$(r/b_{gc} \simeq 0.85)$	$(r/b_{gc} \simeq 0.75)$	$(r/b_{gc} \simeq 0.54)$	$(r/b_{gc} \simeq 0.72)$	$(r/b_{gc} \simeq 0.43)$

[a] Chu (1996)
[b] Chu (1996)
[c] Papanicolaou and List (1987)
[d] Papanicolaou and List (1988)
[e] Papantoniou and List (1989)
Note that C_{max} and C_{min} are the maximum and minimum concentration; the root-mean-square concentration $C_{rms} = \sqrt{\overline{c'^2}}$, where $c' = c - \overline{c}$ is the concentration fluctuation

the visual edge, defined by the 25% maximum concentration contour, corresponds to approximately the concentration half-width defined at $r = B = \sqrt{2}\, b_g$. This location also matches the 50% contour of the intermittency factor. A meaningful definition of jet boundary is convenient for many engineering applications (e.g. study of merging of multiple jets). The advantages of the top-hat profile become even clearer for the more complex case of a bent-over jet in crossflow, when the concentration or velocity profile is clearly not Gaussian.

Problems

6.1 *Jet in Coflow*

The governing equations for an axisymmetric coflowing jet directed in the x-direction can be written in cylindrical co-ordinates (x, r) as:

$$\frac{\partial ru}{\partial x} + \frac{\partial rv}{\partial r} = 0$$
$$\frac{\partial ru^2}{\partial x} + \frac{\partial ruv}{\partial r} = 0$$

where the usual boundary layer approximations applicable to a straight jet have been made, and u, v represent the velocity in the x and r-direction respectively. By multiplying the continuity equation by the coflow U_a and subtracting it from the x-momentum equation, show that the jet excess momentum $M_e = \int uu_e dA = \int u(u - U_a)dA$ is conserved along a coflowing jet. $u_e = u - U_a$ = excess velocity.

6.2 The volume and momentum fluxes of a coflow jet are defined respectively as $Q = \int_o^\infty u dA$, and $M = \int_o^\infty u^2 dA$, where the excess velocity profile is given by Eq. 3 and Fig. 1. Since the integrals Q and M are not bounded, some definition of jet edge is required. If $r = \sqrt{2}b_g$ is defined as the jet edge, show that:

$$Q = \int_0^\infty u dA = \int_0^\infty (u_e + U_a) dA$$
$$\approx \int_0^\infty u_{em} \exp\left(-\frac{r^2}{b_g^2}\right) dA + \int_0^{\sqrt{2}b_g} U_a dA = \pi b_g^2 (u_{em} + 2U_a)$$

where $dA = 2\pi r dr$, $u_e = u_{em} \exp\left(-\frac{r^2}{b_g^2}\right)$ and $u_{em} = \Delta U_g$ is the centerline maximum excess velocity. Similarly show that:

$$M = \frac{\pi}{2} b_g^2 (u_{em} + 2U_a)^2$$

6.3 <u>*Top-hat versus Gaussian profiles*</u>

By referring to Fig. 1, the excess velocity profile of a coflow jet can be represented in terms of either (i) the centerline maximum value and the Gaussian half-width, ΔU_g and b_g respectively (Eq. 3); or (ii) the average excess velocity $\Delta U = U - U_a$ and half-width B of a 'top-hat' profile.

(a) Show by equating the excess volume flux $\int u_e dA$ and the excess momentum flux M_e expressed by the two profiles that:

$$\Delta U = \frac{\Delta U_g}{2}$$
$$B = \sqrt{2} b_g$$

(b) Show that with the above equivalence, the jet volume flux, $Q = \int_0^{B=\sqrt{2}b_g} u dA$, is the same regardless of whichever profile is used. Show that the jet momentum flux, $M = \int_0^B u^2 dA$, can also be equivalently expressed by the two profiles.

(c) If the tracer mass concentration for the top-hat profile is $c = C$ for $r \leq B$, establish the following by equating the tracer mass carried by the excess volume flux, $\Gamma_e = \int u_e c dA$:

$$C = \frac{\lambda^2}{1+\lambda^2} C_m$$

where $\lambda = b_{gc}/b_g$. Show that for $\lambda = 1$, the above relation is consistent with conservation of tracer mass flux, $\pi U B^2 C = \int uC dA = Q_o C_o$.

6.4 *Entrainment coefficient of coflow jet*
Consider the governing equations for a jet in coflow by adopting a top-hat profile:

$$\frac{d}{dx}(\pi B^2 \Delta U) = 2\pi B \alpha \Delta U$$
$$\frac{d}{dx}(\pi B^2 U \Delta U) = 0$$

where U, B are the average velocity and half-width of the coflow jet, $\Delta U = U - U_a$ = excess velocity, and α is the entrainment coefficient. β_s = jet spread rate in stagnant fluid. By expanding the excess momentum equation and using the continuity equation and the jet spreading hypothesis, $dB/dx = \beta_s \Delta U/U$ (Eq. 11), show that the entrainment coefficient in a coflow jet depends only on the relative excess velocity, $\Delta U/U = \Delta U/(U_a + \Delta U)$:

$$\alpha = \beta_s \frac{\Delta U^2}{U(U+\Delta U)}$$

Evaluate the entrainment coefficient for the limiting cases of a strong jet, $\Delta U/U \to 1$ and a weak jet, $\Delta U/U \to 0$.

6.5 *Jet in Counterflow*
Consider a round jet with initial x-velocity u_o and diameter D, discharging into a counterflowing stream of x-velocity $-u_a$. In the presence of the counterflow, the jet is retarded. At some point the jet reaches stagnation at $x = x_p$. The jet effluent issuing into the counterflow occupies a dome shape region. Estimate the penetration distance x_p in terms of the governing parameters.

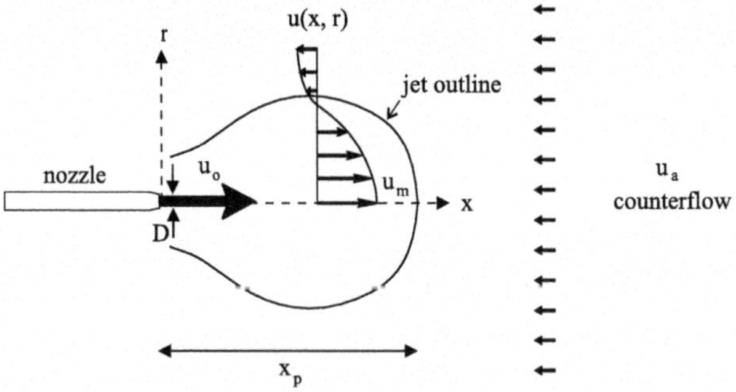

(a) Estimate the jet forward velocity u_j in terms of the jet momentum flux $M_o = u_o^2 \pi D^2/4$ and distance from source x. Assume a strong jet, with $u_o \gg u_a$.
(b) Assuming heuristically that at $x = x_p$, the jet forward velocity balances the counterflow velocity u_a, show that the penetration distance can be estimated by

$$\frac{x_p}{D} \sim \frac{u_o}{u_a}$$

Both the velocity measurements of Belatos and Rajaratnam (1973) and Lam and Chan (1997) show that $x_p/D = 2.6\, u_o/u_a$.

6.6 *Jet Excess Momentum*
The excess velocity $u_e(x, r) = u(x, r) - U_a$ of a jet in coflow is given by (Eq. 3 and Fig. 1):

$$u_e = \Delta U_g \exp\left(-\frac{r^2}{b_g^2}\right)$$

where $\Delta U_g = (u_c - U_a)$ = centerline velocity excess.
For a jet in a counterflow (of magnitude U_a), the excess velocity is $u_e(x, r) = u(x, r) + U_a$, and the centerline velocity excess is $\Delta U_g = (u_c + U_a)$. Show that the excess momentum flux for either (a) a jet in coflow U_a or (b) jet in counterflow current U_a is given by:

$$M_e = \int u u_e dA = \frac{\pi b_g^2}{2}(u_c^2 - U_a^2)$$

Chapter 7
Jet in Crossflow: Advected Line Puffs

When a jet is directed into a crossflow, the interaction of the jet momentum results in a complicated turbulent shear flow. Environmental mixing calculations are often based on a "stillwater dilution", i.e. the receiving environment is assumed to be stationary. However, both experiments and field studies have shown the presence of even a weak crossflow enhances dilution considerably (Wright 1977a; Lee and Neville-Jones 1987a). The prediction of the trajectory and dilution of a jet in crossflow is a considerably more difficult fluid mechanics problem than the case of the coflowing jet. Vortices of various kinds are being generated close to the source, in particular a pair of twin-vortices in the bent-over phase of the buoyant jet. Figure 1 shows a side view of a round jet perpendicular to the cross flow. Figure 2 shows the cross-sectional view of the jet obtained using the laser-induced fluorescence (LIF) technique (Chu and Lee 1996a). Close to the jet source, there are ring vortices characteristic of shear entrainment (similar to jet in stagnant fluid), and wake vortices due to the blockage of the current by the jet itself. In the bent-over phase, however, the jet cross-section is characterised by the vortex pair as seen in the LIF image in Fig. 2. The double vortex induces a large scale entrainment that is manifested in the form of tornado vortices drawn into the jet (Fric and Roshko 1994, Chu and Lee 1996a). As expected, the mixing is greatly enhanced by the vortex pair — an environmentally friendly gesture of nature as effluents as a rule are discharged into moving fluids. We seek a prediction of the turbulent-mean velocity and scalar field of this flow. The theory must be able to capture the shear entrainment regime close to the source (characterized by Gaussian cross-section tracer concentration distribution), the vortex flow in the bent-over jet, as well as the Gaussian-vortex flow transition (Fig. 3).

There is an extensive literature on the jet in crossflow problem. Many experimental studies (e.g. Keffer and Baines 1963; Kamotani and Greber 1972; Rajaratnam 1976; Moussa *et al.* 1977; Andreopoulos and Rodi 1984) have revealed some of the essential

---

Supplementary Information The online version contains supplementary material available at https://doi.org/10.1007/978-3-032-01187-9_7.

J. H. W. Lee et al., *Turbulent Jets and Plumes*,
https://doi.org/10.1007/978-3-032-01187-9_7

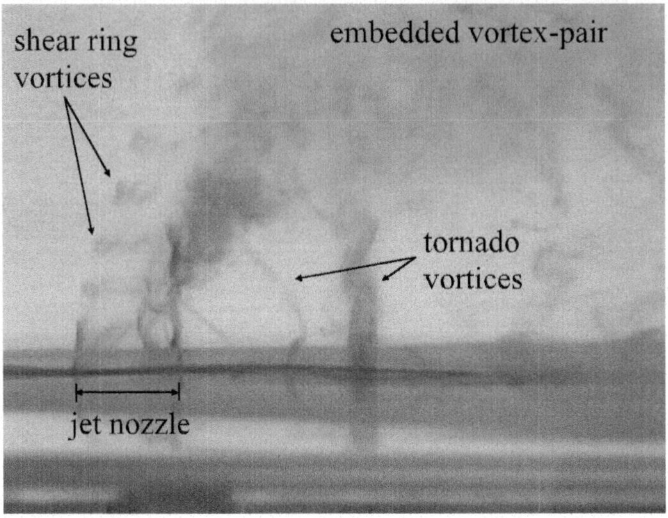

Fig. 1 Momentum jet in crossflow showing the ring vortices near the nozzle, and the tornado-like vortices in the wake drawn down from the jet to touch the floor of the cross flow. Direction of the cross flow in the figure is from left to the right

Fig. 2 Tracer concentration distribution in a cross section of the bent-over jet obtained by Chu and Lee (1996a) using LIF technique

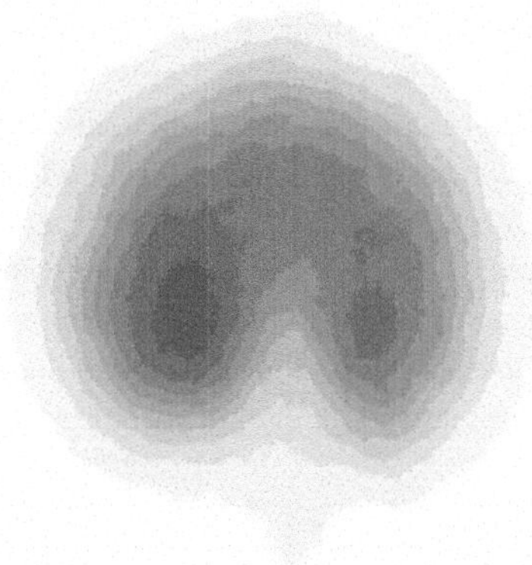

Fig. 3 Asymptotic flow regimes showing the transition from a Gaussian concentration profile in the near field to the vortex-pair profile in the far field

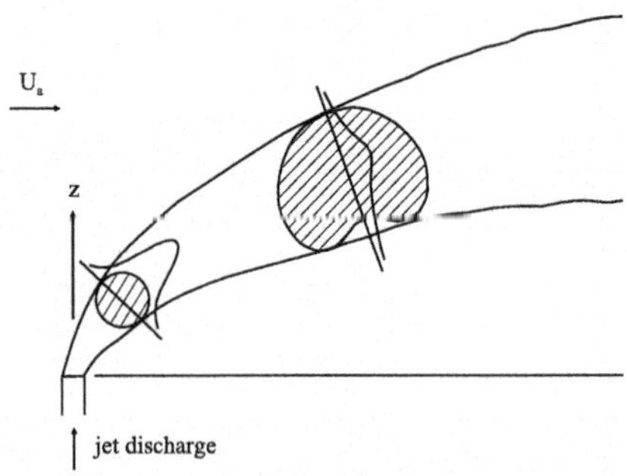

features of the fluid dynamics, among which the most striking is a transition from an initially vertical jet through a bent-over phase during which the jet becomes almost parallel with the main free stream, to a secondary vortex pair flow in the transverse section of the jet. However, most of these studies were concentrated in a region very close to the source, in the order of 5-10 diameters. On the other hand, for environmental mixing studies, the region of "active mixing" is typically in the order of the water depth, which can extend to beyond 100 jet diameters downstream.

To study the mixing of a passive scalar carried in the jet fluid, visual trajectory and limited tracer concentration measurements in the bent-over phase of a momentum-dominated buoyant jet in crossflow have been made (Ayoub 1971; Chu and Goldberg 1974; Wright 1977a). However, most of the concentration measurements were limited to a vertical traverse along the jet centerline. The measured dilution rates also exhibited considerable scatter. On the other hand, numerical studies (Patankar *et al.*1977; Chien and Schetz 1975; Demuren 1983; Sykes *et al.*1986) failed to offer any detailed view of the scalar field, due to the relatively low-resolution and small computational domain used. In the systematic numerical study of Sykes *et al.*(1986), the computational domain extends only 15 jet diameters downstream from the jet source. The computed cross-sectional scalar concentration field fails to reveal any double peak structure. As the only analytical endeavour, a self-similar solution of the bent-over jet in crossflow was obtained by Yih (1981). However, numerical predictions of the scalar field using this free shear layer model are not well-supported by experiments (Lee *et al.*1999). There is scarce data on the scalar field in the bent-over jet. Even the jet boundary could not be clearly defined, and the vortex flow has also not been measured.

In this chapter, we present a comprehensive account of the jet in crossflow. The jet in crossflow problem is examined first using a length scale analysis. In the near field, the effect of the crossflow is shown to be passive. In the far field, the jet behaves essentially as a line puff. In the remainder of the chapter, the emphasis is given to the flow of the advected line puff. Spreading coefficient and added mass coefficient are introduced in a one-dimensional (1D) model as the parameters characterizing the turbulent motion and the irrotational motion surrounding the puff. Two-dimensional (2D) and three-dimensional (3D) turbulence models are developed to compute the turbulent flow in the advected line puff. An extensive series of velocity and tracer concentration measurements obtained in the laboratory are presented. Spreading coefficient and added mass coefficient, the parameters of the 1D model, are evaluated using the 2D and 3D models and the data obtained from the laboratory measurements. The chapter is concluded by a practical example in the final section.

1 Length Scales and Regimes

The problem of the jet in a cross flow is defined schematically in Fig. 4. A round turbulent jet, of diameter D, initial velocity U_{jet}, and tracer concentration C_o is discharging from a nozzle into a steady uniform crossflow U_a. The crossflow is in

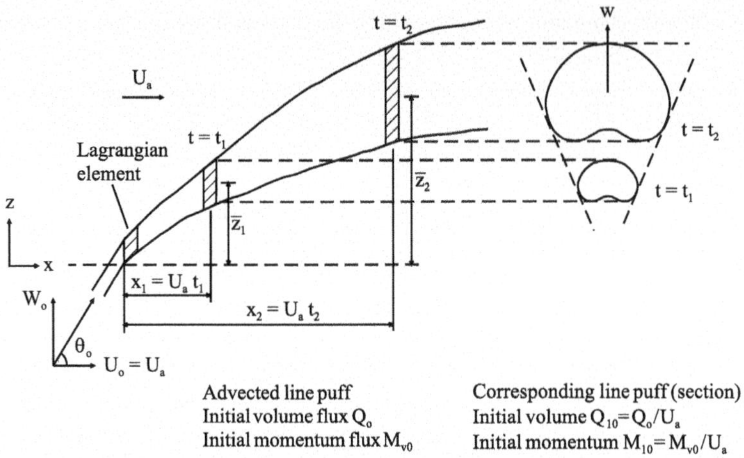

Fig. 4 Schematic of a jet in crossflow showing the Lagrangian element and its cross section in the analog of the advected line puff

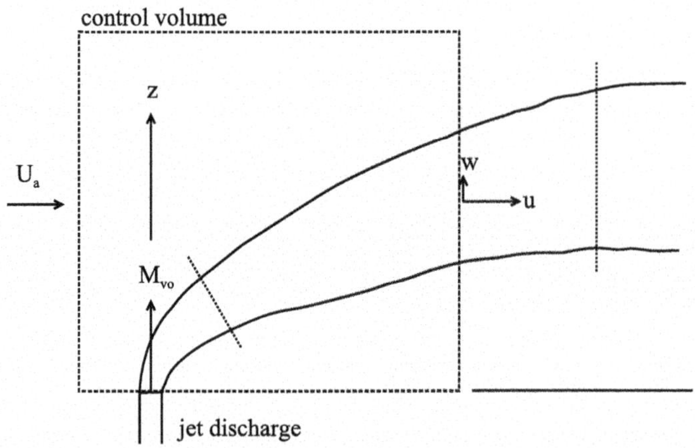

Fig. 5 Control volume for the momentum flux through a vertical cross section of a turbulent jet in crossflow; vertical momentum flux = $M_v = \int w(udA)$, horizontal momentum flux = $M_h = \int u(udA)$

the horizontal x-direction. The jet is oriented at an angle of θ_o to the x-axis. The initial vertical velocity $W_o = U_{jet}\sin\theta_o$. The initial volume flux from the nozzle is $Q_o = U_{jet}\pi D^2/4$. The vertical jet momentum flux is $M_{vo} = Q_o W_o$. The behaviour of a jet is characterized by its momentum flux. The velocity induced by the vertical momentum varies as $w \sim M_{vo}^{1/2} z^{-1}$. As the jet-induced velocity decreases, the jet becomes bent over due to entrainment of ambient fluid with crossflow momentum. At a vertical distance where $w \approx U_a$, or $z \sim M_{vo}^{1/2}/U_a$, heuristically we would expect the jet to be significantly bent over. Thus a crossflow momentum length scale

$$L_{mv} = \frac{M_{vo}^{1/2}}{U_a} \tag{1}$$

can be defined. For $z/L_{mv} = zU_a/M_{vo}^{1/2} \ll 1$, the jet-induced velocity is significantly larger than the ambient current, and the jet is only slightly affected by the current. Thus the mixing is similar to that of a momentum jet in still fluid (Chap. 2). In this region, the jet velocity and concentration are radially-symmetrical, self-similar, and can be well-approximated by Gaussian distributions. By regarding the jet as only advected by the current ($x = U_a t$), the trajectory and dilution relations of the "advected jet" can be obtained from the solutions of a classical pure jet (Wright 1977b; Fischer *et al.* 1979; Chu and Lee 1996a- see Sect. 6). For $z/L_{mv} \gg 1$, the jet is significantly bent over; it is expected that the horizontal exchange of momentum between the jet and the ambient current is complete–i.e. the x-velocity of the jet is approximately equal to U_a. The jet flow in this bent-over phase is similar to that of a line puff. These two asymptotic regimes (advected jet and line puff) are referred to as the *momentum-dominated near field* (MDNF) and *momentum-dominated far field* (MDFF) respectively (Wright 1977a). This chapter concentrates mainly on the MDFF.

1.1 Line Puff Analogy for MDFF

Consider a control volume with outer boundaries far removed from the source, along which the pressure is assumed to be hydrostatic (Fig. 5). By momentum conservation, the vertical jet momentum flux across a vertical section, $M_v = \int_{x-section} u\ w\,dA$, is then equal to the initial jet discharge momentum flux M_{vo}–where u, w are the streamwise (x) and vertical (z) velocities. In the bent-over phase, since $u \approx U_a$, we have $M_v \approx U_a \int w\ dA$. The flow in successive vertical x-sections can then be regarded as equivalent to that of a line puff with initial momentum $M_{lo} = M_{vo}/U_a$ imparted in the $+z$ direction, and initial volume $Q_{lo} = Q_o/U_a$, per unit x-length.

Intuitively, the jet in crossflow can be viewed as issuing line momentum puffs at a steady rate (Fig. 4). As far as the mixing in the bent-over phase is concerned, the action of the jet is to impart a vertical ($+z$) kinematic momentum flux of M_{vo} to the flow streaming by at U_a. In a continuous manner, the jet distributes a vertical momentum of $M_{vo}\Delta t$ to an ambient fluid element of length $U_a\Delta t$. Each of these Lagrangian elements (advected momentum puffs) receives an impulse of $\frac{M_{vo}\Delta t}{U_a\Delta t}$ per unit x-length as they pass by the source. To an observer moving with the crossflow, the ambient environment is stagnant while the turbulent element rises as a line puff. The mixing of the 3D bent-over jet at successive x-sections (x_1, x_2) then correponds to that of an equivalent line puff at corresponding times (t_1, t_2) via a Galilean transformation. This analog, to be confirmed by the experiments, is expected to hold in the bent-over stage of the jet, $z \gg L_{mv}$.

1.2 Similarity Variables for the Line Puff

We consider the mixing created by a cylinder of non-buoyant fluid having initial momentum in an otherwise stagnant fluid. This is illustrated in Fig. 6 for an experiment done upside down (Richards 1965). A cylinder of marked fluid having an initial head is released impulsively from a puff device, creating a dyed line patch with significant initial momentum, and mixes with the surrounding fluid in the water tank. Figure 7 shows a cross-section of the line puff (in the $y - z$ plane) with an initial volume Q_{lo} $[L^2]$ and an initial momentum M_{lo} $[L^3/T]$ per unit length of the line puff. The tracer concentration in the volume initially is C_o. The unsteady flow of the puff is charaterized by vertical coordinate z, half-width b, vertical velocity W, and concentration C. A time scale of the puff

$$t_c = \frac{Q_{lo}^{3/2}}{M_{lo}} = \frac{Q_o^{3/2}}{U_a^{1/2} M_{vo}} \tag{2}$$

is defined for the relative importance of the initial volume Q_{lo} and momentum flux M_{lo}. The related length scale is

$$L_x = U_a t_c = \frac{Q_o^{3/2} U_a^{1/2}}{M_{vo}} \tag{3}$$

As the puff is primarily driven by its initial impulse, the effect of the source volume will be negligible for $t/t_c \gg 1$. Therefore, the puff characteristics is assumed to depend only on M_{lo} and the time t. The following similarity relations for the maximum velocity, spreading rate, trajectory, and scalar dilution of the line puff are

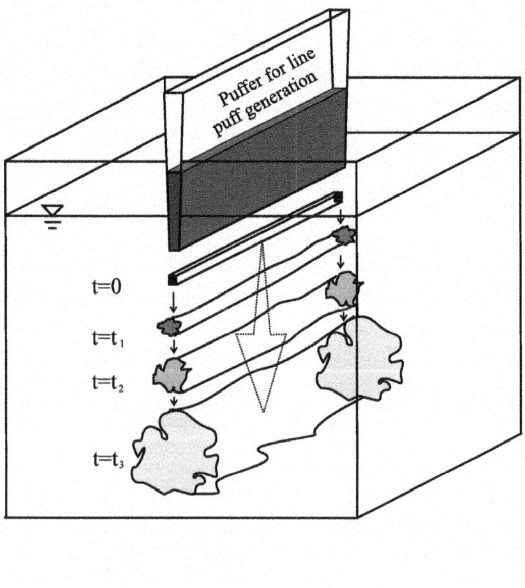

Fig. 6 Line momentum puff generated by releasing fluid impulsively at excess head from a puffer in a tank

obtained based on this assumption:

$$W_m \sim M_{lo}^{1/3} t^{-2/3} \quad \text{or} \quad \frac{W_o}{W_m} \sim (t^*)^{2/3} \tag{4}$$

$$b \sim z \tag{5}$$

$$z \sim M_{lo}^{1/3} t^{1/3} \quad \text{or} \quad z \sim Q_{lo}^{1/2} (t^*)^{1/3} \tag{6}$$

where dimensionless time $t^* = t/t_c$; and W_o and W_m stand for the initial and maximum vertical velocity of the jet respectively. Conservation of tracer mass also implies:

$$S_c = \frac{C_o}{C} \sim \frac{b^2}{Q_{lo}} \sim \frac{M_{lo}^{2/3} t^{2/3}}{Q_{lo}} \sim \frac{z^2}{Q_{lo}} \tag{7}$$

Given the equivalence between successive vertical x-sections of a bent-over jet and the line puff at corresponding times $t = x/U_a$, the following relations can be obtained for the spreading rate, trajectory, and dilution of the 3D bent-over jet in terms of the crossflow momentum length scale L_{mv}.

$$b_{vc} \sim z_c \tag{8}$$

$$\frac{z_c}{L_{mv}} \sim \left(\frac{x}{L_{mv}}\right)^{1/3} \tag{9}$$

$$\frac{S_c Q}{U_a L_{mv}^2} \sim \left(\frac{z_c}{L_{mv}}\right)^2 \tag{10}$$

$$\frac{W_o}{W_m} \sim (x^*)^{2/3} \tag{11}$$

where b_{vc} is the vertical centerline half-width defined by the $e^{-1}C_c$ points, z_c is the vertical location defined by the maximum centerline concentration C_c, $S_c = C_o/C_c$ is referred to as centerline minimum dilution, as illustrated in Fig. 7; and the dimensionless downstream distance $x^* = x/L_x$. For the sake of reference, Eqs. 9 and 10 can be expressed alternatively as:

$$z_c \sim (Q_o/U_a)^{1/2} (x^*)^{1/3} \tag{12}$$

$$S_c \sim z_c^2/(Q_o/U_a) \tag{13}$$

The above similarity relations furnish a useful means of correlating experimental data; the proportionality constants can be derived by best-fitting experimental data or numerical model results.

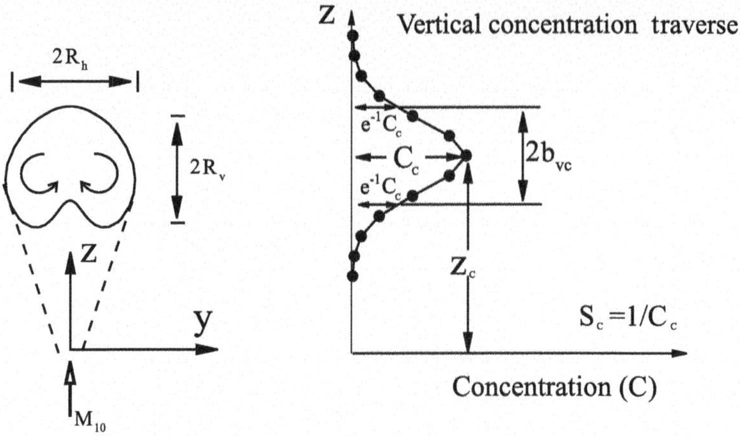

Fig. 7 Definition sketch showing the cross section of the line puff and the concentration profile on the center plane of symmetry

2 1D Model of Line Puff

We begin with the analysis of line puff by a one-dimensional (1D) model. In the absence of body forces, the momentum of the puff, $M_p(t) = \pi R^2 W$ is conserved, where $R = \sqrt{R_h R_v}$ is a characteristic radius of the puff, and W is the average vertical velocity. Application of the general Lagrangian spreading hypothesis (following the puff element or dominant eddy as in Chaps. 4–6) and momentum conservation to the puff in Fig. 7 then gives:

$$\frac{dR}{dt} = \beta_n W \tag{14}$$

$$\frac{d}{dt}((1 + k_n)\pi R^2 W) = 0 \tag{15}$$

where β_n is the puff spread rate and k_n is an added mass coefficient to account for the motion of the irrotational flow outside of the puff. As the puff (the colored patch) moves through the surrounding fluid, energy is expended to form the external flow. The presence of the ambient fluid may be represented by an additional "virtual mass" to the inertia of the puff element (see Prob. 8.3).

Noting that $W = dz/dt$, the average flow or 'global' solution is then given by:

$$R = (\frac{3\beta_n M_{lo}}{(1 + k_n)\pi})^{1/3}\, t^{1/3} \tag{16}$$

$$W = (\frac{M_{lo}}{(1 + k_n)9\pi\beta_n^2})^{1/3}\, t^{-2/3} \tag{17}$$

$$z = (\frac{3M_{lo}}{(1+k_n)\pi\beta_n^2})^{1/3}\, t^{1/3} \tag{18}$$

$$\frac{C_o}{C} = \frac{\pi B^2}{Q_{lo}} = \frac{\pi B^2 U_a}{Q_o} \tag{19}$$

The solution for the puff location translates via a Galilean transformation into the path of the bent-over jet in crossflow:

$$R = \beta_n\, z \tag{20}$$

$$z = (\frac{3}{(1+k_n)\pi\beta_n^2})^{1/3} (\frac{M_{vo}x}{U_a^2})^{1/3} \tag{21}$$

or

$$\frac{z}{L_{mv}} = C_{2p}(\frac{x}{L_{mv}})^{1/3} \tag{22}$$

with $C_{2p} = (\frac{3}{(1+k_n)\pi\beta_n^2})^{1/3}$. Several characteristic parameters of the turbulent-mean puff flow can hence be determined to within a constant without knowledge of the detailed velocity or concentration distributions. There are two model parameters, β_n and k_n, which must be determined independently for a consistent prediction of the puff spread rate, trajectory, and dilution. To obtain these parameters, we turn to numerical computations and measurements of the flow and tracer concentration field of a jet in crossflow.

3 2D Model of Line Puff

Numerical computations have been conducted for the line puff using a two-dimensional (2D) model and for the general problem of jet in crossflow using a three-dimensional (3D) model. The results obtained for the line puff using a two-dimensional k-ϵ turbulence model is presented in this section. The 2D model equations for the line puff is the Reynolds-averaged equations, which for constant density incompressible flow, are:

$$\frac{\partial U_i}{\partial x_i} = 0 \tag{23}$$

$$\frac{\partial U_i}{\partial t} + U_j \frac{\partial U_i}{\partial x_j} = -\frac{1}{\rho}\frac{\partial p}{\partial x_i} + \frac{\partial \tau_{ij}}{\partial x_j} \tag{24}$$

where U_i = fluid velocity in (x, y, z) direction, ρ = density, p = dynamic pressure, τ_{ij} = stress tensor. τ_{ij} is equal to the sum of the viscous and Reynolds-stresses. The eddy-viscosity model (Boussinesq hypothesis) provides the following expression for the stresses:

$$\tau_{ij} = \nu_{eff}(\frac{\partial U_j}{\partial x_i} + \frac{\partial U_i}{\partial x_j}) - \frac{2}{3}k\delta_{ij} \tag{25}$$

where the effective viscosity $\nu_{eff} = \nu + \nu_t$, ν is the molecular viscosity, $\nu_t = C_\mu k^2/\epsilon$ is the turbulent viscosity, k is the turbulence kinetic energy (TKE), ϵ is referred to as dissipation rate (DR) of k. We adopt the standard two-equation $k - \epsilon$ model of Launder and Spalding (1974) for turbulence closure:

$$\frac{\partial k}{\partial t} + U_j \frac{\partial k}{\partial x_j} = \frac{\partial}{\partial x_i}(\frac{\nu_t}{\sigma_k}\frac{\partial k}{\partial x_i}) + \tau_{ij}\frac{\partial U_i}{\partial x_j} - \epsilon \tag{26}$$

$$\frac{\partial \epsilon}{\partial t} + U_j \frac{\partial \epsilon}{\partial x_j} = \frac{\partial}{\partial x_i}(\frac{\nu_t}{\sigma_\epsilon}\frac{\partial \epsilon}{\partial x_i}) + C_{1\epsilon}\frac{\epsilon}{k}\tau_{ij}\frac{\partial U_i}{\partial x_j} - C_{2\epsilon}\frac{\epsilon^2}{k} \tag{27}$$

with the following standard empirical model coefficients: $C_\mu = 0.09$, $C_{1\epsilon} = 1.44$, $C_{2\epsilon} = 1.92$, $\sigma_k = 1.0$, $\sigma_\epsilon = 1.3$ (Rodi 1980). In the above, the turbulent closure is achieved by computing the turbulence kinetic energy locally; the generation, transport, and decay of turbulence can hence be accounted for. TKE is computed by a transport equation for k, while the equation for ϵ is by and large empirical. The eddy viscosity $\nu_t \sim k^2/\epsilon$ is then obtained by dimensional analysis.

In addition to the flow equations, the study of puff characteristics necessitates the calculation of a passive scalar field from the tracer mass conservation equation:

$$\frac{\partial C}{\partial t} + U_j \frac{\partial C}{\partial x_j} = \frac{\partial}{\partial x_j}(\frac{\nu_{eff}}{Sc_t}\frac{\partial C}{\partial x_j}) \tag{28}$$

where Sc_t stands for the turbulent Schmidt number, taken to be 0.75 as found appropriate for a related problem (Sykes *et al.* 1986). Note the above governing equations are applicable for both the 2D line puff and the 3D advected line puff.

3.1 Numerical Simulation of Line Puffs

The nature of the line puff motion can be illustrated by a numerical 'thought experiment'. A finite momentum is imparted to a two-dimensional marked patch of incompressible fluid, and the time evolution of the mean flow and passive scalar field of such a puff is investigated. Figure 8 shows a volume of water initially at rest in a tank of width W_T and depth H. Over a time interval, $t = 0$ to $t = t_s$, an impulse $M_o = M_{lo}$ in the vertical (z) direction is applied to the marked square patch of fluid (of length L_o), which has a uniform tracer concentration C_o and is located sufficiently far away from the solid boundaries. The momentum source gives rise to vorticity. Turbulent flow is produced and is advected in the direction of the imparted momentum while mixing with the ambient fluid. The motion of this turbulent puff (Fig. 9) has been computed using the 2D $k - \epsilon$ model (Lee, Rodi and Wong 1996). The following

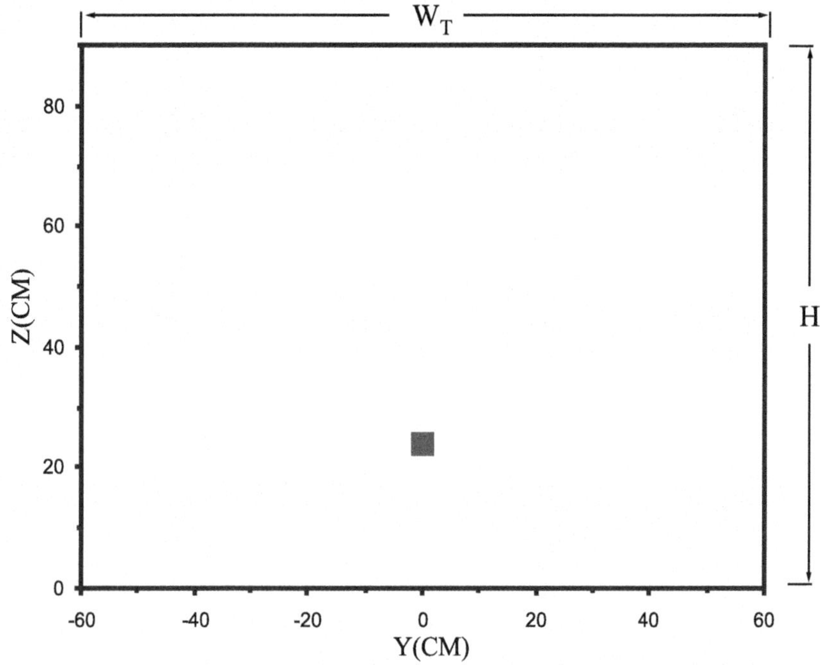

Fig. 8 Numerical experiment of turbulent line momentum puff; an initial impulse is applied in +z direction to marked patch of fluid

parameters are adopted: $W_T = 120\,cm$, $H = 90\,cm$, $t_s = 0.01\,s$, $M_o = 1712\,cm^3/s$, and $L_o = 4\,cm$. The marked patch is initially located at $Z_o = 24\,cm$ from the bottom boundary.

At $t = 0$, zero velocities and pressure are prescribed. The scalar concentration is given the value of $C_o = 1$ inside the marked patch, and zero elsewhere. The puff is defined as the region inside the 0.01 C_m contour, where $C_m(t)$ is the maximum computed concentration, and L defined such that L^2 is the area within that contour. The governing equations are solved using the finite volume method (Patankar 1980). Details of the numerical solution procedure can be found in Lee *et al.*(1996). The main features of the flow and scalar field are summarized below.

The momentum source gives rise to sharp velocity gradients, leading to the formation of two vortices at the lateral edges of the puff (Fig. 10). The force input also generates large positive pressures in front of, and large negative pressures behind the puff. As a result of this pressure interaction, the puff vertical momentum $M_p = \int W dA$ drops to half of the nominal impulse after source introduction. Figure 11 shows that the puff momentum initially decreases, and attains an asymptotic value of 0.5 M_o beyond t* $\approx$ 20. The injection of the momentum puffs into the crossflow requires extra work done to form the ambient irrotational flow; some of the initial impulse is contained in the puff flow, while the rest is carried by the ambient fluid outside of the

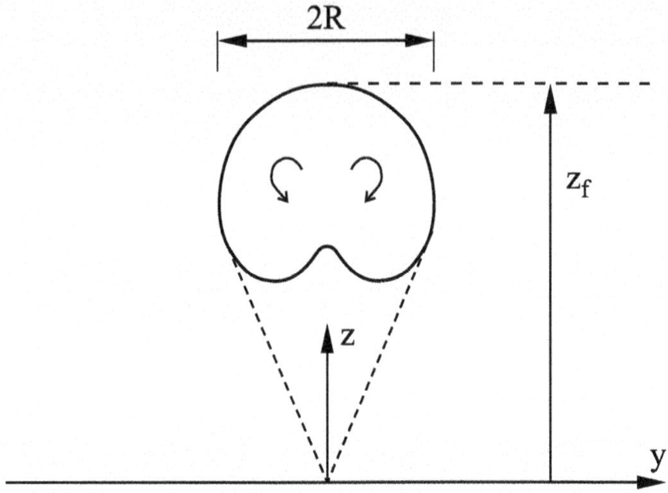

Fig. 9 Definition of puff front and radius

turbulent flow (the added mass effect). The result implies that the line puff motion has an added mass coefficient of $k_n \approx 1$.

The numerical solution shows that the shape of the flow and presssure field, as well as the vorticity, is well-preserved for $t^* \geq 20$. Figure 12 shows clearly the puff flow is self-similar and characterized by a vortex-pair. The computed concentration field and turbulence kinetic energy also display approximate self-similarity as the flow field (Lee *et al.* 1996). From the computed flow and passive scalar field, characteristic puff variables such as the maximum vertical velocity $W_m(t)$ and concentration C_m, elevation z, width R_h, R_v, and aspect ratio R_h/R_v can be determined along with the turbulent kinetic energy k_m. For example, the computed time variation of maximum vertical velocity W_m and front location of the puff z_f are in excellent agreement with the -2/3 and 1/3 power law respectively. The puff front is also linearly proportional to the maximum radius R. It is then possible to determine the dimensionless puff mixing paramters in the self-similar relations from the numerical solution in the asymptotic stage $t^* \gg 1$. These are summarized in Table 1 (SKE-standard $k - \epsilon$ model). Also shown are the ratio of the maximum to mass weighted average vertical velocity, $W_m/\overline{W}$, and the circulation around one half of the puff, defined as $\Gamma = \int_o^H W(0, z)dz$. The properties of the self-similar puff are rather insensitive to the exact initial conditions of the k and ϵ equations assumed. Defining the puff by the 0.05 C_m contour also produced negligible changes in the results. However, one distinct feature of the 2D line puff is the absence of bifurcation of the tracer concentration field in the asymptotic phase; i.e. the aspect ratio of the pear-shaped line puff is approximately one–in contrast to observations of the bent over jet in crossflow. The result is obtained by the standard $k - \epsilon$ model (Lee *et al.* 1996) and also by a modified $k - \epsilon$ model based on ReNormalization Group theory (RNG model in Table 1, Lee

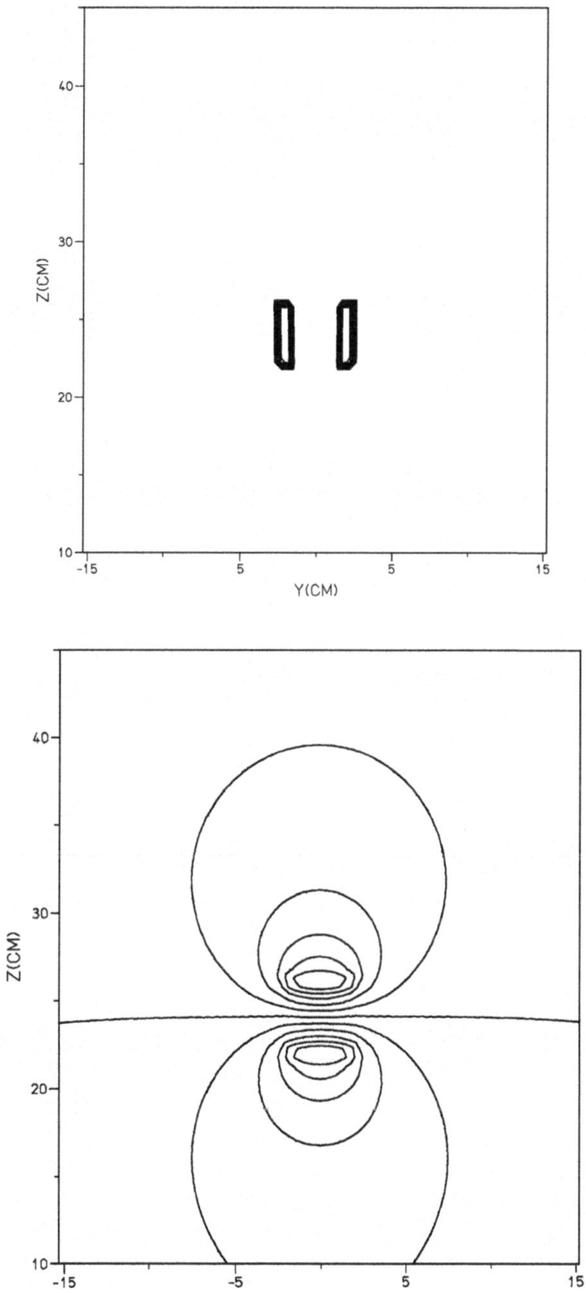

Fig. 10 Initial flow generated at $t = 0.5t_s$. The line momentum source gives rise to vorticity concentrated into two rings (top); (b) pressure field (from Lee, Rodi, and Wong 1996)

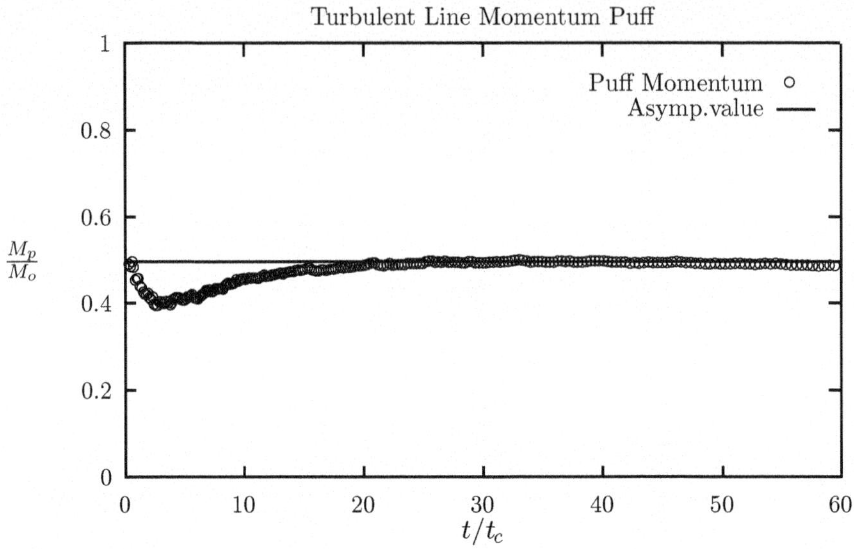

Fig. 11 Time variation of puff momentum M_p/M_o

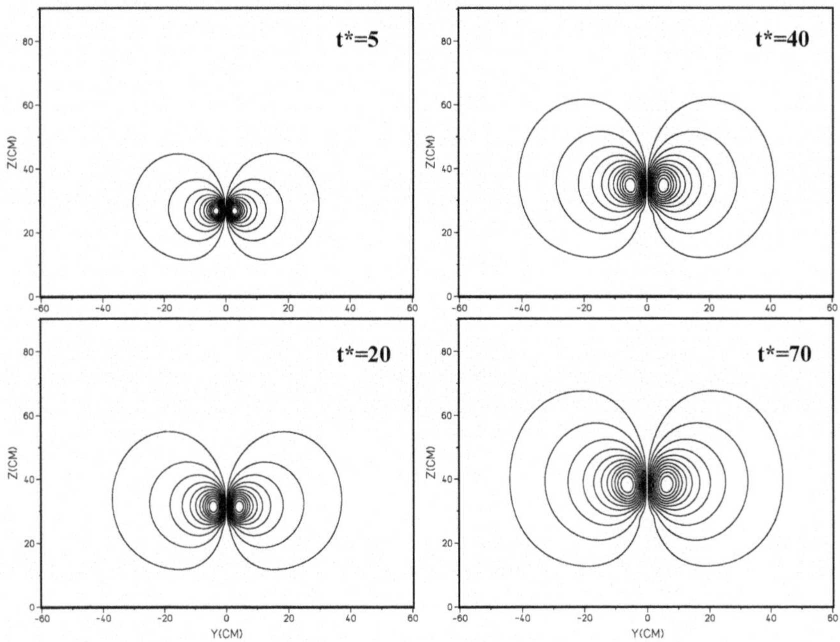

Fig. 12 Computed streamlines of line puff at $t^* = 5, 20, 40, 70$; maximum normalized stream function is $\Psi_m^* \approx 0.18$ (contour interval $\Psi_m^*/12$)

and Chen 1998). This result also shows that a vortex-pair flow does not necessarily imply the bifurcation of the time-averaged concentration field.

4 3D Model of Jet in Crossflow

4.1 The Advected Line Puff

The advected line puff (ALP) has been introduced by Lee and Wong (1993) as an experimental analog for the study of the bent-over jet in crossflow. To best approximate the line puff condition, the round jet is discharged at an angle θ_o so that the initial excess momentum is zero relative to the crossflow. This is done by setting the angle of discharge θ_o such that $U_o = U_{jet}\cos\theta_o = U_a$ as shown in Fig. 4. The jet to ambient velocity ratio is then $W_o/U_a = tan\theta_o$. Since the component of the initial excess momentum in the direction of the cross flow is zero, the vertical motion of the ALP can be viewed in a coordinate system moving with the crossflow. In this moving coordinate system, the ambient is stationary and the tubulent fluid in the ALP rises in the vertical direction essentially as pure line puff. Source conditions are much better controlled in the ALP than the source conditions of the pure line puff produced by a line source of momentum in still fluid. Experimental data obtained in the ALP are much less scattered compared with those obtained in the pure line puff experiment or previous MDFF data of vertical jet in crossflow.

4.2 3D Model of Advected Line Puff

A more accurate picture of the scalar field can be obtained from a three-dimensional $k - \epsilon$ model calculation of the scalar field in a jet in crossflow (Lee *et al.*2002). A 3D numerical experiment of the advected line puff in Fig. 4 can be performed. A near vertical non-buoyant jet is discharged at no excess horizontal momentum into a horizontal crossflow. The scalar field is illustrated for the case of a water jet (diameter 2 cm) discharging in a numerical channel of length 1.2 m, width 0.8 m and depth 0.6 m. The parameters are $U_{jet} = 1.086$ m/s, $U_a = 0.1887$ m/s, and $\theta_o = 80^o$. The computational domain extends to 55 jet diameters downstream. The jet to ambient current ratio is $W_o/U_a = 5.67$.

Figure 13 shows contour plots of the computed scalar field in the plane of symmetry at $y = 0$. The trajectory of the maximum concentration shows a rapid rise to about $z \approx 0.13$ m at $x = 0.1$ m corresponding to $x^* \approx 14$, whereafter the jet rises very slowly, reaching $z \approx 0.27$ m at $x = 1.0$ m corresponding to $x^* \approx 142$. The bent-over phase, during which the scalar contours become almost parallel with the mainstream, takes place essentially in the interval of $10 < x^* < 40$.

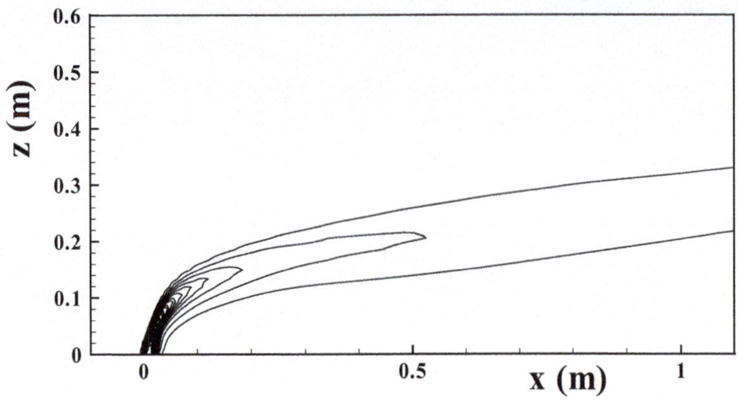

Fig. 13 Computed scalar field in the center plane of symmetry of a jet in crossflow

Figure 14 shows transverse cross sections of the scalar field at different downstream locations. As the salient feature of bent-over jet observed in experiments, a double peak concentration distribution can generally be observed in the kidney-shaped sectional puff, which becomes approximately self-similar for $x^* > 20$.

Characteristics of Advected Line Puff

Analysis of the 3D model results reveals the following characteristics:

(i) *Puff aspect ratio* - As a basic parameter to characterize the sectional puff shape, the puff aspect ratio is defined as the ratio of horizontal half-width R_h over vertical half-width R_v of the contour $C = e^{-1}C_m$, where C_m stands for the maximum concentration over the puff section (Fig. 7). The results show that the puff aspect ratio is roughly equal to 1.20 for the main stage of $20 < x^* < 70$, but reduces, with a minimum value of 1.05, for the earlier and later stages of $x^* \leq 10$ and $x^* \geq 100$. This is consistent with the experimental results of Chu (1996) who found a variation of puff aspect ratio from about 1.05 for the early and later stages to about 1.25 for the bent-over stage, and of Wong (1991) who found a value of 1.23 with probe-based measurements; and comparable to the value of 1.3 suggested by Pratte & Baines (1967) for the developed phase from an analysis of flow visualization.

(ii) *Maximum concentration ratio*—As observed from the concentration contours, the maximum concentration is not located in the center line. It is desirable to correlate the centerline maximum concentration C_c to the sectional maximum concentration C_m, since often only centerline concentration was measured. It is found that the ratio of C_m/C_c varies, from a smaller value of 1.05 for the initial and later stages of $x^* \leq 10$ and $x^* \geq 130$, to a higher value of 1.35 for the central phase of $30 \leq x^* \leq 120$, with an averaged value of 1.2. This is consistent with the range of

1.0–1.5 with an average of about 1.2 found by Chu (1996), and compares well with the range of 1.1 to 1.5 found by Wong (1991).

(iii) Centerline trajectory—Fig. 15 shows the puff trajectory z_c/L_{mv} plotted against downstream distance x/L_{mv}. The results follow the power similarity law given in Eq. 9. Corresponding coefficient is found as $C_{2p} = 1.56$. This value is very close to the experimental values of 1.56 and 1.63 in Wong (1991) and Chu (1996), respectively, and comparable to the value of 1.77 from the buoyant jet experiment of Ayoub (Lee 1989).

(iv) Centerline half-width—Fig. 16 shows the centerline half-width b_{vc} plotted against centerline trajectory z_c. For $z_c/L_{mv} \geq 1.2$, corresponding to $x^* \geq 10$, the results follow the linear similarity law. The corresponding coefficient, referred to as the centerline half-width spreading rate, is found as $C_{1p} = 0.297$. This spreading rate is close to the experimental value of 0.276 by Wong (1991) and 0.299 by Chu (1996).

(v) Centerline dilution—Fig. 17 shows the dimensionless centerline dilution $S_c Q/(U_a L_{mv}^2)$ plotted against centerline trajectory z_c/L_{mv}. The results follow the power similarity law of Eq. 10. The corresponding minimum centerline dilution constant is found as $C_{3p} = 0.488$. No notable effect of virtual origin is observed. This value can be compared to experimental values of 0.409-0.484 (Wong 1991; Chu 1996).

A summary of the self-similar properties of the 3D advected line puff is given in Table 1. The main conclusions of the 3D modelling are as follows:

- The 3D advected line puff (bent-over jet) is longitudinally characterized by a bent-over phase taking place essentially in the interval of $10 < x^* < 40$, during which the fluid flow and scalar contours become almost parallel with the free stream. The flow is approximately self-similar beyond a dimensionless length of $x^* = 20 - 30$.
- Similar to the line puff, the flow in the jet cross-section is characterized by a vortex pair structure, with an added mass coefficient of approximately 1.
- The scalar field is sectionally characterized by a kidney-shaped outline containing a double peak of concentration maxima. The puff aspect ratio (R_h/R_v) and the maximum concentration ratio (C_m/C_c) are both around 1.2.

One practical advantage of a field solution is the determination of the added mass coefficient k_n and the puff spreading rate β_n. The numerical work clearly demonstrates the existence of added mass. The substantial loss of nominal impulse is also supported by the ALP experiments (see next section). The asymptotic value of 0.5 for the puff momentum, a fundamental property, can be interpreted as an added mass coefficient of $k_n = 1$ for the puff motion (Chu 1985;Turner 1986). Half of the nominal impulse accounts for the irrotational flow of the surrounding fluid set up by the moving puff. It is interesting to note that (Table 1) approximately the same value of k_n is obtained in all the 2D and 3D numerical calculations, regardless of the turbulence model. From Table 1, the parameters for the simple Lagrangian model can be inferred–$k_n \approx 1$ and $\beta_n \approx \sqrt{2}\frac{db_{vc}}{dz} \approx 0.4$. The predicted maximum turbulence intensity $u_m/W_m = 0.37$ is also supported by the observed value of 0.42 based on concentration measurements (Chu 1996).

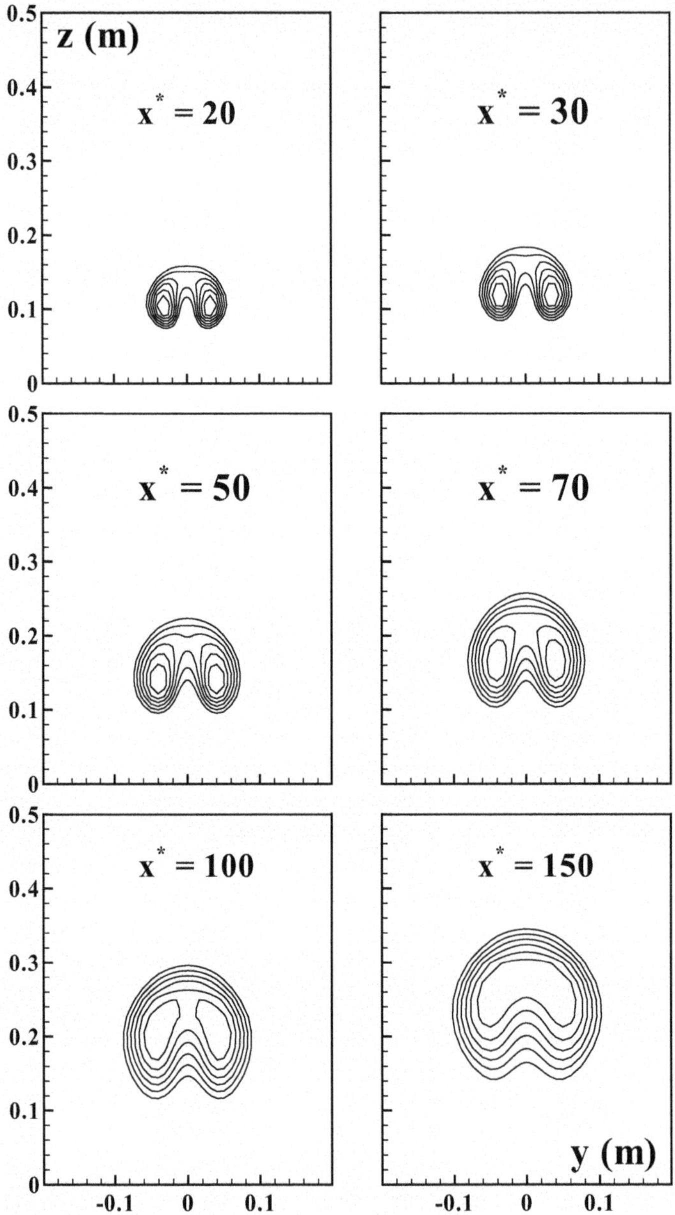

Fig. 14 Computed scalar field at different downstream locations of a bent-over jet in crossflow (contour interval = $(1 - e^{-1})C_m/6$, outmost concentration contour = $e^{-1}C_m$)

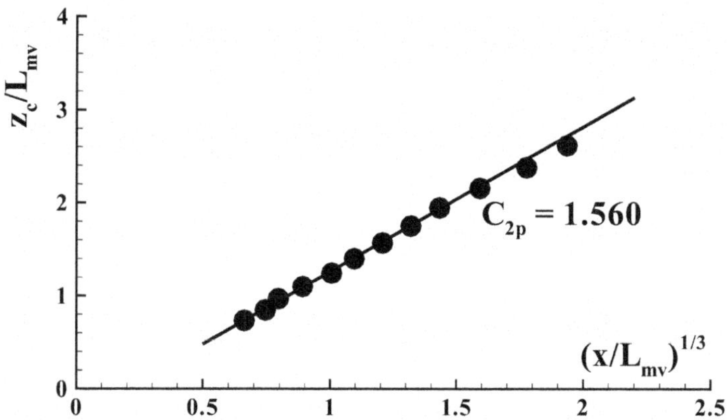

Fig. 15 Puff trajectory z_c/L_{mv} vs downstream distance x/L_{mv}

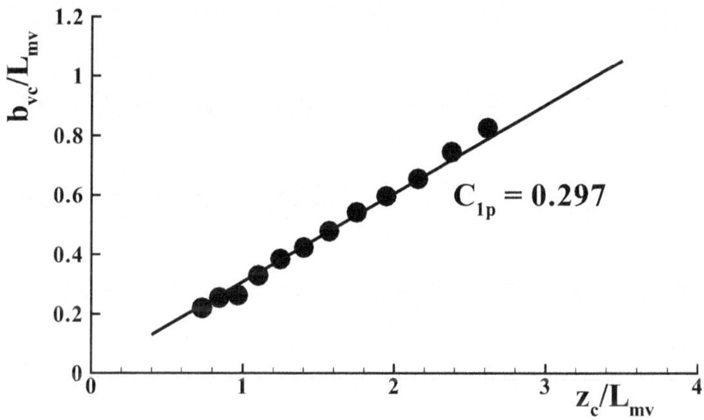

Fig. 16 Centerline half-width b_{vc} vs centerline trajectory z_c

5 Measurements in Advected Line Puffs

Wong (1991) measured the cross section tracer concentration (using salt as tracer) in the ALP using a probe-based techniques. Chu (1996) measured the ALP scalar field using non-intrusive LIF techniques, and the velocity field by LDA. The LIF measurements enable instantaneous and time-average measurements of the cross-section scalar field; from this many parameters of engineering interest (e.g. visual boundary, vertical and horizontal half-width, center of mass, vertical centerline minimum dilution and its location, aspect ratio) can be extracted. In particular, the added mass can be directly determined, and some idea of the turbulence structure gained. As shown in Fig. 18, the puff region is bounded by the puff edge defined by the concentration contour of $C = 0.25C_m$ where C_m is the cross-sectional maximum scalar concen-

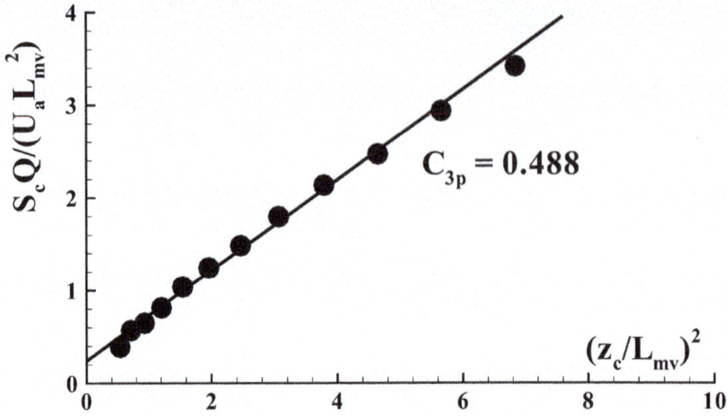

Fig. 17 Centerline dilution $S_c Q/(U_a L_{mv}^2)$ vs centerline trajectory z_c/L_{mv}

Fig. 18 Definition diagram of characteristic properties of an advected line puff

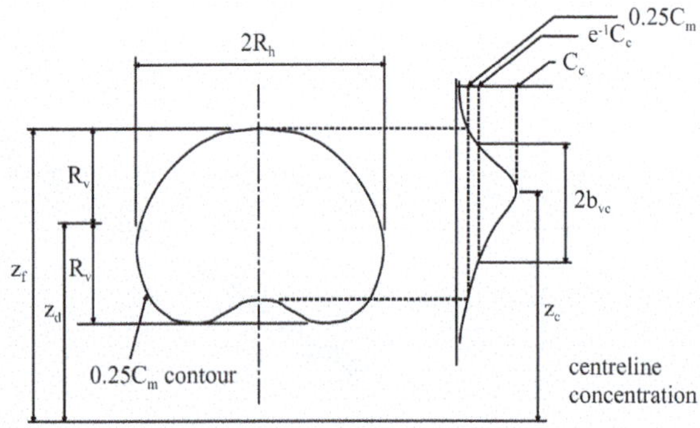

tration. Similar to the coflow jet (Chap. 6), it is found that this edge definition gives good agreement with the visual boundary. $2R_v$,$2R_h$ are the maximum vertical and horizontal width. C_c is the centerline maximum concentration; z_c is the location of C_c. The measured vertical z-velocity across the puff region is integrated to give the actual vertical jet momentum flux M_v, or the momentum of the equivalent line puff $M_p = M_v/U_a$.

Figure 19 shows instantaneous images of the cross-sectional scalar field of ALP. The scalar field is highly intermittent; rotating patches with highly irregular rough boundaries are observed to move around the puff central region. However, unlike a pure jet or coflow jet, a distinct core turbulence structure concentrated near the centre does not exist; the intermittency factor is generally less than unity throughout the entire puff region. The entrainment of the ambient fluid into the core can be clearly observed in the surroundings of the rotating turbulent patches consisting of eddies of different scales. In addition to mixing near the puff front, significant entrainment takes place at the rear bottom of the puff. The engulfing of ambient fluid and traces of fluorescent dye (transported by wake vortices shed near the source) towards the puff

Table 1 Similarity relations of 2D line puff and 3D Advected Line Puff (ALP) derived from numerical solution and laboratory experiment; numbers refer to dimensionless coefficients of equations or ratios.

	Parameter	SKE	RNG	3D JET	Experiment
Puff momentum	M_p/M_{lo}	0.497	0.502	0.48	0.47 (Chu 1996)
					0.3 (vortex ring; Glezer & Coles 1990)
Vertical velocity	$W_m = C_v M_{lo}^{1/3} t^{-2/3}$	1.61	1.90	1.18	1.21 (Chu 1996)
Puff front	$z_f = nR_h$	3.03	3.09		2.6-5.7 (Richards 1965)
					3.1 (Chu 1996)
Trajectory	$z_c = C_{2p} M_{lo}^{1/3} t^{1/3}$	1.70	1.54	1.565	1.44 (Chu 1974)
					1.56 (Wong 1991)
					1.63 (Chu 1996)
Length scale	$L = C_{4p} M_{lo}^{1/3} t^{1/3}$	1.17	1.19		
Centre of mass	$L \sim \overline{z}$	0.75	0.74		0.71 (Chu 1996)
Scalar concentration	$C_o/C_c = C_{3p} z_c^2/Q_{lo}$	0.28	0.28	0.489	0.484 (Wong 1991)
					0.42 (Chu 1996)
	C_m/C_c	1.02	1.02	1.35	1.0-1.56 (Wong 1991)
					1.1-1.5 (Chu 1996)
Centerline half-width	$b_{vc} = C_{1p} z_c$	0.269	0.355	0.274	0.276 (Wong 1991)
					0.294 (Chu 1996)
Aspect ratio	R_h/R_v	1.03	1.00	1.35	1.23 (Wong 1991)
					1.25 (Chu 1996)
					1.3 (Pratte & Baines 1967)
Velocity ratio	$W_m/\overline{W}$	3.18	3.73		3.17 (Chu 1996)
Circulation	$\Gamma/(\overline{W}L)$	2.67	2.71		2.98 (Chu 1996)
Momentum factor	$M_p/(\overline{W}L^2)$	0.69	0.64		0.83 (Chu 1996)
Turbulent viscosity	$\nu_{tm}/(W_m L)$	0.018	0.017		
Turbulent intensity	u_m/W_m	0.39	0.37	0.37	0.42 (Chu 1996)

centre from beneath can be clearly observed. This results in the bifurcation of patches and the counter-rotating vortex-pair flow structure. The time-averaged image of the scalar field is smooth and kidney-shaped. A double peak concentration distribution can generally be observed. The mean scalar field is marked by a lobe-like appendix in the rear; this is related to wake vortices shed near the jet exit, which are re-entrained into the rear of the main vortex-pair downstream (Fig. 1). The self-similar puff flow grows slowly with elevation.

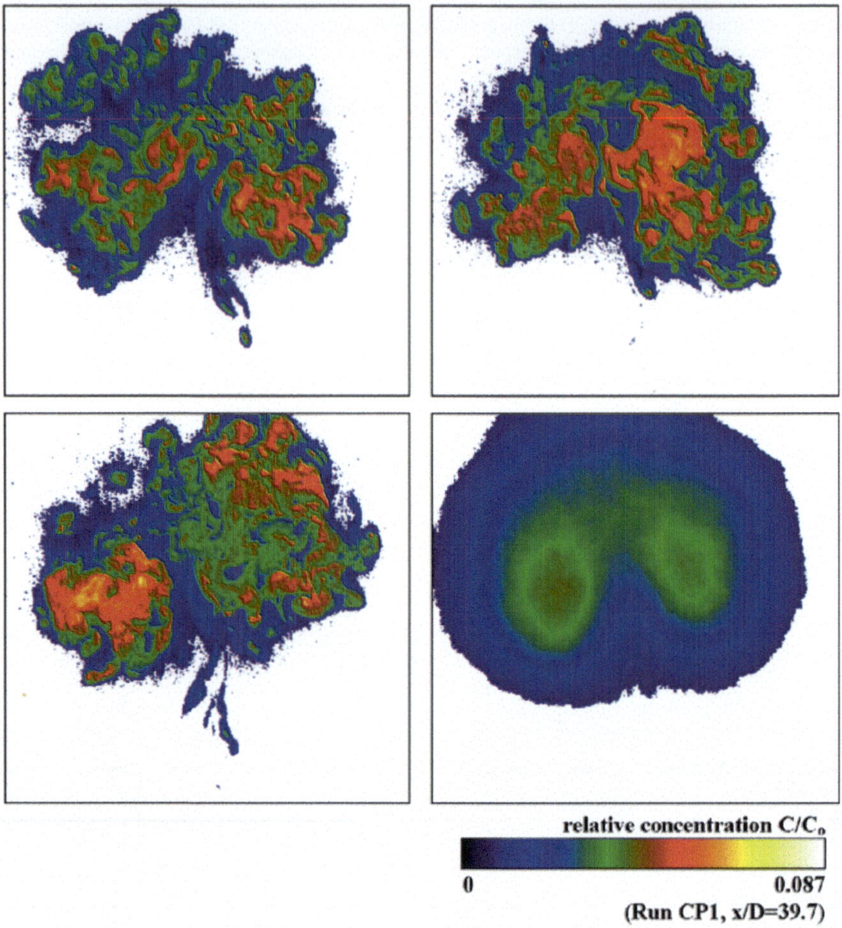

Fig. 19 Typical instantaneous and time-averaged LIF images in the cross-section of the advected line puff ($U_o = 57.5$ cm/s, $U_a = 5.0$ cm/s, $\theta_o = 85°$, $D = 0.75$ cm). Concentration lower than the visual threshold not shown

5.1 Time-Averaged Properties

Figure 20 shows measured centerline half-width b_{vc}, jet trajectory z_c, and centerline minimum dilution S_c (Wong 1991). It can be seen the linear spread rate, the 1/3 trajectory law and the z^2 power dilution law are well supported by the data. Despite the narrow range of z/L_{mv} within which the puff can be measured, the data clearly supports the similarity relations, with $C_{1p} = 0.276$, the trajectory constant $C_{2p} = 1.56$, and the centerline dilution constant $C_{3p} = 0.46$. Similar values have been obtained in the LIF experiments. It is evident from Figs. 14, 18 and 19 that the trajectory z_c defined by the centerline maximum concentration is located generally

Table 2 Summary of experimental constants (visual spreading rate, visual and centreline trajectory, centreline and minimum dilution) of ALP

Previous studies	Advected line puffs (MDFF)				
	Spreading rate	Trajectory		Dilution	
	β	z_d	(z_c)	S_c	(S_m)
	C_{1p}	C_{2p}	(C_{2pc})	C_{3p}	(C_{3pm})
Richards (1965)	0.27[a]	2.07[a]			
Hoult and Weil (1972)	0.42[b]	1.38			
Pratte and Baines (1967)	0.39				
Chan, Lin and Kennedy (1976)		1.44			
Chu and Goldberg (1974)	0.35[b]	1.56			
Wright (1977b)		1.6	(2.1)	0.38	
Chu (1979)	0.34	1.64			
Lee (1989)			(1.77)	0.32	
Wong (1991)	0.43[c]	1.32	(1.56)	0.46	
JETLAG (Lee and Cheung 1990)	0.45[b]	1.3		0.49[d]	
Chu (1996)–shadowgraphy	0.45	1.54			
Chu (1996)–LIF	0.39	1.55	(1.63)	0.41	(0.34)
1D Lagrangian model ($\beta_n = 0.4$ and $k_n = 1$)	0.4[e]	1.44	(1.51)	0.39	(0.33)
				0.43[d]	(0.36[d])

[a] using $R_h/R_v \simeq 1.2$
[b] deduced from entrainment coefficient using $k = 1$
[c] combined the results of visual and concentration half-width spreading rate
[d] based on puff trajectory
[e] selected value

higher than the visual centerline–a typical feature of the pear-shaped concentration structure. The visual trajectory z_d (defined as mid-way between top and bottom boundaries) is found to correspond exactly to the location of the centre of mass; z_c however is on average about 7 % higher than z_d. This result can also be deduced from the numerical model results of the line puff (Problem 8.2). The jet spreads linearly with vertical distance, with $R_v = 0.36\ z_d$ (Fig. 21), and the aspect ratio is $R_h/R_v \approx 1.2$; the puff area is also found to agree perfectly with $A = \pi R_h R_v$. The puff front varies as $z_f = z_d + R_v = 3.52\ R_v$. Using an average aspect ratio of $R_h/R_v = 1.2$, we have $z_f = 3.06\ R_h$; this is contrary to a value of around 2.0 reported earlier by Richards (1965), whose data show a great deal of scatter. In terms of the jet radius in the Lagrangian model, $R = \sqrt{R_h R_v}$, this would translate to a puff spread rate of $dR/dz = \sqrt{2} \times 0.36 = 0.39$. The centerline dilution varies as $S_c \sim z_d^2$, with a dilution constant of 0.46. The ratio of flux-averaged to centerline minimum dilution, $\overline{S}/S_c$, is on average around 1.45. Unlike previous experiments of line puffs (Richards 1965), the constants for the similarity laws can be successfully correlated from the *collective* experimental data. The empirical trajectory and mixing rates obtained by different investigators are shown for comparison in Tables 1 and 2 (Figs. 22 and 23).

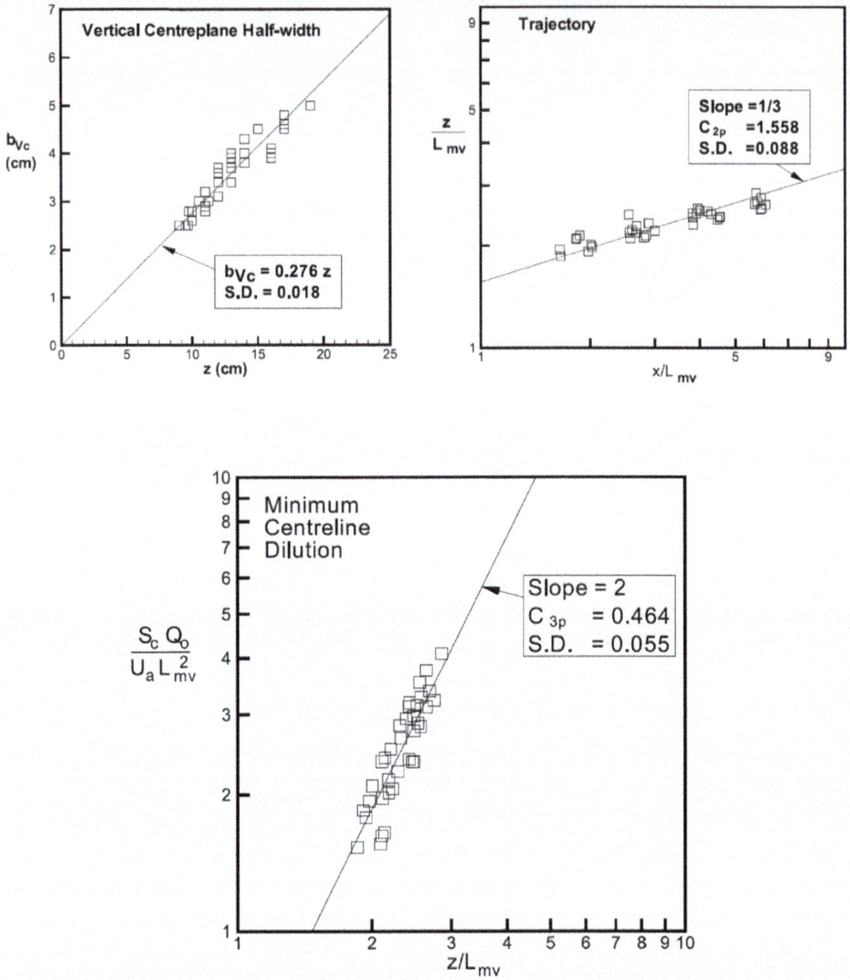

Fig. 20 Comparison of self-similar ALP relations with measured width, trajectory, and dilution in ALP experiments by Wong (1991)

Having defined the location of the ALP, it is possible to map the velocity field within the visually observed puff cross-section. Figure 24 shows a typical three minute-averaged $v - w$ velocity distribution in the jet cross-section. Tracer concentration contours at $C/C_m = 0.05,\ 0.135$ and 0.25 are overlapped on the figure to show the extent of the scalar field. Figure 25 shows the ratio of the measured vertical momentum flux M_p to initial vertical momentum M_{lo} of the equivalent line puff (= ratio of momentum fluxes M_v/M_{vo}) plotted against dimensionless distance $x^* = x/L_x$ $(= t^* = t/t_c)$. In the developed stage, $x^* \geq 75$, the ratio shows a constant value of about 0.42 (s.d. = 0.058). This demonstrates the vertical momentum of the

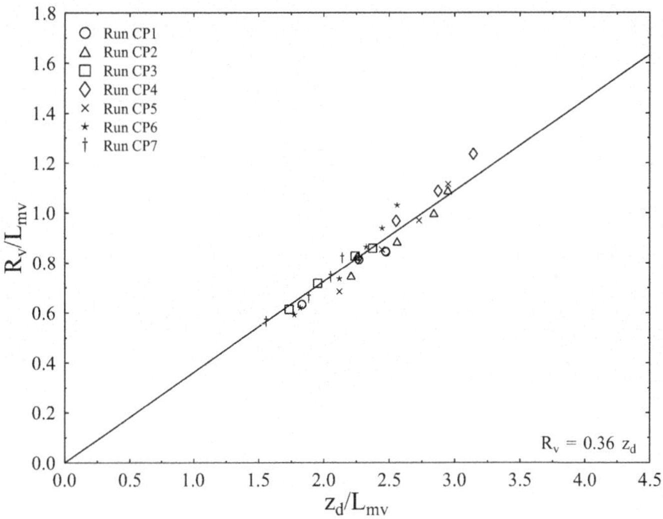

Fig. 21 Puff half-width vs puff trajectory

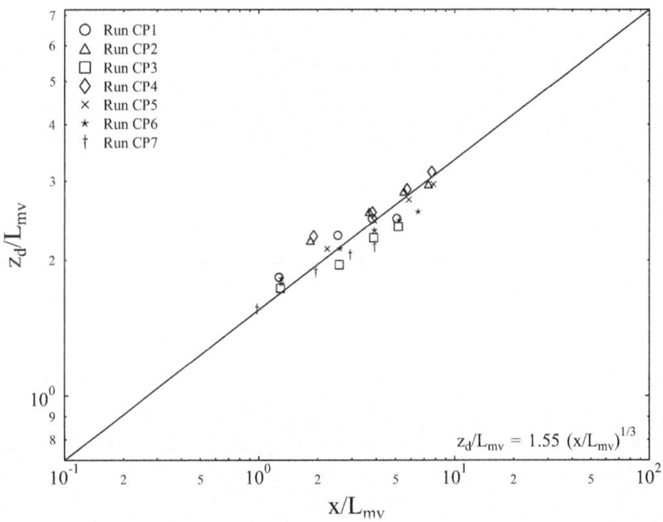

Fig. 22 Puff trajectory vs horizontal distance

advected line puff is conserved once the flow is developed. The vertical momentum contained in the puff region bounded by the puff edge can be estimated to account for only about 91% of the total vertical momentum of the puff. The ratio of total vertical momentum of the advected line puff in the developed phase to the initial vertical momentum is equal to $0.424/0.91 \simeq 0.47$. In other words, the vertical jet momentum flux within the visible jet boundaries has been directly determined to

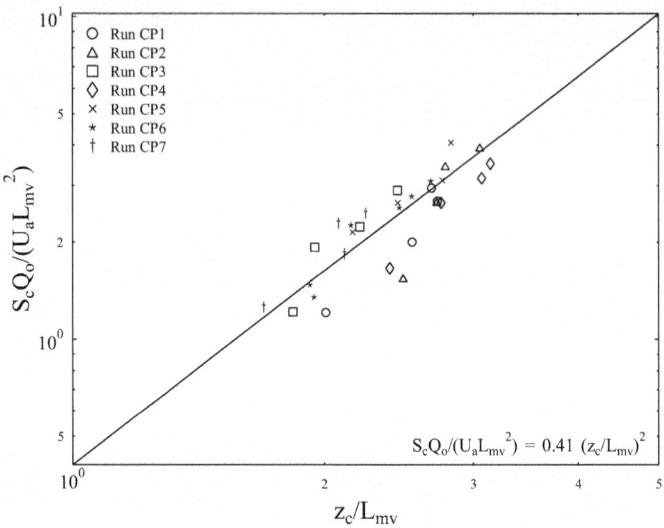

Fig. 23 Minimum centerline dilution vs centerline trajectory

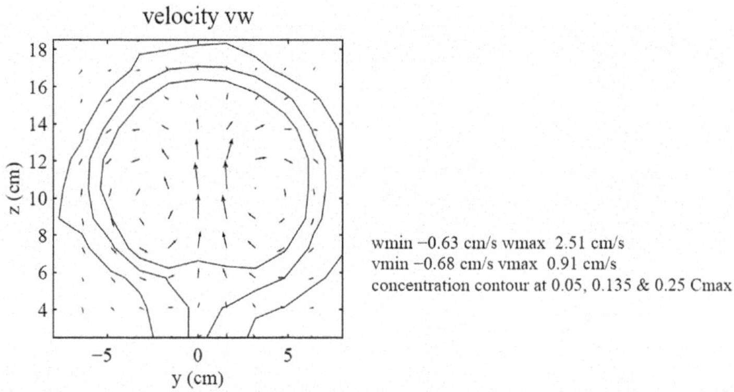

Fig. 24 Typical time-averaged velocity field in vertical section of advected line puff ($U_o = 57.5$ cm/s; $U_a = 5.0$ cm/s; $\theta_o = 85^o$; $D = 0.52$ cm, $t/t_c = 147$)

be approximately half of the jet discharge momentum flux; the existence of added mass has been demonstrated. This experimentally determined puff momentum ratio of about 0.5 can be compared to values of $0.48 - 0.50$ predicted by the turbulence model calculations.

Velocity measurements have also been performed for a vertical jet in crossflow ($\theta_o = 90^o$). The results show good agreement with those of the advected line puff; a momentum loss coefficient of about 0.5 has also been observed (Chu 1996).

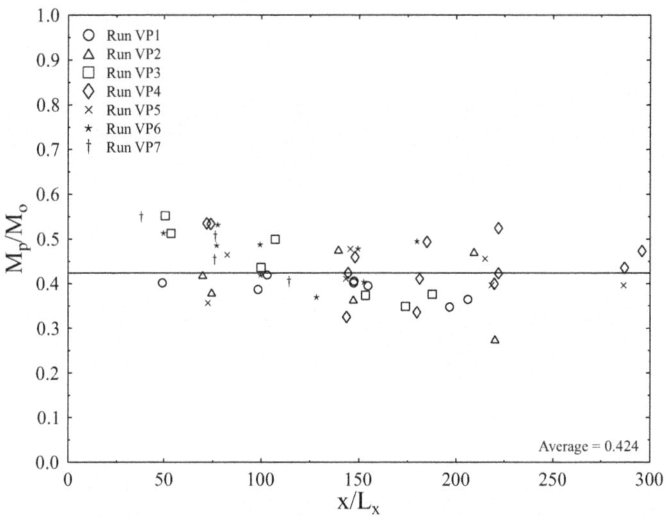

Fig. 25 Ratio of measured puff vertical momentum to initial jet momentum

5.2 *Turbulence Properties*

The measured horizontal and vertical profiles of the rms concentration fluctuation intensity, C_{rms}, are shown in Fig. 26. It is seen the turbulence intensity is much greater than that of the straight jet, with a value of around 0.4 in the core. Figure 26a shows the centerline vertical profile of C_{rms} normalized by the centerline maximum concentration C_c. An asymmetric distribution with double peaks can be noted. The peak in the upper part of the puff is located at $(z - z_c)/b_{vc} \approx 0.75$, with turbulence intensity of 0.35-0.5. The lower peak is located at $(z - z_c)/b_{vc} \approx -0.6$, with intensity in the range of 0.26-0.6. The peak in turbulent intensity in the puff front (upper part) is consistent with turbulence model predictions (Lee *et al.* 1996). The peak in the bottom part appears to reflect the significant entrainment of ambient fluid in the puff rear (cf Figs. 1 and 19). Figure 26b shows the horizontal profile of C_{rms}, normalized by the cross-section maximum concentration C_{cm}, at the elevation of the center of mass of the puff. The turbulent intensity is characterized by two peaks located at $y/b_{vc} \approx \pm 0.8$, with values in the range of 0.3-0.52. In view of such fluctuations, it is not surprising then that the instantaneous maximum concentration can be two times higher than that of the centerline maximum concentration, while the minimum concentration stays close to zero (Fig. 27). Figure 28 shows the horizontal profile of the concentration intermittency at the level of the center of mass of the advected line puff. The intermittency has double peaks located at $y/b_{vc} \approx \pm 0.5$, and has a value of 0.7-1.0 near the centerline. In contrast to a coherent jet core (with an intermittency of one), the measured intermittency shows that parts of the puff core has an intermittency of less than 1; at these locations there is always a probability of ambient fluid present.

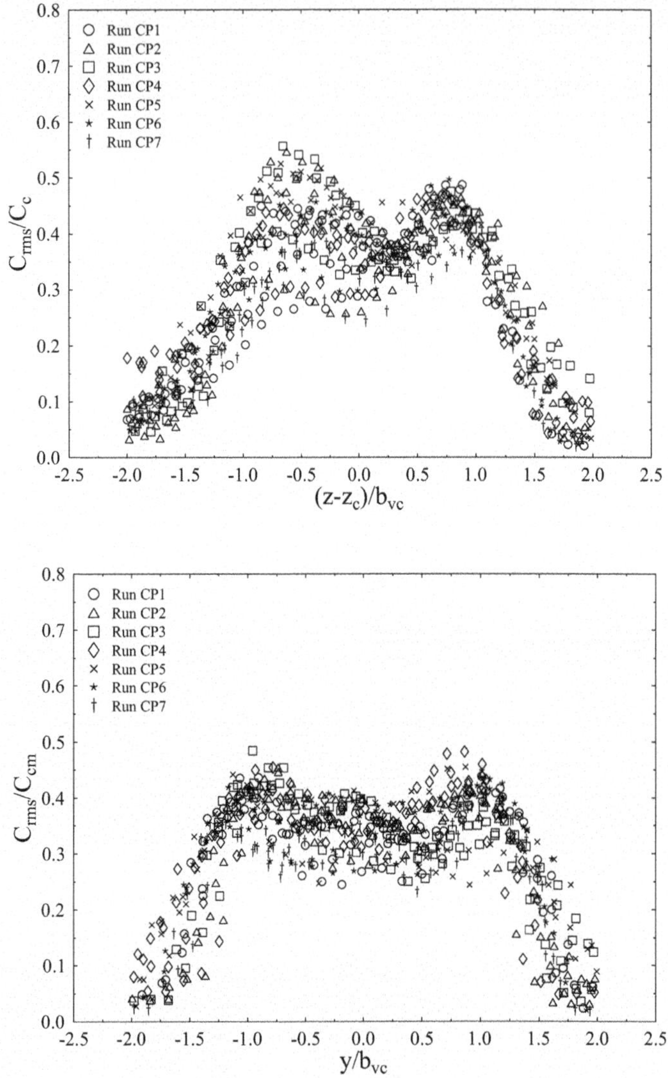

Fig. 26 RMS concentration fluctuation of advected line puff: **a** vertical centerline profile (top); **b** horizontal profile at level of center of mass (bottom)

6 Practical Application

For prediction of jet mixing in a crossflow, the numerical model results can be cast in a more convenient form. For completeness the well-known solution for the MDNF is also included.

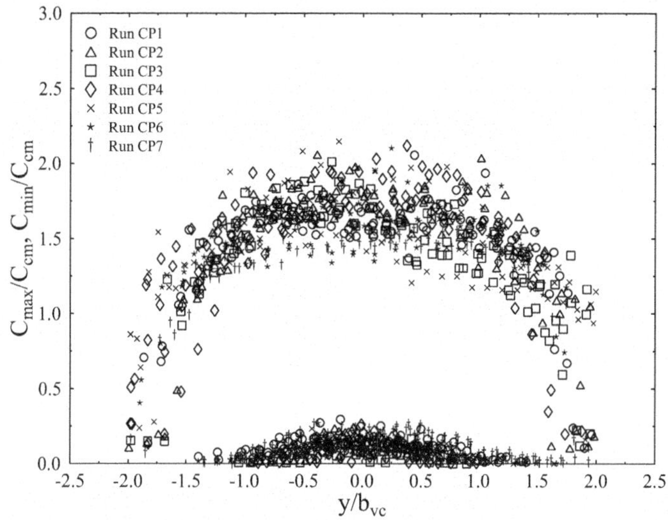

Fig. 27 Horizontal profile of instantaneous maximum and minimum concentration of ALP

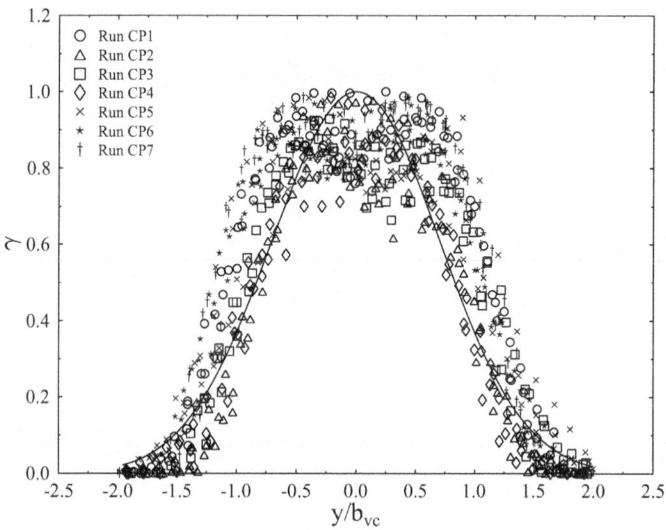

Fig. 28 Horizontal profile of intermittency of ALP

Momentum-dominated Near Field (MDNF)

The classical relations for the characteristic properties of a round jet in stagnant fluid are (Chap. 2):

$$W_m = 7.0\ M_{vo}^{1/2}\ z_c^{-1} \tag{29}$$

$$b_{vc} = 0.114\, z_c \tag{30}$$

$$S_c = 0.16\, \frac{M_{vo}^{1/2}\, z_c}{Q_o} \tag{31}$$

In the MDNF the jet is only weakly advected; mixing is similar to that of a jet in stagnant fluid. Note that $\frac{dz}{dx} = \overline{W}/U_a$, and for a Gaussian profile, the ratio of centerline maximum to average velocity can be shown to be $W_m/\overline{W} = 2$ by equating mass and momentum fluxes. The following equations can then be obtained for the advected jet: $z/L_{mv} \leq 1$**(MDNF):**

$$\frac{z_c}{L_{mv}} = 2.65\, (\frac{x}{L_{mv}})^{1/2} \tag{32}$$

$$b_{vc} = 0.114\, z_c \tag{33}$$

$$\frac{S_c Q_o}{U_a L_{mv}^2} = 0.16\, (\frac{z_c}{L_{mv}}) \tag{34}$$

For example, the trajectory equation, Eq. 32, follows directly from invoking a kinematic relation and using Eq. 29 (with $\overline{W} = W_m/2$). The above simple advected jet model for the MDNF is verified by the detailed LIF experiments of a vertical jet in weak crossflow (Davidson and Pun 1999) which give a trajectory coefficient of 2.68–very close to the value of 2.65 given in Eq. 32. This supports the use of the average velocity in defining the jet trajectory (see also related discussion on coflow jet formulation in Chap. 6 and BDNF in Chap. 8). Jet trajectory predictions by two Lagrangian models based on the use of average velocity are also in excellent agreement with data for a wide range of situations (see Chaps. 9 and 10).

If one thinks of jet mixing as a diffusion process purely effected by small scale eddies (say of the order of 0.1 b), then it may appear that the jet trajectory (defined by the locus of centerline maximum concentration) is determined by the centerline maximum velocity. The experimental results however suggest that the jet mixing is governed by the large scale motion (the 'dominant eddy') which is responsible for the mass transfer between the center region and the edge of the jet, while the many small eddies (carried along by the dominant eddy) smooth the sharp concentration gradients created by the large scale motion. The fluid and scalar material are transported at a range of velocities with the mean jet path defined by the average of these velocities.

According to theory, the MDNF asymptotic flow regime is characterized by $z/L_{mv} \ll 1$; in practical application, however, experiments have shown that $z/L_{mv} \leq 1$ can often be used to delimit the region. It should be noted that none of the asymptotic relations are strictly applicable in the transition region.

Momentum-dominated Far Field (MDFF)

Based on the ALP numerical model and experimental results, the following equations for the advected line puff are recommended: $z/L_{mv} > 1$**(MDFF):**

$$\frac{z_c}{L_{mv}} = 1.56\,(\frac{x}{L_{mv}})^{1/3} \tag{35}$$

$$b_{vc} = 0.28\,z_c \tag{36}$$

$$\frac{S_c Q_o}{U_a L_{mv}^2} = 0.46\,(\frac{z_c}{L_{mv}})^2 \qquad \text{or}$$

$$S_c = 0.46 \frac{U_a\, z_c^2}{Q_o} \tag{37}$$

The jet in crossflow has also been studied using a semi-empirical approach by laboratory and field experiments of a Canadian river (Hodgson and Rajaratnam 1991). It can be shown that the dilution prediction for the asymptotic stage can be expressed as a function of distance downstream (x) from a jet discharge in the following form (Lee 1993):

$$S_c = \frac{C_o}{C_m} = 0.64\,(\frac{W_o\, x}{U_a D})^{2/3} \tag{38}$$

Figure 29 shows the comparison of Eq. 38 with the experiments of Hodgson and Rajaratnam and their field dilution data.

Example 7.1 Consider a single jet discharge into a river with $Q = 0.0138\ m^3/s$. The jet diameter is $D = 0.1$ m. The river depth is $H = 20\,m$, and the river velocity is $U_a = 0.2\,m/s$. The initial density difference (buoyancy) of the effluent is negligible. Estimate the initial dilution, vertical jet velocity, and width at an elevation of $z = 12$ m. Also estimate the corresponding downstream distance of the jet at this location.

Solution: The discharge is modelled as a turbulent jet in crossflow. Compute jet velocity and momentum fluxes and obtain the crossflow momentum length scale L_{mv} as:

$$W_o = \frac{0.0138}{\pi \times 0.1^2/4} = \underline{1.76}\,m/s$$

$$M_{vo} = Q_o W_o = 0.0138 \times 1.76 = \underline{0.0243}\,m^4/s^2$$

$$L_{mv} = \frac{M_{vo}^{1/2}}{U_a} = \frac{\sqrt{0.0243}}{0.2} = \underline{0.78}\,m$$

At an elevation of $z = 12$ m, $z/L_{mv} = 12/0.78 = 15.4 \gg 1$, apply the MDFF equations to obtain the dilution, trajectory, and width.

$$S_c = 0.46 \frac{0.2 \times 12^2}{0.0138} = \underline{960}$$

This occurs at a distance given by the trajectory equation:

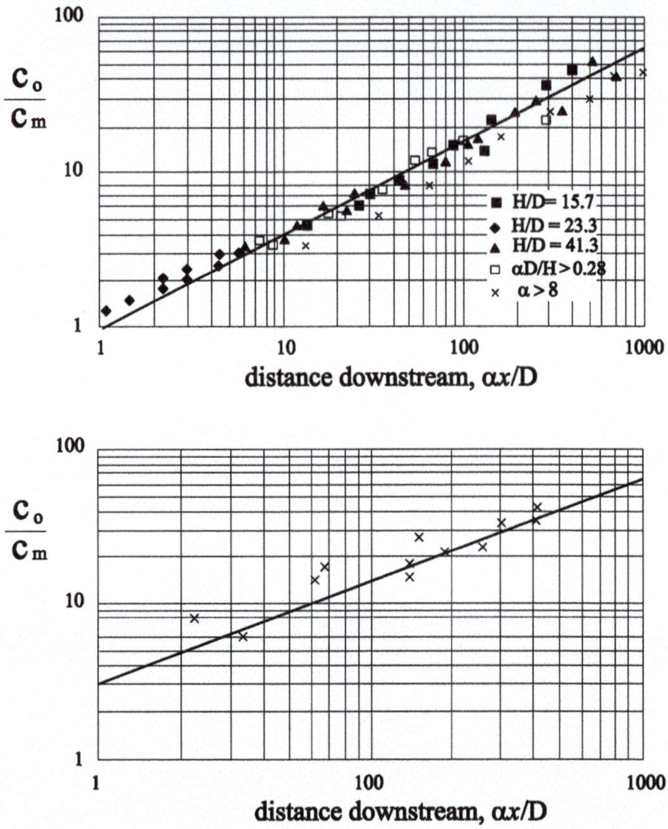

Fig. 29 Comparison of Eq. 38 with the laboratory and field data (upper/lower plot respectively) of Hodgson and Rajaratnam (1992); $\alpha = W_o/U_a$

$$x = 0.78 \times (15.4/1.560)^3 = \underline{750}\ m$$

Note the relatively long distance that it takes for the jet to travel to this elevation. Using $B = \sqrt{R_h R_v} = 0.4z = \underline{4.8}\ m$ and $R_h/R_v = 1.2$, the vertical half-width is $R_v = 4.8/\sqrt{1.2} = \underline{4.4}\ m$, and $R_h = \underline{5.3}\ m$. Typically a good estimate of the visible width of a jet is approximately $\sqrt{2}b$; this suggests that the bent-over jet is quite well-mixed over the cross-section at this location. The vertical velocity at this location can also be shown to have decayed to negligible values:

$$W_m = 1.33 \times (0.0243/0.2)^{1/3}/(750/0.2)^{2/3} = \underline{2.73 \times 10^{-3}}\ m/s$$

The efficiency of mixing in a current ($S_c = 960$) can best be appreciated by computing the stillwater dilution at this elevation (Eq. 31):

$$S_c(U_a = 0) = 0.16 \times \frac{0.0243^{1/2} \times 12}{0.0138} = \underline{21.7}\,!$$

Alternatively, the moving water dilution for this problem can be obtained from the semi-empirical equation based on the field data of Hodgson and Rajaratnam (Eq. 38):

$$S_c = 0.64 \times (\frac{1.76 \times 750}{0.2 \times 0.1})^{2/3} = \underline{1046}$$

Example 7.2 A town discharges a highly toxic flow of $Q_o = 0.442\ \mathrm{m^3/s}$ through a 0.92 m square culvert at a river bank. The river depth is $H = 7.6m$, and the design river velocity is $U_a = 0.0137\ \mathrm{m/s}$. The effluent is just submerged below the surface. Estimate the initial dilution 20 m downstream from the discharge point.

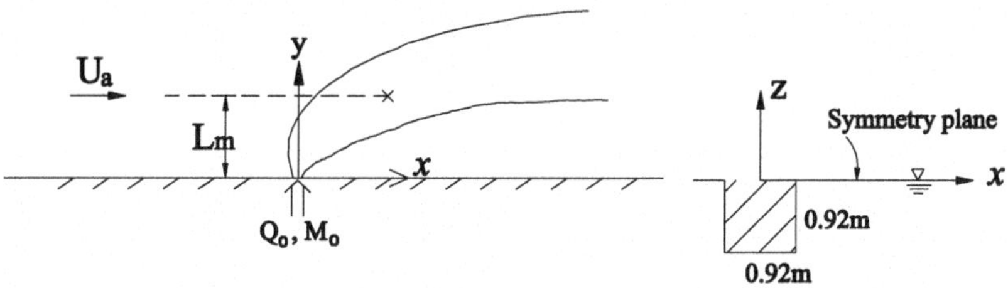

Solution: The toxic effluent is discharged as a surface jet. Assuming zero momentum and mass fluxes across the free surface, the velocity and concentration gradient normal to the free surface is zero. The surface jet can be modelled as one-half of a momentum jet in crossflow, with the free surface approximating a plane of symmetry. The jet velocity, momentum and volume fluxes can be computed to be: $U_o = 0.52\ \mathrm{m/s}$, $Q_o = 2 \times 0.442\ \mathrm{m^3/s}$, $M_o = 0.884 \times 0.52 = 0.46\ \mathrm{m^4/s^2}$ The crossflow momentum length scale is then $L_m = M_o^{1/2}/U_a = 49.5$ m. The equivalent hydraulic diameter is $D = \sqrt{4/\pi \times 2 \times 0.92^2} = 1.47$ m. At $x = 20$ m, use Eq. 35 to obtain $z/L_{mv} = 1.2$. Hence use the MDFF Eq. 38 to obtain: $S_c = 0.64(\frac{0.52}{0.0137}\frac{20}{1.47})^{2/3} = \underline{41}$.

7 Summary

The bent-over phase of a turbulent jet in crossflow has been analysed numerically and experimentally using the advected line puff (ALP) analogy. Numerical computation results obtained using 2D and 3D turbulence models are compared with probe-based as well as non-intrusive LIF and LDA measurements of the advected line puff in the laboratory. Simple equations for mixing analysis are developed. The spread rate and added mass coefficient for the advected line puff are:

$$\text{Spreading rate}: \quad \frac{dR}{dz} = \beta_n = 0.4$$

$$\text{Added mass coefficient:} \quad k_n = 1.0$$

Chapter 8
Plume in Crossflow: Advected Line Thermals

A variety of turbulent flows observed in nature are driven by sources of buoyancy. The interaction of such sources with the surrounding flow is generally characterized as plumes in crossflow. Figure 1 is an example. The plume in crossflow is produced in the laboratory by discharging buoyant water from a pipe of diameter $D = 1$ cm at a velocity $U_o = 0.25$ m/s into a crossflow of velocity $U_a = 0.063$ m/s. The LIF (Laser Induced Fluorescence) images in the figure show the structure of the turbulence and the concentration profile on the plane of symmerty. The initial relative density difference of 2.57% between the source fluid and the crossflow is created by salinity differences ($\Delta\rho_o/\rho_a = 0.0257$). The source densimetric Froude number $Fr = U_o/\sqrt{g\Delta\rho_o/\rho_a D} = 5.0$ is quite small in this case. Buoyancy becomes the dominant effect in a short distance from the source. The buoyant effluent is seen in the figure to contract slightly and then mixed with the surrounding fluid to form the plume in cross flow. Close to the source, the flow behaves essentially as a plume. In the far field, the plume is bent-over to form the advected line thermal. In a co-ordinate system moving with the crossflow, the surrounding fluid is stationary and the plume rises essentially as a line thermal in stagnant fluid. A cross-section of the thermal is a vortex pair as shown in Fig. 2. The average tracer concentration profile in the cross-section of the thermal is similar to that of the advected line puff examined in the previous chapter. While the average flow pattern is similar, the eddying motion in the thermal is more vigorous—the turbulent intensity is much greater when compared with the puff. The study of the thermal is most significant as most of the flows observed in lakes, oceans and the atmosphere are driven by buoyancy. Given sufficient depth for the buoyancy force to act, any buoyant jet in crossflow will tend to behave like plumes and thermals in the ultimate stage of mixing with the surrounding fluid.

Supplementary Information The online version contains supplementary material available at https://doi.org/10.1007/978-3-032-01187-9_8.

J. H. W. Lee et al., *Turbulent Jets and Plumes*,
https://doi.org/10.1007/978-3-032-01187-9_8

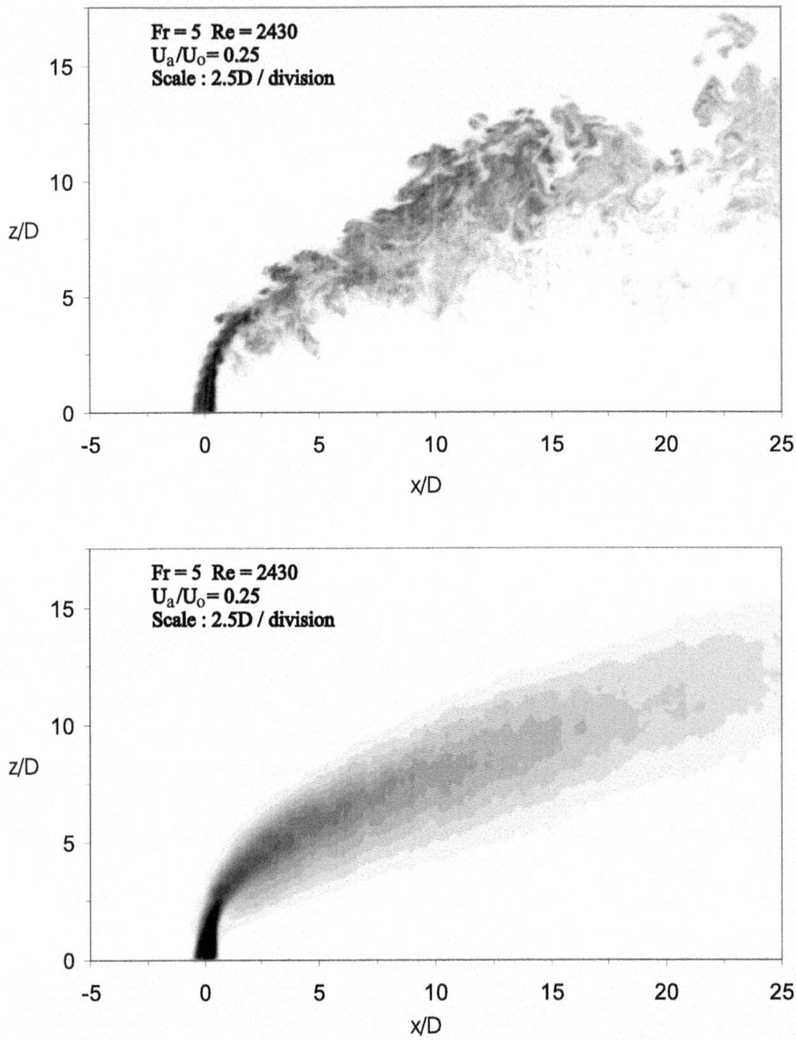

Fig. 1 A vertical plume in crossflow. $Fr = 5$, $U_o/U_a = 4$

Fig. 2 Tracer concentration distribution in cross section of bent-over plume in crossflow ($x/D = 12$, $C_m/C_c = 1.2$)

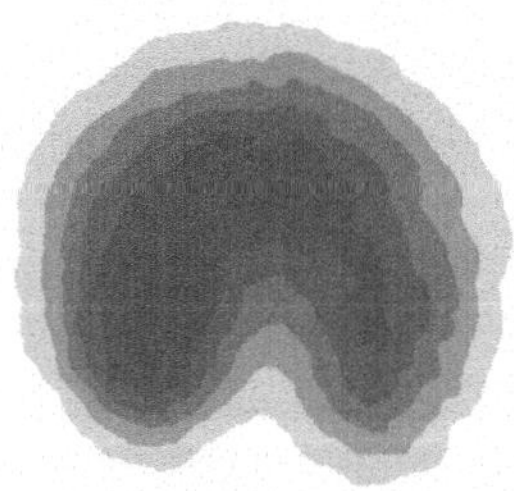

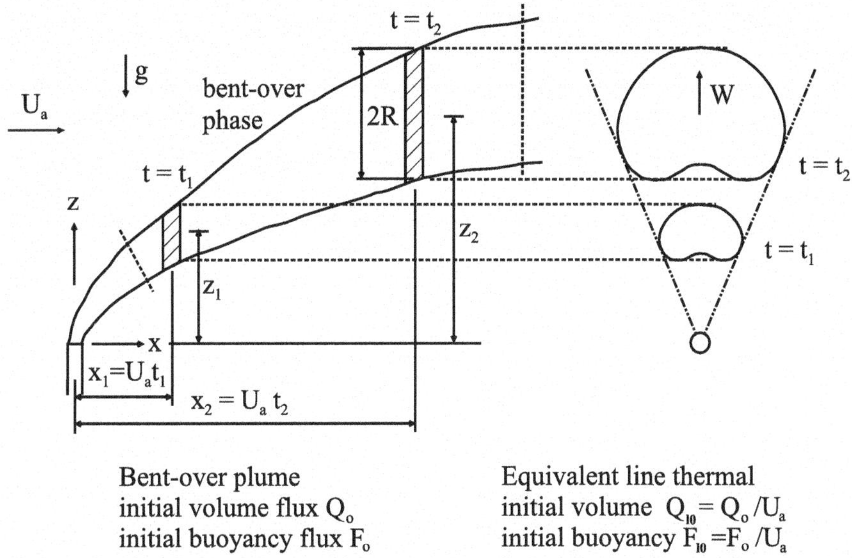

Fig. 3 Schematic of a plume in crossflow showing the Lagrangian element and its cross section in the analog of the advected line thermal

Buoyant plumes in crossflow have been extensively studied. Most of the previous studies are based on probe-based measurements of the trajectory and the dilution on the center plane of symmetry (e.g. Scorer 1959, 1978; Csanady 1965; Chu and Goldberg 1974; Fischer *et al.* 1979). Asymptotic relations of the plume trajectory in the near and far fields have been established from these measurements. Detailed calculations and measurements in the cross-section of a bent-over plume have also been made (e.g. Lilly 1964; Yih 1981; Knudsen 1988; Wong 1991; Wood 1993; Gaskin and Wood 2001; Chen and Lee 2002).

In this chapter, the laboratory data are presented and compared with the results of the computational models. The plume-in-crossflow problem is examined first using a length scale analysis. In the near field, the effect of the discharge buoyancy dominates. In the far field, the plume behaves essentially as a line thermal. In the remainder of the chapter, the emphasis is given to the flow of the line thermal. Spreading coefficient and added mass coefficient are introduced in a one-dimensional (1D) model as the parameters characterizing the turbulent motion and the irrotational motion surrounding the line thermal. Two-dimensional (2D) and three-dimensional (3D) mathematical models are developed to compute the turbulent flow. Measurements and numerical computations on the flow and tracer concentration field of advected line thermals are presented. Spreading coefficient and added mass coefficient, the parameters of the 1D model, are evaluated using the 2D and 3D models and the data obtained from the laboratory measurements.

1 Length Scales and Regimes

The problem of the plume in cross flow is delineated in Fig. 3. The plume is shown to be produced by a vertical round source of diameter D, initial velocity U_o, and tracer concentration C_o discharging into a steady uniform crossflow U_a in the horizontal $+x$ direction. The initial volume flux of the discharge is $Q_o = U_o \pi D^2/4$. The specific buoyancy flux is $F_o = Q_o g_o'$. Neglecting the effect of the initial momentum flux, the behaviour of the plume is determined by its buoyancy flux F_o and the crossflow velocity U_a. Asymptotic regimes of the plume can be defined in terms of the length scale

$$L_b = \frac{F_o}{U_a^3}. \tag{1}$$

The vertical velocity in the plume, W, varies with the height z above the source as a minus one-third power law: $W \sim F_o^{1/3} z^{-1/3}$. As the vertical velocity diminishes with height, the plume is bent over and moves primarily in the horizontal direction of the crossflow. The length scale L_b, defined by the buoyancy flux F_o and the crossflow velocity U_a, is the height of the plume when the buoyancy-induced velocity W has decreased to a value comparable to the velocity of the crossflow ($W \sim U_a$ and $z \sim F_o/U_a^3$). Asymptotic flow regimes are defined by the dimensionless height z/L_b. For dimensionless height $z/L_b = zU_a^3/F_o \ll 1$, the buoyancy-induced velocity W is much greater than the crossflow velocity. In this near-field region, the plume is essentially vertical and the turbulent motion in the region is similar in structure to that of the plume in stagnant fluid, but slightly advected. For dimensionless height $z/L_b \gg 1$, the vertical velocity W is small compared with the horizontally velocity of the crossflow, U_a. The behaviour in this far field is similar to that of an advected line thermal (Scorer 1979; Wood 1993). Wright (1977a) used the terms BDNF (buoyancy-dominated near field) and BDFF (buoyancy-dominated far field) to characterize the "advected plume" and the line thermal regimes respectively. Plume in stagnant fluid was considered in Chap. 3. The transition from BDNF to BDFF will be an example of the general formulation to be described in Chaps. 9 and 10. This chapter concentrates mainly on the BDFF.

1.1 Line-Thermal Analogy for BDFF

The plume in the BDFF can be modeled using a line-thermal analogy. Consider the control volume in Fig. 4, balancing the tracer mass flux of the inflow and the mass flux of the outflow through a cross-section of the plume, we have $Q_o c_o = \int_{x-section} u\, c\, dA$, where u and c are the velocity and the concentration at the cross-section respectively. Assuming a linear equation of state, the buoyancy flux $F(x)$ passing a vertical section is then equal to the initial buoyancy flux, with $F(x) =$

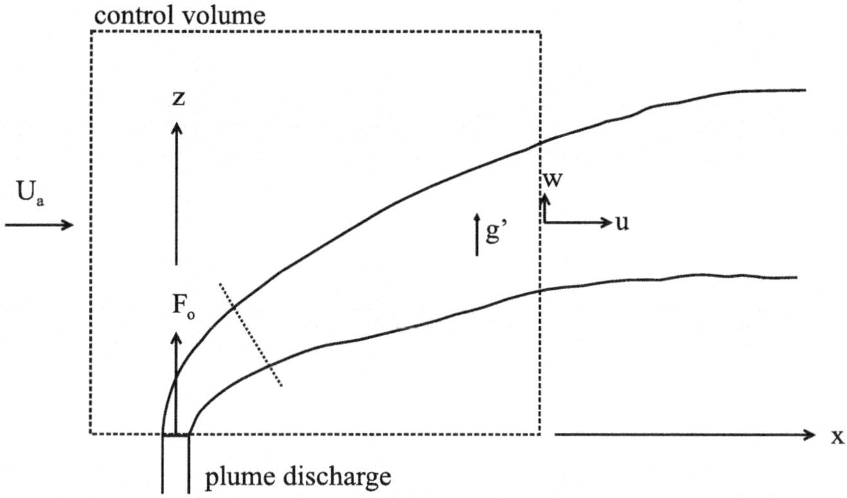

Fig. 4 Control volume for plume in crossflow. The buoyancy flux entering the control volume is F_o. The buoyancy flux through a cross-section perpendicular to the crossflow is $F(x) = \int_A g'(udA)$

$\int_{x-section} u\, g'\, dA = F_o = Q_o g'_o$, where g' = reduced gravity. In the BDFF, $u \simeq U_a$. Hence, $F_o \simeq U_a \int_A g'\, dA$. The flow in successive vertical x-sections can then be approximated by a line thermal with buoyancy $F_{lo} = \int_{x-section} g'\, dA = F_o/U_a$ (per unit x-length) released in the $+z$ direction, and initial volume $Q_{lo} = Q_o/U_a$.

For an observer in a coordinate system moving with the crossflow, the line thermal rises in a stagnant ambient fluid and the plume is issuing the buoyancy to the line element of the thermal at a steady rate. As far as the mixing in the BDFF is concerned, the action of the plume over a period of time Δt is to apportion the source of buoyancy $F_o \Delta t$ to the line thermal over a length equal to $U_a \Delta t$. Therefore, the buoyancy per unit length of the line thermal is F_o/U_a. The plume distributes the buoyancy in a continuous manner to the thermal. The Lagrangian element in the line thermal has a length equal to $U_a \Delta t$ and a buoyancy per unit length of the element equal to $F_o \Delta t / U_a \Delta t$. The bent-over plume at successive cross-sections, located at $x = x_1$ and x_2, corresponds to the line thermal observed at successive times (t_1, t_2), respectively. The space and time relations are $x_1 = U_a t_1$ and $x_2 = U_a t_2$ via the Galilean transformation. This thermal analog of the plume in BDFF is to be confirmed by the experiments and is expected to hold in the bent-over stage of the plume, where $z \geq L_b$. The flow according to the analogy is referred to in the present context as ALT (advected line thermal).

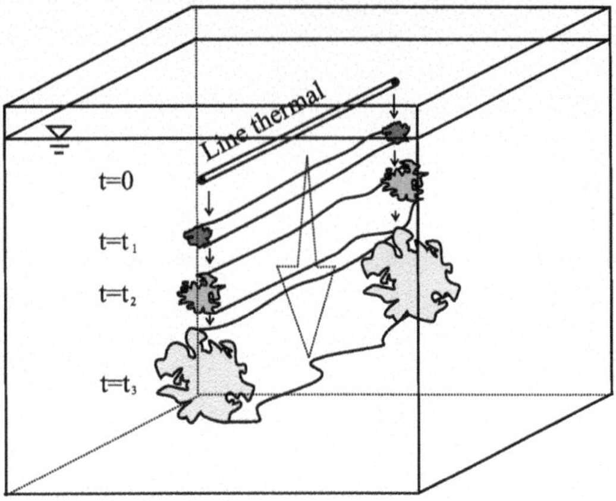

Fig. 5 Line thermal generated by releasing a cylinder of negatively buoyant fluid instantaneously in a tank

1.2 *Similarity Variables for the Line Thermal*

A line thermal is the two-dimensional motion formed when a line of buoyant fluid is released from rest instantaneously. This is illustrated in Fig. 5 for an experiment done upside down (Richards 1963). An isolated mass of dyed negatively buoyant fluid is released from a cylindrical trough extending perpendicularly between parallel vertical walls in a water tank. The release is located sufficiently far away from boundaries.

Consider a line thermal with initial volume Q_{lo} (per unit x-length) and buoyancy F_{lo} at $t = 0$ (Fig. 6). The initial tracer concentration is C_o. Figure 6 shows the two-dimensional unsteady flow in a cross section on the y-z plane of the thermal.

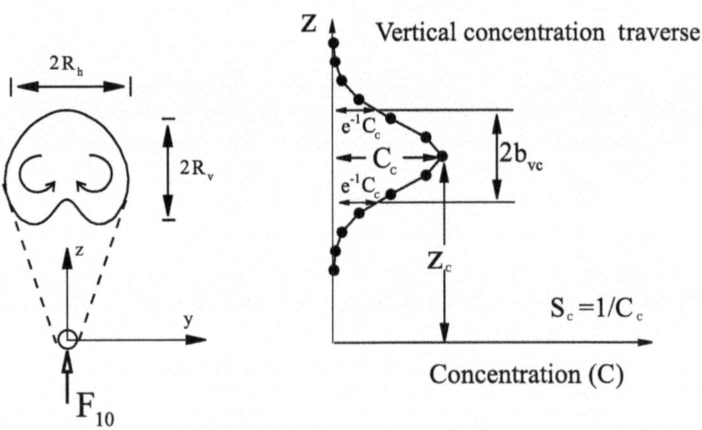

Fig. 6 Definition sketch showing the cross-section of the line thermal and the concentration profile on the center plane of symmetry

The thermal motion is characterized by a vortex pair with its concentration maxima located at the turbulent cores. The geomerty of the thermal is characterized by the radii of the cross section R_v and R_h, the leading edge at the front z_f, the mass center z_d. The motion of the thermal is characterized by its vertical velociy, W, and the concentration profile on the center plane of symmetry by the half-width b_{vc}, the maximum concentration C_c and its location Z_c. The flow in the thermal is driven by the line source buoyancy. The thermal characteristics are function of the buoyancy flux F_{lo} and time t. The initial volume, Q_{lo}, could also be important. The time scale

$$t_c = \frac{Q_{lo}^{3/4}}{F_{lo}^{1/2}} = \frac{Q_o{}^{3/4}}{U_a^{1/4} F_o^{1/2}} \tag{2}$$

and the corresponding length scale

$$L_x = U_a t_c = \frac{(Q_o U_a)^{3/4}}{F_o^{1/2}} \tag{3}$$

are measures of the relative importance of the initial volume Q_{lo} and buoyancy flux F_{lo}. If the flow is self-similar, the following relations for velocity and width are expected from dimensional analysis:

$$\frac{W_m}{W_r} \sim f(\frac{t}{t_c}) \tag{4}$$

$$\frac{b}{Q_{lo}^{1/2}} \sim f(\frac{t}{t_c}) \tag{5}$$

where W_m and W_r stand for the maximum and reference vertical velocity of the thermal respectively. The dimensionless time is $t^* = t/t_c$. The vertical velocity of the thermal is proportional to $W_r = F_{lo}^{1/2}/Q_{lo}^{1/4}$. The effect of the initial volume Q_{lo} is negligible for fully developed thermal at large time. The asymptotic relations for velocity, width and trajectory of the thermal for $t/t_c \gg 1$ are:

$$W_m \sim F_{lo}^{1/3} t^{-1/3} \quad \text{or} \quad \frac{W_m}{W_r} \sim (t^*)^{-1/3} \tag{6}$$

$$b \sim F_{lo}^{1/3} t^{2/3} \tag{7}$$

$$b \sim z \qquad \text{or} \qquad z \sim Q_{lo}^{1/2} (t^*)^{2/3} \tag{8}$$

Conservation of tracer mass also implies:

$$S_c = \frac{C_o}{C} \sim \frac{b^2}{Q_{lo}} \sim \frac{F_{lo}^{2/3} t^{4/3}}{Q_{lo}} \sim \frac{z^2}{Q_{lo}} \tag{9}$$

Given the equivalence between successive vertical x-sections of a bent-over plume and the line thermal at corresponding times $t = x/U_a$, the following relations can be obtained for the spreading rate, trajectory, and dilution of the 3D bent-over plume in terms of the crossflow buoyancy length scale L_b.

$$b_{vc} \sim z_c \tag{10}$$

$$\frac{z_c}{L_b} \sim (\frac{x}{L_b})^{2/3} \tag{11}$$

$$\frac{S_c Q}{U_a L_b^2} \sim (\frac{z_c}{L_b})^2 \tag{12}$$

$$\frac{W_r}{W_m} \sim (x^*)^{1/3} \tag{13}$$

where b_{vc} is the vertical centerline half-width defined by the $e^{-1}C_c$ points, and z_c is the vertical location defined by the maximum centerline concentration C_c. $S_c = C_o/C_c$ is the centerline minimum dilution, as illustrated in Fig. 6; and the dimensionless downstream distance $x^* = x/L_x$. For the sake of reference, Eqs. 11 and 12 can be expressed alternatively as:

$$z_c \sim (Q_o/U_a)^{1/2} (x^*)^{2/3} \tag{14}$$

$$S_c \sim z_c^2/(Q_o/U_a) \tag{15}$$

The proportionality constants in the above similarity relations can be derived by best-fitting experimental data or numerical model results.

2 1D Model of Line Thermal

Similar to the line puff, it is instructive to first analyse the line thermal motion by a one-dimensional model. The formulation is based on the Lagrangian spreading hypothesis and the concept of the dominant eddies described in Chap. 4. The dominant eddies in the thermal rises in a stagnant environment. The elevation of the dominant eddies is z. The rising velocity is

$$W = \frac{dz}{dt}. \tag{16}$$

The size of the dominant eddies is defined by R_v and R_h as shown in Fig. 6. The thermal radius is assumed to increase at a rate proportional to the rising velocity:

$$\frac{dR}{dt} = \beta_n W, \tag{17}$$

where $R = \sqrt{R_h R_v}$ is the radius. The proportionality constant is the spreading coefficient, β_n. Since $W = dz/dt$, it follows the thermal radius increases linearly with height (neglecting initial radius):

$$R = \beta_n z \tag{18}$$

The volume per unit length of the line thermal is $Q_l = \pi R^2 = \pi(\beta z)^2$. The momentum associated with this thermal volume has two parts. The part associated with the turbulent motion is $\pi R^2 W$. The part associated with the irrotational motion is $k_n \pi R^2 W$. The total momentum is

$$M_t(t) = (1 + k_n)\rho\pi R^2 W; \tag{19}$$

the added mass coefficient k_n is the fraction of the momentum in the irrotational surrounding. The thermal momentum increases at a rate equal to the buoyancy force, i.e.,

$$\frac{dM_t}{dt} = \frac{d}{dt}[(1 + k_n)\pi R^2 W] = F_{lo} = \frac{F_o}{U_a} \tag{20}$$

where F_o/U_a is the buoyancy force on unit length of the line thermal. Integrating Eq. 20 with time gives

$$(1 + k_n)M_t = F_{lo}t \tag{21}$$

Further integration with time leads to:

$$R = (\frac{3\beta_n F_{lo}}{2(1 + k_n)\pi})^{1/3} t^{2/3} \tag{22}$$

$$W = (\frac{4F_{lo}}{(1 + k_n)9\pi\beta_n^2})^{1/3} t^{-1/3} \tag{23}$$

$$z = (\frac{3F_{lo}}{2(1 + k_n)\pi\beta_n^2})^{1/3} t^{2/3} \tag{24}$$

Tracer mass conservation results in:

$$\frac{C_o}{C} = \frac{\pi R^2}{Q_{lo}} = \frac{\pi R^2 U_a}{Q_o} \tag{25}$$

The line thermal solution translates via a Galilean transformation into the well-known two-third power trajectory law for a bent-over plume in crossflow:

$$z = (\frac{3}{2(1+k_n)\pi\beta_n^2})^{1/3}(\frac{F_o x^2}{U_a^3})^{1/3} \tag{26}$$

or in dimensionless form:

$$\frac{z}{L_b} = C_{2t}(\frac{x}{L_b})^{2/3} \tag{27}$$

where

$$C_{2t} = [\frac{3}{2(1+k_n)\pi\beta_n^2}]^{1/3} \quad \text{and} \quad L_b = \frac{F_o}{U_a^3}. \tag{28}$$

Since $R = \beta_n z$, and the average dilution $S = C_o/C$,

$$S = \frac{\pi\beta_n^2 z^2 U_a}{Q_o}. \tag{29}$$

where z is given by Eq. 27.

There are two model parameters, β_n and k_n. They can be obtained from numerical computations and measurements of the flow and tracer concentration field of the plume in crossflow. The remainder of this chapter explain how these 1D-model parameters are independently determined for consistent prediction of the spread rate, path, and dilution in the BDFF.

3 2D Model of Line Thermal

3.1 Numerical Simulation of Line Thermals

Numerical calculations are performed for the line thermal using a similar procedure as the numerical experiment for the line puff as described in the previous chapter (see also Lee and Chen 2002). Fully three-dimensional calculations for the plume in crossflow also are conducted (Chen and Lee 2002). The results of the two-dimensional calculations using the k-ϵ model are presented in this section.

Computations are based on the following Reynolds-averaged equations that are derived from the assumptions of (1) small density differences, $(\rho_a - \rho_o)/\rho_a \ll 1$, and (2) linear equation of state, $\rho = \rho_a(1 + \beta C)$:

$$\frac{\partial U_i}{\partial x_i} = 0 \tag{30}$$

$$\frac{\partial U_i}{\partial t} + U_j \frac{\partial U_i}{\partial x_j} = -\frac{1}{\rho_a}\frac{\partial p}{\partial x_i} + \frac{\partial \tau_{ij}}{\partial x_j} + \beta C g_i \tag{31}$$

$$\frac{\partial C}{\partial t} + U_j \frac{\partial C}{\partial x_j} = \frac{\partial}{\partial x_j}\left(\frac{\nu_{eff}}{Sc_t}\frac{\partial C}{\partial x_j}\right) \tag{32}$$

where $U_i = (V, W)$ = fluid velocity in y and z direction respectively, p = dynamic pressure, $\beta = \frac{1}{\rho_a}\frac{d\rho}{dC}$ is a constant fluid property, C = concentration, τ_{ij} = stress tensor, is equal to the sum of the viscous and Reynolds-stresses (Eq. 25 of Chap. 7), and $g_i = (0, -g)$ is the gravitational acceleration vector. The gradient transport analogy (Boussinesq hypothesis) is assumed for the stresses and tracer mass (buoyancy) fluxes. The effective viscosity $\nu_{eff} = \nu + \nu_t$, where $\nu_t = C_\mu k^2/\epsilon$ is the turbulent viscosity. k is the turbulence kinetic energy (TKE), ϵ is the dissipation rate (DR) of k; the effective diffusivity is given by ν_{eff}/Sc_t, where Sc_t is the turbulent Schmidt number taken as 0.7 as commonly recommended. The buoyancy-extended two-equation $k - \epsilon$ model (Rodi 1980) is adopted for turbulent closure:

$$\frac{\partial k}{\partial t} + U_i \frac{\partial k}{\partial x_i} = \frac{\partial}{\partial x_i}\left(\frac{\nu_t}{\sigma_k}\frac{\partial k}{\partial x_i}\right) + \tau_{ij}\frac{\partial U_i}{\partial x_j} + G_b - \epsilon \tag{33}$$

$$\frac{\partial \epsilon}{\partial t} + U_i \frac{\partial \epsilon}{\partial x_i} = \frac{\partial}{\partial x_i}\left(\frac{\nu_{eff}}{\sigma_\epsilon}\frac{\partial \epsilon}{\partial x_i}\right) + \frac{\epsilon}{k}\left(C_{1\epsilon}\tau_{ij}\frac{\partial u_i}{\partial x_j} + C_{3\epsilon}G_b\right) - C_{2\epsilon}\frac{\epsilon^2}{k} \tag{34}$$

where the production of k due to buoyancy is given by:

$$G_b = -g_i \frac{\nu_t}{Sc_t}\frac{1}{\rho}\frac{\partial \rho}{\partial x_i} \tag{35}$$

with the following empirical model coefficients: $C_\mu = 0.09$, $C_{1\epsilon} = 1.44$, $C_{2\epsilon} = 1.92$, $\sigma_k = 1.0$, $\sigma_\epsilon = 1.3$. and $C_{3\epsilon} = C_{1\epsilon}$ when $G_b \leq 0$ (and $C_{3\epsilon} = 0$ when $G_b \geq 0$).

The line thermal flow can be revealed via a numerical experiment. Figure 7 shows a volume of water initially at rest in a tank of width W_T and depth H. At time $t = 0$, a significant buoyancy impulse in the vertical (z) direction is applied to a square patch of fluid (the 'colored' patch) of length L_t. The patch is initially labelled at uniform passive scalar concentration C_o corresponding to density ρ_0 and located sufficiently far away from the solid boundaries. The buoyancy gives rise to fluid acceleration and vorticity; turbulent motion is produced and the thermal moves in the direction of the buoyancy. The following parameters are adopted: $W_T = 1.6$ m, $H = 1.2$ m, $L_t = 0.04$ m. $Q_{lo} = 1.6 \times 10^{-3}$ m^2 and $F_{lo} = 6.43 \times 10^{-5}$ m^3/s^2 - corresponding to an initial density difference of around 0.4 percent. Accordingly, we have $t_c = 1.0\ s$ and $W_r = 0.04\ m/s$. The lower end of the square patch is located at $z = 0.245$ m from the bottom boundary. Both the tank dimensions and initial thermal size are comparable with previous experiments of line thermals and advected line thermals. At $t = 0$ zero velocities and pressure are prescribed; $C_o = 1$ inside the marked patch, and zero elsewhere. The governing equations are solved using the finite volume method; details of the numerical solution can be found in Lee and Chen (2002). The main features of the flow and scalar field are summarized herein.

Figure 9 shows the computed stream function for the line thermal. A vortex pair flow similar to the line puff is clearly revealed. The shape of the flow field is well-preserved for $t^* \geq 10$. Figure 10 shows the computed tracer concentration field. A symmetrical double maxima kidney-shaped structure is formed very early at about $t^* = 3$, and becomes approximately self-similar for $t^* > 10$. This simple kidney-shaped structure is quite different from that for a line puff, for which there is no double peak maxima for the asymptotic phase (Lee and Chen 2002). The concentration maximum C_m is typically as much as 2 times the maximum concentration C_c along the centerline. This value of C_m/C_c is in accord with previous observations in the line thermal (Richards 1963), though higher than the value of 1.7 (Fan 1967) and 1.4-1.6 (Lee and Cheung 1991) obtained for bent-over plumes in crossflow. To characterize the sectional thermal shape, the thermal aspect ratio is defined as the

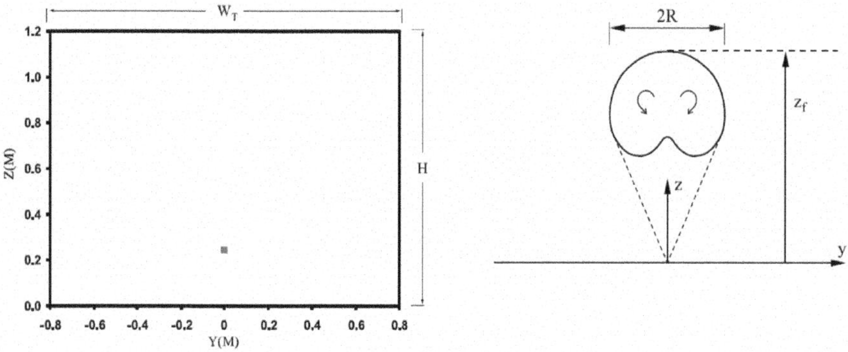

Fig. 7 Numerical experiment of line thermal; an initial buoyancy is applied in the $+z$ direction to the marked patch of fluid

Fig. 8 Definition of line thermal parameters

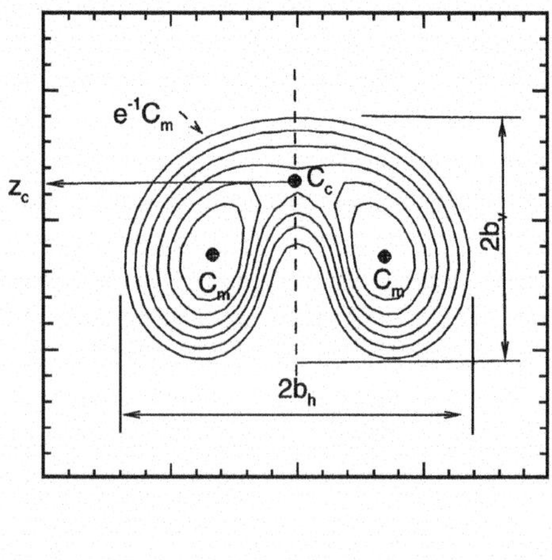

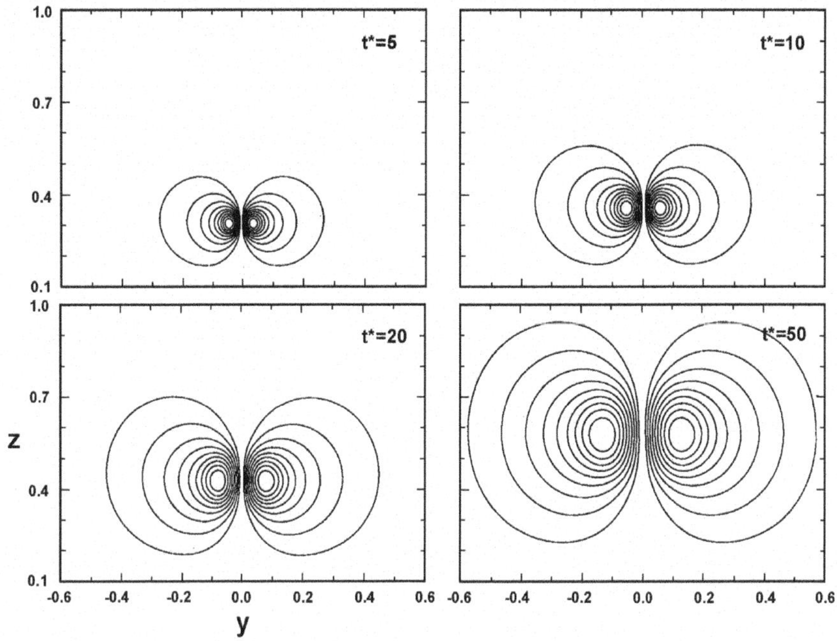

Fig. 9 Computed streamlines of line thermal at $t^* = 5, 10, 20, 50$ (contour interval $\Psi_m^*/10$)

ratio of horizontal half-width b_h to vertical half-width b_v of the contour $C = e^{-1}C_m$ (Fig. 8). The results show that the thermal aspect ratio increases very quickly from its initial value of 1 at source to an asymptotic value of around 1.45 when $t^* > 3$. From the shape of the thermal, it is clear that the location of the centerline maximum concentration (z_c) lies above the visual centerline (z_d, mid-point of upper and lower boundary), $z_c > z_d$ (see also Prob. 8.2).

In Fig. 11, the vertical momentum inside the thermal (as defined by the 0.01 C_m contour), $M_t = \int\int_{x-section} W dydz$, is plotted against time. The best fit of the numerical results gives $M_t = 0.5 F_{lo} t$. This implies that the added mass coefficient of the line thermal is $k_n = 1$ (Eq. 21); half of the momentum is applied to the ambient irrotational flow around the thermal. To investigate self-similarity more fully, a large number of thermal characteristics are computed at each time step. The maximum vertical velocity W_m, trajectory z_c and maximum scalar concentration C_m well satisfies the -1/3, 2/3 and -4/3 similarity laws (Eqs. 6, 8, 9) respectively, for $t^* \geq 10$ (Fig. 12). It is interesting to note that the vertical velocity is roughly constant in the initial phase, with $W_m/W_r \approx 0.5 - 0.55$ in the range of $t^* = 2 - 8$. The linear variation of the vertical location of the thermal front Z_f with the maximum horizontal radius R can also be seen. The thermal is gradually developed from the buoyancy; it can grow to about 20 times its original size by $t^* \approx 80$. The similarity relations in the asymptotic stage $t^* > 10$ are summarized in Table 1. Also shown are a length scale L, defined such that L^2 is the area within the 0.01 C_m contour, the ratio of the maxi-

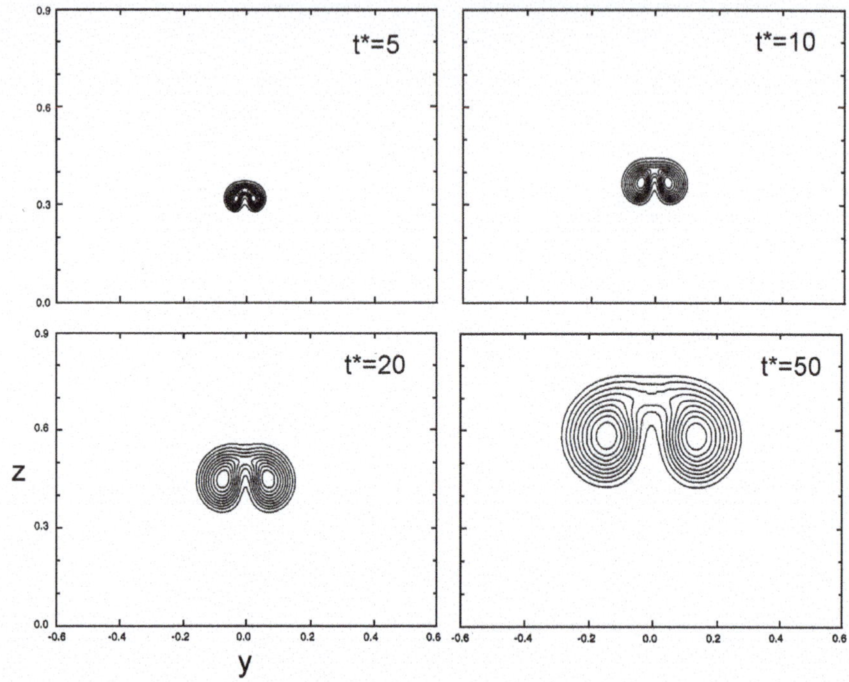

Fig. 10 Tracer concentration field of line thermal

mum to tracer mass-weighted average vertical velocity ($W_m/\overline{W}$), and the circulation around one half of the thermal, defined as $\Gamma = \int_0^H W(0, z)dz$. The vortex pair of the computed line thermal remains connected with time and is never completely divided; the circulation ($\Gamma \sim WL \sim t^{1/3}$) increases with time, in accordance with the measurements of Richards (1963). Similar to the line puff, the derived coefficients in the self-similar relations are not sensitive to the initial conditions assumed over a reasonable range of values.

Figure 13 shows the computed vorticity and turbulent kinetic energy which also display self-similarity for $t^* \geq 10$; a simple pair of vortex eyes can be seen. The computed turbulent kinetic energy shows that the region of high TKE is located towards the front of the patch, where large spatial gradients can be found.

In the self-similar stage, the maximum turbulent intensity is $u_m/W_m \approx 0.39$, where $k_m = 3/2u_m^2$. In the main body of the thermal, the turbulent viscosity, very similar in shape to k, varies spatially by about a factor of 4, with the maximum value $\nu_t/(W_m L)$ = 0.02, and ν_t/ν typically in the order of 100. In Fig. 14, the computed streamlines are shown against concentration contours for the self-similar line thermal. It can be seen the double concentration maxima are located close to the vortex eyes but with a smaller separation distance than that between the vortex eyes. This computed feature is supported by experiments (Fig. 11 of Tsang 1971).

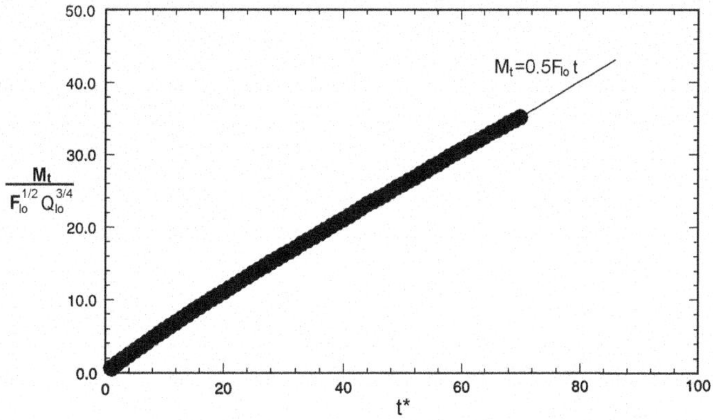

Fig. 11 Time variation of vertical momentum M_t of line thermal

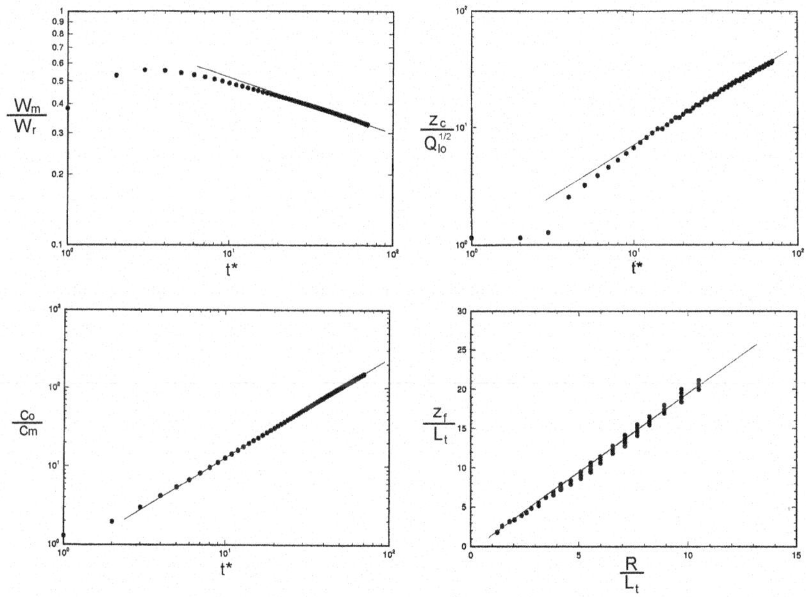

Fig. 12 Time variation of thermal properties: maximum vertical velocity, trajectory, centerline maximum concentration, thermal front location (solid line is best fit self-similar relation)

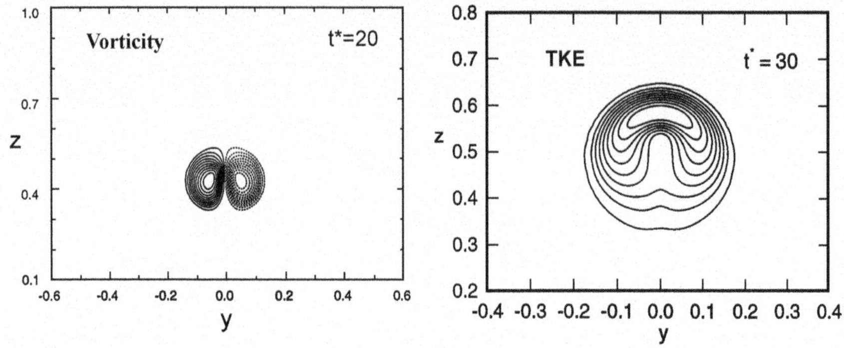

Fig. 13 Vorticity and turbulence kinetic energy of line thermal

Analysis of the computed velocity field reveals the flow outside of the thermal can be modelled by a doublet; the stream function may be approximated by that produced by a circular cylinder having the same cross-section area as the thermal and moving with the velocity of the thermal centre, with strength of 0.53. This is close to the experimental value of 0.522 given by Tsang (1971). The 2D model results and Richards' experiment both suggest that almost all of the fluid presented to the thermal cross-section is entrained into the thermal - this is related to a Projected Area Entrainment hypothesis employed in the Lagrangian model JETLAG (see Chap. 10).

The entrainment coefficient (α) as used in models employing the entrainment assumption (Turner 1986) can be deduced to be $\alpha = 0.46$ (Prob. 8.4). This value of α is very close to the typical value of 0.5 obtained for buoyant forced plumes (Chu and Goldberg 1974), and is similar to the value for a line puff (Lee *et al.* 1996) and a range of $0.25 - 0.67$ obtained by Chu (1985) in an analysis of data for jets in crossflow.

In summary, the computed flow is very similar to the observed mean fluid motion of line thermals, and offers a realistic picture and many insights of the thermal. Csanady (1965) presented an analytical model of the linearized weak thermal by assuming constant diffusivity; the computed flow and scalar field differ greatly from observations. Lilly (1964) solved the full nonlinear equations of motion for a line thermal within confined boundaries, and Yih (1981) presented a semi-analytical model for the bent-over plume in crossflow by assuming $u = U_a$. Both Lilly and Yih essentially adopted an unrealistic eddy viscosity distribution based on a free shear layer hypothesis; the predicted scalar field is not supported by experiments.

Table 1 Similarity relations of 2D line thermal and 3D advected line thermal (ALT) derived from numerical solution and laboratory experiment; numbers refer to dimensionless coefficients of equations or ratios

	Parameter	2D Thermal	3D ALT	Experiment	2D Puff
Added mass coefficient	k_n	1.0	1.0		1.01
Velocity	$W_m = C_v F_{lo}^{1/3} t^{-1/3}$	1.34	1.37		1.61
Thermal front	$Z_f = nR$	2.0		1.3-3.2 (Richards 1963) 2.2-2.5 (Chu 1996)	3.03
Trajectory	$Z_c = C_{2t} F_{lo}^{1/3} t^{2/3}$	0.80	1.30	$0.82t^{*0.767}$ (Wong 1991) 1.19 (Wright 1977) 1.26 (Knudsen 1988) 1.32 (Gaskin 1995)	1.51
Length scale	$L = C_{4t} F_{lo}^{1/3} t^{2/3}$	1.12			1.17
Scalar concentration	$C_o/C_c = C_{3t} Z_c^2/Q_{lo}$	1.14	0.68	0.56 (Wong 1991) 0.41 (Knudsen 1988) 0.44 (Wright 1977) 0.56 (Chu 1979) 0.51 (Cheung 1991)	0.35
	C_m/C_c	2.0	1.5	1.4–1.6 (Cheung 1991) 1.7 (Fan 1967)	1.0
Characteristic radius	$b = C_{1t} Z_c$	0.44	0.43	0.42 (Wong 1991) 0.39 (Chu 1996) 0.39 (Knudsen 1988) 0.42 (Gaskin 1995)	
Centerline halfwidth	$b_{vc} = C_{1t'} Z_c$	0.29			0.269
Aspect ratio	b_h/b_v	1.45	1.43	1–1.38 (Wong 1991) 1.23 (Richards 1963) 1.19 (Gaskin 1995)	1.0
Velocity ratio	$W_m/\overline{W}$	2.63			3.18
Circulation	$\Gamma/(\overline{W}L)$	2.35			2.67
Turbulent Viscosity	$\nu_t/(W_m L)$	0.020			0.018
Turbulent Intensity	u_m/W_m	0.38			0.39

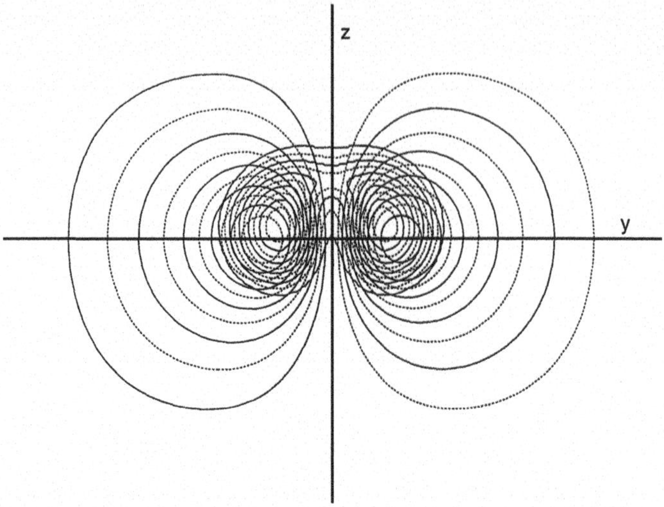

Fig. 14 Computed streamlines shown against concentration contours for $t^* = 30$

4 3D Model of Plume in Crossflow

4.1 The Advected Line Thermal

The flow in the bent-over phase of a buoyancy-dominated jet in crossflow is analogous to that of a line (2D) thermal. An experimental analog of the line thermal is the advected line thermal (ALT) proposed by Knudsen and Wood (1985). Figure 15 shows an advected line thermal produced by discharging a horizontal round buoyant jet with velocity U_o, density ρ_o, and diameter D into a steady coflow (density ρ_a) with the same velocity, $U_a = U_o$. Since there is no initial relative velocity, the flow simulates a vertically rising line thermal (advected at U_a) with initial volume per unit x-length $Q_{lo} = Q_o/U_a$, and buoyancy per unit length $F_{lo} = F_o/U_a$. The flow at two successive vertical sections (x_1, x_2) then correspond to those of the equivalent line thermal at two successive times by a Galilean tranformation. With this method, repeatability of source conditions can be achieved with much less variation than in experiments in which a cylinder of buoyant fluid is released from rest (Richards 1963, Tsang 1971). The data obtained in the latter experiments typically show great scatter. With the advected line thermal, the ensemble-averaged properties at a given time (or x) can be conveniently studied.

The advected line thermal in Fig. 16 is produced by a horizontal heated water jet in a coflow of the same velocity. Flow visualization of the dyed plume clearly shows that close to the source, within $\approx 5D$, a train of buoyancy-generated vortex rings is formed, which become stretched and diffused to form turbulent eddies further downstream. Figure 17 shows an instantaneous and a time-averaged image of the side view of an advected line thermal. The upper boundary of the thermal is highly

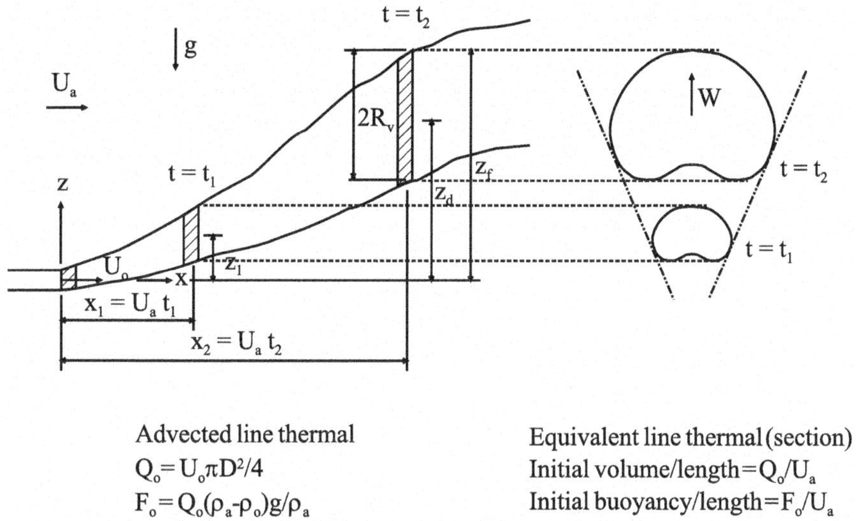

Fig. 15 Schematic diagram of an advected line thermal ($U_o = U_a$)

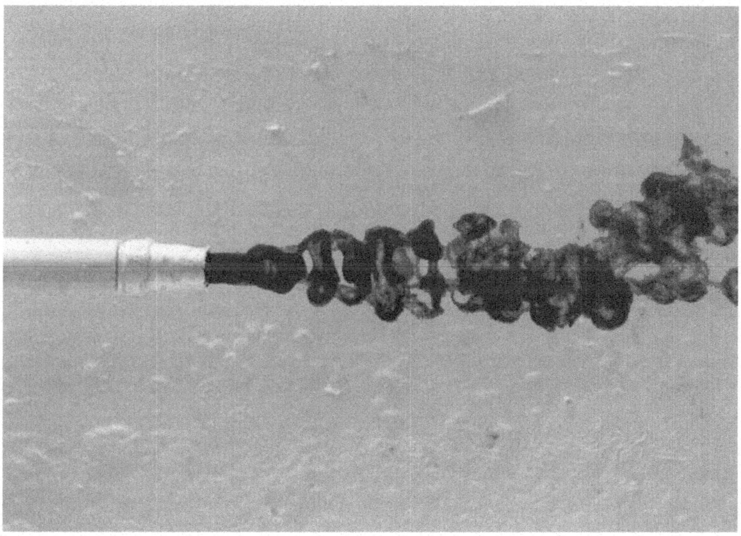

Fig. 16 An advected line thermal produced by a horizontal buoyant jet discharging with no initial excess momentum in a coflow (top view)

irregular and fluctuating, contributed by occcasionally large buoyant elements protruding from the jet proper, while the lower edge is relatively smooth. The path of the advected thermal is initially rather flat, before being bent-over by the crossflow.

4.2 3D Model of Advected Line Thermal

The spatial evolution of an advected line thermal has been studied using the standard two-equation $k-\epsilon$ model with a buoyancy expansion (Chen and Lee 2002). The governing equations outlined in the previous section are solved for the three dimensional flow. A numerical experiment of an advected thermal is performed adopting the following parameters: $U_o = U_a = 0.1\,\text{m/s}$, $D = 1$ cm, $F_o = 1.155 \times 10^{-6}\text{m}^4/\text{s}^3$, and $Q_o = 7.85 \times 10^{-6}\,\text{m}^3/\text{s}$; $(\rho_a - \rho_0)/\rho_a = 1.5 \times 10^{-2}$. The jet densimetric Froude number $Fr = 2.6$. The characteristic length and time scales have values of $L_b = 1.15 \times 10^{-3}$ m, $L_x = 2.45 \times 10^{-2}$ m and $t_c = 0.245$ s, respectively. The computational domain has a length of $L = 1.25$ m, depth $H = 0.4$ m and width $W = 0.3$ m, and extends to $x/D \geq 120$ and $x^* > 50$ downstream of the source. Details of the computation can be found in Chen and Lee (2002).

Figure 18 shows contour plots of the computed scalar field in the plane of symmetry at $z = 0$. The trajectory of the maximum concentration looks very flat on first glance, like a straight line. The process of scalar dilution takes place essentially within 40 diameters downstream, corresponding to $x^* < 15$. The computed thermal trajectory resembles the observed path (Fig. 17).

Figure 19 shows transverse cross sections of the scalar field at different downstream locations. A double peak concentration distribution can be observed in the kidney-shaped thermal, which becomes approximately self-similar for $x^* > 10$. Qualitatively the computed scalar field is very similar to that of the line thermal (Fig. 10), but with slightly more rounded front edges. The double-peak structure is also similar to the numerical results for advected line puffs (Fig. 14). The self-similar relations for the 3D advected line thermal are given in Table 1.

Characteristics of Advected Line Thermal

Analysis of the three-dimensional numerical results reveal the following characteristics:

i) *Thermal aspect ratio*–In Fig. 19, the thermal aspect ratio b_h/b_v increases very quickly from its initial value of 1 at the exit to around 1.43 when $x^* > 3$. This is approximately the same as the value obtained for the line thermal, and comparable with the corresponding experimental result of Wong (1991) who found a variation of thermal aspect ratio from 1.1–1.3, with an average of 1.2. Analysis of thermal boundaries in the classical study of line thermals (Richards 1963) suggests an average aspect ratio of 1.23; analysis of $e^{-1}C_m$ concentration contours measured by Gaskin (1995) in the developed phase of advected line thermals ($Fr \geq 1$) in developed phase suggests an average value of 1.19. This value is also comparable to the value of 1.3 suggested by Pratte & Baines (1967) from an analysis of flow visualization and that found for advected line puffs (Table 1).

ii) *Maximum concentration ratio*–As observed from the concentration contours, the maximum concentration C_m is not located in the centerline. The maximum of

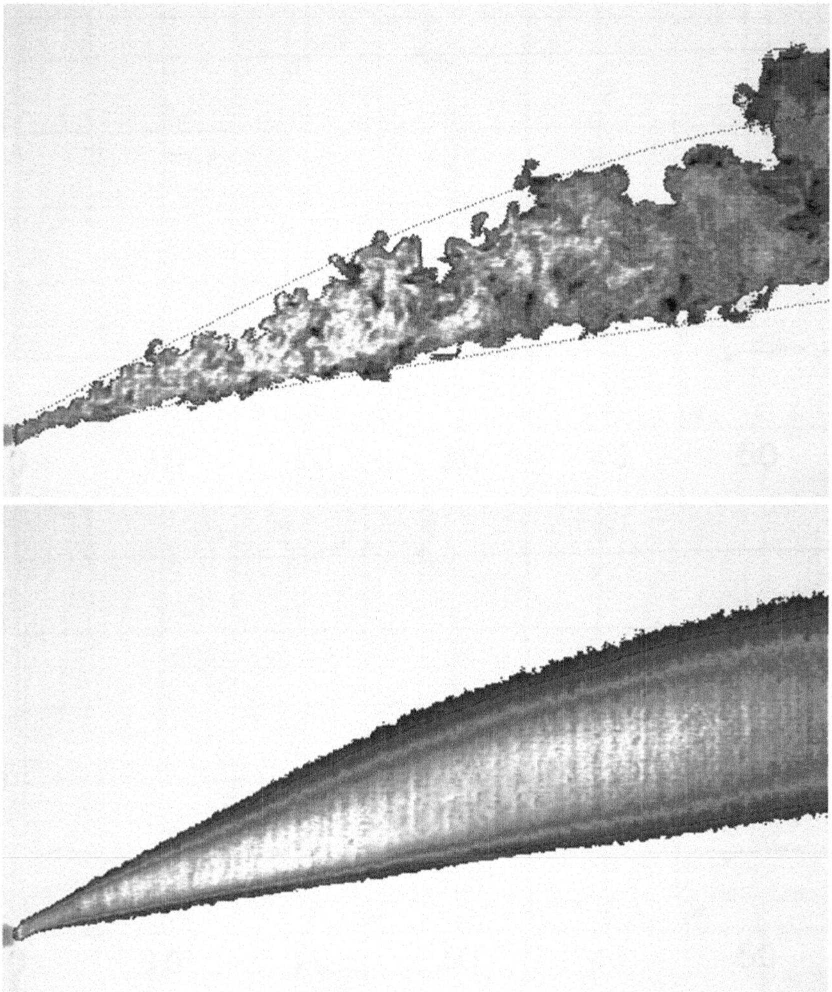

Fig. 17 Instantaneous and time-averaged shadowgraph pseudo-color image (side view) of an advected line thermal ($D = 1.02$ cm; $U_o = 4.4$ cm/s; $U_a = 4.6$ cm/s; $g'_o = 0.1177$ m/s^2)

the ratio of the cross-section to centerline maximum concentration, C_m/C_c, is found to be 1.5, comparable to the value of 1.7 (Fan 1967) and 1.4-1.6 (Lee and Cheung 1991) observed in bent-over plume in crossflow. It is also interesting to note the case of advected line puffs: $C_m/C_c = 1.35$ in the developed phase and values of $1.1 - 1.5$ from experiments.

iii) *Characteristic half-width and trajectory*—Fig. 20 shows the characteristic radius $b = \sqrt{b_h b_v}$ plotted against centerline trajectory z_c; the linear spread is predicted. The width spread rate, $C_{1t} = 0.43$, is close to the value of 0.42 exper-

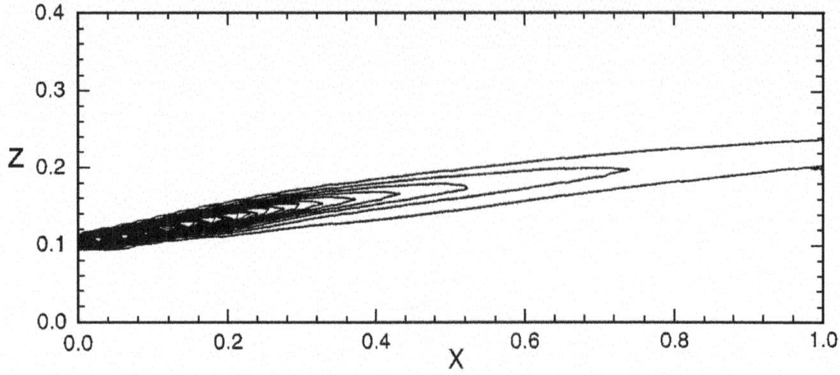

Fig. 18 Computed scalar concentration contours in the plane of symmetry at $y = 0$. Contour interval is 1/100

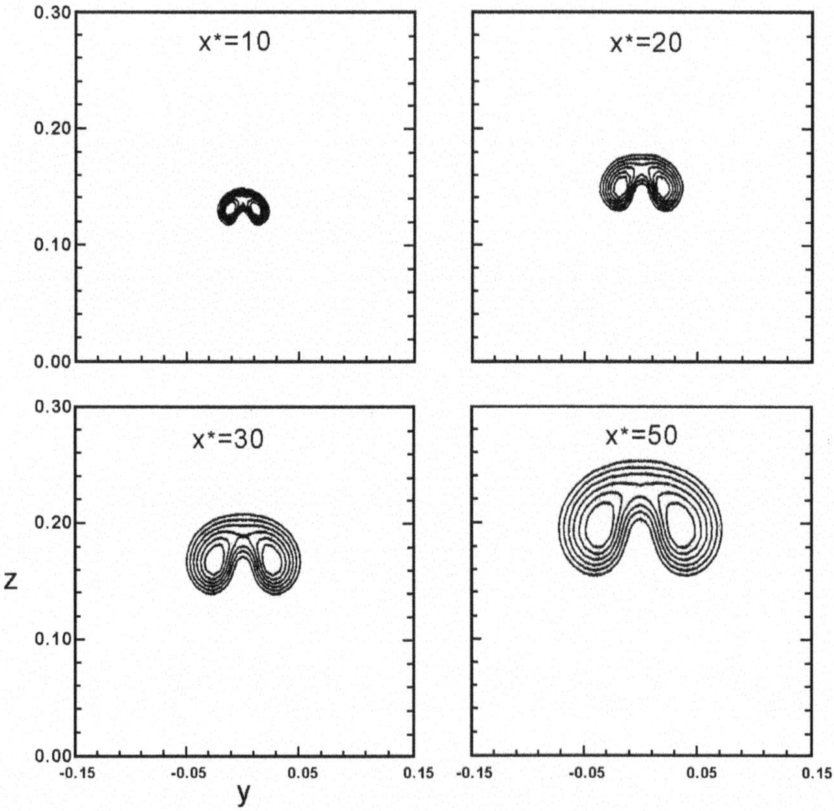

Fig. 19 Computed tracer concentration in different vertical cross-sections of advected line thermal. Contour interval is $(1 - e^{-1})C_m/6$. Outermost concentration contour is $e^{-1}C_m$

imentally found by Wong (1991). It is however notably higher than the theoretical and experimental values of 0.27-0.29 for advected line puffs (defined somewhat differently). Figure 20 also shows the puff trajectory z_c/L_b plotted against downward distance x/L_b. The 2/3 power law is confirmed, with a coefficient of $C_{2t} = 1.30$. This can be compared with experimental values of 1.19 (Wright 1977b), 1.26 (Knudsen 1988), and 1.40 in Wong (1991). The effect of virtual origin is found to be small.

iv) *Centerline dilution*—Fig. 21 shows the centerline dilution $S_c Q/(U_a L_b^2)$ plotted against trajectory z_c/L_b. The results follow the square power similarity law, with a dilution coefficient of $C_{3t} = 0.68$, This value has been obtained by Chu (1979) in his analysis of data of Fan (1967) for the bent-over phase of a buoyant jet, and is comparable to the experimental value of 0.56 in Wong (1991) for advected line thermals. In Chu's analysis, the visual trajectory z_d instead of z_c was used in the correlation. Since $z_c \approx 1.1 z_d$, Fan's data suggests a dilution coefficient of $0.68/1.1^2 = 0.56$.

v) *Vertical momentum and added mass* - Conservation of buoyancy flux (Eq. 21) leads to the following equation for the total vertical momentum flux

$$M_v = \frac{1}{1+k_n} F_o \frac{x}{U_a} \tag{36}$$

where the vertical momentum flux is defined as $M_v = \int\int_{x-section} UW dydz$ and k_n = added virtual mass coefficient. The 3D model computations confirm the above equation, giving a result almost identical to that depicted in Fig. 11. The added mass coefficient is found to be 1.0 for the thermal defined by the $0.1C_m$ contour. Redefining the thermal by the $0.01C_m$, $0.05C_m$, $0.25C_m$ and $e^{-1}C_m$ contours gives values of $k_n = 0.92$, 0.96, 1.04 and 1.09 respectively. In other words, the advected line thermal itself contains only about half of the total vertical momentum resulting from the buoyancy effect; the other half accounts for the approximately irrotational flow of the surrounding fluid set up by the rising thermal. This result is consistent with those found in experiments and numerical studies of advected line puffs and time-dependent line puffs

vi) *Bifurcation of plume in crossflow* –One normally associates the double maxima structure of the scalar field with a buoyant jet in crossflow. Numerical experiments have confirmed that the bifurcation in concentration structure may be absent for cases with small buoyancy–as supported by the experiment of Wong (1991). For example, an advected line thermal with $Fr = 4.12$ with $\Delta\rho_o/\rho_a = 1.5 \times 10^{-3}$ and $U_a = 0.05$ m/s, exhibits only one cross-section concentration maximum centered at the centerline. As the strength of the buoyancy generated circulation can be shown to depend on F_o/U_a, this result seems to be reasonable: for weakly buoyant plumes the circulating flow along and near the thermal centerline is not strong enough to compete with the cross-section turbulent mixing. These cases also result in a smaller aspect ratio of around 1.2.

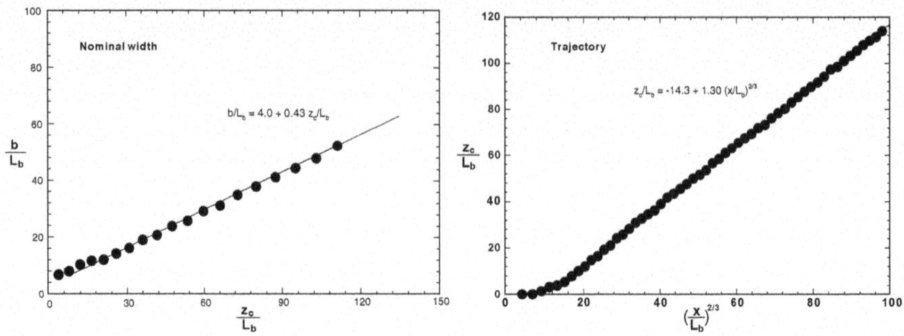

Fig. 20 Width and trajectory of advected line thermal

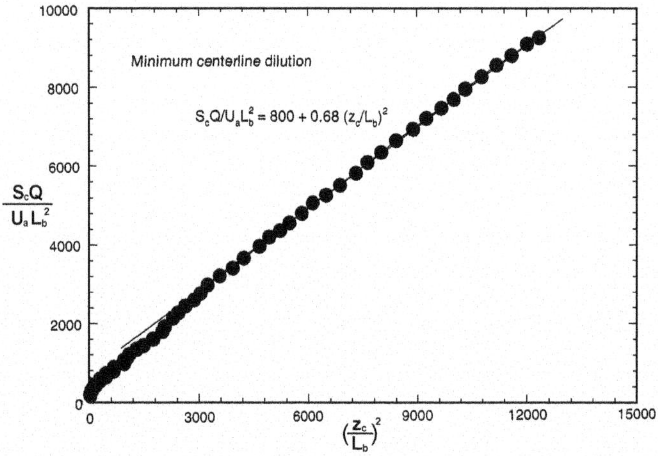

Fig. 21 Dimensionless centerline dilution of advected line thermal

5 Measurements in Line Thermals

5.1 Experiments on Advected Line Thermals

Experiments on advected line thermals have been performed by Knudsen (1988), Wong (1991), and Gaskin (1995). All the results show that the cross-section concentration is often double-peaked, with an aspect ratio of around $b_h/b_v \approx 1.2$, where the half-widths b_h, b_v are defined by the $e^{-1}C_m$ concentration contour. Wong's experiments show clearly the validity of the linear spread relation, with a spreading coefficient of $C_{1t} = 0.42$ (Fig. 22). The thermal trajectory can be described with the 2/3 power law with a trajectory coefficient of $C_{2t} = 1.40$, although the best fit gives $C_{2t} = 0.8$ and a power variation of about 0.767. This can be related to the use of heat as a source of buoyancy. Compared with the case when buoyancy is created by salinity differences (as in Fan 1967 and Knudsen 1988), both numerical models and experiments show that a heated jet with the same initial $\Delta\rho_o$ would take on a lower (more submerged) jet trajectory due to loss of buoyancy (see also Chap. 3). However, the dilution achieved at the same vertical elevation is approximately the same. The minimum centerline dilution follows well the BDFF relation, with a dilution coefficient of 0.56. The ALT has also been studied using video image processing. In these experiments, 250 instantaneous frames from a one-minute video record of the shadowgraph were used to obtain the ensemble-averaged image. The time-averaged visual boundary is defined by a threshold value selected based on visual judgement of thresholding an *instantaneous* image against its visual boundary. The visual centerline trajectory $z_d(x)$ (defined as the mid-point between the upper and lower boundaries), and the corresponding vertical half-width $R_v(x)$ can then be obtained. The time-averaged image reveals a relatively straight initial portion before the jet is bent over (2/3 power law). Figure 23 shows a plot of the centerline trajectory; the initial trajectory can be described by $z_d/D \approx 0.46x/L_x$, before the transition to the developed stage occurs at $x/L_x \approx 10$. The initial linear trajectory is consistent with the approximately constant vertical velocity computed for the line thermal, with $W_m/W_r \approx 0.53$. We note that $dz/dx = \overline{W}/U_a$, and $\overline{W} \approx W_m$ in the initial phase. Using the definition of W_r and t_c and $U_o = U_a$, it can be shown that $W_m \approx 0.53\ W_r$ would have led to an initial trajectory of $z/D = 0.53\sqrt{\pi/4}\ x/L_x = 0.47x/L_x$.

The trajectory coefficient has a value of $C_{2t} = 1.37$. Figure 24 shows that the vertical half-width varies linearly with elevation, $R_v \sim z$, with C_{1t} = 0.48 and 0.39 in the initial and developed phases respectively. Since the upper visual boundary represents the front of the corresponding line thermal, $z_f = z + R_v$, the thermal spreading rate as defined by dz_f/dR_v has a value of 3.11 in the initial phase and 3.57 in the developed phase. As the 3D computation gives an aspect ratio of 1.43, this implies that $z_f \approx 2.2\ R_h$.

Using ($k_n = 1, \beta_n = 0.39$), Eq. 27 would give $C_{2t} = 1.1$; this compares favorably with experimental values of 1.19 and 1.26, derived from an analysis of the data of Wright (1977b) and Knudsen (1988) respectively. In entrainment-based models of buoyant jet in crossflow (e.g. Hoult *et al.* 1969), it can be shown that the entrainment coefficient in the BDFF should be the same as the spreading rate (Prob. 8.5). In previous studies, the entrainment coefficient α_o was typically determined from the observed jet trajectory in the BDFF (2/3 power relation, Eq. 27) without accounting for added mass. It can be seen that the entrainment coefficient $\alpha_o = \beta\sqrt{1+k}$ so determined will not match the observed spreading coefficient, an anomaly noted earlier (Hoult and Weil 1972). Using the reported entrainment coefficient, the spread rate observed in these studies can be inferred and given in Table 2.

5.2 *Concentration Measurements*

Detailed concentration measurements in the bent-over phase of plumes in crossflow (BDFF) as well as advected line thermals (ALT) have been made in a number of studies. The nature of the measurements are summarized below along with the con-

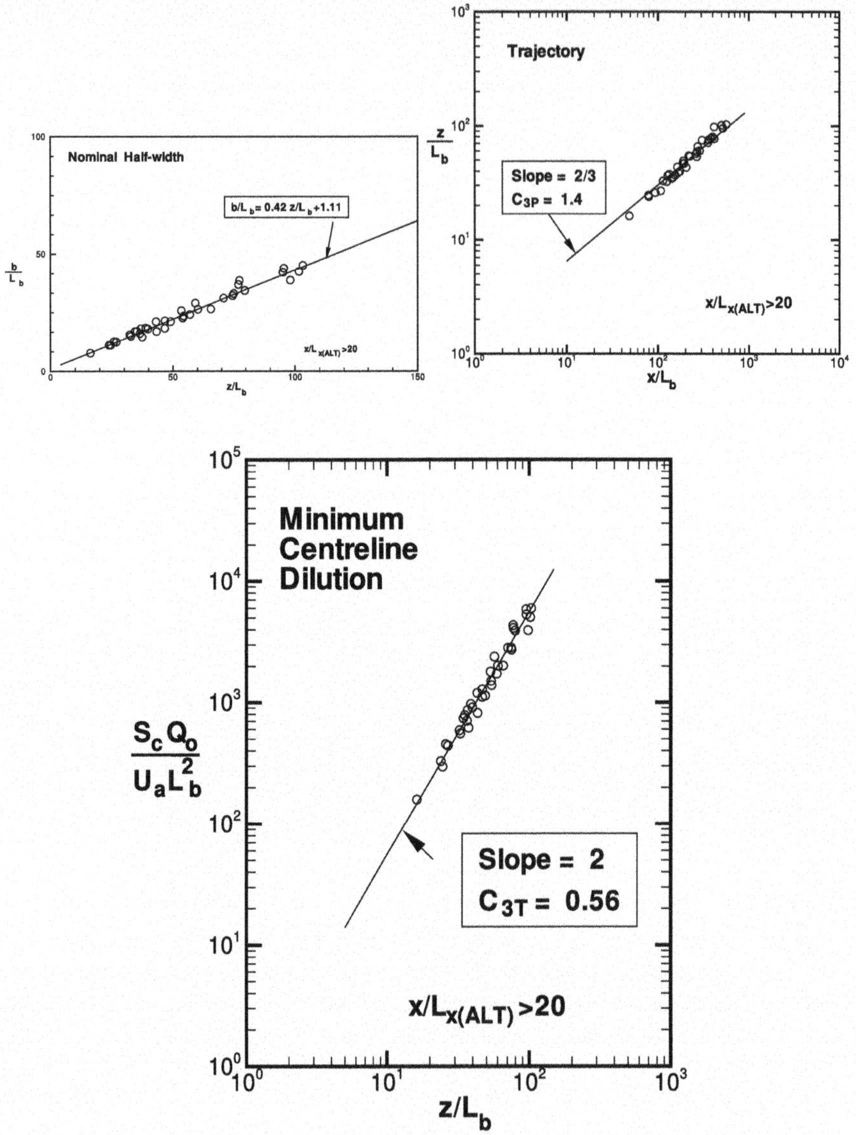

Fig. 22 Observed visual trajectory, width, and dilution of the advected thermal (Wong 1991)

centration spread rate C_{1t}, trajectory coefficient C_{2t} and centerline minimum dilution constant C_{3t} in the BDFF so obtained. The key values can also be found in Table 2.

Fan 1967 Vertical buoyant (saline) jet in crossflow; centerline concentration (conductivity) normal to jet axis in the vertical plane of symmetry; concentration and visual trajectory (z_c, z_d and selected cross-section measurements. Momentum-

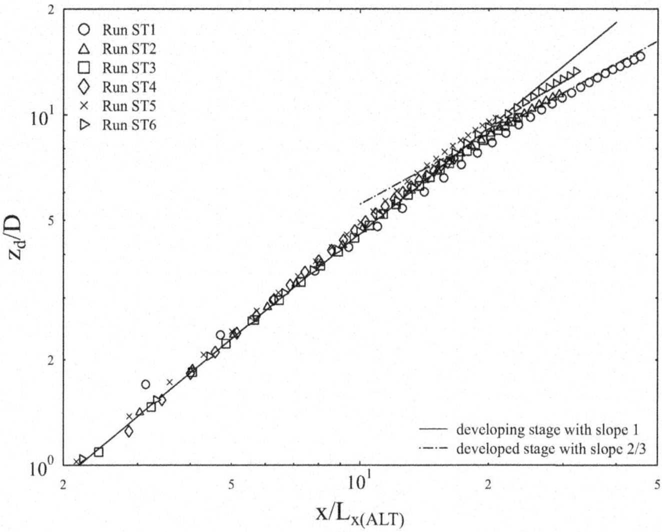

Fig. 23 Observed visual trajectory of the advected thermal: **a** $z \sim x$ for initial stage; **b** $(z/L_b) \sim (x/L_b)^{2/3}$ in developed phase

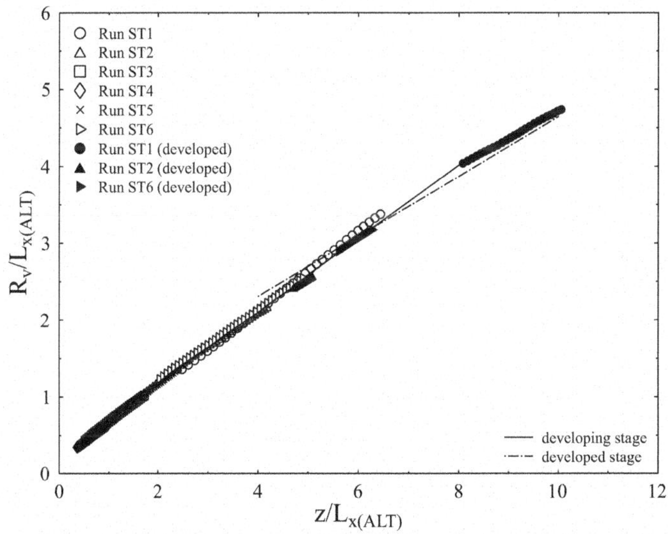

Fig. 24 Spreading rate of advected line thermal: linear spread of visual vertical half-width versus elevation

dominated jets ($Fr = 10 - 80$, $K = U_o/U_a = 4 - 16$). Analysis of Fan's data gives $C_{1t} = 0.34$ (visual), $C_{3t} = 0.56$.

Ayoub 1971 Horizontal buoyant jet in coflow and crossflow; centerline concentration (conductivity) normal to jet axis in the vertical plane of symmetry; trajectory (z_c, z_d) and selected cross-section measurements. Momentum-dominated jets with 3D trajectories ($Fr = 15 - 90$, $K = 5 - 20$).

Wright 1977 Towed vertical saline jet in uniform and stratified still ambient; centerline concentration (conductivity) along a vertical traverse in the plane of symmetry; visual trajectory. Extends to weak crossflow regime ($K \geq 4$). Analysis of all data with $z/L_b \geq 1$ gives $C_{2t} = 1.19$ (visual) and $C_{3t} = 0.44$. Data with $z/L_b \geq 2$ has an average dilution constant of $C_{3t} = 0.51$,

Knudsen 1988 Advected line thermal; towed horizontal saline plume in uniform still ambient. Cross-section (conductivity) measurements in six experiments. $C_{1t} = 0.36$, $C_{2t} = 1.39$, $C_{2t} = 1.1$ (visual), $C_{3t} = 0.41$.

Wong 1991 Advected line thermal; heated horizontal thermal plume in coflow. Cross-section temperature measurements (3-minute average) on a 17×14 grid and visual trajectory. A total of 42 cross-sections of 12 thermals. Digital analysis of video-recorded dyed plume. $C_{1t} = 0.42$, $C_{2t} = 0.8$ (best fit 0.767 power law), $C_{2t} = 1.4$ (2/3 power law), $C_{3t} = 0.56$. Analysis of video records of dyed thermals gives $C_{1t} = 0.43$ (assuming aspect ratio of 1.2), and $C_{2t} = 1.37$ (z_d).

Cheung 1991 Vertical and horizontal buoyancy-dominated heated jets in crossflow. ($Fr = 2 - 10$, $K \geq 2$). Transverse and normal centerline temperature profiles and visual trajectory particularly in BDNF-BDFF transition. Detailed cross-section measurements for horizontal buoyant jet in crossflow. $C_{3t} = 0.51$.

Gaskin 1995 Advected line thermal; towed saline jet in uniform still ambient. Cross-section conductivity measurements; Instantaneous and time-averaged LIF measurements of longitudinal centerline section of thermal; instantaneous LIF data on cross-section. PIV measurement of external entrainment flow. $C_{1t} = 0.41$, $C_{2t} = 1.32$ (z_d).

Pun 1999 Advected jet and plumes; towed saline jet in uniform still ambient. Detailed LIF data in centerline section of vertical buoyant jet in weak crossflow. $C_{2t} = 1.35$ (z_d). Data on MDNF and BDNF.

5.3 *Summary of Experimental Data*

In the experiments described above, different measurement methods have been used; e.g. the concentration half-width has been defined from the visual half-width (between the upper and lower boundaries), from a vertical concentration traverse, or a nominal radius based on the extreme widths of the $e^{-1}C_m$ concentration contour. Different periods of time-averaging have also been used. Nevertheless, the following points can be made:

- The nominal radius based on the $e^{-1}C_m$ contour ranges from 0.36 (Knudsen 1988) to 0.42 (Wong 1991). In Knudsen's experiments, if we define the concentration half-width as λb, and assume (as in puffs) a visual radius of $\sqrt{2}b$, then the theoretical spread rate of the ALT is $C_{1t} = \sqrt{2}/1.2 \times 0.36 = 0.424$ where the concentration to velocity width ratio has been assumed to be $\lambda = 1.2$. Digital image processing of the video records of advected line thermals (Chu 1996) gives a spread rate of $C_{1t} = 0.39$ for the visual vertical half-width. Assuming an aspect ratio of $b_h/b_v = 1.2$, this gives a nominal visual half-width spread rate of $C_{1t} = \sqrt{1.2} \times 0.39 = 0.43$. For the measured ALT concentration half-width of $C_{1t} = 0.42$, the corresponding visual radius can also be interpreted as $C_{1t} = 0.42/1.2 \times \sqrt{2} = 0.50$
- The visual trajectory constant is quite stable, in the range of $C_{2t} = 1.1 - 1.37$; the digital analysis of ALT visual trajectory gives a value of 1.37. These values are generally lower than the trajectory defined by z_c.
- The experiments of buoyant jet in a current suggests a centerline dilution constant of $C_{3t} = 0.51 - 0.56$, while the towed jet experiments give a generally lower value of $C_{3t} = 0.41 - 0.51$. The variability is partly related to the sensitivity of the dilution constant to choice of z (z_d or z_c) and that not all data lie in the fully developed BDFF. The relatively higher dilution constant measured with a stationary jet in a current may be affected by the ambient turbulence.
- In the bent-over phase, the average dilution is given approximately by

$$\overline{S} = U_a \frac{\pi B^2}{Q_o} = 0.80 \frac{U_a z^2}{Q_o} \tag{37}$$

where $B = 0.5z$ is adopted. For the simple jet/plume, the ratio of average to centerline dilution is 1.7. For advected line puffs, a ratio of average to centerline dilution constant of $\overline{S}/S_c = 1.45$ has also been measured. Using the turbulence model results for the line thermal and line puff, the ratio $\overline{S}/S_c$ can also be determined to be around 1.7 (Prob. 8.6). A value of 2.08 has also been determined from Fan's (1967) data (Chu and Goldberg 1974). Since the experimental value for the ALT is unknown, adopting a value of $\overline{S}/S_c = 1.7$ results in:

$$S_c = 0.46 \frac{U_a z^2}{Q_o} \tag{38}$$

This value is seen to lie within the range of measured dilution constants. It is also consistent with field measurements of initial dilution at sea outfalls after accounting for the surface layer (see Chap. 10).
- The overall numerical and experimental results suggest that a value of $C_{1t} \approx 0.4 - 0.5$, $C_{2t} \approx 1.3$, and $C_{3t} \approx 0.46$ can be reasonably adopted for the BDFF.

The LIF data on instantaneous flow of advected line thermals (Gaskin 1995; 2001) revealed a highly variable and complex three-dimensional structure. The lead-

Table 2 Summary of experimental constants (visual spreading rate, visual and centerline trajectory, centerline and minimum dilution) of ALT

Previous studies	Advected line thermals (BDFF)				
	Spreading rate	Trajectory		Dilution	
	β	z_d	(z_c)	S_c	(S_m)
	C_{1t}	C_{2t}	(C_{2tc})	C_{3t}	(C_{3tm})
Richards (1963)	0.62[a]	0.90[a]			
Tsang (1971)	0.42[a]	1.09[a]			
Hewett, Fay and Hoult (1971)	0.50[b]		(0.98)		
Hoult and Weil (1972)	0.42[b]	1.10			
Chu and Goldberg (1974)	0.35[b]	1.24			
Wright (1977)		$0.85\,(L_m/L_b)^{1/6}$	$(1.4\,(L_m/L_b)^{1/6})$	0.41	
Chu (1979)	0.34	1.30		0.68	(0.41)
Lee and Neville-Jones (1987a, 1987b)			(0.94)		(0.32)
Cheung (1991)				0.51	
Knudsen (1988)	0.42[c]	1.10	(1.39)	0.41	
Wong (1991)	0.51[d]	1.36	(1.40)	0.56	(0.53)
Gaskin (1995)	0.49[c]		(1.32)		
JETLAG (Lee and Cheung 1990)	0.45[b]	1.1		0.45[e]	
Chu (1996)–shadowgraphy	0.43	1.37			
1D Lagrangian model ($\beta_n = 0.4$ and $k_n = 1$)	0.4[f]	1.14	(1.20)	0.39	(0.33)
				0.43[e]	(0.36[e])

[a] using $R_h/R_v \simeq 1.2$
[b] deduced from entrainment coefficient using $k = 1$
[c] deduced from concentration half-width
[d] average of visual and concentration half-width spreading rate
[e] based on thermal trajectory
[f] selected value

ing (top) edge of the thermal is hydrodynamically unstable and is seen to result in random protrusions into the ambient flow (see Fig. 17). This results in much greater concentration fluctuations than the advected line puff. The ratio of concentration fluctuation to the maximum time-mean concentration (C_{rms}/C_m) can have a maximum value of 0.8 towards the leading edge, while the maximum instantaneous concentration can be up to 3.5-4 times C_m maximum time-averaged value. In contrast, the corresponding figures for the advected line puff are approximately 0.4-0.5 and 2.0 respectively (Fig. 26 and 27). The intermittency factor is however broadly similar for both thermals and puffs. The formation and transport of 3D 'sub-thermal' structures in the far field of the ALT has been discussed by Gaskin and Wood (2001).

6 Buoyant Jet in Crossflow

Since in general a buoyant jet has both momentum and buoyancy, whether a discharge is momentum or buoyancy dominated depends on L_m/L_b. If $L_m/L_b \ll 1$, the effect of buoyancy is exerted over a much greater distance than momentum, and the discharge is said to be buoyancy-dominated. Conversely, for $L_m/L_b \gg 1$, the initial momentum is relatively more important than buoyancy before the jet is bent-over, and the discharge is said to be momentum-dominated. Since the initial jet momentum loses its effect within a distance of the order of $L_s = M_o^{3/4}/B_o^{1/2}$ (the momentum length scale in stagnant fluid, Chap. 3), for any discharge the effect of buoyancy will govern asymptotically. In other words, the buoyancy-dominated far field (BDFF) is the ultimate flow regime of any discharge. In Lagrangian terms, given enough time (at large distances), no matter how weak the initial buoyancy is, the buoyancy effect will be felt. It can be demonstrated from analysis of experimental data or numerical model predictions that the behaviour of a momentum-dominated jet in crossflow may be described by a MDNF-MDFF-BDFF sequence of flow regimes (Prob. 8.7); similarly it is likely the behaviour of a buoyancy-dominated jet in crossflow can be described by a MDNF-BDNF-BDFF sequence (Wright 1977; Fischer *et al.* 1979). In general the important flow regimes are the flows very close to and very far from the source; the transition is usually not conveniently dealt with by length scales and is better treated using mathematical models. Note also that the three length scales L_s, L_m and L_b are not independent, with $L_s/L_b = (L_m/L_b)^{3/2}$.

6.1 *Line Thermals and Puffs*

Line puffs and thermals are in many ways similar types of free shear flows. Line puffs are generated by an instantaneous source of momentum, while thermals are generated by an instantaneous source of buoyancy. Both are vortex-pair like flows which move through and mix with the surrounding fluid, carrying the ambient fluid with them; the added mass coefficient for both flows is approximately 1. The mixing and turbulence intensity is greatest towards the front of the puff/thermal. The dilution constants and maximum dimensionless eddy viscosities are also comparable. There are however significant differences in the computed scalar concentration field. In the case of line puffs, the motion is dominated by the initial impulse. In the relatively short-lived asymptotic phase, the aspect ratio of the puff is about 1, and (contrary to experiments of jet in crossflow) no concentration peaks are predicted. The line thermal, on the other hand, is a motion that develops gradually from the buoyant impulse, and the flow has a clearly bifurcated double concentration maxima.

The results indicate that the unsteady line thermal is a reasonably good approximation to the bent-over plume in crossflow. The bifurcation of the scalar field in a typical buoyant plume in crossflow is predicted; the computed C_m/C_c and aspect ratio of 1.5 and 1.43 respectively can be compared to measured values of $C_m/C_c = 1.2 - 1.7$

and $b_h/b_v \approx 1.2 - 1.4$. It appears that the detailed structure of the scalar field would depend critically on the initial conditions as well as the location of the section (i.e. whether in the BDNF-BDFF transition or in the developed phase). In addition, in experiments the distinct bifurcation is observed only with relatively large initial buoyancy in a weak to moderate crossflow. For experiments with weaker buoyancy and/or stronger current, such a bifurcation is not observed, and $C_m \approx C_c$ (Wong 1991, Cheung and Lee 1991). The absence of double concentration peaks may also be related to the turbulence in the ambient flow (Gaskin 1995). A summary of the experimental results for advected line thermals and 2D thermals is also given in Table 2. Additional results on the BDFF concerning laboratory and field data will be dealt with in Chap. 10.

Line puffs and thermals are useful analogs for correlating experimental data of buoyant jet in crossflow. Nevertheless, the 3D ALP numerical computations (Lee, Chen and Kuang 2002) show that the excess x-velocity, $u_e = u - U_a$ is not negligible. Although the overall excess x-momentum is conserved (=zero), the distribution of u_e is highly non-uniform; maximum values of u_{em}/U_a around 0.3 are found even in the bent-over phase. The excess x velocity only becomes negligible for $x^* \sim 100$. Hence the representation of a section of the bent-over jet in crossflow (or advected line puff) by a line puff moving at the ambient velocity has certain limitations. This appears to explain the inability of the line puff to simulate the bifurcation of the scalar field. Such a limitation does not exist with the line thermal. The 3D ALT computations (Chen and Lee 2002) reveal that at $x^* = 8$ the horizontal excess velocity differs negligibly from the ambient velocity, with the maximum excess velocity $u_{em} \approx 0.05U_a$.

Although line thermals are evidently better mixing agents than line puffs, it may appear strange on first sight that they have similar spreading and dilution laws in terms of the vertical distance z. The similar dilution characteristics seem to suggest that buoyancy is not important–perhaps a surprising result. This apparent anomaly is clarified in an example below (Example 8.3).

6.2 Application Example

For prediction of plume mixing in a crossflow, the numerical model results can be cast in a more convenient form. For completeness the well-known solution for the BDNF is also included.

Buoyancy-dominated Near Field (BDNF)

The classical relations for the characteristic properties of a round plume in stagnant fluid are (Chap. 3):

$$W_m = 4.71\, F_o^{1/3}\, z_c^{-1/3} \tag{39}$$

$$b_{vc} = 0.11\, z_c \tag{40}$$

$$S_c = 0.096 \frac{F_o^{1/3} z_c^{5/3}}{Q_o} \approx 0.1 \frac{F_o^{1/3} z_c^{5/3}}{Q_o} \tag{41}$$

In the BDNF the jet is only weakly advected; mixing is similar to that of a plume in stagnant fluid. Similar to the advected jet (Chap. 7), the following equations can then be obtained for the advected plume:

$z/L_b \ll 1$ **(BDNF):**

$$\frac{z_c}{L_b} = 2.36\,(\frac{x}{L_b})^{3/4} \tag{42}$$

$$b_{vc} = 0.11\, z_c \tag{43}$$

$$\frac{S_c Q_o}{U_a L_b^2} = 0.1\,(\frac{z_c}{L_b})^{5/3} \tag{44}$$

The above simple advected plume model for the BDNF is verified by the detailed LIF experiments of a vertical plume in weak crossflow (Pun and Davidson 1999) which confirms the validity of the $x^{3/4}$ trajectory law and give a trajectory coefficient of 2.57. This supports the use of the average velocity in defining the plume trajectory (see also related discussion on coflow jet formulation in Chap. 6 and MDNF in Chap. 7).

According to theory, the BDNF asymptotic flow regime is characterized by $z/L_b \ll 1$; for many engineering application estimates. Concentration measurements have shown the transition occurs in the range of $z/L_b = 0.1 - 1$ (Cheung 1991); the LIF trajectory data show a transition at $z/L_b \approx 0.5$. For practical purposes, a transition at around $z/L_b \approx 1$ can often be assumed; the transition issue will be further discussed in Chap. 10.

Buoyancy-dominated Far Field (BDFF)

Based on the ALT numerical model and experimental results, the following equations for the advected line thermal are recommended:

$z/L_b > 1$ **(BDFF):**

$$\frac{z_c}{L_b} = 1.3\,(\frac{x}{L_b})^{2/3} \tag{45}$$

$$b_{vc} = 0.4\, z_c \tag{46}$$

$$\frac{S_c Q_o}{U_a L_b^2} = 0.46\,(\frac{z_c}{L_b})^2 \quad \text{or}$$

$$S_c = 0.46 \frac{U_a\, z_c^2}{Q_o} \tag{47}$$

Note that the dilution constant in the vortex flow in the bent-over phase is the same for both the MDFF and the BDFF.

Example 8.1 Vertical buoyant jet in crossflow–BDFF
Consider a vertical buoyant jet in crossflow in the laboratory as shown in Fig. 1 , with the following discharge and ambient flow data: $W_o = 0.25$ m/s, $U_a = 0.063$ m/s, $D = 0.01$ m, $\Delta\rho_o/\rho_a = 0.0257$. Estimate the elevation, width, and dilution of the buoyant jet at a distance of $x = 0.12$ m downstream.

Solution: The densimetric Froude number of the jet can be determined to be $Fr = \underline{5}$. First compute the jet volume, momentum, and buoyancy fluxes as:

$$Q_o = 0.25 \times \pi \times 0.01^2/4 = 0.196 \times 10^{-4}\ m^3/s$$

$$M_o = Q_o W_o = 0.491 \times 10^{-5}\ m^4/s^2$$

$$F_o = Q_o g_o' = 0.494 \times 10^{-5}\ m^4/s^3$$

The crossflow momentum and buoyancy length scales are then: $L_m = L_{mv} = \frac{M_o^{1/2}}{U_a} = \underline{0.0352}$ m, and $L_b = \frac{F_o}{U_a^3} = \underline{0.0198}$ m. Since $L_m/L_b = 1.78$, the two crossflow length scales are comparable to each other. At $x = 0.12$ m, the jet trajectory is located in the BDFF, with

$$\frac{z}{L_b} = 1.3\,(\frac{x}{L_b})^{2/3}$$

or $z = 1.3 \times 0.12^{2/3} \times (0.0198)^{1/3} = \underline{0.086}$ m. Since $z/L_b > 1$ and $z/L_m > 1$, use the BDFF Eq. 46 to estimate the width, $w = 2B = 2 \times 0.4 \times 0.085 = 0.068$ m. The centerline dilution at this elevation is given by Eq. 47, with $S_c = 0.046 \times 0.063 \times 0.085^2/(0.196 \times 10^{-4}) = \underline{10.7}$. The minimum dilution in the cross-section is probably less by 20-30 percent. Since $B = \sqrt{b_h b_v}$, assuming an aspect ratio of $b_h/b_v = 1.2$, the vertical and horizontal half-widths can be estimated as: $b_v = B/\sqrt{1.2} = \underline{0.062}$ m, and $b_h = 1.2 \times b_v = \underline{0.075}$ m. Note that the above calculations assume a buoyant jet in an unbounded crossflow. The effect of the free surface on the dilution is treated in Chap. 10.

Example 8.2 Plume in weak crossflow BDNF/BDFF transition
Consider a laboratory vertical heated plume in crossflow with the following characteristics: $W_o = 0.1075$ m/s, $U_a = 0.0175$ m/s, $D = 0.0075$ m, $\Delta\rho_o/\rho_a = 0.012$ (corresponding to a jet discharge temperature of $T_o = 56^o$ C, and $\Delta T_o = 21.8^o$ C). Estimate the dilution at a depth of $z = 0.05$ m $z = 0.136$ m above the source.

Solution: The densimetric Froude number of the jet is $Fr = 3.62$. The jet volume, momentum, and buoyancy fluxes can be determined as: $Q_o = 0.475 \times 10^{-5}$ m^3/s,

$M_o = 0.511 \times 10^{-6}$ m^4/s^2, and $F_o = 0.5592 \times 10^{-6}$ m^4/s^3. The jet momentum length scale is $l_s = M_o^{3/4}/F_o^{1/2} = 0.026$ m. The crossflow momentum and buoyancy length scale are respectively $L_m = 0.0408$ m, and $L_b = 0.104$ m.

This is a case of $L_m \sim L_b$, with $L_m/L_b = 0.39$. At $z = 0.05$ m, since $z/L_b \approx 0.5$, it is natural to use the BDNF relation Eq. 41 to compute the dilution of the advected plume, with $S_c = 0.1 \times (0.5592 \times 10^{-6})^{1/3} \times 0.05^{5/3}/(0.475 \times 10^{-5}) = \underline{1.07}$. At $z/D = 0.05/0.0075 = 6.7$, the buoyant jet is just beyond the zone of flow establishment, and hence the center of the plume is not much diluted. In this case, however, the prediction based on such a reasoning would under-estimate the actual observed value by a factor of 6!

What is at fault? We note that at $z = 0.05$ m, $z/L_b < 1$, and $z/l_s = 1.9$, suggesting that the flow is buoyancy-dominated at this depth. At the same time, $z > L_m$ and hence the momentum-induced velocity has already decayed to less than the ambient velocity value, although the flow is still influenced by the discharge buoyancy. If we assume a vortex flow, the dilution (for either the MDFF or BDFF) would be given by $S_c = 0.46 \times 0.0175 \times 0.05^2/(0.475 \times 10^{-5}) = \underline{4.7}$, much closer to the measured dilution of 6.0. In situations when we are dealing with transitions, it is often difficult to estimate dilutions based purely on length scales. Alternatively the width and dilution in the transition can be predicted by a 1D Lagrangian model for the case of a general buoyant jet with both discharge momentum and buoyancy (Chaps. 9 and 10). Similarly, application of the BDFF equation gives the centerline dilution at $z = 0.136$ m as $S_c = \underline{31.4}$.

Example 8.3 Advected line thermals versus advected line puffs

Consider a vertical discharge with the following characteristics: $W_o = 1.0$ m/s, $U_a = 0.2$ m/s, $D = 0.1$ m, $\Delta\rho_o/\rho_a = 0.025$. Use Eq. 27 to determine the BDFF trajectory. Similarly, use Eq. 18 to determine the MDFF trajectory for a corresponding nonbuoyant discharge. Estimate the half-width and dilution at $z = 5$, 10 m, and the corresponding travel times.

Solution: The densimetric Froude number of the buoyant jet is $Fr = 6.5$. The jet volume, momentum, and buoyancy fluxes can be determined as: $Q_o = 7.85 \times 10^{-3}$ m^3/s, $M_o = 7.85 \times 10^{-3}$ m^4/s^2, and $F_o = 1.926 \times 10^{-3}$ m^4/s^3. Noting that $F_{lo} = F_o/U_a$, and $M_{lo} = M_o/U_a$, Eq. 27 (with $k = 1, \beta = 0.4$) then gives $z = (1.492 F_{lo})^{1/3} t^{2/3}$ for the line thermal, and similarly Eq. 18 gives $z = (2.98 M_{lo})^{1/3} t^{1/3}$ for the line puff. The trajectories are sketched in the attached figure (ignoring the near field).

The figure shows clearly that although the width and dilution achieved at the same z vertical distance above the jet are the same for the thermal and the puff, the time taken to rise to 5-10 m is 10-30 times longer for the puff. The buoyant thermal is clearly a much more vigorous mixing agent than the momentum puff; the ratio of the half-width at a given time for the thermal and puff is $B_r = R_{thermal}/R_{puff} = (0.5\frac{F_o t}{M_o})^{1/3}$. On simplification, we have $B_r = (0.5\frac{g_o' t}{W_o})^{1/3}$. The relative mixing efficiency of thermal versus puff is determined by the ratio $g_o' t/W_o$ i.e. the buoyant velocity to the

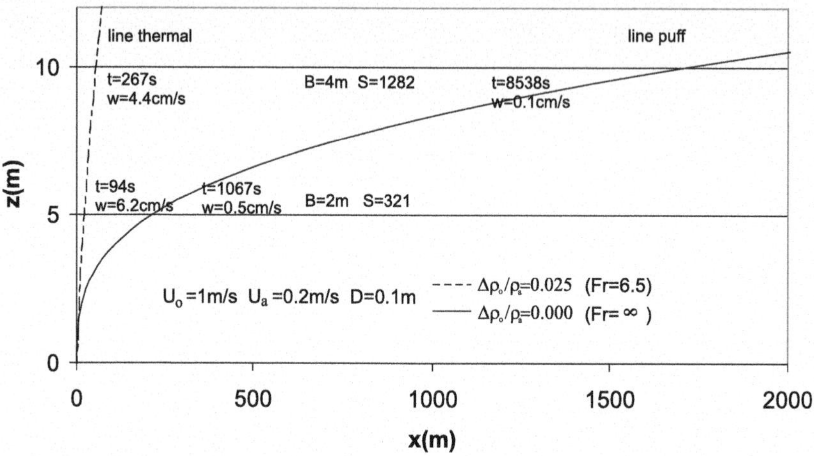

Fig. 25 Comparison of line thermal and line puff characteristics for plume/jet in crossflow

intial velocity. Since this ratio quickly exceeds unity, the thermal can be seen to be a much more efficient mixing device than puffs when interpreted in Lagrangian manner (Fig. 25).

6.3 *Case Study*

Innovative design of chlorine jet for disinfection

The Hong Kong Harbour Area Treatment Scheme (HATS) is a sewage collection and treatment scheme that serves a population of around 3 million in the urban areas of Hong Kong, and the largest wastewater treatment plant of its kind in the world. The sewage receives Chemically Enhanced Primary Treatment (CEPT), and the treated sewage flow of 1.4 million m^3/day (16 m^3/s on average) is discharged via a 1.2 km long submarine outfall diffuser into western Victoria Harbour. To protect the bacterial water quality of nearby bathing beaches, a concentrated 10% chlorine solution (sodium hypochlorite with a specific gravity of 1.2) is used directly to disinfect sewage. The dense chlorine solution is injected into the treated sewage flow in the form of multiple dense jets (Fig. 26). Field observations have revealed large spatial and temporal fluctuations of residual chlorine and *E. coli* in the chlorinated CEPT effluent and hence bacterial quality of the HATS effluent into coastal waters. It should be noted that excess chlorine in the SCISTW effluent is also harmful to the environment.

The mixing of sewage with chlorine (and hence the chlorine demand) in a small beaker in the laboratory is very different from the mixing of a chlorine jet in a sewage flow inside a treatment plant. To study the mysterious chlorine consumption and test different jet dosing designs, experiments in a 1:2-scale sectional physical model are

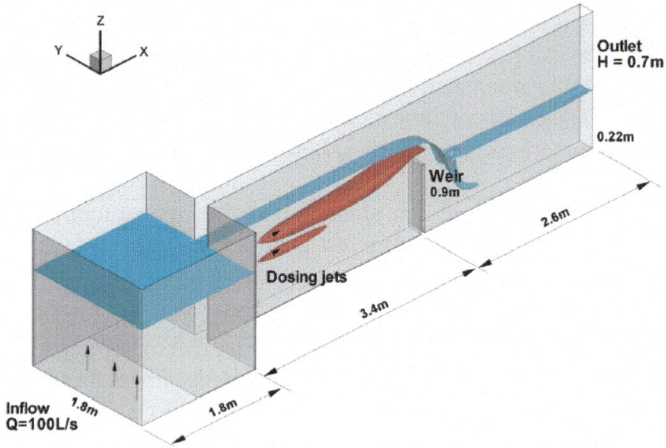

Fig. 26 CFD simulation of chlorine jet in field scale physical model in sewage treatment works (Chan et al. 2017)

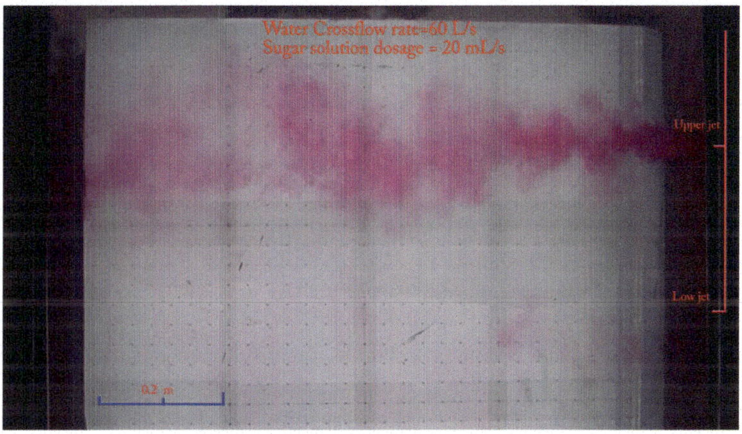

Fig. 27 Observed initial mixing of a chlorine dosing jet in a turbulent sewage coflow (flow direction from right to left); the dense jet was simulated with dyed sugar solution (Lee et al. 2017)

carried out using the on site sewage and chlorine. It was found that the injected chlorine is rapidly consumed close to the jet source before effective disinfection can take place. To improve the mixing of chlorine jet dosing system, an integral jet model based on the theory of coflow jet (Chap. 6), advected line thermal (Chap. 8), ambient turbulence (Chaps. 2 and 6), and second-order chemical kinetics has been developed to guide and screen dosing unit design proposals (Chan et al. 2017). Visualization of the chlorine jet flow using dyed sugar solution is given in Fig. 27. Three-dimensional CFD model was then developed to study the detailed mixing and distribution of residual chlorine in the sewage treatment works.

Mixing is an important aspect in any reactor process inside a wastewater treatment plant. It is found the chlorine consumption depends on the source chlorine concentration in a highly nonlinear manner: the higher the chlorine concentration, the greater the consumption. Hence it is of utmost importance to minimize the *time of exposure* of sewage to high chlorine concentrations. Both experiments and theoretical modeling suggest that the chlorine demand can be significantly reduced by around 10–15% by discharging a pre-diluted chlorine solution (to 2.5%) with a four-jet dosing unit in the sectional model. The proposed optimization was fully supported by a set of 50 field experiments using real sewage and chlorine. In addition, with optimal jet design and placement, field operation has demonstrated chlorine savings of up to 30 percent (Lee et al. 2023). Through understanding the interaction between jet mixing and chlorination kinetics, significant savings of chemicals and energy has been achieved - thus contributing to carbon neutrality with fluid mechanics.

7 Summary

To an observer moving with the ambient flow, a section of a bent-over plume in crossflow rises like a line thermal. Turbulence model computations of unsteady 2D line thermals and steady 3D advected line thermals show that the flow is self-similar beyond a dimensionless time of around $t^* = 10$. The thermal is characterized by a double-vortex flow, and typically a kidney-shaped cross-section scalar field with off-centered double concentration maxima. The aspect ratio of the thermal is around 1.2-1.4. The cross-sectional maximum concentration can be up to 40-60 percent greater than the centerline maximum concentration. Both the predictions and the experiments show that the line thermal is a very good analogy of the bent-over plume (BDFF). The existence of added mass is confirmed. A review of measurements of thermals is presented. The predicted relations for time-averaged trajectory, width, and dilution are well-supported by experimental data. It is demonstrated that the mixing is governed by the large scale flow structure, and the path is defined by the average rather than the centerline velocity.

The instantaneous flow of the advected line thermal is highly variable and much more complex than the time-averaged structure. Thermals exhibit much greater concentration fluctuation than jets, plumes, and puffs. Thermals are by nature much more efficient mixers than puffs. Surprisingly, the Eulerian spreading and dilution relations are very similar. For both the advected thermal (BDFF) and the advected puff (MDFF), numerical computation and direct measurements have confirmed the linear spread with vertical distance above source. The recommended values for the added mass and spreading coefficients are:

$$k_n = 1 \quad \text{and} \quad \beta_n = 0.4 \quad \text{for line thermals and puffs} \tag{48}$$

Problems

8.1 *Consider the waste gas jet discharge in Example 3.7 in Chap. 3, with* $w_o =$ 5 m/s, $\Delta\rho_o/\rho = 0.034$, *and* $D = 0.2$ *m. Estimate the dilution of the waste plume at* $z = 10$ *m in a cross wind of a)* $U_a = 1$ m/s, *and b)* $U_a = 3$ m/s. *Compare with the plume dilution in a stagnant atmosphere. What is the downstream location of the plume at this height?*

8.2 *Concentration versus visual trajectory of jet in crossflow*
Consider the predictions of the concentration trajectory z_c *and maximum vertical velocity* W_m *of the line puff derived from the* $k - \epsilon$ *model (Table 1):*

$$z_c = 1.7 M_{lo}^{1/3} t^{1/3}$$
$$W_m = 1.61 M_{lo}^{1/3} t^{-2/3}$$

a) The visual trajectory (center of mass) z_d *is given by* $dz_d/dt = \overline{W}$, *where* $\overline{W}$ *is the average velocity. Noting that* $W_m/\overline{W} = 3.18$, *determine the ratio of* z_c/z_d. *Is your result in accord with observations?*

b) Using the numerical results for the line thermal in Table 1, obtain z_c/z_d *for a plume in crossflow.*

8.3 *Added mass of line puff/thermal*
The virtual added mass of a line puff or thermal can be likened to that of a 2D cylinder. Consider a circular cylinder of radius a and mass per unit length M, *moving at a velocity* $U(t)$ *in the x-direction through a fluid of density* ρ_a. *The otherwise stagnant fluid will be displaced by the cylinder and a flow pattern is formed. The stream function* $\psi(r, \theta)$ *of the instantaneous irrotational flow outside of the cylinder at any time is (see e.g. Lamb 1932):*

$$\psi = -\frac{Ua^2}{r} \sin\theta$$

where a=radius of cylinder, and (r, θ) *are the radial and angular co-ordinates (with respect to the centre of the sphere at time t, and the* $x-$ *direction respectively). The radial and tangential velocities at any point in the flow are then:*

$$V_r = -\frac{1}{r}\frac{\partial\psi}{\partial\theta} = U\frac{a^2}{r^2}\cos\theta$$
$$V_\theta = \frac{\partial\psi}{\partial r} = U\frac{a^2}{r^2}\sin\theta$$

Show that the total kinetic energy of the external flow is:

$$E_{flow} = \int_a^\infty \frac{1}{2}\rho_a q^2 2\pi r dr = \frac{1}{2} M_a U^2$$

where $q^2 = V_r^2 + V_\theta^2$, *and* $M_a = \rho_a(\pi a^2)$ *is the virtual added mass of the moving cylinder. The total kinetic energy required to move the cylinder is then:*

$$E = \frac{1}{2}(M + M_a)U^2$$

If F *is the external force acting on the cylinder in the* $x-$ *direction, then the rate at which* F *does work must equal to the rate of increase of the total kinetic energy,* $FU = \frac{dE}{dt}$. *Hence show that:*

$$F = (M + M_a)\frac{dU}{dt}$$

i.e. the cylinder experiences a resistance of $-M_a\frac{dU}{dt}$ *due to the presence of the fluid.*

For line puffs and thermals, with $\rho \approx \rho_a$, *and* $M \approx M_a$, *if the above equation is assumed, we obtain:*

$$F = (M + M_a)\frac{dU}{dt} = (1 + k)M\frac{dU}{dt}$$

where $k = 1$ *is the added mass coefficient. However, unlike a circular cylinder, the interface between the puff/thermal element and the ambient fluid is not a solid boundary; it is a moving fluid interface across which there is significant entrainment (particulary at the rear). The puff/thermal is also kidney-shaped and the flow around it is not separated. In view of these complexities the added mass coefficient must be determined by a full numerical calculation of the time-dependent flow and passive scalar field (Figs. 11 of Chap. 7 or Fig. 11) or by direct measurement (Fig. 25).*

8.4 *Entrainment coefficient of line thermal*
Consider the following typical entrainment model for a line thermal or puff: the rate of change of volume of the thermal is proportional to the inflow due to turbulent entrainment:

$$\frac{d}{dt}(\pi B^2) = 2\pi B(\alpha \overline{W})$$

where $B = 0.44\, z_c$ *is the characteristic radius of the thermal and* $\overline{W}$ = *average vertical velocity. The entrainment velocity is assumed to be proportional to the thermal velocity, with* α=*entrainment coefficient.*

a) Using the turbulence model predictions of the concenstration trajectory z_c, *maximum vertical velocity* W_m, *and* $W_m/\overline{W}$ *for the line thermal (Table 1), show that* $\alpha \approx 0.46$. *Obtain a similar value for the entrainment coefficient of the line puff using Table 1.*

b) If we assume the line thermal to be an analog for the bent-over plume in crossflow, the BDFF, then $dB/dt = U_a dB/dx$. Noting that $dz/dx = \overline{W}/U_a$, show that $B = \alpha z$–i.e. in this case the entrainment coefficient is the same as the spreading rate.

8.5 *Entrainment-based model for bent-over buoyant jets*
An entrainment-based model for the line thermal (BDFF) or line puff (MDFF) of a bent-over buoyant jet in crossflow can be formulated in Eulerian form.

a) For a buoyant jet in crossflow, the entrainment relation is often stated as the sum of two parts, one due to shear entrainment, and one due to the entrainment of the crossflow:

$$\frac{dQ}{ds} = 2\pi B[\alpha_s(U - U_a \cos\phi) + \beta' U_a \sin\phi]$$

where $Q = V\pi B^2$ = jet volume flux, s=natural streamwise co-ordinate, α_s = shear entrainment coefficient, and β' = crossflow entrainment coefficient, and ϕ=local jet inclination to x-axis. In the bent-over phase, $V \approx U_a$, and the second crossflow term dominates. Noting that $\sin\phi = dz/ds$, show that $B = \beta' z$. The use of the this model together with momentum conservation leads to the 2/3 power trajectory law, Eq. 27.

b) In experiments, the BDFF trajectory coefficient of a plume in crossflow has been determined to be $C_{2t} = 1.1$, and the observed vertical half-width is $B_v = 0.35\ z$, where C_{2t} is given by Eqs. 27 and 28. If added mass is neglected, determine the spreading rate from the trajectory observation. Compare your result with the observation and give an interpretation of the data.

8.6 *Average to centerline minimum dilution for line thermal*
For the line thermal, the volume (per unit length) of the thermal (defined within the $0.01C_m$ concentration contour) is given by L^2. The average dilution is hence given by $\overline{S} = L^2/Q_{lo}$, where the terms have their usual meaning. On the other hand, the centerline minimum dilution is given by $S_c = 1.14z_c^2/Q_{lo}$. Using the turbulence model predictions of z_c and L given in Table 1 show that $\overline{S}/S_c \approx 1.7$ for the line thermal. Obtain a similar value for the line puff using Table 1.

8.7 *MDFF-BDFF trajectory*
Consider a vertical buoyant jet discharge with $w_o = 4$ m/s, $U_a = 0.2$ m/s, $D =$ 0.1 m, and $\Delta\rho_o/\rho_o = 0.005$. Using the model VISJET or equivalent, predict the buoyant jet trajectory. By plotting the predicted trajectory on a log-log scale, identify the slopes for the asymptotic flow regimes (if any) for the MDFF (1/3) and BDFF (2/3). Comment on your results. How would your results change if $\Delta\rho_o/\rho_o = 0.025$?

8.8 *Line plume in stratified crossflow*
The behaviour of a line plume in a stratified crossflow can be effectively studied using a Lagrangian model using the principles outlined in Chaps. 3, 5, and 8 (Tate, P.M. and Middleton, J.H. , "The height of rise and dilution of a two dimensional plume

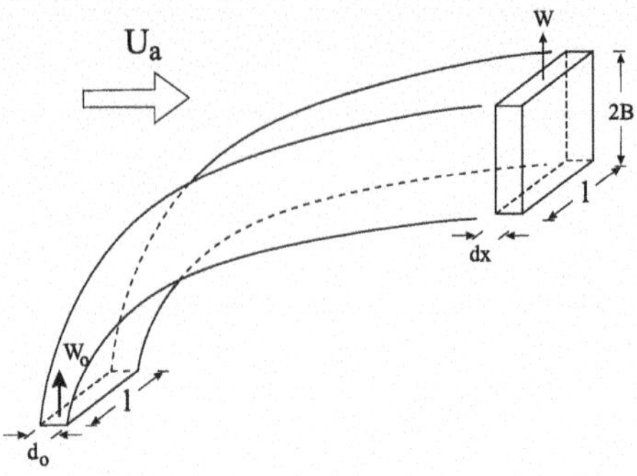

Fig. 28 A line plume in stratified crossflow

in a stratified crossflow", Sydney Water Board internal report, 1990, unpublished). The discharge from an ocean outfall of length L is sometimes modelled as a line source of buoyancy. The discharge and buoyancy flux per unit diffuser length of this line plume are $Q_o = Q_s/L$, $F_o = Q_s(\rho_a(0) - \rho_o)g/L$ *respectively, where* Q_s*=total source flow from diffuser.*

Consider a line plume in a perpendicular stratified crossflow of velocity U_a *and density profile* $\rho_a(z)$*. In a moderate to strong current, when the plume velocity (*$\sim (F_o/\rho)^{1/3}$*) is less than the ambient current (or* $F = U_a^3/(F_o/\rho) \leq 1$*), it is expected the line plume is quickly bent over by the crossflow and the x−velocity in the curtain of mixed sewage is approximately* U_a*. Figure 28 shows an element of this line plume, characterized by length* dx*, vertical half-width* B*, volume (*$dx2B$*), buoyancy* $\Delta\rho g = (\rho_a - \rho)g$*, and vertical velocity* W*.*

a) By using the Lagrangian spreading hypothesis, and applying momentum and buoyancy conservation, show that the line plume satisfies the following equations:

$$\frac{dB}{dt} = \beta W$$
$$\frac{dM}{dt} = \frac{d}{dt}(\rho 2WB) = (\Delta\rho g 2B) = F$$
$$\frac{dF}{dt} = \frac{d}{dt}(\Delta\rho g 2B) = (\rho 2WB)\left(\frac{g}{\rho}\frac{d\rho_a}{dz}\right)$$
$$= -N^2 M$$

where M, F *= vertical momentum of and buoyancy force acting on the plume element,* N*=stratification frequency, and* β*=spreading or entrainment coefficient (Hint: see Eqs. 60–61, and Eq. 8).*

b) For a linear stratification, N*=constant, show that the buoyancy force* $F(t)$*, and the trajectory* $z(t)$ *of the line plume element are given by:*

$$F(t) = \frac{F_o}{U_a} \cos(Nt)$$

$$z^2 = -(\frac{F_o}{\rho \beta N^2 U_a})[\cos(Nt) - 1]$$

c) At the equilibrium height of rise z_e, the buoyancy $B = 0(Nt = \pi/2)$ [at the maximum rise height z_m the vertical momentum $M = 0(Nt = \pi)$]. The average dilution is given by $\overline{S} = U_a(2B)/Q_o$. Show that z_e and $\overline{S}$ are given by:

$$z_e = \frac{1}{\sqrt{\beta}}(\frac{F_o}{\rho U_a})^{1/2} N^{-1} = 1.29(\frac{F_o}{\rho U_a})^{1/2} N^{-1}$$

$$\overline{S} = 2\sqrt{\beta} U_a^{1/2} \frac{(F_o/\rho)^{1/2}}{Q_o} N^{-1} = 1.55 U_a^{1/2} \frac{(F_o/\rho)^{1/2}}{Q_o} N^{-1}$$

where $\beta = 0.6$ has been used. As a comparison, the experiments of a line plume in a stratified crossflow (Roberts et al.1989) give for the equilibrium level of the line plume:

$$z_e = 1.5(\frac{F_o}{\rho U_a})^{1/2} N^{-1}$$

for $0.1 < F < 100$, where $F = U_a^3/(F_o/\rho)$ is a crossflow parameter. The minimum dilution in the spreading layer is given by:

$$S_m = 0.67 \frac{(F_o/\rho)^{1/3} z_e}{Q_o} F^{1/6}(2.19 F^{1/6} - 0.52)$$

It should be noted that the first term in the Roberts dilution equation has the same form as that given by the Lagrangian model, with a coefficient of 1.46.

8.9 *Ocean outfall in stratified current*

Consider an ocean outfall with total discharge flow $Q_s = 10$ m^3/s, diffuser length $L = 800$ m, and initial relative density difference of 0.025. Estimate the equilibrium level and dilution S in the waste field in a stratified current of $U_a = 0.2$ m/s, and stratification frequency $N = 0.01$ s^{-1}. Compare the estimates obtained from the Lagrangian model with that using the Roberts formula.

Chapter 9
General Lagrangian Formulation

The development of jets, plumes, puffs or thermals (JPPT) in coflow and crossflow, and in a surrounding environment of non-uniform density, have been the topics treated at length in previous chapters. The interaction of the JPPT with the density stratification in the environment was considered in Chap. 5. The effects of uniform coflow and uniform crossflow were examined in Chaps. 6, 7 and 8. In the present chapter, the combination of all the effects that have been examined as limiting cases are included in a general formulation. The concept of dominant eddies, the method of excesses, and the hypotheses of entrainment and spreading are revisited in a general context. The general formulation, for non-uniform coflow and crossflow and for non-uniform density stratification, is carried out here following the Lagrangian method by Lee and Cheung (1990) and Chu (1994). The equivalent formulation using the Eulerian method is given in a more recent work by Chu and Lee (1996).

1 Elemental Volume

The Lagrangian formulation is based on mass balance in an *elemental volume* and the concept of *dominant eddies*. In puffs and thermals, the elemental volume is the dominant eddy occupying the entire puffs/thermals volume. The formulation for the puffs and the thermals is naturally a Lagrangian problem. Jets and plumes can also be formulated using the Lagrangian method if an elemental volume is correctly defined for the problem. Figure 1 shows the elemental volume in the general problem of a buoyant jet in a non-unform crossflow of changing direction. The elemental volume is defined by the dominant eddies that are extending across the full width of the

Supplementary Information The online version contains supplementary material available at https://doi.org/10.1007/978-3-032-01187-9_9.

J. H. W. Lee et al., *Turbulent Jets and Plumes*,
https://doi.org/10.1007/978-3-032-01187-9_9

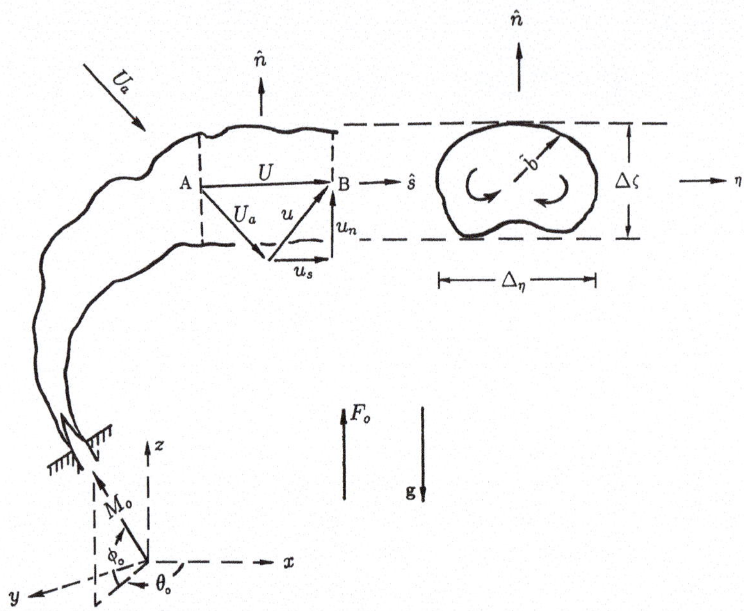

Fig. 1 Buoyant jet in a non-uniform crossflow of changing direction. The three-dimensional trajectory is produced as the buoyancy force acts in a general direction relative to the cross flow. The elemental volume, AB, is defined by the discharge of momentum and buoyancy at the source over a period of one time unit

turbulent region and moving along the path of the buoyant jet. The volume is defined along the path over the length AB as marked in the figure. The orientation of AB is along the path of motion. The magnitude and direction of AB is determined by the velocity vector, $\tilde{\mathbf{U}}$, tangential to the path of dominant eddies. The dominant eddies over the length AB are produced by momentum and buoyancy injection at the source over a period of one time unit. With such definition of the elemental volume, the momentum and buoyancy initially associated with AB are equal in magnitude of the momentum and buoyancy fluxes at the source, M_o and F_o, respectively.

As the dominant eddies move through their surrounding, materials are drawn into the eddies from the surrounding. The result is a continuous increase of materials in the elemental volume. This increase of materials in the elemental volume does not change the *excess* value in the volume if the excess is defined as the concentration of the materials relative to its value in the environment. The following section explains how the problem is conveniently formulated by the *method of excess*.

2 Method of Excesses

Due to entrainment, the 'mass' and volume in an elemental volume increase continuously. The volume is not fixed in space but moves with the velocity of the dominant eddies. The conventional fluid-mechanic methods such as the method of the control volume and the method of the material volume are not suitable for the present formulation. We overcome the difficulty associated with the entrainment using the method of excess. The concentration relative to its value in the ambient environment is the concentration *excess* $(c - c_a)$, where c_a is the ambient concentration. The mass-balance equation for the excess is derived by considering the mixing between the fluid in the elemental volume and the entrained ambient fluid.

The block diagram in Fig. 2 shows the mass-balance. The fluid in the elemental volume of carrying mass $\tilde{\mathcal{M}}$ and concentration $\tilde{c}$ is mixed with a small carrying mass $\delta\tilde{\mathcal{M}}$ of the ambient fluid of concentration c_a. As the fluid in the environment is drawn into the dominant eddies in the elemental volume, tracer 'mass' is exchanged between the eddies and its environment. For the present formulation, the concentration $\tilde{c}$ and c_a are intensive variables and are defined as the tracer 'mass' per unit mass of the carrying fluid. With this definition, the tracer 'mass' in the elemental volume is $\tilde{c}\tilde{\mathcal{M}}$. The tracer 'mass' of the entrained fluid entering the elemental volume is $c_a\delta\tilde{\mathcal{M}}$. After the entrain mixing, the mass concentration becomes $\tilde{c} + \delta\tilde{c}$ and the carrying mass becomes $\tilde{\mathcal{M}} + \delta\tilde{\mathcal{M}}$. Equating the tracer 'mass' of the system before and after the entrainment,

$$c_a\delta\tilde{\mathcal{M}} + c\tilde{\mathcal{M}} = (\tilde{c} + \delta\tilde{c})(\tilde{\mathcal{M}} + \delta\tilde{\mathcal{M}}), \tag{1}$$

which gives the mass-balance equation

$$\delta[(\tilde{c} - c_a)\tilde{\mathcal{M}}] = -\tilde{\mathcal{M}}\,(\delta c_a). \tag{2}$$

The tracer 'mass' concentration in this equation can represent either the concentration of scalars, such as salinity and heat, or concentration of vectors such as linear and angular momentum.

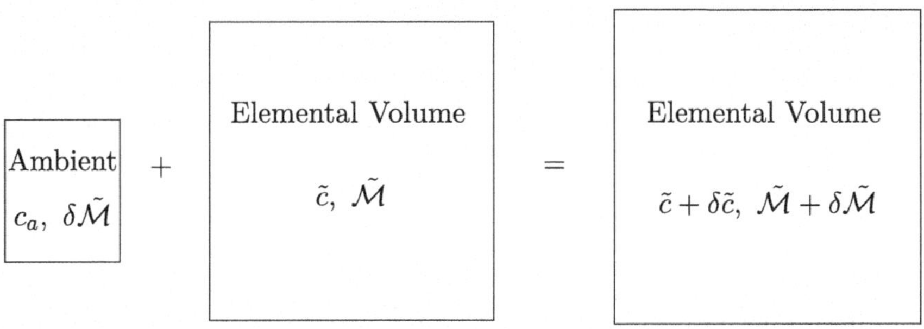

Fig. 2 Mixing of fluid in an elemental volume of carrying mass $\tilde{\mathcal{M}}$ and concentration $\tilde{c}$ with a small carrying mass $\delta\tilde{\mathcal{M}}$ of the ambient fluid of concentration c_a

Momentum Equation If the tracer 'mass' is linear momentum, the concentration of the momentum will be the velocity vector, which is momentum per unit mass of the carrying fluid. With this usage, the Eq. 2 becomes the momentum equation, which is

$$\delta[(\tilde{\mathbf{U}} - \mathbf{U}_a)\tilde{\mathcal{M}}] = -\tilde{\mathcal{M}}(\delta \mathbf{U}_a) \tag{3}$$

if momentum is conserved. However, the momentum of the system may change due to the buoyancy force. The equation for momentum changed by the impulse of the buoyancy force is

$$\delta[(\tilde{\mathbf{U}} - \mathbf{U}_a)\tilde{\mathcal{M}}] = -\tilde{\mathcal{M}}(\delta \mathbf{U}_a) + \tilde{\mathbf{B}}\delta t \tag{4}$$

that, in a more compact form, is as follows:

$$\delta \tilde{\mathbf{m}} = -\tilde{\rho}\tilde{V}\delta\tilde{\mathbf{U}}_a + \tilde{\mathbf{B}}\delta t \tag{5}$$

where $\tilde{\mathbf{m}} = \tilde{\rho}\tilde{V}\tilde{\mathbf{u}}$ is the momentum *excess* and $\tilde{\mathbf{B}}$ is the buoyancy force vector. The velocity excess, $\tilde{\mathbf{u}} = \tilde{\mathbf{U}} - \mathbf{U}_a$, is the velocity of the dominant eddies relative to the ambient. The density of the carrying fluid in the elemental volume is $\tilde{\rho}$. The mass of the carrying fluid in the elemental volume of volume $\tilde{V}$ is $\tilde{\mathcal{M}} = \tilde{\rho}\tilde{V}$. The tilde symbol, $\tilde{}$, denotes the averages over the dominant eddies. The excess momentum $\mathbf{m}$ is modified at a rate proportional to the buoyancy force, $\tilde{\mathbf{B}}$. In the absence of buoyancy force ($\tilde{\mathbf{B}} = 0$), the excess momentum would be constant if the ambient velocity is constant ($\delta\tilde{\mathbf{U}}_a = 0$).

Buoyancy Equation The buoyancy force is not a constant if the fluid in the environment is density stratified. The basic equation for the variation of the buoyancy is

$$\delta\tilde{\mathbf{B}} \simeq g\hat{\mathbf{k}}\tilde{V}\delta\rho_a. \tag{6}$$

where $\tilde{\mathbf{B}} = B\hat{\mathbf{k}}$ and $B = g(\rho_a - \rho)\tilde{V}$; g is the gravity constant, $\hat{\mathbf{k}}$ is the base vector pointing in the opposite direction to the force of gravity, and $(\rho - \rho_a)$ is the density of the fluid relative to the reference density ρ_a. This is the same buoyancy-variation equation given by Eq. 8 in Chap. 5. Note that $\tilde{B}$ is positive if $\rho < \rho_a$ and $\delta\tilde{B}$ is positive if $\delta\rho_a > 0$.

General Form of the Equation for the Excess Following similar procedure, it is possible to obtain excess equations for any tracer mass in the flow (e.g. nutrient, suspended sediments, fuel, toxic fume, smoke, heat and moisture). These equations are similar in form and all can be expressed in terms of the excesses as follows:

$$\delta[(\tilde{c} - c_a)\tilde{\rho}\tilde{V}] = -\tilde{\rho}\tilde{V}\delta(c_a) - \mathcal{K}\delta t \tag{7}$$

where $[(\tilde{c} - c_a)\tilde{\rho}\tilde{V}]$ is the concentration excess of the tracer mass, $\tilde{\rho}\tilde{V}$ is the mass of the carrying fluid; $\tilde{c}$ is the tracer concentration (tracer mass per unit mass of the eddies), c_a the tracer concentration of the ambient fluid, and $\mathcal{K}$ the 'mass' dissipation rate. By definition, excesses are zero in the environment. Entrainment of materials from the environment do not change the values of the excesses in the elemental volume. Without the source term, the excesses are constant if the ambient concentration is constant; i.e.,

$$\delta[(\tilde{c} - c_a)\tilde{\rho}\tilde{V}] = 0 \text{ if } \delta(c_a) = 0 \text{ and } \mathcal{K} = 0. \tag{8}$$

Analogous relations may be obtained for momentum and buoyancy excesses.

3 Spreading Hypothesis

The formulation is completed with a turbulent closure equation. In a one-dimensional formulation, this closure equation is based on an entrainment or spreading hypothesis. A number of variations in the hypothesis has been proposed. The great majority of the reported models are derived from an Eulerian point of view; many are not entirely consistent with all the experimental data. In the present formulation, the radius of the dominant eddies, $\tilde{b}$ or R, is assumed to increase at a rate in proportional to the relative speed, $|\tilde{\mathbf{u}}|$, as follows:

$$\frac{D\tilde{b}}{Dt} = \beta|\tilde{\mathbf{u}}| \tag{9}$$

where $\tilde{\mathbf{u}} = \tilde{\mathbf{U}} - \mathbf{U_a}$ is the velocity excess. In this formulation, D/Dt is a Lagrangian operator for time differentiation in a Lagrangian reference frame moving with the dominant eddies. Since $|\tilde{\mathbf{u}}| = Ds/Dt$,

$$\frac{D\tilde{b}}{Dt} = \beta\frac{Ds}{Dt}. \tag{10}$$

where s is the migration distance of the eddies relative to the surrounding fluid. The radius of the dominant eddies, $\tilde{b}$, increases at a rate in proportional to this migration distance, s, according to this hypothesis. The spreading rate is

$$\beta = \frac{D\tilde{b}}{Ds}. \tag{11}$$

In the general formulation for buoyant jet in a non-uniform crossflow (see Fig. 1), the excess velocity vector $\mathbf{u}$ is resolved into a tangential component u_s in the direction parallel to $\mathbf{U}$ and into a normal component u_n in the direction perpendicular to $\mathbf{U}$. The entrainment of fluid into the dominant eddies is known to dependent on the orientation of elemental volume. Lee and Cheung (1990) and Lee et al. (2000) in their Lagrangian formulation of the buoyant jet problems have introduced the terms *shear-induced* entrainment and *vortex* entrainment to associate the process with the tangential and normal components of the excess velocity. The shear-induced entrainment is the dominant mechanism in the near field while vortex entrainment is dominant in the far field. There is yet no satisfactory model to handle the transition from the shear-induced entrainment to the vortex entrainment. The formulation of the entrainment hypothesis can be either based on an additive hypothesis, or a maximum hypothesis, as follows:

$$\frac{D\tilde{b}}{Dt} = \beta_s|u_s| + \beta_n|u_n|, \tag{12}$$

$$\frac{D\tilde{b}}{Dt} = Max[\beta_s|u_s|, \beta_n|u_n|] \tag{13}$$

Lee and Cheung (1990) found the maximum hypothesis gives slightly better results. Chu (1994) and Dehghani (1996) however were able to obtained satisfactory model agreement with experimental data using the additive model. A heuristic theory to account for the transition from the shear entrainment to the vortex entrainment regime is given in Eq. 36 in Chap. 10 (Lee et al.2000). Chu (1994) has recommended a value of the shear-induced coefficient $\beta_s \simeq 0.17$ based on the data from the limiting case of the jets and plumes in stagnant fluid. When used together with an additive spreading hypothesis, the value recommended for the vortex coefficient is $\beta_n \simeq 0.34$. Analysis of data obtained for advective line puff and advective thermal in the previous chapters has lead to a higher votex coefficient of $\beta_n \simeq 0.4$. The value of the spreading coefficients are critically dependent on the definitions of $\tilde{b}$ and $\tilde{\mathbf{u}}$. Our recommendation is to define the boundary of the dominant eddies at the location where the turbulence intermittency $\gamma = 50$ %, so that $\tilde{b}$ (or R) is the radius of the region defined by this boundary and $\tilde{\mathbf{u}}$ is the average of the velocity in this region. As noted in the previous chapters, unlike the Eulerian coefficient, the Lagrangian coefficient β_s is approximately constant for straight jets and plumes as well as coflowing jets. The following sections show how the Lagrangian coefficient is obtained from the entrainment hypothesis introduced in a reference frame moving with the dominant eddies of the turbulent flow.

3.1 Projected Area Entrainment

Besides the spreading hypothesis (Eqs. 12 and 13), an alternate model is the Projected Area Entrainment (PAE) hypothesis (Lee and Cheung 1990; Lee et al. 2000). The PAE model was introduced to explain the process of the vortex entrainment (Fig. 3). The entrainment into the elemental volume is assumed in this model to be equal to the ambient-flow intercepted by the 'windward' face of the volume.

For a line puff or a line thermal, the volume per unit length of the line element is πR^2; the frontal projection area per unit length is $A_p = 2R_h$; the flow into the

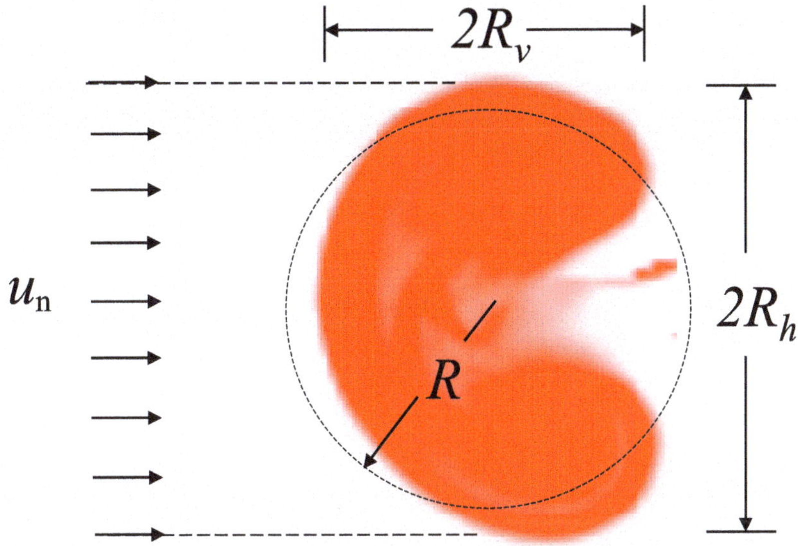

Fig. 3 Projection Area Entrainment (PAE) hypothesis and Surface Area Entrainment (SAE) hypothesis. R (or $\tilde{b}$) is the radius of an equivalent cylindrical or spherical volume; u_n is the component of the velocity relative to the dominant eddy

frontal projection area on the wind-ward face is $\mathcal{A}_p u_n = 2R_h u_n$. Equating the rate of volume change to the wind-ward interception,

$$\frac{D}{Dt}(\pi R^2) = 2R_h u_n \tag{14}$$

which gives the spreading rate

$$\frac{DR}{Dt} = \frac{R_h}{\pi R} u_n = \beta_n u_n, \tag{15}$$

and the normal component of the coefficient

$$\beta_n = \frac{R_h}{\pi R}. \tag{16}$$

Note that R is the radius of the equivalent cross section while $2R_h$ and $2R_v$ are the widths of the dominant eddy normal to and along the ambient flow direction, as shown in Fig. 3. Hence, $\pi R^2 \simeq \pi R_h R_v$ (i.e., $R = \sqrt{R_h R_v}$) and the value of β_n is determined by Eq. 16:

$$\beta_n = \frac{1}{\pi}\sqrt{\frac{R_h}{R_v}} \simeq 0.35. \tag{17}$$

This value of the spreading coefficient, obtained directly from PAE hypothesis, is remarkably close to the value recommended for line puffs/thermals by Chu (1994), and close to the experimental value of around 0.4 for advected line puffs and thermals (Chaps. 7 and 8). The above calculation was based on the experimental and theoretical value of $R_h/R_v = 1.2$ (Chap. 8).

Similar calculation can be made for the asymmetric puffs/thermals. For this case the volume is $\mathcal{V} = \frac{4}{3}\pi R^3$; the frontal projection area is $\mathcal{A}_p = \pi R_h^2$. The mass balance for the elemental volume is

$$\frac{D}{Dt}(\frac{4\pi}{3}R^3) = (\pi R_h^2)u_n \tag{18}$$

This PAE model gives a spreading rate

$$\frac{DR}{Dt} = \beta_n u_n = \frac{R_h^2}{4R_v^2}u_n \tag{19}$$

for the axisymmetric volume and a spreading coefficient

$$\beta_n = \frac{1}{4}(\frac{R_h}{R_v})^2 \simeq 0.36 \tag{20}$$

if $R_h/R_v \simeq 1.2$. This spreading coefficient obtained by PAE model for the round puffs/thermals is comparable to the value of 0.34 for the line puffs/thermals. The round-puff data obtained by Richards (1965) has a spreading rate varying from $\beta_n = 0.25$ to 0.8.

3.2 *Surface Area Entrainment*

Alternatively, the entrainment rate can be assumed to be proportional the 'surface area' instead of the projected area. This SAE (Surface Area Entrainment) model gives identical spreading coefficient for round and line puffs/thermals that is in better agreement with experimental data. The mass balance equation of the SAE model is

$$\frac{D}{Dt}(\mathcal{V}) = \beta_n(\mathcal{A})u_n. \tag{21}$$

The entrainment rate is assumed to be proportional to the surface area $\mathcal{A}$ and velocity excess u_n. For line puffs/thermals, the surface area is $\mathcal{A} = 2\pi R$, and the volume is $\mathcal{V} = \pi R^2$, per unit length of the line element. Equation 21 becomes

$$\frac{D}{Dt}(\pi R^2) = \beta_n(2\pi R)u_n. \tag{22}$$

For round puffs/thermals, the volume is $\mathcal{V} = \frac{4}{3}\pi R^3$, the surface area is $\mathcal{A} = 4\pi R^2$, and the equation is

$$\frac{D}{Dt}(\frac{4\pi}{3}R^3) = \beta_n(4\pi R^2)u_n. \tag{23}$$

Both Eq. 22 for the line puffs/thermals, and Eq. 23 for the round puffs/thermals, give the same spreading rate

$$\frac{DR}{Dt} = \beta_n u_n. \tag{24}$$

The coefficient β_n now represents the effective fraction of the surface area (or effective fraction of velocity) for turbulent entrainment. The values of the spreading coefficients for both the line puffs/thermals and round puffs/thermals are the same according to SAE hypothesis/model. The SAE model is conceptually simple and therefore most adaptable to complex flows encountered in practical application. Since the entrainment rate is directly proportional to the surface area, the volume of turbulent fluid with greater the surface area will have greater rate of entrainment into the volume.

The SAE hypothesis also leads to a naturally simple rule for the treatment of *'merging'*. Merging occurs as multiple parts of the jets and plumes are occupying the same space and in paths of the jets and plumes with large curvature. In such complex situation encountered in practical engineering applications, the natural approach would be to simply adapt the SAE model assuming that the entrainment rate is proportional to the surface area. Computation algorithms already have been developed to find the surface area in highly complex geometry (see Chap. 10). The entrainment rate in the overlapping region is reduced as effective surface area is reduced due to merging.

3.3 Shear Entrainment

The shear entrainment hypothesis is also consistent with the SAE model as can be demonstrated by examining jets/plumes in stagnant fluid. In a Lagrangian reference frame moving with the dominant eddies, the cross sectional area of the jets/plumes increases at a rate proportional to the surface area and the shear velocity. For round jets/plumes, the volume per unit length of the jet/plume element is $\mathcal{V} = \pi\tilde{b}^2$ and the surface area per the same unit length is $\mathcal{A} = 2\pi\tilde{b}$. The mass balance between volume increase in the element and rate of entrainment is

$$\frac{D}{Dt}(\pi\tilde{b}^2) = \beta_s(2\pi\tilde{b})u_s \tag{25}$$

For plane jet, the volume per unit length is $\mathcal{V} = 2\tilde{b}$ and the surface area per unit length is $\mathcal{A} = 2$. Hence,

$$\frac{D}{Dt}(2\tilde{b}) = \beta_s(2)u_s \tag{26}$$

Both Eqs. 25 and 26 give the same equation

$$\frac{D\tilde{b}}{Dt} = \frac{\partial \tilde{b}}{\partial t} + u_s \frac{\partial \tilde{b}}{\partial s} = \beta_s u_s. \tag{27}$$

which becomes

$$\frac{d\tilde{b}}{ds} = \beta_s \tag{28}$$

in steady flow. Data of both plane and round jets/plumes are consistent with this spreading rate (Eq. 28) derived from SAE model. The shear-entrainment coefficient for *steady flow* is $d\tilde{b}/ds = \beta_s = 0.17$ (Chu, 1994), if $\tilde{b}$ is defined at the laminar and turbulent interfaces where the intermittency factor is 50%.

3.4 Summary

To summarize, our recommendation is to use the SAE model in the general formulation. For steady flow, and the additive hypothesis adopted, the values of the β-coefficients are $\beta_s = 0.17$ and $\beta_n = 0.34$ for shear-entrainment and vortex-entrainment, respectively.

4 Puffs and Thermals

The formulation for the puffs and the thermals is carried out as a prelude to the general formulation of the more difficult problem to be presented in the next section. The problem of the puffs and thermals is simple because a single dominant eddy occupies the entire volume of a puff or a thermal. The round thermal is considered here as the first example using the method of excess.

Example 9.1 *A 1.0 m*^{3} *of dredge spoil contaminated with 10.0 kg of cyanide is accidentally released from a ship into a river. The density of the dredge spoil is 1100 kg/m*^{3}*. The river velocity relative to the ship is 8 m/s. An axisymmetrical thermal is formed as the cloud of the spoil mixes with the river water. (a) Find the momentum excess of the thermal upon entering the water surface. (b) Find the dilution and cyanide concentration after the thermal has reached a depth of 5 m below the water surface.*

Solution: Example 5.2 is considered again with the effect of the crossflow included. (a) Upon entering the water surface, the velocity of the spoil (1100 kg) is increased suddenly from zero to 8 m/s. The momentum excess also changes suddenly. At time $t = 0$, the x-component of the momentum excess is $m_x(0) = 1100\text{kg} \times (0 - 8)\text{m/s} = \underline{-8800}$ kg-m/s; the z-component $m_z(0) = 0$. (b) The subsequent variation of the excess momentum is determined by the momentum equations, Eq. 5:

$$\delta \tilde{m}_x = 0 \tag{29}$$

$$\delta \tilde{m}_z = B \delta t \tag{30}$$

The horizontal momentum excess stays constant. The vertical momentum excess changes at a rate proportional to the downward buoyancy force, which is $B = (1100 - 1000) \times 9.81 = 981$ N. Integrating these excess momentum equations once with time, gives

$$\tilde{m}_x = (1+k)\rho(\frac{4\pi R^3}{3})\frac{d\xi}{dt} = -8800, \tag{31}$$

$$\tilde{m}_z = (1+k)\rho(\frac{4\pi R^3}{3})\frac{d\zeta}{dt} = 981\, t. \tag{32}$$

where $(d\xi/dt, d\zeta/dt)$ are the velocity excess. The added-mass coefficient is $k = 0.5$ if the shape of the thermal is assumed to be spherical. The absolute displacements (dx, dz) are the relative displacements $(d\xi, d\zeta)$ plus the contribution from the crossflow. If the crossflow is in the x-direction,

$$dx = U_a\, dt + d\xi,\ dz = d\zeta. \tag{33}$$

The radius of the thermal is assumed to increase at a rate proportional to the velocity excess

$$\frac{dR}{dt} = \beta_n |\tilde{\mathbf{u}}| = \beta_n \sqrt{(\frac{d\xi}{dt})^2 + (\frac{d\eta}{dt})^2}. \tag{34}$$

Hence, the radius increases linearly with distance along the path of the thermal as follows:

$$dR = \beta_n \sqrt{(d\xi)^2 + (d\zeta)^2}. \tag{35}$$

The spreading coefficient $\beta_n = 0.34$. Equations 31, 32 and 35 are numerically integrated to give R, ξ, and ζ. Figure 4 shows dilution ratio, and the path of the thermal, in a coordinate system moving with the ambient cross flow. The thermal moves backwards relative to the ship due to the negative excess momentum of—8800 Kg-m/s. At 5 m depth, the dilution ratio is $S = \mathcal{V}_o/\mathcal{V} = \underline{192}$, which may be compared with the dilution ratio $S = 21$ for the same thermal at the same depth without the

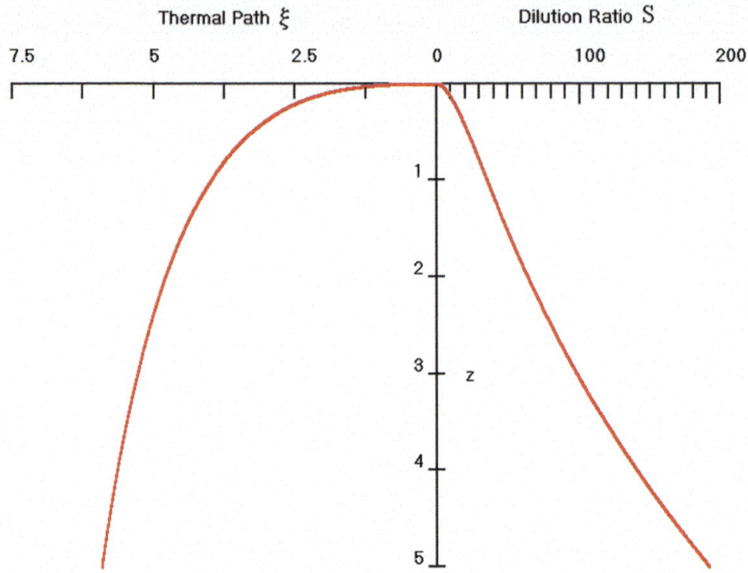

Fig. 4 Path and dilution ratio of the round thermal in a coordinate system moving with the crossflow (Example 9.1)

crossflow. The effect of the crossflow is significant. It enhances the dilution by nine-fold. The volume of the thermal at this depth is 192 m^3 and the cyanide concentration is $\tilde{c} = 10/192 = \underline{0.0521}$ kg/m^3 (i.e, 52.1 mg/L).

5 Buoyant Jet in Crossflow

The formulation for jets and plumes in crossflow follows similar procedure as for the puffs and thermals except that the *elemental volume* must be defined so that the initial condition can be correctly related to the momentum flux and the buoyancy flux at the source. The sketch in Fig. 5 shows how such an elemental volume is defined. By definition, the elemental volume AB is the part of the jet and plume that has a length equal to speed $|\tilde{U}|$. Such an elemental volume is produced by injection of momentum and buoyancy at the source for a period of one time unit. The dominant eddy at one end of the element, A, are released from the source a unit of time earlier than the time when the dominant eddy at other end of the element, B, are released from the source. By defining the elemental volume this way, the momentum and buoyancy associated with the elemental volume is equal to the momentum flux M_o and the buoyancy flux F_o at the source. The interaction between the adjacent elements may be ignored so the formulation for the jets and plumes produced by the continuous source is similar to the method as described in the previous section for puffs and thermals.

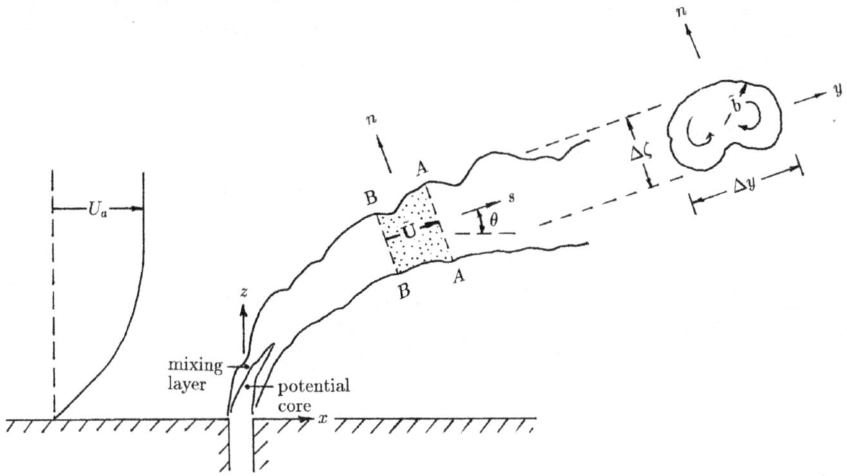

Fig. 5 Elemental volume AB defined by the velocity of the dominant eddies, **U**, along the path of a buoyant jet in a non-uniform cross flow. The momentum flux and the buoyancy flux at the source are M_o and F_o, respectively; the crossflow velocity is $\mathbf{U_a}$

The formulation for jets and plumes in a non-uniform crossflowis by far the most complex problem being considered in this monograph. Due to the possible change in the direction of the cross flow, the buoyant jet follows in general a three-dimensional trajectory. Although the governing equations are the same, much details regarding the geometry is required before the actual numerical computations can be carried out for properties such as the jet trajectory and dilution ratio. Figure 1 shows the buoyant jet and the elemental volume AB produced by the momentum and buoyancy flux, M_o and F_o, at a general orientation relative to the crossflow and direction of the buoyancy force. The direction of the absolute velocity vector $\tilde{\mathbf{U}}$ is in a direction along the path of the buoyant jet. This velocity is equal to the sum of the velocity of the crossflow $\mathbf{U_a}$ and the relative velocity $\tilde{\mathbf{u}}$ as shown in the figure. The geometry of the flow in the zone of flow estabilishment (ZFE) is affected by the development of the potential core and the formulation for the ZFE near the source is different from the formulation for ZEF far away from the source. Therefore the problems for the two regions are considered separately. The flow in the ZEF will be considered first. The problem of the ZFE will be examined in a subsequent section.

5.1 Zone of Established Flow

The calculation in the ZEF is conducted using a Lagrangian coordinate moving with the cross flow. A plume elemental volume AB is selected so that the initial momentum and initial buoyancy associated with the volume is equal to momentum flux M_o and buoyancy flux F_o. This is the elemental volume due to the discharge of momentum and buoyancy at the source for a period of one time unit. The length of this element is equal to the absolute velocity $\tilde{U}$ along the path of the buoyant jet as shown in Fig. 1. The vector tangential to the path of the jet is

$$\tilde{\mathbf{U}} = \tilde{U}\hat{\mathbf{s}}. \tag{36}$$

The excess velocity vector is $\tilde{\mathbf{u}} = \tilde{\mathbf{U}} - \mathbf{U_a}$. It is resolved into two components, u_s and u_n, as follows:

$$\tilde{\mathbf{u}} = \tilde{u}_s\hat{\mathbf{s}} + \tilde{u}_n\hat{\mathbf{n}} \tag{37}$$

where $\hat{\mathbf{s}}$ and $\hat{\mathbf{n}}$ are unit vectors, and $u_s = \tilde{\mathbf{u}} \cdot \hat{\mathbf{s}}$ and $u_n = \tilde{\mathbf{u}} \cdot \hat{\mathbf{n}}$ are the velocity components, in the tangential and normal directions to the path, respectively. The tangential unit vector is calculated using the formula

$$\hat{\mathbf{s}} = \frac{\tilde{\mathbf{U}}}{|\tilde{\mathbf{U}}|} \tag{38}$$

while the normal vector using

$$\hat{\mathbf{n}} = \frac{\tilde{\mathbf{u}}_n}{|\tilde{\mathbf{u}}_n|} = \frac{\tilde{\mathbf{u}} - \tilde{\mathbf{u}}_s}{|\tilde{\mathbf{u}} - \tilde{\mathbf{u}}_s|} \tag{39}$$

The vector $\tilde{u}$ is resolved into components so that the radius of the elemental volume, $\tilde{b}$, is assumed to spread at a rate dependent on the orientation of the element as follows:

$$\frac{\delta\tilde{b}}{\delta t} = \beta_s|\tilde{u}_s| + \beta_n|\tilde{u}_n|. \tag{40}$$

An additive assumption is employed for the two basic entrainment mechanisms. The spreading coefficient β_s is the rate associated with the shear entrainment and the coefficient β_n is the rate associated with vortex entrainment as previously described. The two coefficients are known to have very different values. The spreading coefficients in the tangential shearing,

$$\beta_s = 0.17, \tag{41}$$

is assumed to be the same as the straight jets and plumes in stagnant environment. The spreading coefficient associated with the vortex pair in the direction perpendicular to the path is selected to fit with the experimental data. The best-fit value is

$$\beta_n = 0.34. \tag{42}$$

This spreading coefficient of the vortex entrainment is the same value as the coefficient recommended previously for the line thermal.

The incremental changes in buoyancy and momentum are governed by the buoyancy and momentum equations:

$$\delta \tilde{F} = \tilde{g}\tilde{V}\delta\rho_a, \tag{43}$$

$$\delta\tilde{\mathbf{m}} = -\rho_a\tilde{V}\delta\mathbf{U}_a + \tilde{\mathbf{F}}\delta t \tag{44}$$

where $\tilde{\mathbf{F}} = \tilde{F}\hat{\mathbf{k}}$ is the buoyancy flux vector. The buoyancy terms $\tilde{B}$ in Eqs. 6 and 5 becomes buoyancy flux $\tilde{F}$, since the initial buoyancy in the elemental volume AB is produced by the flux released at the source over a period of one time unit.

Since the elemental volume is produced over a period of one time unit, the length of the volume is $\tilde{U}$ and the volume $\tilde{V} = \pi\tilde{b}^2\tilde{U}_s$ where U is velocity of the dominant eddies along the path of the jet. The momentum associated with the elemental volume, $\tilde{V}$, has two parts. Part of the momentum comes from the turbulent part of the jet and part of it from the induced irrotational motion surrounding the turbulent jet. The part with the turbulent fluid is $\tilde{V} = \pi\tilde{b}^2|\mathbf{U}|$. The added-mass effect due to the irrotational motion depends on the orientation of the jet relative to the cross flow. A pair of added-mass coefficients is introduced to account for the relative motion in the tangential and normal directions perpendicular to the path of the crossflowas follows:

$$m_s = \rho(1+k_s)\tilde{V}u_s, \quad \text{and} \quad m_n = \rho(1+k_n)\tilde{V}u_n \tag{45}$$

where (m_s, m_n) are the components of the excess momentum vector $\tilde{m}$ and (u_s, u_n) are the components of the excess velocity in the tangential and normal directions, respectively.

The solutions for the five unknowns, $\tilde{b}, \tilde{u}_s, \tilde{u}_n, \tilde{U}_s, \theta$, are obtained by numerical integration of Eqs. 40, 43, and 44 using the algebraic relations imposed by Eq. 45. The added mass coefficient selected for the tangential shearing is the same as the one for the free jet, i.e.,

$$k_s = 0.18. \tag{46}$$

The added-mass coefficient for relative motion normal to the path is

$$k_n = 1.00. \tag{47}$$

which is assumed to be the same as the value for acceleration of a circular cylinder normal to the flow.

5.2 *Potential Core in the ZEF*

The entrainment process in the potential core region near the source is determined by the size and velocity of the dominant eddies in the mixing layer. The size of the eddies in the mixing layer is defined by the outer and inner radius, R and r, as shown

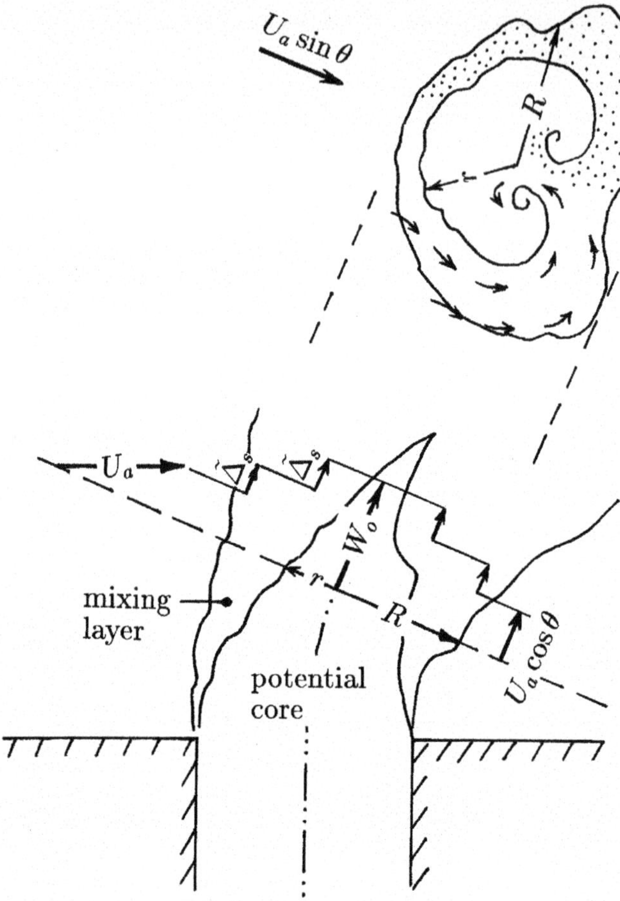

Fig. 6 The development of the mixing layer in the potential core

in Fig. 6. A mixing layer develops around the core due to the exchange of momentum between the fluid in the core and the cross flow. The inner radius, r, decreases while the outer radius, R, increases with distance, $\tilde{s}$, from the source. The inner radius reduces to zero at the end of the potential core. The calculation for the flow in the potential core follows a procedure as given below.

The fluid in the core is lighter than the fluid in the cross flow. The increase in velocity due to buoyancy is determined by the Bernoulli equation:

$$W^2 - g_o' z = W_o^2 \tag{48}$$

where W_o and $g_o' = g(\rho_a - \rho)/\rho_a$ are the velocity and the reduced gravity at the source, respectively.

The dominant eddies in the mixing layer move along the shear layer in the tangential direction. The velocity of the eddies is the average of the velocities in the core and the cross flow: $\tilde{U}_c = \frac{1}{2}(W + U_{as})$. In the tangential direction, the difference in velocity across the mixing layer is $\tilde{\Delta}_s = \frac{1}{2}(W - U_{as})$. In normal direction, the difference is $\tilde{\Delta}_n = U_{an}$. The components of the crossflowvelocity in tangential and normal directions are U_{as} and U_{an}, respectively. The part of the fluid entrained into the mixing layer due to the tangential shearing is $\beta_s \tilde{\Delta}_s 2\pi(R + r)$ while the part due to shearing in the normal direction is $\beta_n \tilde{\Delta}_n 2\pi(R + r)$. The total entrained volume divided by the circumference, $2\pi(R + r)$, gives the entrainment velocity $v_e = \beta_s \tilde{\Delta}_s + \beta_n \tilde{\Delta}_n$. The thickness of the mixing layer, $(R - r)$, is assumed to increase at a rate proportional to the entrainment velocity:

$$\frac{\delta(R - r)}{\delta t} = \tilde{U}_c \frac{\delta(R - r)}{\delta s} = \beta_s \tilde{\Delta}_s + \beta_n \tilde{\Delta}_n. \tag{49}$$

The change in momentum in the potential core follows a relation

$$\delta\tilde{m} = -\rho \tilde{Q} \delta \mathbf{U}_a + \tilde{B} \delta t \tag{50}$$

that is similar to the one in the ZEF, except that the volume flux and the body force terms due to buoyancy now have slightly different expressions:

$$\tilde{Q} = \pi r^2 W + \pi(R^2 - r^2)\tilde{U}_c, \tag{51}$$

$$\tilde{B} = \pi r^2 W \mathbf{g}' + \pi(R^2 - r^2)\tilde{U}_c \frac{\mathbf{g}'}{2}. \tag{52}$$

The excess momentum, $\tilde{m}$, is again resolved into s- and n-components as before. The s-component is

$$\tilde{m}_s = 2\rho\pi r^2 W \tilde{\Delta}_s + \rho(1 + k_s)\pi(R^2 - r^2)\tilde{U}_c \tilde{\Delta}_s. \tag{53}$$

The calculation for the potential core ends when $r = 0$. The variation of the velocity in the potential core, δW, is determined by

$$W \delta W - g_o' \delta z = 0, \tag{54}$$

which is obtained by differentiation of Eq. 40. The variation δr is

$$\delta \tilde{m}_s = f_1 \delta r + f_2, \tag{55}$$

which is obtained by the differentiation of Eq. 50. Hence,

$$\delta r = \frac{\delta \tilde{m}_s - f_2}{f_1}, \tag{56}$$

where

$$f_1 = 4\pi\rho r W\tilde{\Delta}_s + 2\pi\rho(1+k_s)(R-r)U_c\tilde{\Delta}_s,$$
$$f_2 = -\pi\rho[2r^2 W\delta W + r^2 W\delta U_{as} - r^2 U_{as}\delta W]$$
$$-\frac{1}{4}\pi\rho(1+k_s)[8RU_c\tilde{\Delta}_s + (R^2-r^2)(2W\delta W - 2U_{as}\delta U_{as})]. \quad (57)$$

The deflection of the jet is affected by entrainment of ambient fluid into the mixing layer surrounding the core. The process in mixing layer is particularly important for jets and plume with low exit-to-crossflow velocity ratio. The calculation for the mixing layer follows essentially the same procedure as in the ZEF. The variation $\delta\tilde{m}$ is determined by Eq. 55, the variation δW by Eq. 54, the variation $\delta(R-r)$ by Eq. 49, and δr by Eq. 56.

Numerical calculations using the above formulation for jets and plumes of non-unform crossflow were conducted by Dehghani (1996). The results obtained using the standard coefficients (Eqs. 41, 42, 46 and 47) were very consistent with experimental observations for a wide range of crossflow conditions. Some of the numerical results obtained for non-buoyant jets discharged in a direction normal into a crossflow of constant density is presented in this chapter. Figures 7, 8 and 9 show the boundary of

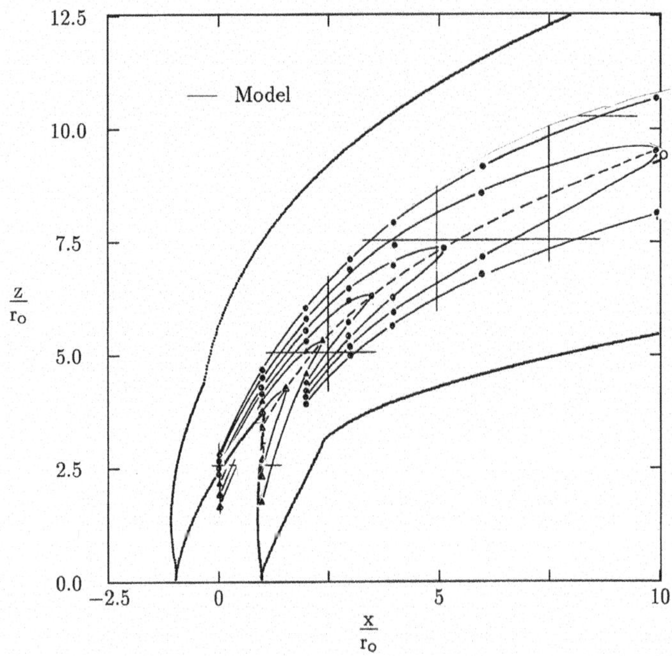

Fig. 7 Jet in the region near the potential core; $W_o/U_a = 2.37$. The contours of constant velocity are reproduced from Chassaing et al. (1974)

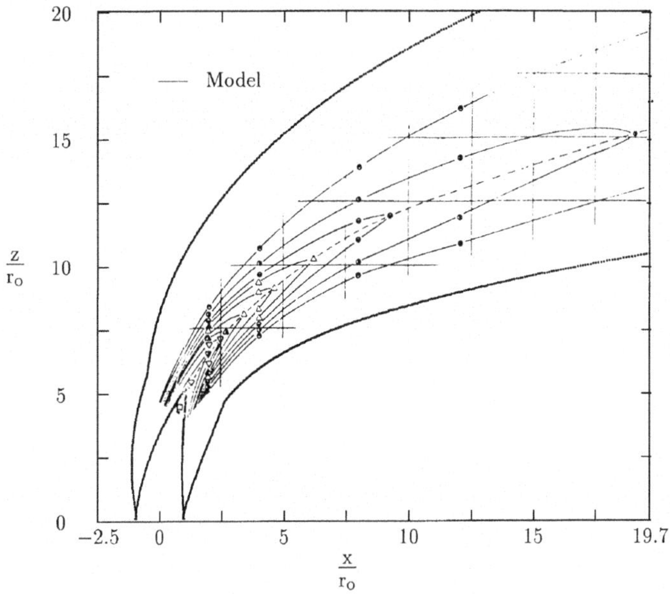

Fig. 8 Jet in the region near the potential core; $W_o/U_a = 3.95$

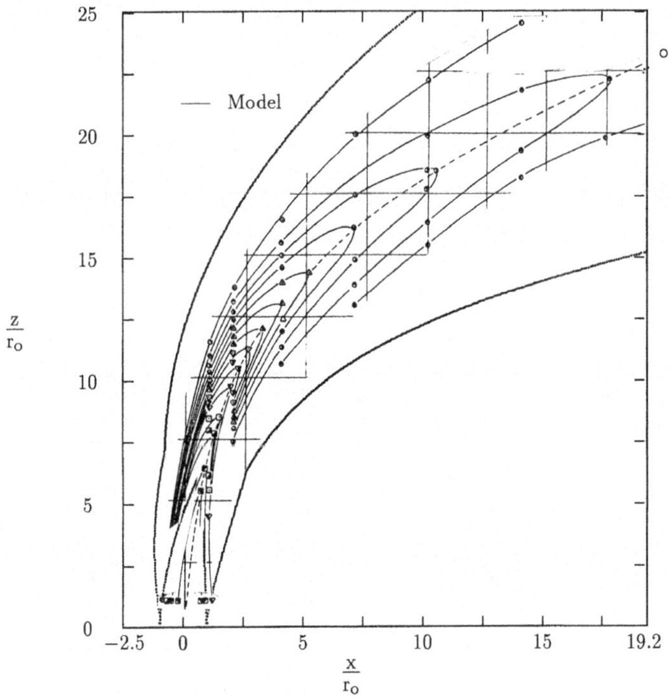

Fig. 9 Jet in the region near the potential core; $W_o/U_a = 6.35$

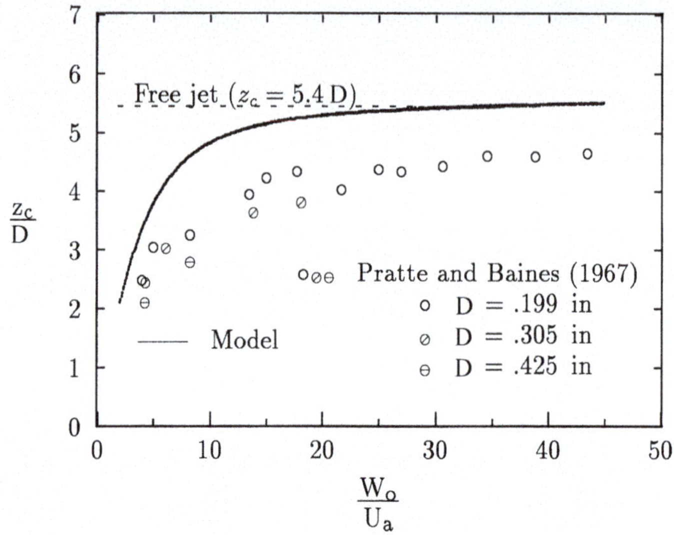

Fig. 10 The length of the potential core obtained from the model and comparison with experimental observation

the deflecting jet in the potential core region, obtained for exit-to-crossflow velocity ratio $W_o/U_a = 2.37, 3.95$, and 6.35, respectively. The contour lines of constant velocity obtained for the same exit-to-crossflow velocity ratios by Chassaing et al. (1974) are superimposed on the same figure for comparison with the numerical results obtained by the Lagrangian model. The development of the potential core can be seen in these figures quite accurately predicted by the model. At the end of the potential core, $r = 0$, $R = \tilde{b}$, $\tilde{\Delta}_s = \tilde{u}_s$ and $\tilde{\Delta}_n = \tilde{u}_n$. Figure 10 plotted the length of the potential z_c (normalized by the diameter of the pipe D) for a range of exit-to-crossflowvelocity ratio varying from $W_o/U_a = 2$ to 45. The data by Pratte and Baines (1967) was obtained by detecting the change of velocity at the end of the core using a hot-film anemometer. The observations follow the same trend as the model, with generally somewhat lower values than predicted. The asymptotic value ($z_c = 5.4D$) for the high exit-to-crossflow velocity ratio is in agreement with the observation in straight jets (see the data of Albertson et al. in Chap. 2).

Figure 11 shows the top and bottom boundaries of the turbulent jet in cross flow. The data are obtained by flow visualization of smoke jets in a wind tunnel (Pratte & Baines, 1967). The lines are the model predictions. Obtained in the ZEF far away from the source, these predictions by the Lagrangian model are also in close agreement with the experimental observation.

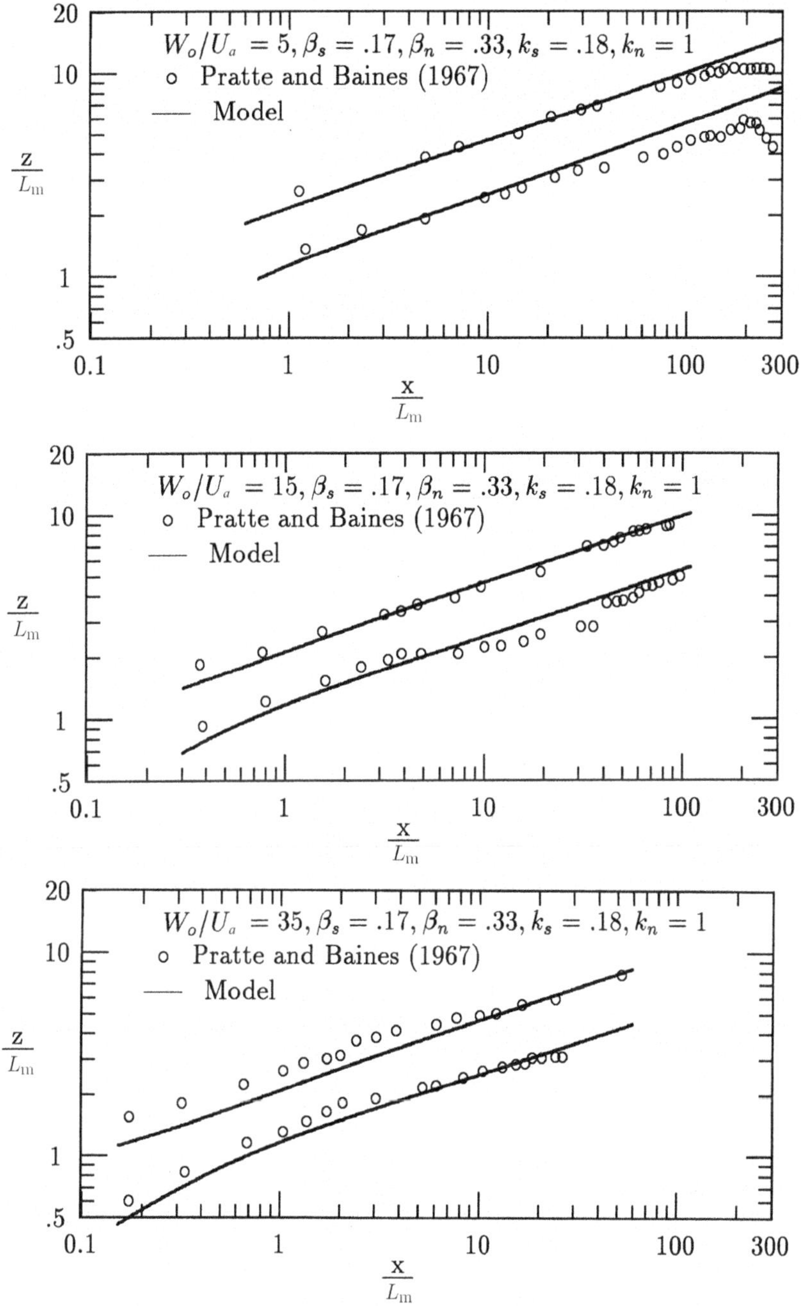

Fig. 11 The development of the jet in the far field region; **a** $W_o/U_a = 5$, **b** $W_o/U_a = 15$, **c** $W_o/U_a = 35$. The set of curves in each figure denote the upper and lower boundaries of the turbulent jet. The experimental data of Pratte and Baines (1967) were obtained by flow visualization of smoke jets in a wind tunnel. $L_m = \sqrt{M_o/\rho}/U_a$ = momentum length scale

6 Summary

The Lagrangian method of the jets, plumes, puffs or thermals (JPPT) is based on two sets of parameters—the spreading coefficients (β_s, β_n) and the added mass coefficients (k_s, k_n). The spreading coefficients are the parameters that characterize the turbulent motion within the dominant eddies. The added mass coefficients on the other hand are to account for the effect of the irrotational motion surrounding the turbulent flow. These parameters are introduced as required by the one-dimensional (along the path) formulation of the problem.

The formulation of the general problem is described in this chapter primarily based on the works of Chu (1994) and Dehghani (1996) which adopts a Lagrangian spreading hypothesis for turbulent closure. A parallel development of the Lagrangian method was made earlier by Cheung (1991) and Lee and Cheung (1990). The Lagrangian jet model JETLAG by Lee and Cheung (1990) and Lee et al. (2000) is based on the project-area entrainment (PAE) hypothesis. A heuristic theory is adopted to treat the shear to vortex entrainment transition. The model has been extensively validated against basic laboratory data and applied to a variety of practical engineering problems. An interactive three-dimensional virtual reality model (VISJET) using JETLAG as the model engine will be given in the next chapter of this monograph.

With the advance of computational method and turbulence modelling, turbulence in jets, plumes, puffs and thermals has increasingly been simulated by two- and three-dimensional models. Techniques such as k-ϵ model and LES model have provided greater details of the local process. The fine scales of the turbulence often are simulated to match directly with the experimental observation. (See Chaps. 5, 7 and 8 for examples of k-ϵ modelling and LES modelling results.) Nevertheless, the one-dimensional simulation method described in this monograph will continue to be significant in engineering analysis for its simplicity and its ability to focus on the overall physical process and the dominant structure of the turbulent flow.

Problems

9.1 A jet is discharged into a river with a flow rate $Q = 0.0138$ m^3/s and a direction perpendicular to the crossflowof the river. The jet diameter is $D = 0.1$ m. The river velocity is $U_a = 0.2$ m/s. Find (a) the excess momentum $\mathbf{m}$ and (b) the tangential and normal components of the excess velocity, u_s and u_n, of the jet immediately after the jet enters the river. See Eq. 45 for the definitions of the excess momentum and excess velocity. Use $\beta_s = 0.17$ and $\beta_n = 0.34$ in your calculation.

9.2 The momentum of a buoyant plume in a cross wind is changed by the buoyancy force in the vertical direction and by the variable wind in the horizontal direction. Approximate the wind profile by the following formula

$$U_a = U_\infty[1 - \exp(-\frac{z}{\delta})].$$

in which U_∞ is the velocity at the edge, and δ the displacement thickness, of the atmospheric boundary layer; z is the elevation above ground.

(a) Write down the momentum, buoyancy and spreading equations for an elemental volume moving with the dominant eddies. Explain how the variable wind enters as an effect to the equations.

(b) Specify the initial condition for a point source of buoyancy flux F_o in the cross wind. Describe how the dilution and the path of the plume are calculated in a computer program. List the steps in the numerical calculation.

9.3 A computer program is developed to calculate the buoyant jet in a crossflow of non-uniform velocity and non-uniform density stratification. The numerical calculations are carried out using an elemental volume following the motion of the dominant eddies. In a Cartesian coordinate system, the absolute and relative positions of the elemental volume are x_i and ξ_i, respectively. The incremental displacement of the volume over a period of time dt is

$$dx_i = \tilde{U}_i dt = (U_{ai} + \tilde{u}_i)dt; \quad i = 1, 2, 3$$

where U_{ai} is the velocity of the cross flow; $\tilde{U}_i = dx_i/dt$ and $\tilde{u}_i = d\xi_i/dt$ are the abosolute and relative velocities of the dominant eddies, respectively.

(a) Show that the components of the unit vector in the tangential and normal direction to the path of the motion are, respectively,

$$s_i = \frac{U_i}{\sqrt{U_\ell U_\ell}}; \quad i = 1, 2, 3$$

$$n_i = \frac{u_{ni}}{\sqrt{u_{n\ell} u_{n\ell}}}; \quad i = 1, 2, 3$$

where $u_{ni} = u_i - (u_\ell s_\ell)s_i$. The repeated index denotes the inner product, which is sum over all three components in the Cartesian coordinates.

(b) If the momentum deficit vector is m_i, show that the components of the momentum deficit vector in the tangential and normal directions are $m_s = m_\ell s_\ell$ and $m_n = m_\ell n_\ell$.

(c) The components of the velocity in the normal and tangential directions are

$$u_s = \frac{m_s}{\rho(1+k_s)\tilde{V}} \quad \text{and} \quad u_n = \frac{m_n}{\rho(1+k_n)\tilde{V}},$$

respectively. Explain why these velocity components, u_s and u_n, are calculated in the computer program.

Chapter 10
Numerical Modelling and Field Application

Jets and plumes are often encountered in environmental engineering applications. Wastewater disposal into coastal waters is an example. In this chapter, we examine a number of data sets obtained in field investigations, and show how numerical model and asymptotic formulae are developed to guide the interpretation of these data.

The problems of wastewater disposal are often analyzed by dimensional consideration in asymptotic flow regimes. Simple formulae for the jet trajectory and dilution in each of the asymptotic flow regimes are obtained by length scale analysis, and flow classification furnishes a means of correlating laboratory and field data. Basic asymptotic flow regimes of a turbulent buoyant jet in a stratified current have been examined: straight jets and plumes, effect of buoyancy and ambient density stratification, strong and weak coflowing jet, puffs and thermals. Relations derived from the length scale analysis are useful in the limiting cases. However, for many practical problems we are faced with the prediction of mixing in the transition between regimes. For example, a strongly buoyant jet in a weak crossflow (e.g. the 95 percentile value of the ambient current speed) is often a design condition. Dilution predictions in the BDNF-BDFF transition, a difficult problem (see later section), are hence needed. As a second example, prediction of the trapping level of a submerged sewage field (which may determine whether the swimmers in a nearby beach will be protected) may be based on a measured vertical ambient density stratification that is neither linear nor two-layered. Another commonly encountered situation is when the direction of momentum is not aligned with the direction of buoyancy, and the jet trajectory is three-dimensional—e.g. in shallow coastal waters the effluent is often discharged horizontally into a perpendicular alongshore tidal current (horizontal jet in crossflow). In these and similar situations, effective use of length scales is much more difficult; the results can sometimes be unreliable or misleading. For engineering application, a general mathematical model for predicting the characteristic jet

---

Supplementary Information The online version contains supplementary material available at https://doi.org/10.1007/978-3-032-01187-9_10.

J. H. W. Lee et al., *Turbulent Jets and Plumes*,
https://doi.org/10.1007/978-3-032-01187-9_10

properties is required (e.g. Muellenhoff et al. 1985). As pointed out by Wood (1993), such a model should be validated in the asymptotic flow regimes. A discussion of representative mathematical models and their pertinent features can be found in Lee et al. (1987c) and Chu and Lee (1996a).

We begin in this chapter by examining first the characteristics of the field data sets in their asymptotic regimes. This will be followed by discussing the transition problem between the regimes. We will examine the so called BDNF-BDFF regimes and compare the predictions with both laboratory and field data. The problem of mixing in a weak current and the necessity for the solution by a Lagrangian numerical model (JETLAG) is discussed. The JETLAG computational model embodies many of the concepts and shares the spirit of the Lagrangian approach outlined in the previous chapter. However, JETLAG uses a projected area entrainment hypothesis (shown to be consistent with the spreading hypothesis in the last chapter) and a heuristic theory to treat the shear to vortex entrainment transition. The model has been extensively validated against laboratory and field data, and successfully applied in a wide range of practical problems. The implementation of JETLAG and the visualization of the plume geometry (using the supplied VISJET software) are illustrated by a series of examples.

1 Initial Dilution of Buoyant Plumes in a Current: The BDNF and BDFF

Figure 1 shows the general case of a horizontal buoyant jet discharge; without loss of generality the ambient current U_a is assumed to flow in the x-direction. The trajectory is shown both for the vertical (x–z) buoyancy plane, and the horizontal (x–y) momentum plane. Based on the concepts outlined in Chap. 8, it is possible to delineate the Buoyancy-dominated Near Field (BDNF) and the Buoyancy-dominated Far Field (BDFF) regimes for a buoyancy jet discharge (e.g. coastal sewage effluent). The relevant equations can be summarized as:

<u>Buoyancy-dominated Near Field (BDNF)</u> $z/L_b = zU_a^3/F_o \ll 1$

$$Dilution: \quad \frac{SQ_o}{U_a L_b^2} = C_1 \left(\frac{z}{L_b}\right)^{\frac{5}{3}} \quad \text{or} \quad \frac{SQ_o}{F_o^{\frac{1}{3}} z^{\frac{5}{3}}} = C_1 \tag{1}$$

$$Trajectory: \quad \frac{z}{L_b} = C_2 \left(\frac{x}{L_b}\right)^{\frac{3}{4}} \tag{2}$$

<u>Buoyancy-dominated Far Field (BDFF)</u> $z/L_b = zU_a^3/F_o \gg 1$

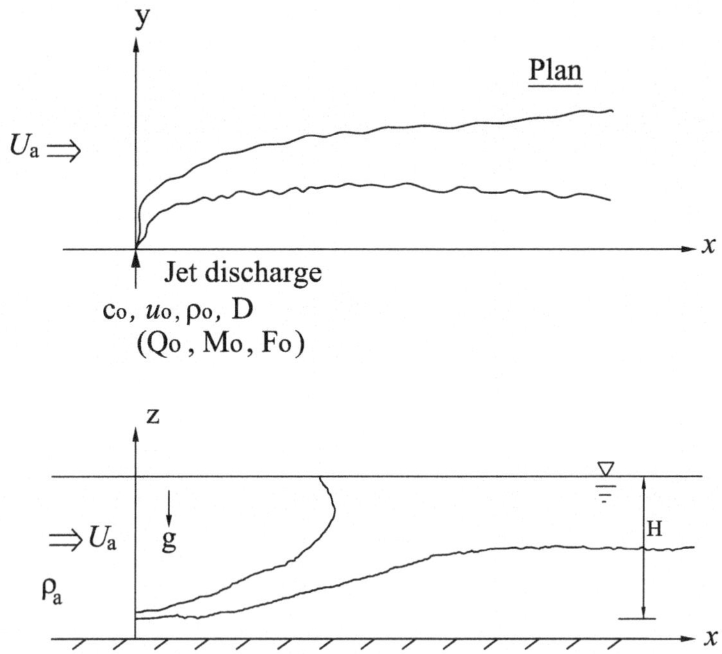

Fig. 1 A horizontal buoyant jet in a crossflow

$$\textit{Dilution}: \quad \frac{SQ_o}{U_a L_b^2} = C_3\left(\frac{z}{L_b}\right)^2 \quad \text{or} \quad \frac{SQ_o}{F_o^{\frac{1}{3}} z^{\frac{5}{3}}} = C_3\left(\frac{z}{L_b}\right)^{\frac{1}{3}} \tag{3}$$

$$\textit{Trajectory}: \quad \frac{z}{L_b} = C_4\left(\frac{x}{L_b}\right)^{\frac{2}{3}} \tag{4}$$

where $L_b = F_o/U_a^3$ is the crossflow buoyancy length scale. Here z and S are the characteristic vertical location (defined e.g. by location of maximum concentration, or by the center of mass) and dilution. Q_o, F_o are the discharge volume and specific buoyancy fluxes respectively.

According to this classification, jet behaviour is determined by the parameter $H/L_b = HU_a^3/F_o$. Whether a given elevation H is in the BDFF or BDNF depends on the ambient current U_a and the buoyancy flux F_o. The weaker the ambient current and the stronger the buoyancy flux, the more likely a given vertical location on the jet trajectory will lie in the near field. Thus in a coastal situation, as the tidal current varies, the free surface can alternately lie in the BDNF and BDFF. Figure 2 shows an example of an actual field condition for each case (Hastings outfall, UK); the plume geometry is generated by a Lagrangian model (to be discussed in the next section). The buoyant plume is portrayed for a situation (i) around tidal slack (surfacing plume in BDNF), and (ii) when the current gains strength (surfacing plume

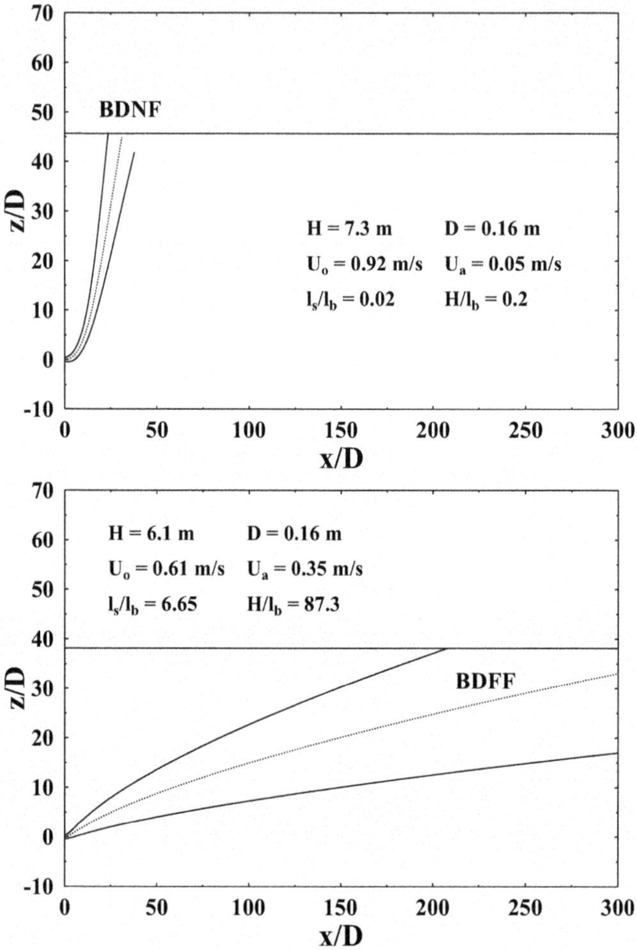

Fig. 2 Example of an actual field condition for BDNF and BDFF (Hastings outfall, UK)

in BDFF). (Note: The reader should be cautioned that the terminology of near and far field in this context is different from the customary usage of near/far field to distinguish the region close to the effluent (and dynamically affected by the buoyant discharge) from the region further away and dominated by ambient advective and decay processes.)

The above length scale analysis has been verified in the laboratory (e.g. Wright 1977a; Lee and Cheung 1991; Knudsen 1988; Wood 1993). Figure 3 shows the correlation of measured centerline dilution of a vertical buoyancy-dominated heated jet as a function of z/L_b. In these experiments the tracer concentration was measured in the centerline plane, either along a horizontal traverse (for near vertical jets) or a vertical traverse (bent-over jets). The location z_c of the maximum concentration

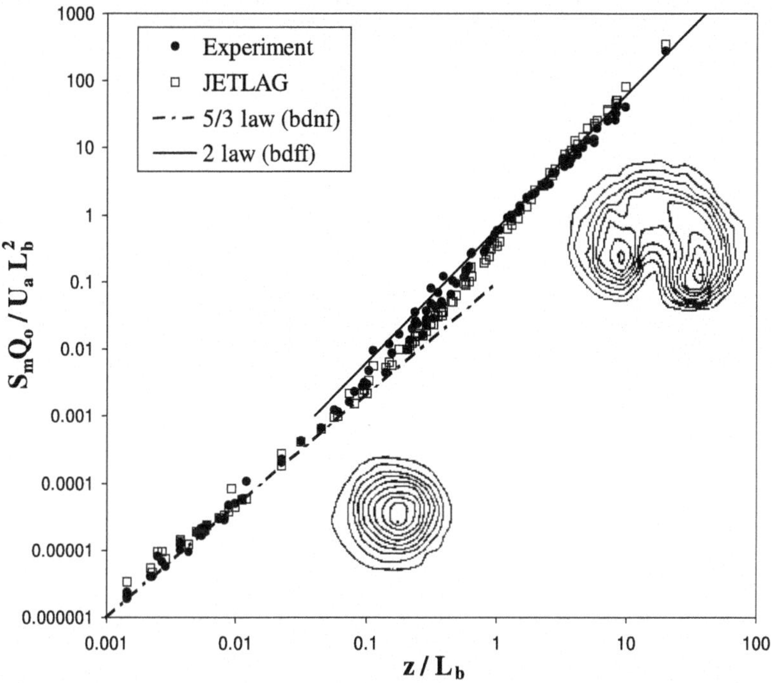

Fig. 3 Comparison of predicted and measured dimensionless centerline dilution as a function of z/L_b for vertical heated jet in crossflow (Cheung 1991)

C_m in such a profile can be obtained, from which the centerline minimum dilution, $S_m = C_o/C_m$, can be determined. It can be seen that the dilution relations for both the limiting cases of a BDNF (weakly advected plume, with a Gaussian cross-section concentration), and the BDFF (advected thermal, with a vortex cross-section) are well-supported by the experiments, with a BDNF/BDFF dilution constant of 0.10 and 0.51 respectively. The data also shows that the BDNF/BDFF transition occurs at $z/L_b \approx 0.1 - 1$.

Figure 4 shows the results of length scale analysis of all available field data of initial dilution collected at a number of sea outfalls in the United Kingdom and in Southeast Florida. In each of these field experiments, the initial dilution refers to the minimum dilution (inferred from the concentration of a tracer) in the "sewage boil" of a surfacing plume. The UK field data are measured for single port discharges (Sidmouth, Bridport, Gosport) or widely spaced ports along a multiport diffuser (Hastings); the discharge can be characterized as a horizontal buoyant jet in a perpendicular unstratified crossflow (Lee and Neville-Jones 1987a). The in-situ dilution measurements are based on radioactivity counts or Rhodamine WT dye concentrations. The field data for the Hastings Outfall (Bennett 1981, 1983) and the Southeast Florida ocean outfalls (SEFLOE experiments, Proni et al. 1994) are derived from a large number of dye injection experiments. The surface dye concentration was tracked with a con-

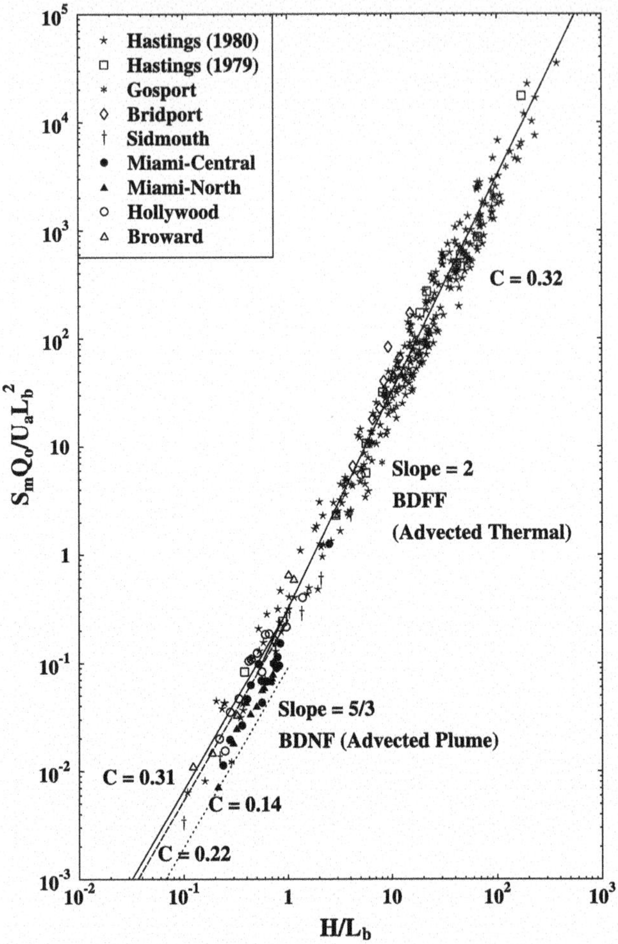

Fig. 4 Comparison of initial dilution prediction equation with field data

tinuous flow fluorimeter, and the minimum dilution was inferred from the maximum dye concentration in the crossing of the boil or from grab samples taken from the observed boil. In addition, part of the Florida dilution data was based on salinity derived from CTD (conductivity-temperature-depth) measurements. Ambient stratification was present in some of the Florida experiments; but apparently the dilution was too low as surface boils were always observed. Together with the measured port discharge, initial density difference, ambient velocity, and water depth above the jet discharge, the dilution constant can be derived for each field condition. The Florida ocean outfall data complement the UK data very well; the discharges are horizontal buoyancy-dominated jets located in deeper waters and characterized by $H/L_b \leq 3$. Most of the data are from single-port outfalls (Hollywood, Broward); the data for multiport outfalls (Miami-Central and Miami-East) are shown only for comparison but not used.

In Fig. 4, the dimensionless initial dilution (minimum surface dilution) in moving water is plotted against the dimensionless discharge depth H/L_b. It is seen that the measured dimensionless dilution (which spans eight decades) for $H/L_b \gg 1$ is well-supported by the BDFF relation (z^2 dilution law in advected thermal), with an average dilution constant of $C = 0.32$ for $H/L_b \geq 1$. The statistical variabilities of the BDFF dilutions have also been studied (Lee and Neville-Jones 1987b); the median value of the BDFF constant is 0.27. Note that the correlation embodies the effect of the surface layer thickness; the dilution constant is hence somewhat lower than that obtained from the laboratory correlations. Although the field data were mainly for horizontal jets in a perpendicular crossflow, the difference of vertical vs horizontal discharges could not be distinguished against the background scatter in the field data.

For the buoyancy-dominated limit, the relatively few field data for $H/L_b \leq 0.1$ can be described by the BDNF equation with a dilution constant of $C = 0.14$. Even without accounting for the surface layer thickness in a near-stagnant water, the BDNF dilution theoretically should only have a value of 0.091. However, the observed dilution is much greater than the stillwater value. This may reflect additional mixing in the boil, and effects not accounted for by the theory (e.g. wave effects, and current non-uniformity). The Florida data is seen to be consistent with the Hastings data. In the BDNF-BDFF transition, $H/L_b = 0.1 - 1$, the BDNF dilution constant has an average value of 0.26. The collective data for $H/L_b \leq 0.5$ gives a dilution constant of 0.24. The measured dilution for the Miami outfalls are in general lower; this has been attributed to plume merging (Proni et al. 1994) and the data have not been included in the analysis.

The results of the data analysis can be summarized in the following initial dilution equations which can be used to estimate in general initial dilution of buoyant wastewater discharges in a current:

Buoyancy-dominated Near Field (BDNF) $H/L_b = HU_a^3/F_o \leq 0.5$

$$Dilution: \quad \frac{SQ_o}{U_a L_b^2} = 0.26\,(\frac{H}{L_b})^{\frac{5}{3}} \quad \text{or} \quad S = 0.26\,\frac{F_o^{\frac{1}{3}} H^{\frac{5}{3}}}{Q_o} \tag{5}$$

Buoyancy-dominated Far Field (BDFF) $H/L_b = HU_a^3/F_o \geq 0.5$

$$Dilution: \quad \frac{SQ_o}{U_a L_b^2} = 0.32\,(\frac{H}{L_b})^2 \quad \text{or} \quad S = 0.32\,\frac{U_a H^2}{Q_o} \tag{6}$$

Example 10.1 *For a vertical sewage discharge with jet velocity $u_o = 1.2$ m/s, D=0.125 m, H=14 m, estimate the initial dilution in moving water for a design ambient current of $U_a = 0.05$ m/s and 0.2 m/s respectively. Assume a relative density difference of 0.025 at the discharge point.*

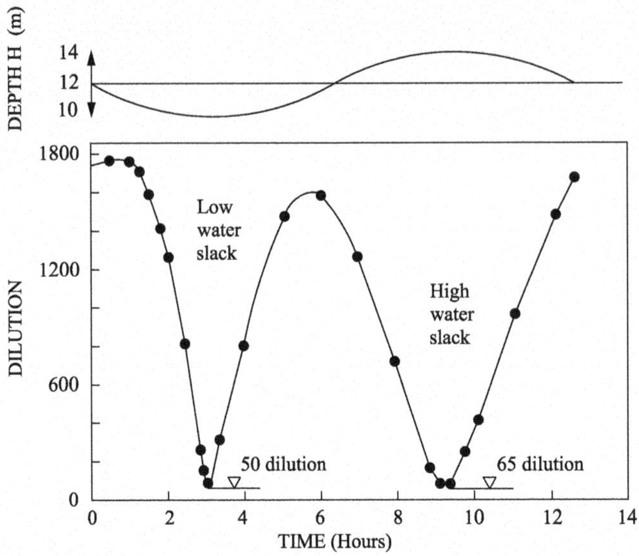

Fig. 5 Huge variation of the initial dilution over a tidal cycle. Dilution ratio changes from 50 at low water slack to nearly 1800 at flood tide

Solution: To find the dilution in a current, first compute the volume flux $Q_o = 1.2 \times \frac{\pi}{4} \times 0.125^2 = 0.01473\ \mathrm{m^3/s}$, and the buoyancy flux $F_o = 1.2 \times \frac{\pi}{4} \times 0.125^2 \times 9.81 \times 0.025 = 3.612 \times 10^{-3}\ \mathrm{m^4/s^3}$. For $U_a = 0.05$ m/s, $HU_a^3/F_o = 0.48$, therefore Eq. 5 with an average dilution constant of 0.26 can be used to compute the BDNF dilution $S_m = 0.26 \times (3.612 \times 10^{-3})^{1/3} \times 14^{5/3}/0.01473 = \underline{221}$.

For $U_a = 0.2\,m/s$, $HU_a^3/F_o = 30.7$, the plume lies in the BDFF when it surfaces. Use Eq. 6 to obtain $S_m = 0.32 \times 0.2 \times 14^2/0.01473 = \underline{852}$.

Compared to a stillwater dilution of 85 (with BDNF dilution coefficient of ≈ 0.1), it can be seen the initial dilution is increased considerably even in the presence of a weak current. In coastal locations with adequate currents, Eqs. 5 and 6 give much more realistic prediction of dilution than the stillwater dilution. In typical situations, the travel time of a buoyant plume to the free surface is of the order of a minute; while the time scale of tidal variation is of the order of an hour (Prob. 3.1). Thus in a tidal situation, the initial mixing problem can often be treated as a quasi-steady problem. Figure 5 illustrates the typical variation of initial dilution over a tidal cycle at a sea outfall.

1.1 *The BDNF-BDFF Transition*

The above field data suggests that even in a weak current the dilution can be about 3 times greater than the still water value (with a dilution constant of approx. 0.1).

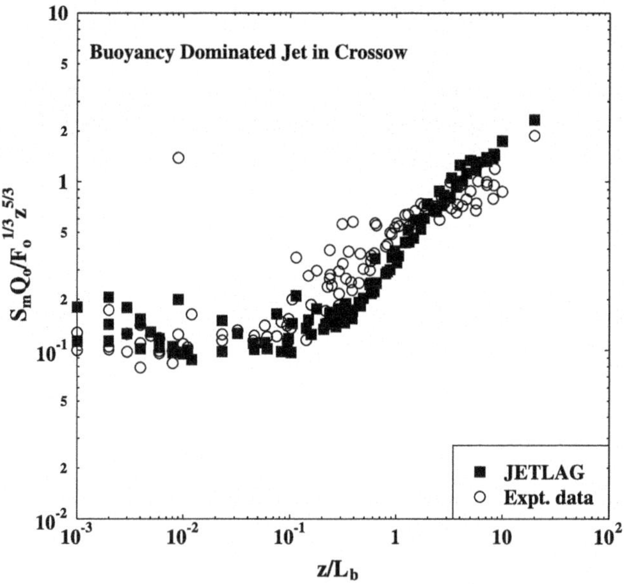

Fig. 6 Dimensionless dilution of buoyant plume in weak current. JETLAG predictions (solid symbol) are compared with experimental data (open symbol)

At first sight this seems to be paradoxical as one would expect the BDNF dilution to be asymptotically the same as that of a stillwater plume. In the field, in the great majority of time, there is an ambient current. In most cases, the dilution departs from the stillwater value at a z/L_b of the order of 0.1. This can be noted in the centerline dilution measurements of laboratory experiments (Wright 1977a; Lee and Cheung 1991) and is apparently related to the change in jet structure from the Gaussian to the vortex type as the ambient current is increased. Figure 6 shows the measured centerline minimum dilution relative to the stillwater dilution for a vertical buoyancy-dominated jet in a weak crossflow (Lee and Cheung 1990; Cheung 1991). In this figure it is clearly seen that the dilution can significantly increase even in a weak current. However for $z/L_b \leq 0.2$, the stagnant plume dilution value is recovered. The prediction of mixing in this BDNF-BDFF transiton is a difficult problem; it will be further discussed in the next section.

2 JETLAG—A Lagrangian Buoyant Jet Model

For environmental impact assessment, outfall design studies, and post-operation monitoring, it is desirable to have a computer model that is able to predict the initial mixing of buoyant wastewater discharges in a current. This is needed, for example, to help define mixing zones outside of which water quality objectives are legally enforced. For discharges into shallow coastal waters, water quality objectives simply cannot be achieved if the effect of a tidal current (which is present for most of the time) is not taken into account. In order to carry out satisfactory risk assessment, the model must give reliable predictions over a wide range of conditons (jet orientation, ambient current and stratification). On the other hand, few models can treat satisfactorily jets with three-dimensional trajectories, such as oblique buoyant jets or rosette-shaped jet groups discharging from modern ocean outfall risers. This section describes a Lagrangian model for buoyant jets with three-dimensional trajectories. The model combines the current version of the Lagrangian model JETLAG with computer graphics techniques to give virtual reality simulations of merging jets issuing from an ocean outfall. In the following the Lagrangian entrainment formulation is first presented. Representative comparisons of model predictions with experimental data are then presented for flow situations that are difficult to model but often encountered in practice. The use of computer graphics for visualizing and assisting the model predictions is then highlighted.

2.1 *Overview of Model*

The Lagrangian model JETLAG (Lee and Cheung 1990) predicts the mixing of buoyant jets with three-dimensional trajectories. Conceptually, the present JETLAG model is a generalization of a previous treatment of jets with a two-dimensional trajectory (Winiarski and Frick 1976; Frick 1984), and is developed after extensive testing against laboratory data. The model does not, strictly speaking, solve the usual Eulerian governing differential equations of fluid motion and mass transport. Instead, the model simulates the key physical processes expressed by the governing equations. The unknown jet trajectory is viewed as a sequential series of "plume-elements" which increase in mass as a result of shear-induced entrainment (due to the jet discharge) and vortex-entrainment (due to the crossflow)—while rising by buoyant acceleration and being sheared over by the crossflow. The model tracks the evolution of the *average* properties of a plume element at each step by conservation of horizontal and vertical momentum, conservation of mass accounting for entrainment, and conservation of tracer mass/heat. The vortex entrainment is determined by a heuristic Projected Area Entrainment (PAE) hypothesis originally proposed by Frick (1984) for buoyant jets with 2D trajectories, while pressure drag is ignored. A justification for the PAE hypothesis based on fundamental principles has been advanced in the previous chapter. A discussion of Frick's (1984) model can be found in Cheung and Lee (1996).

Figure 7 shows a round buoyant jet directed into a uniform horizontal cross flow of velocity U_a. Without loss of generality, the ambient current is assumed to flow along the x-axis. The major parameters governing the flow are the jet volume flux $Q_o = V_o \pi D^2/4$, the kinematic momentum flux $M_o = Q_o V_o$, and the specific buoyancy flux $F_o = Q_o g'_o = Q[(\rho_a - \rho_o)/\rho_a]g$, where V_o, ρ_o are jet discharge velocity and density, respectively, and g is gravitational acceleration. The initial jet discharge

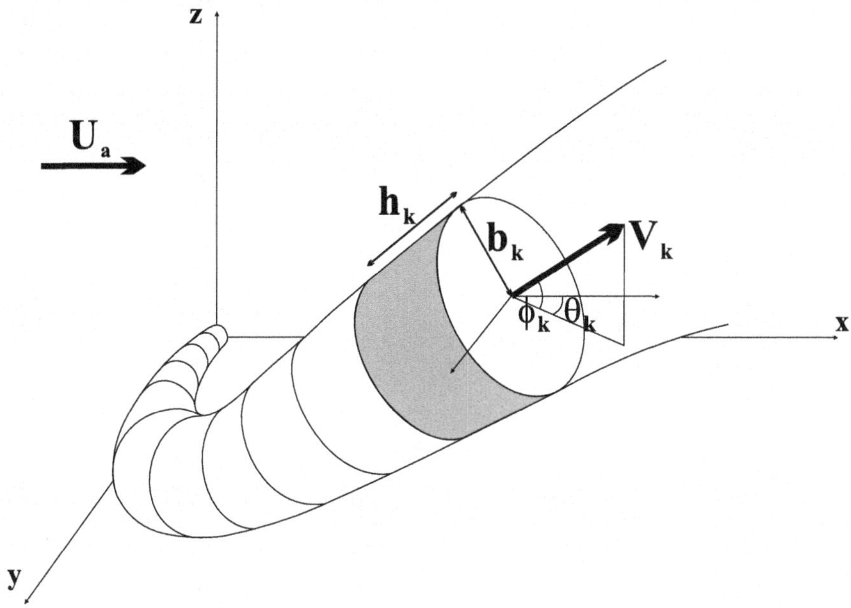

Fig. 7 General Lagrangian model for buoyant jet with three-dimensional trajectories; schematic diagram of jet trajectory traced out by Lagrangian plume elements

angle is (ϕ_o, θ_o), where ϕ_o is vertical discharge angle with respect to the horizontal $x - y$ plane, and θ_o is horizontal discharge angle with respect to x-axis. The objective is to predict the unknown plume geometry and the characteristic dilution and velocity in the jet cross section.

2.2 Basic Concepts of Lagrangian Model

The unknown plume trajectory is viewed as consisting of a sequential series of plume elements (Fig. 7). Each plume element, which can be thought of as a section of a bent cone, is characterized by its location, average velocity, pollutant concentration, tem perature and salinity, width, and thickness. We examine the changes in the properties of a plume element at discrete time steps of Δt.

The turbulent entrainment of the ambient fluid into the plume element is calculated at each step. Based on the increase in elemental mass, the momentum, energy, and tracer mass conservation equations can be solved in their integral form to give the velocity, density, and concentration at the next step. The kinematic and geometrical relations are then used to give the trajectory and relevant dimensions. The evolution of a single plume element is then equivalent to a solution of the properties along the unknown jet trajectory in the three-dimensional steady flow. By defining the thickness of a plume element as proportional to the local jet velocity, the Lagrangian method is similar to a time integration along the jet trajectory. The formulation, however, allows a certain flexibility in accounting for the detailed jet geometry in the determination of the vortex entrainment. The initial characteristics of the element at the source can be clearly related to the discharge parameters.

2.3 Model Formulation

At the kth step consider a plume element located at (x_k, y_k, z_k) with velocity (u_k, v_k, w_k) where (u, v) is the horizontal velocity, and w is the vertical velocity; $V_k = \sqrt{u_k^2 + v_k^2 + w_k^2}$ is the magnitude of the velocity (Fig. 7). The temperature, salinity, and density are denoted by T_k, S_k and ρ_k. The jet axis makes an angle of ϕ_k with the horizontal plane, and θ_k is the angle between the x-axis and the projection of the jet axis on the horizontal plane. The half-width or radius of the plume element is b_k; h_k is the thickness/length, defined as proportional to the magnitude of the local jet velocity, $h_k \propto V_k$. The mass of the plume element is then given by $M_k = \rho_k \pi b_k^2 h_k$.

Given the increase in mass due to turbulent entrainment, ΔM_k, the plume element characteristics at the next step are obtained by applying conservation of mass, horizontal and vertical momentum, energy, and tracer mass to the discrete element.

- Mass:

$$M_{k+1} = M_k + \Delta M_k \tag{7}$$

$$M_{k+1} = \rho_{k+1} \pi b_{k+1}^2 h_{k+1} \tag{8}$$

- Concentration and density:

$$S_{k+1} = \frac{M_k S_k + \Delta M_k S_a}{M_{k+1}} \tag{9}$$

$$T_{k+1} = \frac{M_k T_k + \Delta M_k T_a}{M_{k+1}} \tag{10}$$

$$\rho_{k+1} = \rho(S_{k+1}, T_{k+1}) \tag{11}$$

$$c_{k+1} = \frac{M_k c_k + \Delta M_k c_a}{M_{k+1}} \tag{12}$$

- Horizontal momentum:

$$u_{k+1} = \frac{M_k u_k + (\Delta M_k) U_a}{M_{k+1}} \tag{13}$$

$$v_{k+1} = \frac{M_k v_k}{M_{k+1}} \tag{14}$$

- Vertical momentum:

$$w_{k+1} = \frac{M_k w_k + M_{k+1} \left(\frac{\Delta\rho}{\rho}\right)_{k+1} \mathrm{g}\Delta t}{M_{k+1}} \tag{15}$$

$$(HVEL)_{k+1} = (u_{k+1}^2 + v_{k+1}^2)^{1/2} \tag{16}$$

$$V_{k+1} = (u_{k+1}^2 + v_{k+1}^2 + w_{k+1}^2)^{1/2} \tag{17}$$

- Thickness/radius:

$$h_{k+1} = \frac{V_{k+1}}{V_k} h_k \tag{18}$$

$$b_{k+1} = \left(\frac{M_{k+1}}{\rho_{k+1}\pi h_{k+1}}\right)^{1/2} \tag{19}$$

- Jet orientation:

$$\sin\phi_{k+1} = \left(\frac{w}{V}\right)_{k+1} \tag{20}$$

$$\cos\phi_{k+1} = \left(\frac{HVEL}{V}\right)_{k+1} \tag{21}$$

$$\sin\theta_{k+1} = \left(\frac{v}{HVEL}\right)_{k+1} \tag{22}$$

$$\cos\theta_{k+1} = \left(\frac{u}{HVEL}\right)_{k+1} \tag{23}$$

- Location:

$$x_{k+1} = x_k + u_{k+1}\Delta t \tag{24}$$

$$y_{k+1} = y_k + v_{k+1}\Delta t \tag{25}$$

$$z_{k+1} = z_k + w_{k+1}\Delta t \tag{26}$$

$$\Delta s_{k+1} = V_{k+1}\Delta t \tag{27}$$

- Initial condition:

$$(u, v, w)_o = (V_o cos\phi_o cos\theta_o, V_o cos\phi_o sin\theta_o, V_o sin\phi_o) \tag{28}$$

$$(b, h)_o = (0.5D, 0.5D) \tag{29}$$

$$(S, T, \rho, c)_o = (S_o, T_o, \rho_o, c_o) \tag{30}$$

$$\Delta t_o = 0.1 \times \frac{h_o}{V_o} \tag{31}$$

In these equations, pressure drag is tacitly neglected, and all of the entrained ambient flow is assumed to have velocity $U_a\hat{i}$. The initial location of the plume element is set at the discharge location. It can be readily shown that a consistent formulation can be obtained if the initial plume element velocity, radius, density, salinity, and pollutant concentration, are set equal to the jet discharge values (see Prob. 10.7 for a heuristic proof of the Lagrangian formulation).

By the nature of the formulation, there is no distinction between the zone of flow establishment (the potential core) and the zone of established flow of the buoyant jet; only the average properties in the jet cross section are represented. In effect, the model postulates a top-hat profile for both the velocity and the concentration. The formulation, however, can be readily extended to include the potential core in accordance with the concepts laid out in Chap. 4; such a model also gives satisfactory predictions similar to those in Figs. 7 and 9 in Chap. 4.

2.4 Shear and Vortex Entrainment

The crux of the plume model is the computation of turbulent entrainment for the general situation of an arbitrarily-inclined plume element in a crossflow. The increase in mass of the plume element is due to turbulent entrainment of the ambient flow. Close to the discharge point, or in a very weak currrent, shear-induced entrainment dominates. In the bent-over phase, however, the vortex entrainment of the crossflow dominates. The increase in mass of the plume element at each step, ΔM, is computed as a function of two components: the shear entrainment due to the relative velocity between the plume element and the ambient velocity in the direction of the jet axis, ΔM_s, and the vortex entrainment ("forced" entrainment) due to the ambient crossflow, ΔM_f. The formulation of the shear and vortex entrainment is first separately presented below. The computation of the total entrainment from these components is discussed in a later section.

Shear entrainment :

In the current model, the shear entrainment ($\Delta M_s = E_s$) at each time step k is computed as (Fig. 7):

$$E_s = 2\pi \alpha_s b_k h_k \Delta U \, \Delta t \tag{32}$$

$$\alpha_s = \sqrt{2}\,(0.057 + 0.554 \sin\phi_k / F_l^2)\,(\frac{2V_k}{\Delta U + V_k}) \tag{33}$$

where V_k = jet velocity, $\Delta U = |V_k - U_a \cos\phi_k \cos\theta_k|$ is the relative jet velocity in the direction of the jet axis, and b_k, h_k are the radius and thickness of the plume element (Fig. 7); α_s and F_l are the entrainment coefficient and the local jet densimetric Froude number respectively. This shear entrainment expression can be shown to predict correctly the limiting case of straight jets and plumes, and a jet in coflow (Eqs. 45 in Chap. 3 and 44 in Chap. 6).

Projected Area Entrainment (PAE)

The vortex entrainment due to the crossflow is modelled using the Projected Area Entrainment (PAE) hypothesis (Frick 1984, Lee and Cheung 1990, Cheung and Lee 1996). Experimental observations (Chu and Goldberg 1974, Chu 1977) have shown that the transfer of horizontal momentum is complete beyond a few jet diameters. We assume that all the ambient flow on the windward side of the plume is entrained into the plume element—i.e. the entrainment due to the crossflow (the vortex pair entrainment in the far field) is equal to the ambient flow intercepted by the 'windward' face of the plume element. The vortex entrainment of the ambient flow into an arbitrarily inclined plume element, the incremental increase in mass at each step $\Delta M_f (= E_f)$, can be formulated and results in the following PAE expression (Lee and Cheung 1990):

$$\begin{aligned}\Delta M_f &= \rho_a U_a \,[2 b_k h_k \sqrt{1 - \cos^2\theta_k \cos^2\phi_k} \\ &\quad + \pi b_k \Delta b_k \cos\phi_k \cos\theta_k \\ &\quad + \frac{\pi b_k^2}{2} \Delta(\cos\phi_k \cos\theta_k)]\, \Delta t \\ &= E_p(\text{projection}) + E_w(\text{growth}) + E_c(\text{curvature}) \\ &= \rho_a U_a (A_p + A_w + A_c)\, \Delta t \end{aligned} \tag{34}$$

An initial estimate of ΔM_f can be obtained as:

$$\begin{aligned}\Delta M_f &= \rho_a U_a b_k h_k \,[2\sqrt{\sin^2\phi_k + \sin^2\theta_k - (\sin\phi_k \sin\theta_k)^2} \\ &\quad + \pi \frac{\Delta b_k}{\Delta s_k} \cos\phi_k \cos\theta_k \\ &\quad + \frac{\pi b_k}{2} \frac{(\cos\phi_k \cos\theta_k - \cos\phi_{k-1} \cos\theta_{k-1})}{\Delta s_k}]\, \Delta t \end{aligned} \tag{35}$$

Based on the initial estimate of ΔM_f, the properties of the plume element at the next step $k + 1$ can be computed (see Eqs. 7–27). A revised estimate of $\Delta b/\Delta s$ and $\Delta(\cos\phi \cos\theta)/\Delta s$ can be calculated using the values at step k and $k + 1$ [e.g., $\Delta b/\Delta s = (b_{k+1} - b_k)/\Delta s_k$]. An improved estimate $(\Delta M_f)_k$ can then be obtained from Eqs. 34 and 35, and the procedure repeated until convergence is achieved. In all the results presented here, typically two iterations are required for convergence of the key variables (with a relative tolerance of 10^{-4}) at each step. The time step

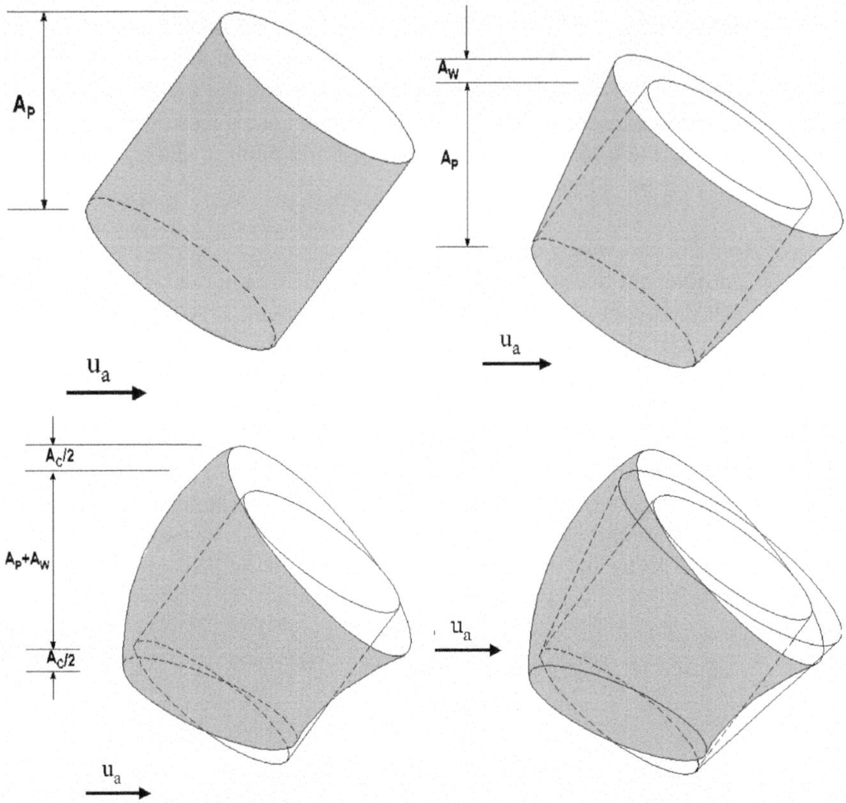

Fig. 8 In the current model, the Illustration of the three contributions of the Projected Area Entrainment (PAE): projected area term A_p, increase in area due to plume growth A_w, and correction in area due to plume curvature A_c, and the sum of all three terms

Δt can be fixed or variable; it is chosen via a "predict-correct" procedure to attain a prescribed fractional change in mass (typically of the order of 1%) at each step. It should be emphasized that the ambient density ρ_a takes on local values; thus ambient stratification can be readily handled.

In Eqs. 34 and 35, there are three contributing terms to the projected area: they are respectively the entrainment due to the projected plume area normal to the cross flow—the projection or cylinder term E_p, a correction term due to the increase in plume width E_w, and a correction term due to plume curvature E_c; the total projected area entrainment is $E_f = E_p + E_w + E_c$. Figure 8 illustrates the meaning of the respective projected area terms pictorially for normal situations for a current flowing from the left. It can be shown that each term has its importance over different parts of the trajectory, so that none of the terms can be neglected. It is noted that the first term has been rewritten to minimize roundoff error, and derivatives are backward differenced from values at the present and previous steps. For jets with two-dimensional trajectories, $\theta_k = 0$, Eq. 34 reduces to the PAE expression given by Frick (1984).

In developing this general vortex entrainment formulation, a wide variety of flow situations have been considered. Equation 34 is generally valid without having to make special provisions—i.e. the signs and angles take care of themselves properly; it is not necessary (in fact erroneous) to force each term to be positive. The projection term A_p is always positive. However, there are situations when the growth term A_w or curvature term A_c can be negative. For example, $A_w \leq 0$ for $\phi_k \geq 90^o$, when a plume element is directed against the ambient current (oblique jets or horizontal jets with $\theta_o \geq 90^o$). A detailed discussion of the physical interpretation of the individual terms for different cases can be found in Cheung and Lee (1999).

The total entrainment

The total entrainment can be obtained from a maximum hypothesis, $\Delta M = Max(\Delta M_s, \Delta M_f)$; alternatively an additive hypothesis, $\Delta M = \Delta M_s + \Delta M_f$ can be used. Comparison with basic data shows that the maximum hypothesis in general gives better results. However, the use of this hypothesis may give unreasonable predictions for a weak current. When U_a is small, e.g. around tidal slack, or when Q_o is large, it is expected that shear entrainment dominates. However, the relative jet velocity decreases as the ambient current increases - thus leading to a decrease in entrainment with increasing crossflow, a result that is not borne out by experiments of plumes in weak crossflow (Lee and Cheung 1991). Neither of these two approaches can reproduce satisfactorily the initial mixing data in the bdnf-bdff transition (Fig. 3). A general modelling framework is needed to handle the transition from the shear entrainment (advected jet/plume) regime to the vortex entrainment (advected puff/thermal) regime.

2.5 Formulation for Near-Far Field Transition

Consider the general situation of a jet/plume in a weak crossflow, with the jet axis making an angle of ψ_k with the crossflow (Fig. 9); the ambient current in the plane of the jet cross-section is then $U_a \sin \psi_k$. The general entrainment formulation, which models the near-far field transition (or weak to strong current), is as follows:

$$\Delta M = E_s \frac{(\pi - \varphi_k)}{\pi} + E_f \sin \varphi_k \tag{36}$$

where φ_k is a "separation angle" which delineates the relative importance of shear and vortex entrainment. φ_k is computed from the maximum radial shear entrainment velocity and the ambient velocity according to:

$$\cos(\varphi_k) = Min(\frac{V_r(max)}{U_a \sin \psi_k}, 1) \tag{37}$$

Fig. 9 Weakly bent-over buoyant jet in crossflow

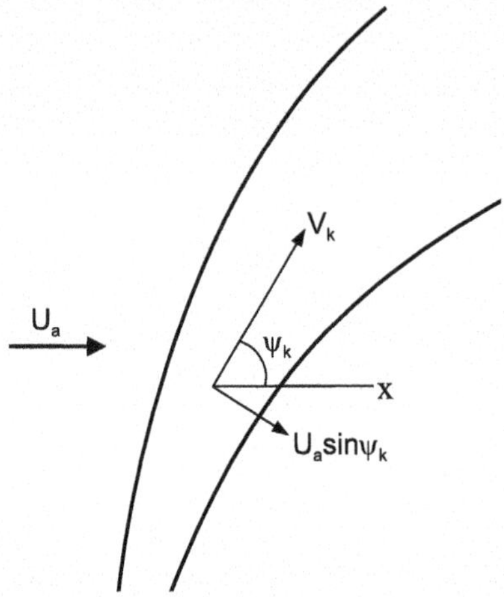

For a round jet or plume in stagnant fluid, the longitudinal jet velocity is self-similar and approximately Gaussian; the radial entrainment velocity field V_r can be obtained from continuity as a function of the local centerline velocity; since the spreading rate of straight jets and plumes is approximately the same $db/ds \approx \sqrt{2}(0.109)$, it can be shown that $V_r(max) = 0.421\alpha_s \Delta U_g$, and occurs at a radial position of $r/b_k \approx \sqrt{2}$ (Problem 10.8). where $\Delta U_g = 2\Delta U$ is the centerline excess velocity.

The above formulation is motivated by the work of Gaskin (1995) and Gaskin et al. (1995). In a weak crossflow, both theory and experiments suggest the irrotational entrainment flow in the plane of the jet cross-section can be modelled as the sum of the radial entrainment flow of a straight jet/plume in stagnant fluid, V_r, and the uniform ambient flow defined by $U_a \sin \psi_k$. This provides a heuristic general framework to piece together E_s and E_f. Figure 10 shows the external flow outside of the jet obtained in this way (for convenience we define the jet/plume boundary by the location of maximum entrainment velocity, $r = \sqrt{2}b_k$). Consider the radial velocity *into* the plume element along the centerline axis; after superposition the velocity on the windward side of the plume element will be reinforced, $V_r + U_a \sin \psi_k$, while that on the leeward side will be decreased, $V_r - U_a \sin \psi_k$. Note that the turbulent entrainment which accounts for mixing is the inflow into the plume element. If the net radial velocity all around the element boundary is directed inwards, the total entrainment will be exactly equal to the shear entrainment E_s, since the ambient velocity U_a contributes no net entrainment into the element when integrated around the circumference. The critical transition point, when total entrainment exceeds shear entrainment, is given by $U_a \sin \psi_k = V_r(max)$, where $V_r(max)$ can be computed locally in a general way. To see the possibility of a zero radial inflow on the plume boundary, we take the maximum radial shear entrainment velocity and equate to the

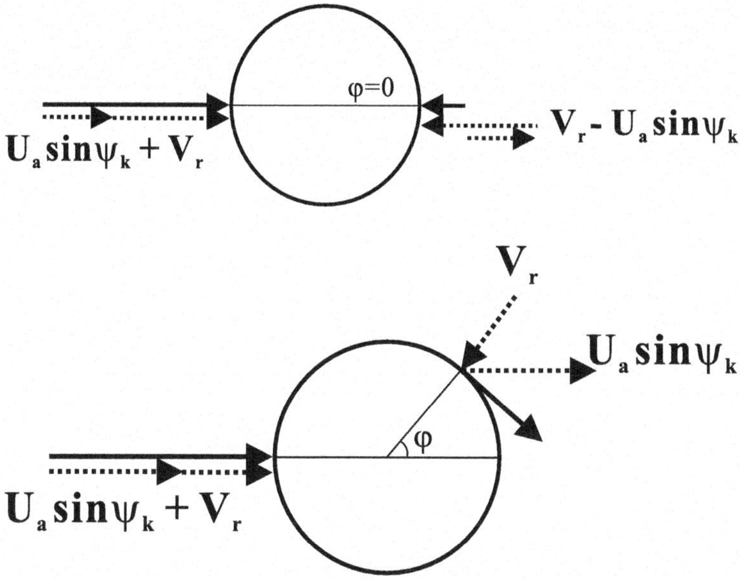

Fig. 10 Net radial velocity at jet boundary (entrainment flow in jet cross-section); **a** total entrainment = shear entrainment, $E = E_s$ (top); **b** $E \geq E_s$, with stagnation point at angle φ

radial component of the ambient velocity (Fig. 10):

$$\frac{U_a \sin \psi_k}{\alpha_s \Delta U_g} \cos \varphi_k = 0.421 \tag{38}$$

The above equation shows that, for a given ΔU_g the minimum ambient velocity U_a^* to give a 'stagnation point' (point of zero normal velocity) on the plume boundary is $\frac{U_a^* \sin \psi_k}{\alpha_s \Delta U_g} = 0.421$. For $U_a \geq U_a^*$, Eq. 38 can be used to locate the "stagnation point", i.e. the angle φ_k (Fig. 10). The corresponding stream function for such a flow is illustrated in Fig. 11. Equations 36 and 37 show that as $U_a \to 0$, $\varphi_k \to 0$, $\Delta M \to E_s$; i.e. the entrainment is entirely due to shear entrainment, while as $U_a \to \infty$, $\varphi \to \pi/2$—i.e. entrainment is almost entirely due to crossflow (E_f typically much greater than E_s). The two symmetrical limiting streamlines that pass through the stagnation points give the width of the upstream flow that gets entrained into the plume element and hence the total entrainment flow.

To summarize, the main consequence of this hypothesis is that for an ambient current up to a value defined by ψ_k and ΔU_g, the total entrainment is entirely due to the shear entrainment E_s. Beyond this critical limit, the entrainment increases to the vortex entrainment value E_f. It can be shown that (Problem 10.9) the entrainment flow q_e can be written as (Gaskin 1995):

$$\frac{q_e}{2\pi\alpha\Delta U_g b_g} = \frac{\pi - \varphi_k}{\pi} + \frac{2U_a \sin \psi_k r^*}{2\pi\alpha\Delta U_g b_g} \sin \varphi_k \tag{39}$$

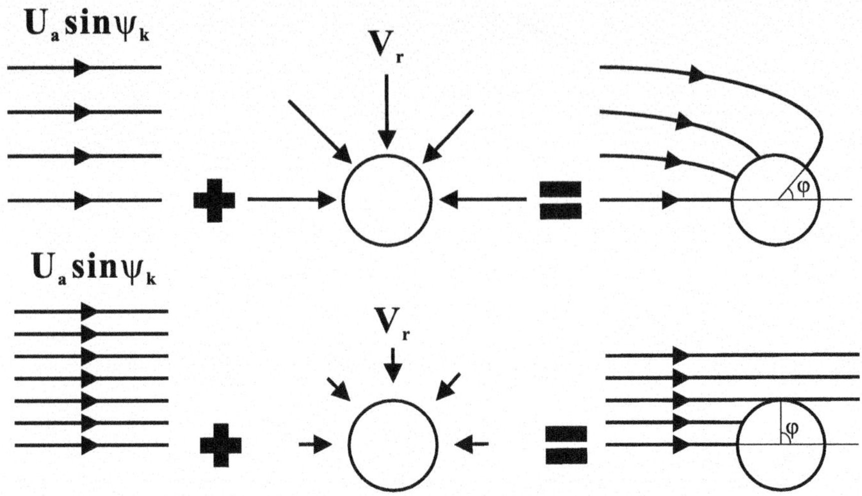

Fig. 11 Entrainment flow pattern for **a** jet in weak crossflow (top); and **b** jet in strong crossflow (bottom)

where $r^* \approx \sqrt{2}b_k$ is the radial location of the plume boundary. Since $E_s = 2\pi\alpha\Delta U_g b_g$, and the PAE entrainment is $E_f = 2U_a \sin\psi_k r^*$, the above equation can be replaced by the equivalent form, Eq. 36, which is amenable to general calculation. Note that Eqs. 36 and 37 are generally applicable to jets with three-dimensional trajectories. It should be emphasized that the ultimate test of the heuristic theory outlined is validation against experimental data.

An update to the transition formulation

Consider the original transition formulation Eq. 36, in the limiting case of an advected line thermal, with $\varphi = \pi/2$, it gives a result that is slightly larger than the desired limit of $\Delta M \approx E_f$, with $\Delta M = E_s/2 + E_f$. Extensive model testing against benchmark laboratory data suggests that model performance can be optimized with a modified implementation of Eq. 36:

$$\Delta M = E_s \frac{(\pi - \varphi_k)}{\pi} \tanh\left[10\left(\frac{\pi}{2} - \varphi_k\right)\right] + E_f \sin\varphi_k \tag{40}$$

which ensures that the turbulent entrainment is due to only vortex entrainment as $\varphi_k \to \pi/2$ in the far field.

For the majority of cases, the predictions of both the present model (Eq. 40) and the earlier version (Eq. 36 from First Edition) give similarly good results (More details are given in Lee et al. 2008). This is expected as both equations are practically equivalent in the far-field because $E_f \ll E_s$. However the new transition formulation is more robust with exactly correct asymptotic behavior, and gives overall slightly better results. The improved model has been incorporated to the VISJET software that is accompanied with this monograph.

2.6 Comparison of Model Predictions with Laboratory Data

Using the general formulation, the turbulent entrainment into the Lagrangian plume element can be computed at each step. The change in plume properties and the plume trajectory can then be obtained. JETLAG reproduces the correct behaviour of (i) a round buoyant jet in stagnant or near stagnant fluid, and (ii) an advected line puff/thermal in a bent-over momentum/buoyancy dominated jet. The model has been validated against experimental data by different investigators for: straight jets and plumes, vertical buoyant jet and dense plume in crossflow, oblique momentum jet in crossflow, horizontal buoyant jet in coflow; horizontal buoyant jet in crossflow; vertical buoyant jet in stratified crossflow; coflow and counterflowing momentum jets; buoyant plumes in weak current. We present herein only model-data comparisons for representative cases, including some that have proved difficult for integral models in general (e.g. Muellenhoff et al. 1985; Wood 1993).

Figure 12 shows the comparison of predicted dilution of a horizontal buoyant jet in still fluid with the classic data of Cederwall (1968); the model is well-supported by the data. Predictions of velocity and concentration for a vertical buoyant jet in stagnant fluid (Fig. 13) are also in good agreement with the data of Papanicolaou and List (1988). Figure 14 shows the comparison of predicted and observed trajectories and concentration of a horizontal buoyant jet in crossflow (experimental data from Cheung 1991); both the jet paths in the horizontal and vertical planes and tracer concentration are well-predicted. Figure 3 shows the predicted centerlinedilution for a plume as a function of $z/l_b = zU_a^3/F_o$, where z= elevation above source, and F_o=jet buoyancy flux. In a very weak current, $z/l_b \ll 1$, it is seen that the dilution is given by that of a straight plume, while significant increases in dilution can be achieved even in a weak current, $z/l_b \approx 0.1 - 1$. The present prediction represents a significant improvement over alternative formulations which do not consider the transition problem. Figure 15 shows representative comparison of predictions of jets in stratified crossflow with the experiments of Wright (1977b). A stratified crossflow is often encountered in the field situation. It can be seen the model predictions of final equilibrium level (trap level indicated by circles) compare very well with the experimental data (solid horizontal bar at right of figure). Finally, Fig. 16 shows a comparison of predicted maximum plume rise height with data of Wright (1977b), and Fig. 17 compares predicted centerline jet velocity in a counterflow with data of Lam and Chan (1997); the model predictions are in reasonably good agreement with experimental data. Other cases of buoyant jet in still water, a coflowing jet, vertical jet in crossflow, dense plumes, and oblique jets have been reported (Lee and Cheung 1990; Lee et al. 2000). In addition to comparison of model predictions with data of individual experiments, numerical experiments simulating a momentum-dominated jet and a buoyancy-dominated jet have been reported (Lee et al. 1987c).

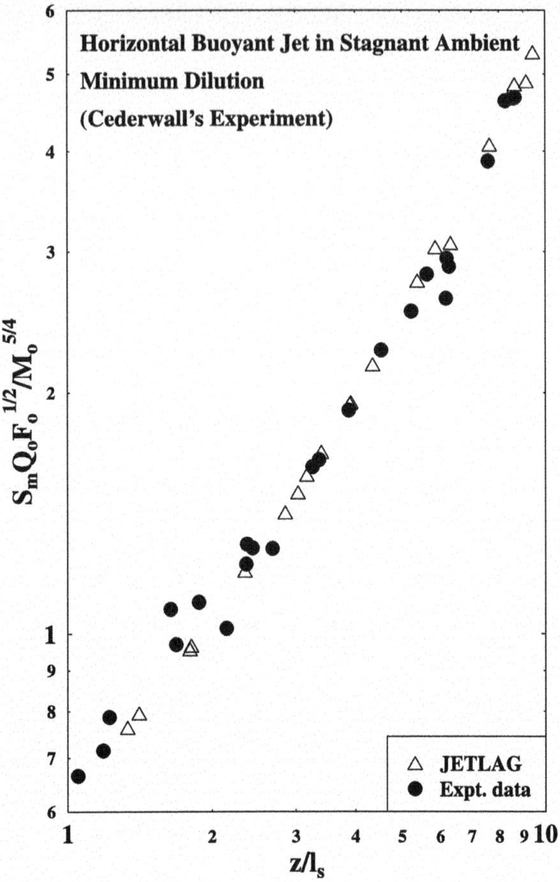

Fig. 12 Comparison of predicted dilution with measurements of a horizontal buoyant jet in still fluid

The model predictions of jet trajectory and dilution reproduce the power laws as given by Fischer et al. (1979) for the limiting flow regimes; the value of the trajectory and dilution constants derived from the numerical calculations compare favorably with experiments. The reader is reminded that in all the calculations reported herein, none of the model parameters is changed.

3 Field Application and Verification

3.1 Post-Operation Monitoring of Sydney Outfall

Sydney, Australia is serviced by three major ocean outfalls. Since around 1991, primary-treated domestic effluent is discharged into the coastal water through three offshore outfalls at North Head, Bondi, and Malabar (Gordon and Fagan 1991; Wood

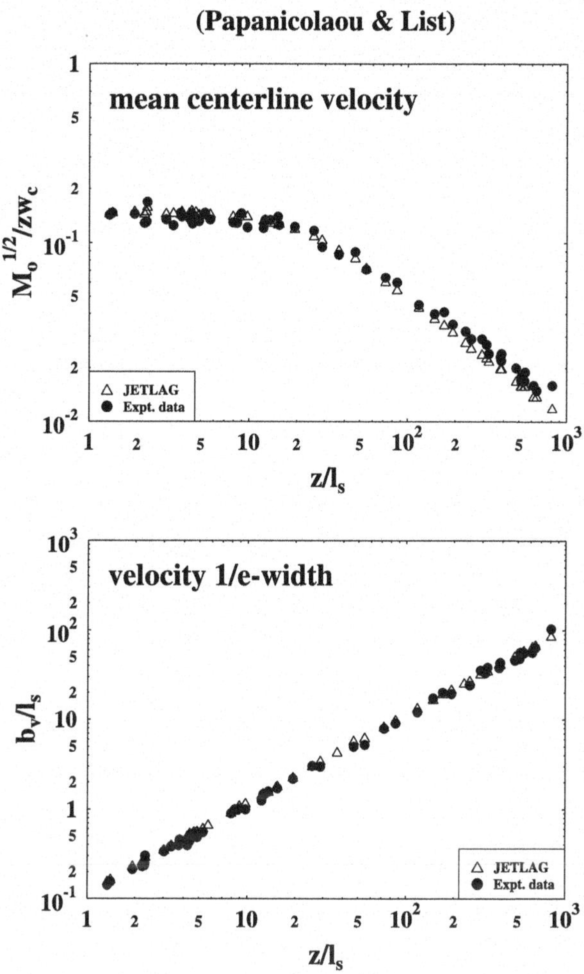

Fig. 13 Comparison of predicted and measured velocity and concentration for a vertical buoyant jet in stagnant fluid

et al. 1993). An environmental monitoring program was developed to examine the post-operation performance of the outfalls. The objective was to (i) assess the impact of the outfalls on the marine environment, and (ii) to compare the operation of the outfalls against design assumptions and provide data for design of future outfalls. In view of the high level of natural variability of the physical environment, it was necessary to use a suite of initial dilution and circulation models to provide a framework for interpretation of field data. The model JETLAG was applied to many field situations and have been verified from the field data obtained by the large scale experiments—"... with JETLAG proving to be the most versatile and useful model in most instances" (Gordon and Fagan 1991). It was also concluded that "the min-

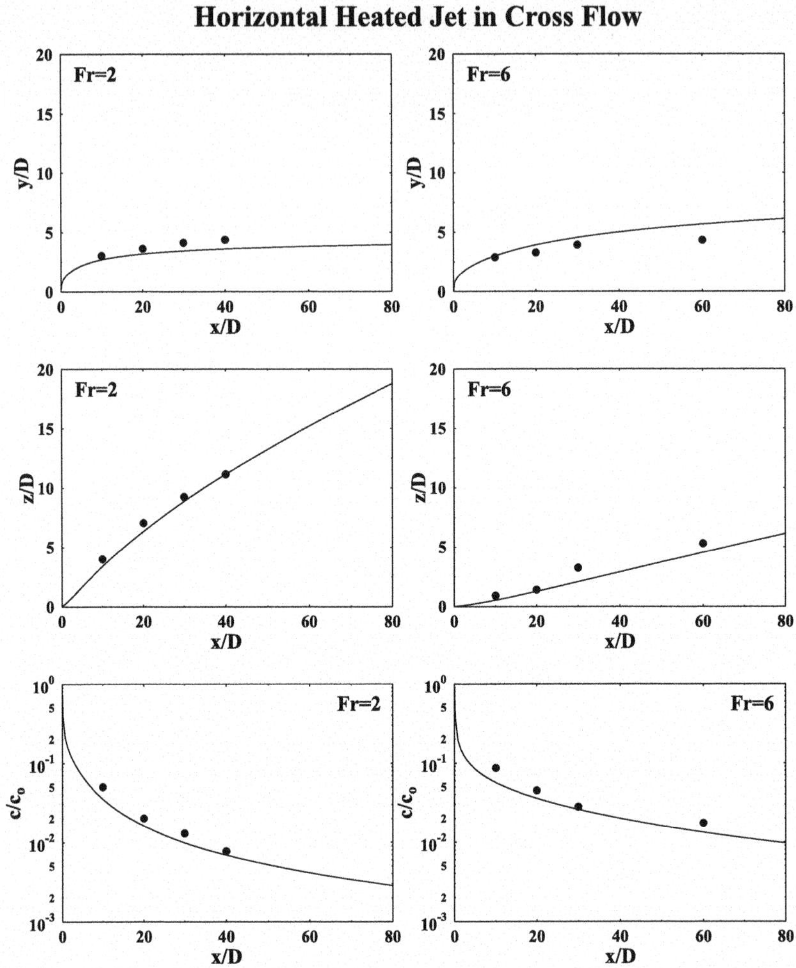

Fig. 14 Comparison of predicted (solid line) jet trajectory (horizontal $[x, y]$ and vertical $[x, z]$ planes), and tracer concentration with data of horizontal buoyant jets in crossflow (Cheung 1991)

imum dilution estimates obtained from the field work were typically twice those obtained from the plume model. This appears to be due to the additional dispersion resulting from the fluctuating ambient current which may not have been included in the laboratory experiments used to calibrate these models". The verified model was used during the outfall commissioning phase and also in the post-commissioning phase to provide temporal and spatial information on water quality. This information was integrated with the regular monthly water quality measurements. Figure 18 shows the computed plume for the Malabar outfall at different times of the year, for a typical ambient current of 0.1 m/s. The 3.6 km long outfall (diameter 3.5 m) consists of shore normal tunnel with a 775 m long diffuser section at the seaward end. A total

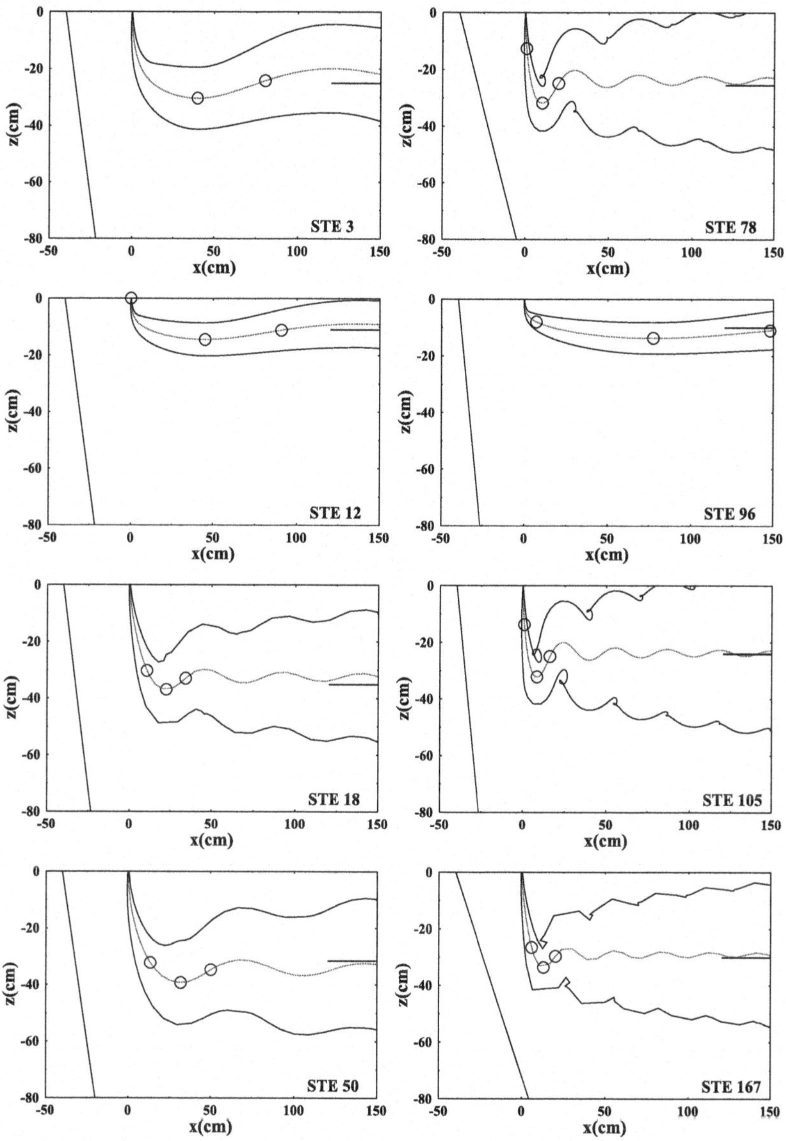

Fig. 15 Comparison of predicted and observed terminal level of a vertical buoyant jet in stratified crossflow (Experiment of Wright 1977b; observed level indicated by horizontal bar)

sewage flow of $Q_o = 2.3$ m^3/s is discharged through 28 risers at a depth of 80 m. Each riser is fitted with a "gas-burner" type head, discharging 4–6 horizontal jets (typical D=0.1 m). Because of the relatively large depth (77.5 m), it is acceptable to model the discharges from each riser by an equivalent buoyant jet with the same volume, momentum and buoyancy flux. It is seen that in June, the water is unstrat-

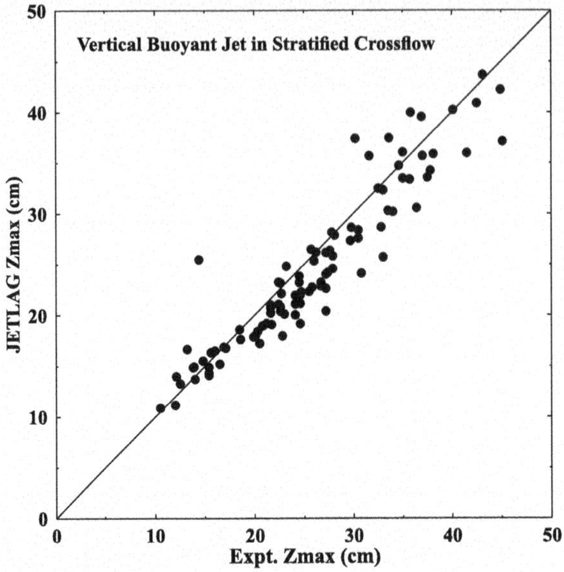

Fig. 16 Comparison of model predictions and data of maximum rise height of a vertical buoyant jet in stratified crossflow (Wright 1977b)

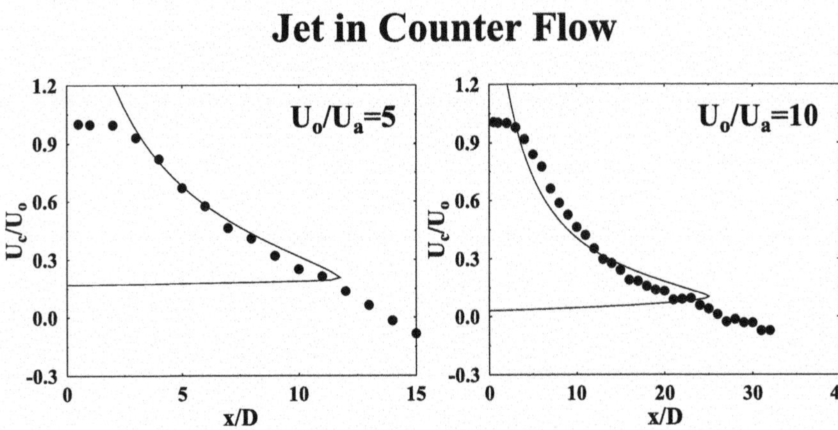

Fig. 17 Comparison of model predictions (solid line) and data (symbols) of centerline velocity of counterflowing jet (Chan and Lam 2018)

ified, and the plume surfaces with a dilution which can exceed 1000. In February however, there is a linear temperature stratification in the surface layer, with a 4.8 ^oC differential over the upper 50 m. The effect of the temperature stratification always results in a trapped plume; it is also seen that in the presence of a thermocline (6 ^oC jump over 10 m), the sewage field is trapped at about the level of the thermocline.

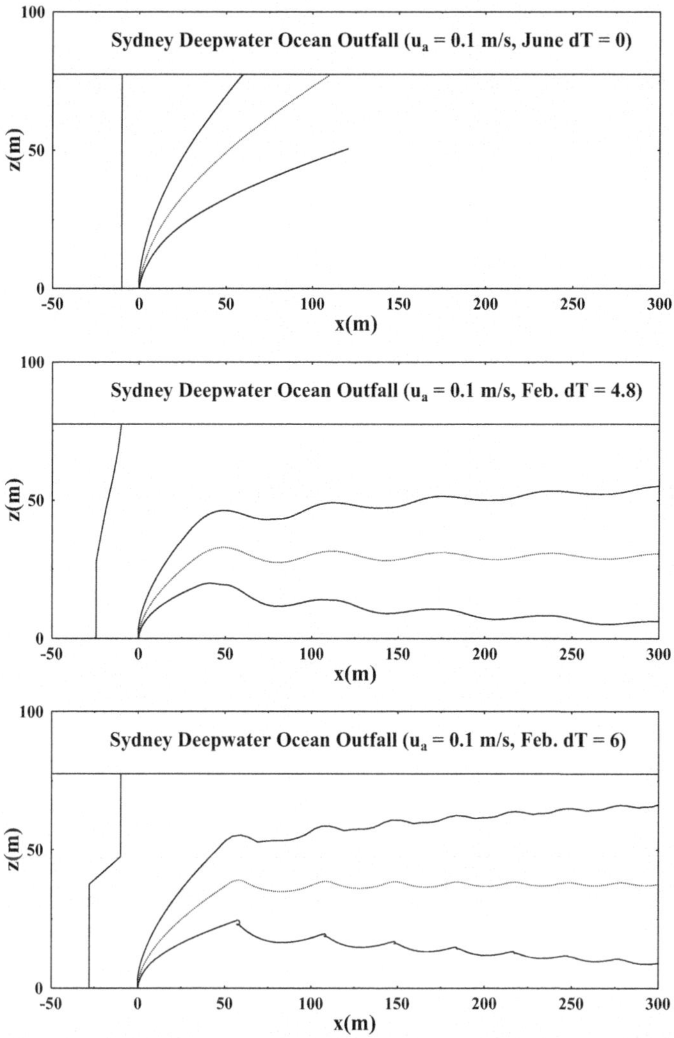

Fig. 18 Computed ocean outfall plume behaviour: Sydney Ocean Outfall, Australia

3.2 Field Verification at North West New Territories Outfall, Urmston Road, Hong Kong

The North West New Territories outfall in Hong Kong consists of a 2.6 km long submarine pipeline into the Urmston Road tidal channel. A 600 m diffuser section at the offshore end contains 30 risers at a depth of about 20 m. Each diffuser riser has two discharge ports (port diameter 0.3 m), discharging horizontally in opposite directions. Several post-operation monitoring surveys of the outfall were carried

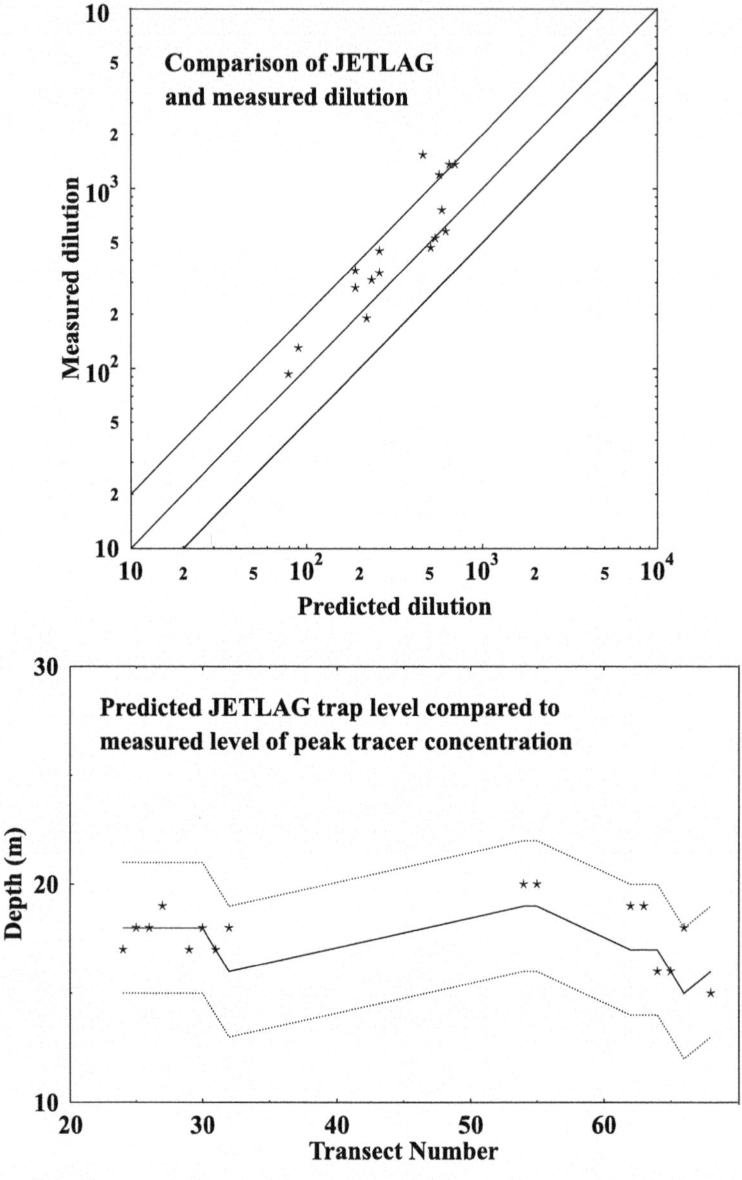

Fig. 19 Comparison of model predictions of dilution and trap height with field data at North West New Territories sewage outfall, Urmston Road, Hong Kong

out using dye and radio-isotope tracers, CTD (conductivity-temperature-depth) and acoustic doppler current profiling (ADCP); details can be found in Horton et al. (1997). The measurements enable the determination of the initial dilution and trap height. Figure 19 shows the comparison of measured average dilution against the predictions by JETLAG for a wet season survey of July 1995. In the field experiments, the density stratification was evident, with up to a 15 ppt and 4 ^{o}C difference in salinities and temperatures respectively through the water column. The riser flow was in the range of 0.05–0.12 m^3/s; $U_a \approx 0.3 - 0.8 m/s$. Initial dilutions varied between 50 to over 1000, with a trap centerline depth of 16–19 m. Higher dilutions were associated with faster currents and lower discharges. Given the variabilities in the field conditions and the complexity of the phenomenon the overall agreement of the prediction with data is quite satisfactory. All model dilution estimates were well within a factor of two of the measured dilutions as shown on the figure. The predicted plume trapping height (with estimated upper and lower plume boundaries) is compared with the measured vertical location of peak tracer concentration at the same horizontal position. The observed trapping level was well within the effluent jet/plume as predicted by JETLAG.

3.3 Environmental Impact Assessment of the Hong Kong SSDS Ocean Outfall

As part of the Strategic Sewage Disposal Scheme (SSDS) of Hong Kong, it was proposed to discharge sewage from a population of about 3 million (after advanced primary treatment) through a 15 km long tunnelled outfall into the southern coastal waters of Hong Kong. Preliminary studies have resulted in an initial diffuser design with 20 risers (spaced at 60 m) each with six to eight jets discharged horizontally like a rosette (Howard et al. 1991). For outfall design and impact assessment, it was necessary to predict mixing under the entire range of ambient and discharge conditions. For example, the design 95 percentile current speed was 0.04 m/s, while velocities up to 0.6 m/s must also be considered. For this application, the limiting water quality objective that dictated the initial dilution requirement was the allowable un-ionised ammonia nitrogen concentration (0.021 mg/L). Figure 20 shows an example of the predicted plume for a horizontal jet in coflow ($\phi_o = 0$, $\theta_o = 0$) and a horizontal jet in crossflow ($\phi_o = 0$, $\theta_o = 90^o$) for a summer condition, for $u_a =$ 0.1 and 0.2 m/s. The riser and jet diameters are 3 m and 0.25 m respectively; the design flow is $Q_t = 23.1$ m^3/s. The jet densimetric Froude number is $Fr = 16.3$; the temperature and salinity vary respectively from 23 ^{o}C and 34.5 ppt at the sea bed to 28 ^{o}C and 25 ppt at the surface. It is seen that in all cases the plumes are trapped beneath the surface. For the stronger current, the increased dilution results in a greater submergence of the sewage field. Figure 22 shows the computed buoyant sewage plumes issuing from the proposed Hong Kong Ocean outfall diffuser risers in the form of rosette jet groups. The complex turbulent buoyant shear flow, with jet configurations ranging from a coflow jet, oblique jet, to a counterflow buoyant jet, can clearly be seen. The computed plume trajectories enable a meaningful definition of the initial mixing zone.

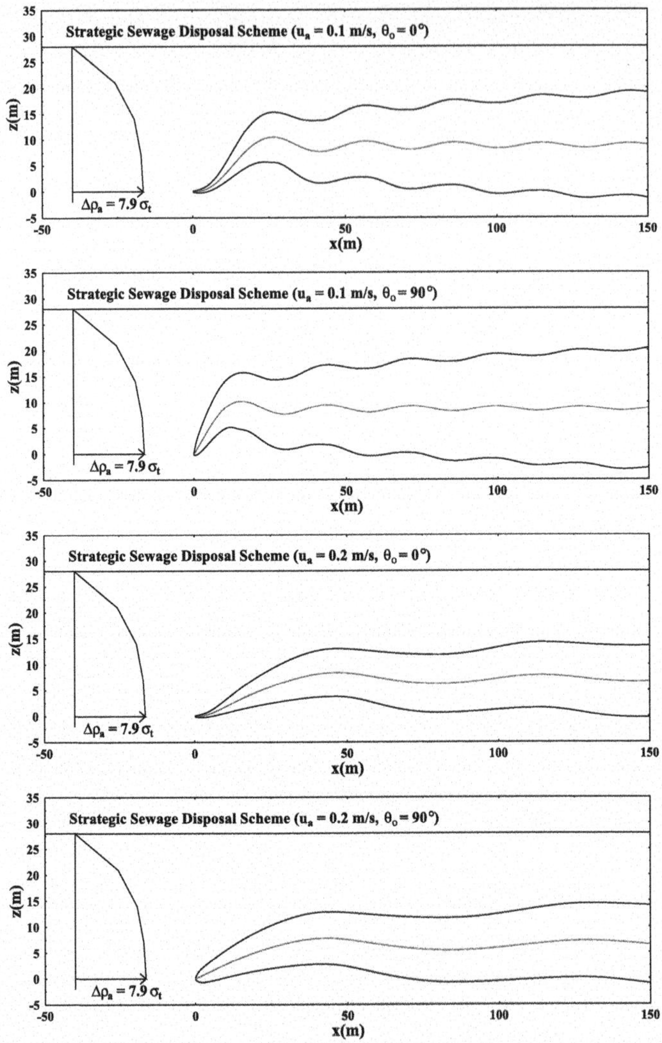

Fig. 20 Computed plume trajectory of buoyant jets at different angles from the SSDS outfall

3.4 *VISJET Interactive Virtual Reality Model*

The Lagrangian model JETLAG has been further developed by incorporating three-dimensional computer graphics techniques. The new VISJET software package is a PC-based interactive model that gives virtual reality displays of the predicted plume. Figures 21 and 22 are samples of the VISJET graphic output for a multi-jet discharge from rosette diffuser. Views of the jets and the diffuser beneath the ocean surface can be obtained interactively; the user can manipulate the computer interface to exam-

Fig. 21 Submerged diffuser below the ocean surface (top). The closeup view of the same diffuser (bottom) is obtained interactively using VISJET's three-dimensional graphic rendering routines. The graphics also shows the air and water interface at the far end of the image on the top of the figure

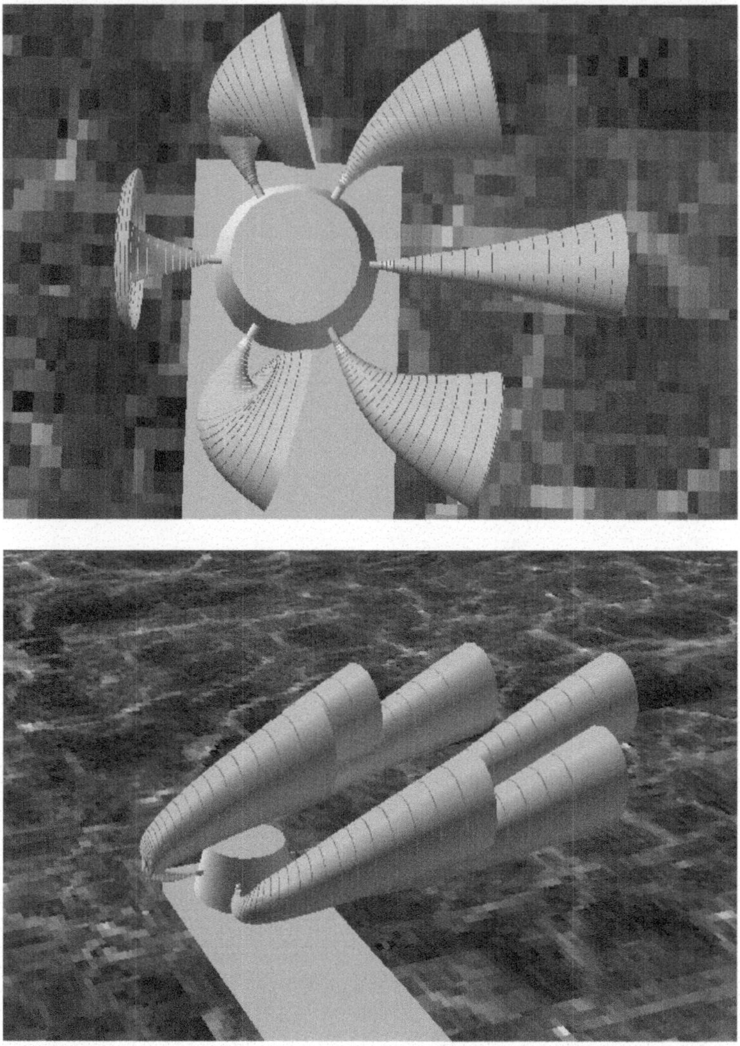

Fig. 22 Merging of buoyant jets above a rosette-shaped ocean outfall riser. Top and side views of the jets are obtained interactively using the VISJET graphic rendering routines. A sequence of cross-sectional slices along the path of each buoyant jet is shown

ine the flow closeup from different orientations relative to the directions of the jets and the crossflow. The buoyant jets discharged in multi-directions interact with the crossflow to form a complex three-dimensional pattern. The predicted concentration and dilution are obtained directly simply by pointing the cursor at the location where quantitative information is desired. When the receiving water is density-stratified, VISJET depicts the predicted trapping level of the plume and spreading of the plume fluid in a horizontal layer. The interrogative computer graphics also provides the determination of the degree of merging between the bent-over jets from a multi-jet group in any downstream cross-section plane. The composite dilution of the jet group can hence be determined.

The VISJET software and application examples are attached with this monograph. VISJET can also be downloaded from
http://www.aoe-water.hku.hk/visjet.

4 Other Specialized Applications

While this chapter has focused on examples of environmental discharges because of the drive to implement such systems in the past decades, the general theory and models introduced in this monograph are equally applicable to other important problems related to climate change, water scarcity, energy, land use, and ecosystem restoration. In this section we have selected a few specialized applications for interested readers to appreciate the broader scope of application of the theory of jets and plumes.

Dense jet—Desalination and LNG regasification

Desalination is gaining popularity in recent years as an alternative or key source of water supply to address the challenges of water scarcity, which is expected to be intensified in future due to climate change. As a by-product of the desalination processes, a concentrated brine solution (e.g. 50 ppt–70 ppt, 1035 kg/m^3–1050 kg/m^3) has to be returned to nearby water bodies. It is typically discharged upwards at an inclined angle via a submarine outfall to minimize environmental impact. As the brine discharge is denser than the surrounding seawater (1025 kg/m^3), the jet will eventually fall back onto the sea floor where it continues to spread horizontally as a bottom density current.

Desalination dense jet discharge is fundamentally different from positively buoyant discharge because its vertical momentum is opposite to the buoyancy direction, leading to a distinct feature—'detrainment' of the jet. Opposite to jet entrainment, detrainment is the process in which the dense fluid is leaving the main dense jet structure, reducing the volume and buoyancy flux of the jet. Lai and Lee (2012) studied a dense jet discharging into a stagnant ambient with a wide range of discharge angles, and found that JETLAG conservatively under-predicts the maximum terminal rise height by 10–15% and dilution by 30% because the 'detrainment' is not accounted for. However, with an addition of an ambient current, as is observed in most engineering applications, the detrainment effects on the jet diminished (Lai and Lee 2014). Defining a crossflow Froude number $F = U_r F_r$, where $U_r = U_a/U_o$ is the ambient-to-jet velocity ratio, Lai and Lee (2014) found that when $F \geq 0.8$, the dense jet becomes significantly bent-over during its ascent phase; the jet mixing is dominated by vortex entrainment and the detrainment has less effects. When $F \geq 2$, the detrainment ceases to have any effect on the jet behavior. The jet trajectory in both the horizontal momentum and buoyancy planes are well predicted by JETLAG.

The environmental and ecological impact due to the elevated salinity close to the seabed resulting from the dense jet discharge is a key concern in desalination. In addition, the risk of the saline spreading layer recirculating back to its nearby intake, which increases the desalination cost, often needs to be assessed. All these should be assessed not only with an integral jet model, but also with a numerical shallow water circulation model in which the mixing of a saline spreading layer can be predicted (Choi et al. 2016). The coupling of an integral jet model with the far-field shallow water circulation model will be discussed in Chap. 11.

Dense jet is also seen in the by-products of the Liquefied Natural Gas (LNG) regasification process, in which the very cold liquefied natural gas is converted back to gaseous state for energy generation. The LNG is 'warmed' by the intake seawater in the process, and the cooled seawater (typically 4° less than the ambient) is discharged back to the sea using an outfall, resulting in a dense jet. The dense jet, however, is normally discharged downwards with the vertical momentum in the same direction as the buoyancy—the mixing can be treated in the same way as an "inverted" positively buoyant jet discharge (i.e. no detrainment). The mixing of other common by-products such as sodium hypochlorite for minimizing bio-fouling involves coupling an integral jet model with chemical kinetics—an example in wastewater treatment application can be found in Yang et al. (2019).

Sediment jet—Wastewater disposal and land reclamation

Sediment jets are seen in wastewater disposal because of the presence of total suspended solids (TSS) in the wastewater. Untreated domestic wastewater typically contains 100 mg/L–400 mg/L of TSS. Depending on the degree of treatment, TSS can be reduced to 30 mg/L–60 mg/L before discharging to the receiving water body via an outfall. Such concentration of TSS (e.g. volume fraction $<1\%$) does not contribute significantly to the buoyancy of the wastewater, so the wastewater buoyancy can be considered as conserved even the TSS can 'fall-out' from the jet partly due to its settling velocity (Chap. 4).

The fall-out of sediment leads to the formation of 'sediment/sludge' banks at the seabed at the vicinity of an outfall, which causes environmental and ecological concerns (Chan 2013). To predict the sediment deposition, Chan and Lee (2014) developed a general model for the prediction of sediment deposition from buoyant jets in stagnant water by coupling JETLAG with a general particle-tracking model to represent the sediments. Deposition of the sediment has been successfully predicted.

Sediment jets are also seen in land reclamation, which large amount of sand is dumped into water to create land mass, and is commonly seen in coastal cities facing the shortage of land for urban development. Different from the TSS that is laden with the wastewater buoyant jet, such discharge has a large sediment volume fraction (e.g. volume fraction $\approx 60\%$), and the buoyancy the jet is generated by the sediment itself. The mixing of such 'hyperconcentrated' sediment jets can also be predicted based on two-phase integral modelling (Lai et al. 2016) following the ideas of jet and plume theory introduced in the previous chapters. The fall out of sulphur bearing deposits from deep sea hydrothermal vents ('black smoker' plumes, see Problem 5.6 in Chap. 5) is also an important problem in earth sciences.

Bubble plumes—Ecosystem restoration and oil well blowouts
Dissolved oxygen (DO) is a key water quality parameters in many water bodies, and is important to sustain an ecosystem. In particular, the stable thermal stratification in the summer is often associated with oxygen depletion in the hypolimnion of eutrophic lakes (Chap. 5), because the stratification prevents vertical mixing with the upper epilimnion with higher DO concentration. Bubble-plume diffusers are increasingly used to oxygenate the hypolimnion of lakes and reservoirs to restore the water quality (Toledo et al. 2021).

Bubble plume is another example of particle-driven plume, in which the buoyancy of the flow is from the air bubbles. Although complex air bubble dynamics and phase interaction can exist (e.g. Socolofsky and Adams 2002; Lai and Socolofsky 2019), first-order bubble plume integral modelling (Milgram 1983; Sun and Faeth 1986) has continued to find important applications in a wide range of areas (such as oil well blowouts) due to its simplicity and effectiveness. More specific applications of oil and gas plumes can be found in Yapa et al. (1999).

5 Role of Computational Fluid Dynamics (CFD)

With the recent advancements in computing power and rapid CFD software developments, accessibility to CFD has been much improved, and its usage is almost certain to increase in future. CFD is considered in a broad sense here, which includes the solvers of 3D Navier-Stokes equations such as the Reynolds-averaged Navier-Stokes (RANS) equations, as well as its variants such as the shallow water equations. Many natural and engineering problems involve unsteady flows with boundary effects, or require consideration of detailed distribution of jets and plumes' mixing and turbulence characteristics—in such cases the steady state integral modelling of jets and plumes that are free from boundaries are no longer sufficient, and have to rely on numerical solutions. These can ranges from small scale problems such as sediment particle tracking under the effects of buoyant jet turbulence (Chan 2013), to large scale problems such as mixing zone analysis of a submarine outfall under time-varying tidal current flows near the coastline.

5.1 *Near-Field CFD Simulation*

While CFD is very powerful in providing solution to many problems, it should be cautioned that the results from CFD can be unreliable if the model is not properly setup (such as boundary conditions) and calibrated. There are also limitations to each type of CFD solvers and should be chosen carefully to ensure it is applicable to the problem. For example, in the commonly used standard two-equation $k - \epsilon$ turbulence closure model that is used for estimating the turbulent eddy viscosity μ_t, it is well known that the spreading rate of a round jet is over-predicted by 20% with $db/dx \approx 0.14$ (e.g. Lai 2009; Chan 2013). Tuning the turbulence model parameters to $C_{1\epsilon} = 1.2$, $C_{2\epsilon} = 1.6$, $C_\mu = 0.11$, $\sigma_k = 1.0$, and $\sigma_\epsilon = 1.3$ (Lai and Socolofsky 2018) can improve the solution, or alternatively the realizable $k - \epsilon$ model can be used. Examples detailed model setup can be found in Lai (2009) or Chan (2013).

When using CFD to predict buoyant jet flows, it is also important to ensure the buoyancy effects on the turbulence kinetic energy k and dissipation rate ϵ have been accounted for in the model (both standard and realizable $k - \epsilon$), i.e. the buoyancy extended $k - \epsilon$ model (Rodi 1980). Details of the applying the standard $k - \epsilon$ model can be found in e.g. Kuang et al. (2006). The three-dimensional flow of jets and plumes is obtained by solving the RANS in Cartesian coordinates (x, y, z). Temperature difference is used as a source of buoyancy for some buoyant jet simulations. The Boussinesq approximation is adopted in buoyant flow as the density difference is small. This results in the following governing equations:

$$\frac{\partial U_i}{\partial x_i} = 0$$

$$U_j \frac{\partial U_i}{\partial x_j} = -\frac{1}{\rho}\frac{\partial p}{\partial x_i} + \frac{\partial \tau_{ij}}{\partial x_j} + \frac{\rho - \rho_a}{\rho_a} g_i$$

$$U_j \frac{\partial h}{\partial x_j} = \frac{\partial}{\partial x_j}\left(\frac{\nu_t}{\sigma_h}\frac{\partial T}{\partial x_j}\right) + U_i \frac{\partial p}{\partial x_i} + \tau_{ij}\frac{\partial U_i}{\partial x_j}$$

$$U_i \frac{\partial k}{\partial x_i} = \frac{\partial}{\partial x_i}\left(\frac{\nu_t}{\sigma_k}\frac{\partial k}{\partial x_i}\right) + G_k + G_b - \epsilon$$

$$U_i \frac{\partial \epsilon}{\partial x_i} = \frac{\partial}{\partial x_i}\left(\frac{\nu_t}{\sigma_\epsilon}\frac{\partial \epsilon}{\partial x_i}\right) + C_{1\epsilon}\frac{\epsilon}{k}(G_k + (1 - C_{3\epsilon})G_b) - C_{2\epsilon}\frac{\epsilon^2}{k}$$

where U_i is the velocities in the x, y, and z direction; ρ is the density of the fluid; ρ_a is the density of the ambient fluid; p is the dynamic pressure; τ_{ij} is the stress tensor; $g = (0, 0, -g)$ is the gravity; $h = C_p T$ is the static enthalpy of the fluid; T is the temperature of the fluid; C_p is the specific heat capacity of the fluid; σ_h is the turbulent Prandtl number for energy; k is the turbulent kinetic energy; ϵ is the dissipation rate of k; $\nu_t = \frac{C_\mu k^2}{\epsilon}$ is the turbulent viscosity; G_k and G_b are the production of k due to shear and buoyancy respectively.

$$G_k = \nu_t\left(\frac{\partial U_i}{\partial x_j} + \frac{\partial U_j}{\partial x_i}\right)\frac{\partial U_j}{\partial x_i}$$

$$G_b = g_i \frac{\nu_t}{\rho \sigma_h} \frac{\partial \rho}{\partial x_i}$$

$$\tau_{ij} = \mu_t \left(\frac{\partial U_j}{\partial x_i} + \frac{\partial U_i}{\partial x_j} \right) - \frac{2}{3} k \delta_{ij}$$

The constants in the standard $k - \epsilon$ model that are optimized for jet and plume applications are given by Lai and Socolofsky (2018), and $\sigma_h = 0.7$. In carrying out any CFD simulations, it is always a good practice to first compare CFD predictions with established jets/plumes experimental data or model to ensure the reliability of the results, such as those in the earlier chapters, before proceeding to more complex cases.

The recent advances in computational power have enabled the more computationally intensive Large Eddy Simulations (LES) to be carried out within a reasonable amount of time. Studies of LES of a dense jet include that from Zhang et al. (2016), Jiang et al. (2019), and more recently by Law et al. (2023), enabling a better understanding of the turbulence characteristics and eddy structures of a dense jet. For a more comprehensive review of the recent developments and applications of CFD in outfall discharges, readers are referred to Mohammadian et al. (2023).

5.2 *Far-Field Circulation Model*

In many engineering applications such as the mixing zone analysis of of outfalls and assessing re-circulation risks of an intake-outfall pair, the buoyant jet trajectory and mixing can be well-predicted by a validated integral model or CFD simulations in the vicinity of the discharge ('near field'). Near field models can resolve down to scales of the order of the jet diameter; typical time and length scales of the plume rise are in the order of minutes and water depths respectively (see Fig. 27 in Chap. 11).

On the other hand, in the 'far field', the effluent is passively transported by ambient currents (such as tidal current) and further diluted by turbulent diffusion; typical time and length scales are in the order of hours and kilometres. In practice, the far-field simulation can cover the whole coastal city with length scale of tens or hundreds of kilometres, and simulation period of months if it involves other long-term processes such as water quality studies. A circulation model which solves the shallow water equations with hydrostatic approximation (for the vertical momentum equation) is typically used for the far-field.

In the near-far field transition ('intermediate field'), the dynamics depends on the interaction of the near field plume mixing and the ambient flow, which can result in changes in background concentration, gravitational spreading, and modifications in ambient stratification (e.g. Baines and Turner 1969; Wallace and Wright 1984; Lee and Cheung 1986; Wong 1986; Roberts 1979). Successful predictions in the intermediate field requires the near-field buoyant jet dynamics to be resolved in a far-field model, which has been a challenge because of the large disparity in length and time scales. Even if one attempts to refine the far field model grid size to the near-field length scale, the hydrostatic pressure approximation still fails to cater for the buoyant jet dynamics with strong vertical acceleration. In Chap. 11, the near-far field coupling problem will be re-visited, and an approach to resolve this challenging problem will be introduced.

Problems

10.1 One approach to correlate the initial dilution in moving water, S_M, is to first compute the stillwater dilution S_s, using e.g. the Cederwall Equation, and then to correlate the ratio of the moving water to stillwater dilution (with dimensionless parameters such as the ambient to jet velocity ratio, $\frac{u_a}{u_j}$, and depth to jet diameter ratio, z/D). The following are three such semi-empirical equations which are based on field or laboratory data: (i) the Agg and Wakeford Equations based on sampling of tracer concentration at sewage boils; (ii) the Hydraulics Research Station (HRS) equation based on heated jet experiments in the laboratory; and (iii) the Bennett equation based on dye tracing studies at the Hastings Outfall in the UK:

$$\frac{S_M}{S_s} = 12.794\,(\frac{u_a}{u_j})^{0.938} \qquad \text{(Agg \& Wakeford 1972)}$$

$$\frac{S_M}{S_s} = 1 + 6(\frac{z}{D})\,(\frac{u_a}{u_j})^{1.5} \qquad \text{(HRS 1977)}$$

$$\frac{S_M}{S_s} = 1.296\,\frac{u_a^{0.682} z^{1.183}}{Q_o^{0.955}} \qquad \text{(Bennett 1983)}$$

Using the above equations, compute the initial dilution in moving water for two cases: (i) a single port outfall with diameter $D = 0.5$ m, discharge jet velocity $u_j = 0.6$ m/s, depth $H = 10$ m, and ambient velocity $u_a = 0.1$ m/s; (ii) a jet discharging from a multiport diffuser with widely spaced jets, with $D = 0.1$ m, $u_j = 1$ m/s, $H = 20$ m, and $u_a = 0.1$ m/s. Assume the same initial relative density difference for both cases, $\frac{\Delta\rho_o}{\rho_a} = 0.026$. Comment on your results.

10.2 A horizontal round buoyant jet with initial jet velocity $u_o = 1$ m/s, diameter $D = 0.15$ m, and relative density difference $\Delta\rho_o/\rho_a = 0.025$ discharges in depth of $H = 10$ m. Compute the initial dilution (a) in still water; and (b) in an unstratified crossflow of (i) 0.02 m/s, (ii) 0.1 m/s, and (iii) 0.5 m/s. Comment on your results.

10.3 *Outfall Mixing Anomaly in Weak Current*

An ocean outfall discharges a total flow of $Q_t = 23.1$ m^3/s into coastal water of 22 m depth. There are 21 widely spaced risers on the diffuser section. There are six horizontal jets on each riser (D=0.25 m). Initial mixing calculations for two discharge flows have been performed for a weak current $u_a = 0.04\ m/s$ and a design

Discharge flow (m^3/s)	23.1	10.4
Predicted neutral buoyancy level z_t (m)	7.7	8.4
Predicted average dilution S	31.3	27.8

stratification condition. The results show surprisingly that the initial dilutions for the two drastically different discharge flows are very similar.
(a) Is the above result reasonable? Assuming a stagnant ambient condition and $\Delta\rho_o/\rho = 0.025$, and adopting the Cederwall equation or otherwise, compute the jet densimetric Froude number and the average dilution for the two discharge flow conditions at $z = 7.7$ m and $z = 8.4$ m respectively. Comment on the dependence of predicted dilution on the discharge flow.
(b) Use the VISJET model to obtain the plume trajectory and average dilution for the following jet discharges: $\theta_o = 30, 90, 120°$. What is the maximum lateral penetration distance of the perpendicular discharge ($\theta_o = 90°$)? If the riser spacing is 60 m, will there be interference between the jets from adjacent risers at the surface or trap level. The effluent salinity and temperature are 2 ppt and 20°C respectively. Assume $u_a = 0.04$ m/s, and the following ambient profile:

Depth below surface (m)	Salinity (ppt)	Temperature (°C)
0.0	25.0	28.0
5.5	29.5	27.0
11.0	33.0	25.0
16.5	34.0	23.0
22.0	34.5	23.0

10.4 *Buoyant plume in stratified crossflow*
It is often puzzling that the predicted initial dilution of a round buoyant jet in a stratified crossflow appears to be rather insensitive to the discharge flow. This apparent insensitivity is examined as follows. Consider a buoyancy-dominated jet with specific buoyancy flux $F_o = Q_o g'_o$, discharging into a stratified crossflow U_a. Assume the receiving water is linearly stratified, with the stratification parameter $\epsilon = -\frac{g}{\rho_a}\frac{d\rho_a}{dz}$. If the mixing of the buoyant plume in crossflow is assumed to be analogous to that of an advected line thermal of initial buoyancy F_o/U_a, then the maximum height of rise z_{max} can be assumed to depend on the line thermal buoyancy F_o/U_a and the stratification parameter ϵ.
(a) Show that the maximum height of rise is a function of the jet buoyancy flux, ambient current, and ambient stratification parameter:

$$z_{max} \sim \left(\frac{F_o}{U_a \epsilon}\right)^{1/3}$$

(b) Assuming the dilution equations for a weak and strong current, i.e. the BDNF and BDFF equations respectively (Eqs. 5 and 6), show that the dependence of the dilution at the trap level for a given discharge flow of Q_o is:

$$S_t = S(z_{max}) \sim \frac{g_o'^{8/9}}{(\epsilon U_a)^{5/9} Q_o^{1/9}}$$

for the buoyancy-dominated near field (BDNF). It is seen that the dilution at the trap level is not as sensitive to current or port discharge as the unstratified situation ($S \sim F_o^{1/3} H^{5/3} / Q_o \sim (g_o')^{1/3} H^{5/3} / Q_o^{2/3}$). Similarly, show that the initial dilution at trap level for the buoyancy-dominated far field (BDFF) is given by:

$$S \sim \frac{U_a^{1/3} g_o'^{2/3}}{\epsilon^{2/3} Q_o^{1/3}}$$

10.5 (a) For the dense jet in Prob. 3.7, use VISJET to determine the jet trajectory, width, terminal height of rise, and dilution for $\phi_o = 30, 45, 60°$. Assume the jet is located at 1 m from the sea bottom.
(b) For $\phi_o = 60°$, use VISJET to determine the trajectory, maximum height of rise, and dilution of the dense jet in a crossflow $U_a = 0.1$ m/s.

10.6 Wastewater from a small town next to a popular beach (frequented by wind surfers in the dry season) is discharged in the form of a single horizontal round jet in coastal water of 12 m depth. The idea is to discharge the jet at high velocity from a rock cliff face away from the beach area as much as possible. Located at 1 m from the bottom, the jet diameter is $D = 0.036$ m and the discharge flow is $Q_o = 0.02$ m^3/s (see Prob. 5.8). Based on field measurements, the average ambient conditions in the dry season are as follows:

Depth (m)	S (ppt)	T(°C)	U_a (m/s)
1	30.1	21.2	0.1
3	30.1	21.2	0.08
5	30.1	21.2	0.06
7	30.2	21.1	0.05
9	30.5	20.9	0.04
11	30.6	20.7	0.03

Use VISJET to predict the trajectory of this horizontal buoyant jet in a crossflow, with a 3D trajectory. Determine the offshore penetration distance and whether the jet is well-trapped beneath the free surface. Assume effluent characteristics of S = 2 ppt and T = 20°C.

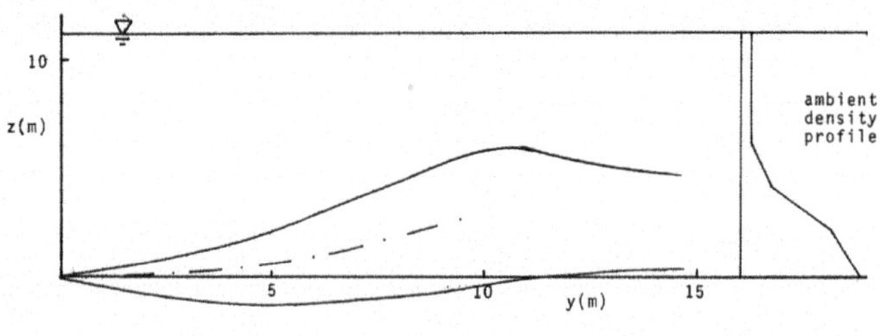

10.7 *Lagrangian model formulation*

The Lagrangian jet model (JETLAG) formulation in Eqs. 7 to 27 (and its correspondence to the Eulerian formulation) can be best illustrated using a vertical buoyant jet in uniform stagnant ambient fluid. Using top-hat profiles, the Eulerian equations for conservation of mass and momentum are, respectively:

$$\frac{d}{dz}\int \rho w \, dA = \frac{d}{dz}(\rho A W) = \rho_a Q_e = \rho_a 2\pi B(\alpha W)$$
$$\frac{d}{dz}\int \rho w^2 dA = \frac{d}{dz}(\rho A W^2) = \Delta\rho g A$$

where $B, W, \Delta\rho = \rho_a - \rho$, and $A = \pi B^2$ are respectively the plume radius, average velocity and density deficit, and plume area respectively. Q_e is the entrainment flow per unit length along the plume axis, and α=entrainment coefficient.

Noting that $W = dz/dt$, we can cast the above equations in terms of *time*:

$$\frac{d}{dt}(\rho A W) = \rho_a 2\pi B(\alpha W)W$$
$$\frac{d}{dt}(\rho A W^2) = \Delta\rho g A W$$

Suppose we multiply each of the above equations by an arbitrary (constant) time interval parameter Δt^*, we would have

$$\frac{d}{dt}[\rho A(W\Delta t^*)] = \rho_a 2\pi B(\alpha W)(W\Delta t^*)$$
$$\frac{d}{dt}[\rho A W(W\Delta t^*)] = \Delta\rho g A(W\Delta t^*)$$

In the above equations, if we interpret $h = W\Delta t^*$ as a thickness of a plume element, then the equations can be discretized as:

$$\frac{\Delta M_k}{\Delta t} = \frac{\Delta}{\Delta t}(\rho_k A_k h_k) = \rho_a \alpha W_k 2\pi B_k h_k$$
$$\frac{\Delta M_k W_k}{\Delta t} = \frac{\Delta}{\Delta t}(\rho_k A_k h_k W_k) = \Delta\rho_k g A_k h_k$$

where $M_k = \rho_k A_k h_k$ is the plume element mass. This is effectively our Lagrangian model using just the shear entrainment (Eq. 32). Note that the time step Δt is a free parameter (dictated by numerical stability and accuracy considerations). This is chosen to achieve a prescribed percentage increase in mass of plume element (typically one percent). With this framework, $h_k \propto W_k$, the volume of the plume element represents the volume flux. The initial choice of plume element thickness,

$h_o \sim D$, is however immaterial and will not affect the solution of the plume sectional properties.

In this Lagrangian framework, we are tracking the changes of a plume element mass that is released from the source over a time interval (Δt^*). The consecutive plume elements are assumed to be independent and do not mix with each other. Note that there is no need for the Boussinesq approximation, $\Delta\rho/\rho \ll 1$ (see also Chap. 3). For example, for the same absolute initial density difference $|\Delta\rho_o|$, small differences in the behaviour of positively or negatively buoyant jets can be simulated. Since added mass is not explicitly included in the formulation, it can be shown the bent-over jet or plume width predicted by JETLAG is related to the top-hat width B by $B' = \sqrt{1+k_n}B \approx \sqrt{2}B$ (cf Eqs. 22 in Chap. 7 and 27 in Chap. 8, and Prob. 8.5).

10.8 *Radial entrainment velocity distribution*
Consider a vertical jet or plume in stagnant fluid. The shear entrainment is the radial entrainment flow caused by the velocity gradient between the jet and the surroudings. Since the axial velocity $w_m(s = z)$ is known (Table 3 in Chap. 2), the entrainment flow per unit axial length into the jet/plume at any level s is known, $\frac{dQ}{ds} = 2\pi r V_r(r)$, where $V_r(r)$ is the radial inflow velocity at any radial position r. By symmetry, the entrainment flow field in the normal cross-section of the jet/plume is given by a 2D sink flow of strength $m = 2\pi r V_r(r) = 2\pi\alpha\Delta U_g b_g$.

(a) By axisymmetry, show from the continuity equation the following expression for the entrainment velocity at any radius $r = r_1$:

$$\frac{d}{ds}\left(\int_o^{r_1} w(s,r)2\pi r dr\right) = 2\pi r_1 V_r(r_1)$$

(b) Assuming Gaussian profile for $w(s,r)$, show that the radial entrainment velocity is given by:

$$\frac{d}{ds}\{\pi\Delta U_g b^2(1 - e^{-(r/b)^2})\} = 2\pi r V_r(r)$$

Hence show that $V_r(r)$ can be obtained from the local centerline velocity excess ΔU_g and Gaussian half-width $b = b_g$ as:

$$\frac{V_r(r)}{\alpha\Delta U_g} = \frac{(1 - e^{-(r/b)^2}) - A(r/b)^2 e^{-(r/b)^2}}{r/b}$$

where $A = \frac{1}{\alpha}\frac{db}{ds}$. The attached figure shows the radial variation of the entrainment velocity.

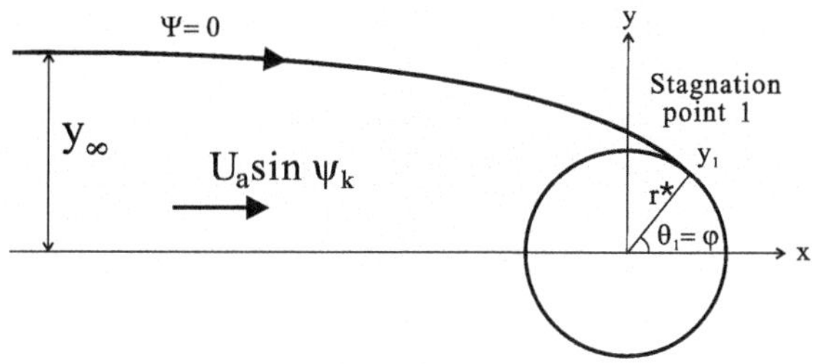

Fig. 23 Entrainment flow pattern in cross-section of jet in crossflow (normal to jet axis)

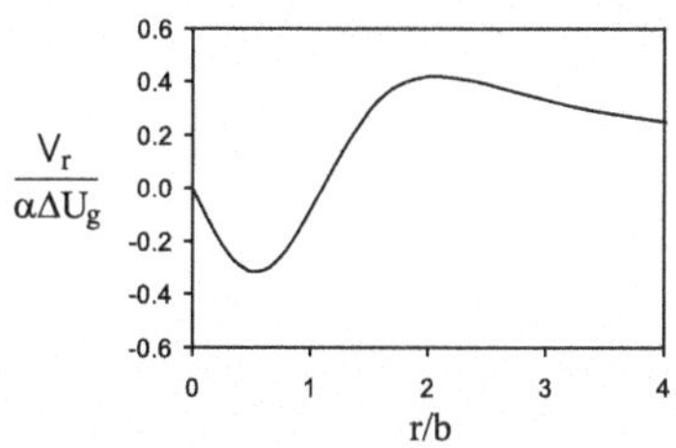

(c) If $\alpha = 0.056$, and $db/ds \approx 0.109$, $A = 1.94$, show that the radial entrainment velocity has a maximum value of 0.421 at $r/b_g = 2.071$.

10.9 *External flow of a bent-over jet in crossflow*
Consider a slightly bent-over vertical buoyant jet in a weak horizontal crossflow. The jet axis is inclined at angle ψ_k to horizontal. Experiments show that the flow outside of the jet is the sum of the shear entrainment and the crossflow velocity $U_a \sin \psi_k$ (Figs. 11 and 23). In the cross-section perpendicular to the jet axis, outside of the jet, it can be shown from continuity considerations that $\frac{\partial r V_r}{\partial r} + \frac{\partial V_\theta}{\partial \theta} = 0$, where V_r, V_θ are the radial and tangential velocities. Hence a 2D stream function $\Psi(x, y) = \Psi(r, \theta)$ can be defined with

$$V_r = \frac{1}{r}\frac{\partial \Psi}{\partial \theta};\ V_\theta = -\frac{\partial \Psi}{\partial r}$$

(a) By super-imposing the radial entrainment velocity (Prob. 10.8) and the crossflow velocity, show that the 2D stream function is given by:

$$\Psi(r, \theta) = -\alpha \Delta U_g b f(r)(\theta - \theta_1) + U_a \sin \psi_k (y - y_1)$$

where $f(r) = (1 - e^{-(r/b)^2}) - A(r/b)^2 e^{-(r/b)^2}$, and $A = \frac{1}{\alpha}\frac{db}{ds}$. $(\theta_1 = \varphi_k, y_1 = r \sin \theta_1)$ are the angular and y-coordinate of the stagnation point.

For the limiting streamline that passes the stagnation point on the plume boundary, $r = r^* \approx 2b_g = \sqrt{2} b_k$, show that $(r \to \infty, \theta \to \pi)$

$$\frac{y_\infty - y_1}{b} = \frac{\alpha \Delta U_g}{U_a \sin \psi_k}(\pi - \theta_1)$$

where y_∞= half-width of streamtube far upstream.

(b) Since the entrainment flow into the plume element is $q_e = 2U_a \sin \psi_k y_\infty$, show that

$$q_e = 2(\alpha \Delta U_g b(\pi - \theta_1) + U_a \sin \psi_k \, y_1)$$

where $\theta_1 = \varphi_k$, and $r_1 = r^*$. Hence the relation:

$$\frac{q_e}{2\pi\alpha\Delta U_g b} = \frac{\pi - \varphi_k}{\pi} + \frac{2U_a \sin \psi_k \, r^*}{2\pi\alpha\Delta U_g b} \sin \varphi_k$$

As $2\pi\alpha\Delta U_g b_g$ and $2U_a \sin \psi_k \, r^*$ are respectively the shear and vortex entrainment (Fig. 11), this leads to the general entrainment computation (Eq. 36):

$$E = E_s \frac{\pi - \varphi_k}{\pi} + E_f \sin \varphi_k$$

where the 'separation angle' φ_k can be dynamically computed (Eq. 37).

Chapter 11
Mixing of Multiple Buoyant Jets

A general theory to predict the mixing characteristics of a single buoyant jet has been developed in previous chapters. In many applications, however, multiple buoyant jets are discharged in close proximity as a jet group and can result in markedly different behaviour than a single buoyant jet. For example, partially treated wastewater from coastal cities are often discharged as rosette-shaped jet groups from a submarine outfall to optimize mixing; air pollution sources induced by a heat island effect are considered as multiple buoyant plumes (Baines and Keffer 1974); electronic equipment and humans in an office enclosure are also analyzed as thermal plume groups (Linden 1999).

The complex nature of multiple buoyant jet mixing is illustrated in Fig. 1, which is a physical model study of the condenser cooling water discharge of the Setubal power station in Portugal. It consists of an array of straight buoyant jets discharging in parallel. The jets can interact kinematically when the jets merge—the merging of the jets will reduce their surface area available for ambient entrainment, leading to a reduction in dilution. The jets can also interact dynamically due to the jet-induced external flow and pressure—the jet group in Fig. 1 can be seen to contract with the individual jets deflecting inwards, further affecting the jet group mixing. A general theory to account for the multiple buoyant jet interaction is needed to accurately predict the jet group mixing characteristics.

1 Multiple Jets in Stagnant Ambient

For discharges with large initial momentum relative to buoyancy (as seen in some cooling water discharges), the problem can be simplified as multiple momentum jet

Supplementary Information The online version contains supplementary material available at https://doi.org/10.1007/978-3-032-01187-9_11.

J. H. W. Lee et al., *Turbulent Jets and Plumes*,
https://doi.org/10.1007/978-3-032-01187-9_11

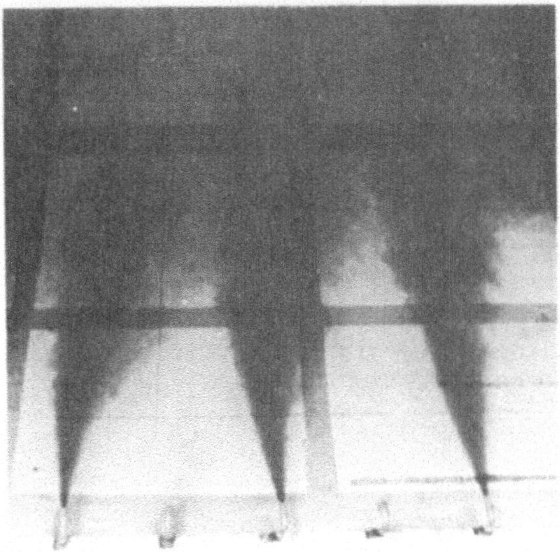

Fig. 1 Merging of horizontal three-dimensional jets from a submerged multiport diffuser; top view (jet nozzle spacing 19.5 cm, water depth 6.1 cm, floor grid size 30.5 cm) (Lee and Jirka 1980)

discharges. For an infinitely long array of jet discharge in an unbounded ambient, the dynamic interaction among jets can also be neglected and each jet will have a straight trajectory.

The definition sketch of the problem is given in Fig. 2, with multiple round jets with spacing s (Fig. 2) discharging into the $+x$-direction. Figure 3 shows the process of jet merging. Initially, before merging the jet velocity is the same as that of a single free jet with a Gaussian distribution. As the jets spread by turbulent entrainment and begin to merge, their velocity fields 'overlap'; the velocity distribution is no longer Gaussian and eventually becomes uniform like a plane jet when the jets are 'completely' merged. Early attempts to determine the jet velocity interaction simply adopted linear superposition of the jet velocity u of each individual jet; however this lacks theoretical support and leads to incorrect predictions.

1.1 Momentum Superposition

Instead of superimposing u, it can be shown that the jet velocity squared u^2 should be superimposed to predict the interacting velocity (Lai and Lee 2012). In the multiple jet discharge shown in Figs. 2 and 3, each jet has an initial momentum flux M_o. By momentum flux conservation, at a given distance x the momentum of each jet is also $M = M_o$. The jet group momentum flux M_{NJ} is hence equal to the sum of momentum flux of the individual jets:

$$M_{NJ} = \int u_{NJ}^2(x, y, z)dA = M_1 + M_2 + \cdots + M_N = NM_o \tag{1}$$

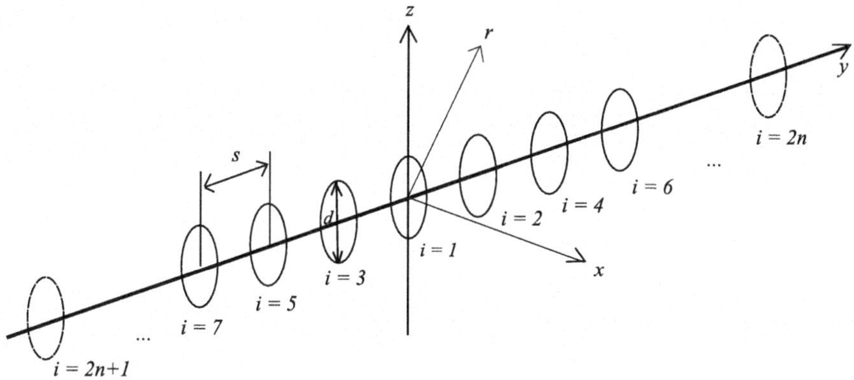

Fig. 2 An array of jets discharging into the $+x$-direction

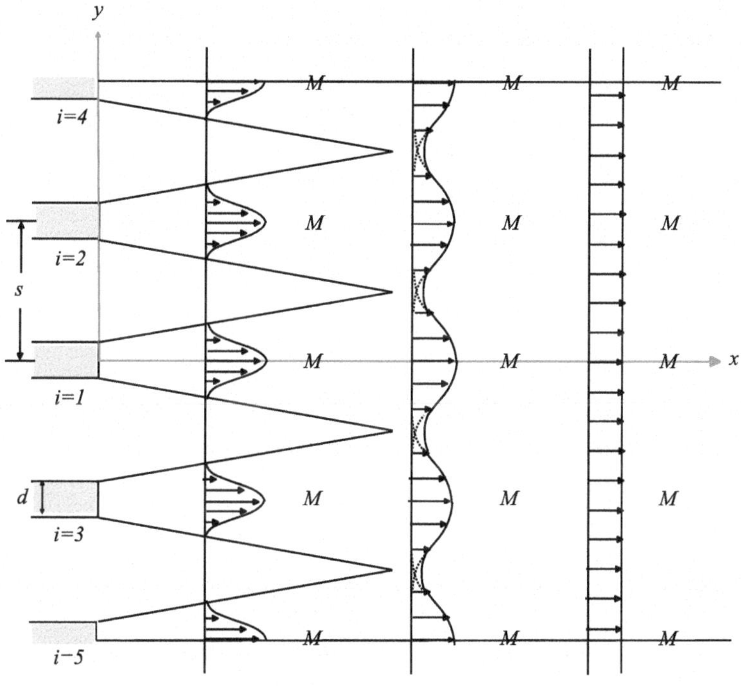

Fig. 3 An array of merging momentum jets

where the area integral is taken over the (y, z) plane and $N = 2n + 1$. Then:

$$\int u_{NJ}^2(x, y, z)dA = \int u_1^2 dA + \int u_2^2 dA + \cdots + \int u_N^2 dA \tag{2}$$

where $u_i(x, y, z)$ is the velocity field induced by the i-th independent jet, and $u_{NJ}(x, y, z)$ is the velocity field of the jet group. A solution to Eq. (2) at (x, y, z) is:

$$u_{NJ}^2(x, y, z) = u_1^2 + u_2^2 + \cdots + u_N^2 \tag{3}$$

which provides the theoretical basis for superimposing u^2. Alternatively, using the Reichardt's hypothesis the governing momentum equation can be shown to be linear in u^2—hence the validity of linear superposition of momentum (e.g. Hinze 1975 or Problem 2.11).

Example 11.1 *Multiple round jets from a power station*
Consider a cooling water discharge from a power station with negligible density difference. Suppose the cooling water is discharged horizontally via a diffuser as three round pure jets with nozzle diameter $D = 0.5$ m, initial velocity $u_o = 2.0$ m/s, and jet spacing $s = 5$ m into an unbounded stagnant ambient (Fig. 4). Determine the centerline velocity of the jet group at (i) $x = 5$ m and (ii) $x = 50$ m from the diffuser.

Solution: (i) The centerline velocity of each individual round jet at $x = 5$ m is given by $u_m(x) = 6.2u_o(x/D)^{-1} = 6.2(2.0)(5/0.5)^{-1} = 1.24$ m/s (Chap. 2, Table 2). To determine the jet group centerline velocity, first determine the momentum contribution $u(x, r)^2$ from each of the jet at $x = 5$ m (Fig. 4):

$$u_1(5, 0)^2 = [u_m(5)e^{-[0/(0.114\times5)]^2}]^2 = [u_m(5)]^2 = 1.5376 \text{ m}^2/\text{s}^2$$

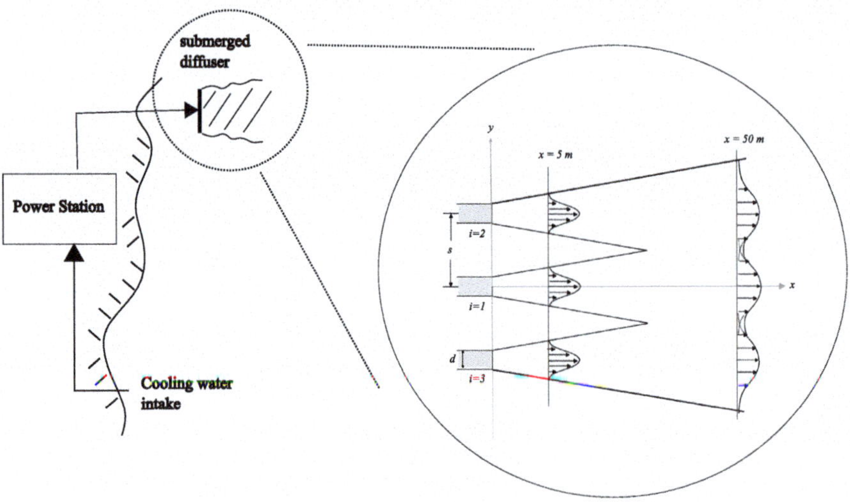

Fig. 4 Cooling water discharged from a power station as multiple round jets in Example 11.1 (not to scale)

$$u_2(5,-5)^2 = [u_m(5)e^{-[-5/0.114(5)]^2}]^2 = 0.0\ \mathrm{m^2/s^2}$$

$$u_3(5,5)^2 = [u_m(5)e^{-[5/0.114(5)]^2}]^2 = 0.0\ \mathrm{m^2/s^2}$$

The jet group momentum is then $u_{NJ}^2 = u_1^2 + u_2^2 + u_3^2 = 1.5376 + 0.0 + 0.0 = 1.5376\ \mathrm{m^2/s^2}$, and jet group centerline velocity $u_{NJ} = \sqrt{1.5376} = \underline{1.24\ \mathrm{m/s}}$—same as that of a single free jet, suggesting the jets have not merged. In the above equations, jet spreading rate of $\beta = 0.114$ has been used (Chap. 2).

(ii) At $x = 50$ m, the centerline velocity of each individual round jet is given by $u_m(50) = 6.2(2.0)(50/0.5)^{-1} = 0.124$ m/s. The momentum contribution $u(x,r)^2$ from each of the jet at $x = 50$ m are given by:

$$u_1(50,0)^2 = [u_m(50)e^{-[0/(0.114\times5)]^2}]^2 = [u_m(50)]^2 = 0.0154\ \mathrm{m^2/s^2}$$

$$u_2(50,-5)^2 = [u_m(50)e^{-[-5/0.114(50)]^2}]^2 = 0.00330\ \mathrm{m^2/s^2}$$

$$u_3(50,5)^2 = [u_m(50)e^{-[5/0.114(50)]^2}]^2 = 0.00330\ \mathrm{m^2/s^2}$$

The jet group momentum is then $u_{NJ}^2 = u_1^2 + u_2^2 + u_3^2 = 0.0154 + 0.00330 + 0.00330 = 0.022\ \mathrm{m^2/s^2}$, and jet group centerline velocity $u_{NJ} = \sqrt{0.022} = \underline{0.148\ \mathrm{m/s}} > 0.124$ m/s of a single free jet because of the jet interaction.

If velocity is additive, the jet group velocity is given by $u_{NJ} = u_1 + u_2 + u_3 = 0.124 + 0.0574 + 0.0574 = 0.239$ m/s, which will over-predict the jet group velocity significantly!

2 Multiple Plumes in Stagnant Ambient

For buoyant jet group with large buoyancy and negligible momentum, such as thermal plumes from electronic equipment (Linden 1999), they can be considered as pure plume group. Partially treated wastewater discharged from a submarine multiport diffuser can also be treated as a pure plume group when buoyancy becomes dominant (Chap. 3). To predict the plume group's mixing characteristics such as velocity and concentration, a different approach from momentum superposition is required as the momentum flux of a plume continually changes due to the effect of buoyancy (see Chap. 3, Table 1).

2.1 Kinetic Energy Superposition

Proof of Kinetic Energy Superposition Approach for Plumes

The theoretical basis of the superposition approach for plumes can be derived from basic plume theory (Lai and Lee 2012). First, consider the mechanical energy conservation equation of a round buoyant jet in cylindrical coordinates (z, r), where z is the vertical and r is the radial coordinates:

$$\frac{\partial u^3}{\partial z} + \frac{1}{r}\frac{\partial r u^2 v}{\partial r} = 2ug' - \frac{2u}{r}\frac{\partial (r\overline{u'v'})}{\partial r} \tag{4}$$

where $u(z, r)$ and $v(z, r)$ are the turbulent-mean z and r-velocity respectively, g' is the local reduced gravity, and $u'$and v' are the turbulent velocity fluctuations in the z- and r-direction respectively. Equation (4) can be obtained from the governing continuity and momentum equations.

The dependence of the entrainment coefficient α on the local jet densimetric Froude number $Fr_L = u_m/\sqrt{g'_m b_g}$ is $\alpha = \alpha_j + (\alpha_p - \alpha_j)(Fr_p/Fr_L)^2$, where α_j and α_p are the entrainment coefficients of a momentum jet and plume respectively. u_m and g'_m are the velocity and reduced gravity at the centreline of the buoyant jet respectively, and b_g is the Gaussian jet half-width (Chaps. 3 and 4, and Problem 3.17). It can be shown that $\alpha \to \alpha_j$ for momentum jets, $g'_m \to 0$, and that all buoyant jets tend to be plumes far away from the source, with $Fr_L \to Fr_p$, which is the asymptotic plume densimetric Froude number, $\alpha \to \alpha_p$. For a round plume, $Fr_p = \sqrt{\frac{5\lambda^2}{4\alpha_p}} = 4.6$ for $\alpha = 0.085$ and $\lambda = 1.2$ (A recent summary of α and λ can be found in Kaminski et al. 2005) respectively.

Specifically, it can be shown that the integral form of Eq. (4) can be written as (Fox 1970; Lee and Jirka 1981; Lai and Lee 2012):

$$\frac{\partial}{\partial z}\int u^3 dA = \frac{2\pi\lambda^2}{1+\lambda^2} u_m g'_m b_g^2 - \frac{2}{3}\pi u_m^3 \alpha_j b_g \tag{5}$$

Adopting the classical solution of $u_m = 4.71 F_o^{1/3} z^{-1/3}$ and $b_g = 0.105z$ for a pure plume (Rouse et al. 1952; Table 1 in Chap. 3), and assuming Gaussian velocity and buoyancy profiles we have:

$$\frac{2\pi\lambda^2}{1+\lambda^2} u_m g'_m b_g^2 = 2F \tag{6}$$

and adopting $\alpha_j = 0.057$ we obtain

$$\frac{2}{3}\pi u_m^3 \alpha_j b_g \approx 1.3F \tag{7}$$

where F is the plume buoyancy flux which is conserved. Substituting Eqs. (6) and (7) into Eq. (5), the plume kinetic energy flux can be related to the buoyancy flux as:

$$\frac{\partial E}{\partial z} \sim \frac{\partial}{\partial z} \int u^3 dA \approx 2F - 1.3F = 0.7F \tag{8}$$

and thus

$$\int u^3 dA \approx kFz \tag{9}$$

in a plume, where $k \approx 0.7$. Now by the conservation of buoyancy flux, at a given level z the plume group buoyancy flux F_{NJ} is equal to the sum of buoyancy flux of the NJ non-interacting individual plumes (Fig. 5):

$$F_{NJ} = F_1 + F_2 + \cdots + F_N \tag{10}$$

$$kF_{NJ}z = kF_1z + kF_2z + \cdots + kF_Nz \tag{11}$$

Assuming the kinetic energy flux of a plume group can be expressed in terms of the buoyancy flux in the same way as a single plume, we have:

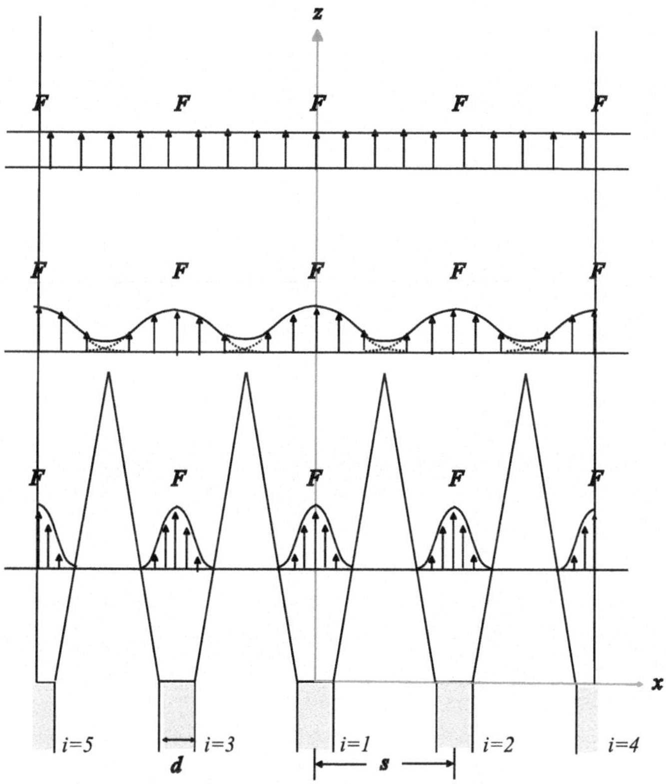

Fig. 5 An array of merging pure plumes

$$\int u_{NJ}^3(x, y, z)dA = \int u_1^3 dA + \int u_2^3 dA + \cdots + \int u_N^3 dA \tag{12}$$

A solution to Eq. (12) is:

$$u_{NJ}^3(x, y, z) = u_1^3 + u_2^3 + \cdots + u_N^3 \tag{13}$$

The above shows heuristically that the plume velocity cubed u^3 at each point of a given level z of individual plumes can be superimposed to give u^3 of the plume group. More generally, numerical experimentation (Lai and Lee 2012) shows the jet group velocity for a buoyant jet with both momentum and buoyancy (say $1 < Fr < 40$) can be obtained by interpolation between the jet-induced and the plume-induced jet-group velocity (obtained from superposition of u^2 and u^3 respectively), in accordance with the local densimetric Froude number (similar to the computation of the local entrainment coefficient—see Problem 11.2).

With the jet or plume group velocity u determined, uc at each point of a given level z can be superimposed (by mass conservation) to determine the concentration field $c_{NJ}(x, y, z)$ of the jet or plume group:

$$u_{NJ}c_{NJ} = u_1c_1 + u_2c_2 + \cdots + u_Nc_N \tag{14}$$

$$c_{NJ} = \frac{u_1c_1 + u_2c_2 + \cdots + u_Nc_N}{u_{NJ}} \tag{15}$$

Equivalently, this can also be shown by extending the Reichardt's hypothesis to compute the concentration field of a jet group (Hodgson et al. 1999).

Example 11.2 *Multiple round plumes from a submarine outfall*
Consider a partially treated wastewater discharged from a multiport diffuser with $NJ = 5$ ports, port spacing $s = 2$ m, and port diameter $d = 0.1$ m, into a receiving water of depth $H = 20$ m. The initial jet velocity $u_o = 1.0$ m/s and concentration $c_o = 1.0$ mg/L, and relative density difference $\Delta\rho_o/\rho_a = 0.025$ (Fig. 6). Find the plume group centerline (i) velocity and; (ii) concentration at height $h = 20$ m from the discharge.

Solution: The flowrate, momentum and buoyancy flux from each individual port are given by $Q_o = \pi(0.1)^2/4 \times 1.0 = 0.007854$ m^3/s, $M_o = Q_o u_o = 0.007854$ m^4/s^2, and $F_o = Q_o g' = 0.007854 \times 9.81 \times 0.025 = 0.001926$ m^4/s^3. The momentum length scale $l_s = 0.007854^{3/4}/0.001926^{1/2} = \underline{0.6}$ m. At $h = 15$ m, $h/l_s = 25 >> 1$ and the use of plume solution is well justified (Chap. 3).

(i) Centerline velocity of the round plume at $h = 15$ m (Table 1 in Chap. 3) is $u_m = 4.71F_o^{1/3}z^{-1/3} = 4.71 \times 0.001926^{1/3} \times 20^{-1/3} = \underline{0.216}$ m/s.

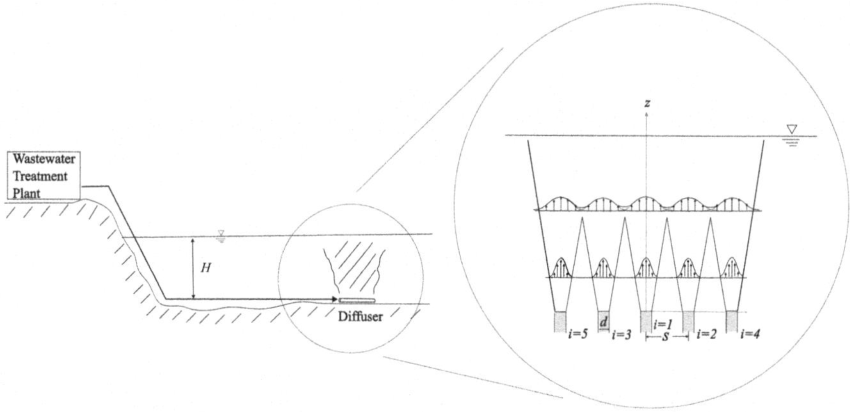

Fig. 6 Partially treated wastewater discharged as multiple round plumes of Example 11.2 (not to scale)

To determine the centerline velocity of the plume group, we can apply the superposition approach. First determine the kinetic energy distribution $u(z, r)^3$ from each of the plume at $z = 20$ m:

$$u_1(20, 0)^3 = [u_m(20)e^{-[0/(0.105\times 5)]^2}]^3 = [u_m(20)]^3 = 0.0101\ \text{m}^3/\text{s}^3$$

$$u_3(20, 2)^3 = [u_m(20)e^{-[2/0.105(20)]^2}]^3 = 0.0006645\ \text{m}^3/\text{s}^3$$

$$u_5(20, 4)^3 = [u_m(20)e^{-[4/0.105(20)]^2}]^3 = 1.90 \times 10^{-7}\ \text{m}^3/\text{s}^3$$

$$u_2(20, -2)^3 = u_3(20, 2)^3 = 0.0006645\ \text{m}^3/\text{s}^3$$

$$u_4(20, -4)^3 = u_5(20, 4)^3 = 1.90 \times 10^{-7}\ \text{m}^3/\text{s}^3$$

In the above equations, a plume spreading rate of $\beta = 0.105$ has been adopted. The plume group centerline velocity is then:

$$u_{NJ} = (u_1^3 + u_2^3 + u_3^3 + u_4^3 + u_5^3)^{1/3} = \underline{0.225\ \text{m/s}}$$

If one superimposes the plume velocity u instead, the plume group centerline velocity is given by $u_{NJ} = (u_1 + u_2 + u_3 + u_4 + u_5) = 0.309$ m/s, significantly overpredicting the velocity!

(ii) The plume group centerline concentration can be also be determined by the superposition approach. At $z = 20$ m, the concentration of a single round plume is given by (Chap. 3): $c_m = 10.46 Q_o c_o F_o^{-1/3} z^{-1/3} = 0.243$ mg/L. The concentration contribution at the plume group centerline by each plume are (with spreading rate $\lambda\beta = 0.105 \times 1.19 = 0.125$):

$$c_1(20, 0) = c_m(20)e^{-[0/(0.125\times 20)]^2} = c_m(20) = 0.243 \text{ mg/L}$$

$$c_3(20, 2) = c_m(20)e^{-[2/(0.125\times 20)]^2} = 0.128 \text{ mg/L}$$

$$c_5(20, 4) = c_m(20)e^{-[4/(0.125\times 20)]^2} = 0.0188 \text{ mg/L}$$

$$c_2(20, -2) = c_3(20, 2) = 0.128 \text{ mg/L}$$

$$c_4(20, -4) = c_5(20, 4) = 0.0188 \text{ mg/L}$$

The plume group concentration is then:

$$c_{NJ} = (u_1c_1 + u_2c_2 + u_3c_3 + u_4c_4 + u_5c_5)/u_{NJ} = \underline{0.334 \text{ mg/L}}$$

If one simply superimposes the plume concentration c instead, the plume group centerline concentration is given by $c_{NJ} = (c_1 + c_2 + c_3 + c_4 + c_5) = 0.537$ mg/L, significantly over-predicting the concentration!

2.2 Experiments and Numerical Simulations

The superposition approach for jets has been validated experimentally by Knystautas (1964). The velocity field of multiple round jets at different spacing was measured and compared with the predictions by superimposing u^2. The agreement between predictions and observation for both the centerline velocity decay and the traverse velocity profile were satisfactory.

Pani and Dash (1983) followed a similar approach and used the principle of superposition (of u^2) and found good agreement between predictions and experiments of multiple jets in stagnant ambient. Later Pani and Dugad (2002) extended the application of Reichardt's hypothesis to the case of three-dimensional (rectangular) outlets by introducing two different length scales in the y and z directions, which further supported the validity of the approach. Hodgson et al. (1999) and Wang and Davidson (2003) later showed that the concentration field can be predicted by superimposing the tracer mass flux of the jet group. The approach has also been extended to multiple jets in coflow; details can be found in Pani et al. (2009).

A systematic study of multiple plume interaction was carried out by Yannopoulos and Noutsopoulos (2006), with the centerline velocity and traverse concentration profile of a plume group measured. By assuming the kinetic energy flux or u^3 can be superimposed, they were able to predict both the transverse concentration and centerline velocity decay of the plume group well.

A detailed numerical study using the FLUENT CFD code was carried out by Lai and Lee (2012) to study the mixing of a plume group. It was shown that the kinetic energy flux of a twin plume discharge (NJ = 2) is close to that obtained by summing the kinetic energy flux of each individual plume along z. This supports $\int u_{NJ}^3 dA = \int u_1^3 dA + \int u_2^3 dA$ and thus the validity of kinetic energy superposition $u_{NJ}^3 = u_1^3 + u_2^3$. The numerical solution also demonstrates that $E \approx kFz$ ($k \approx 0.7$) for both the individual plume and the plume group. By superimposing u^3 for a row of round plumes, Lai and Lee (2012) also showed that the plume group velocity approaches that of a plane plume after merging, showing the correct asymptotic behaviour of the superposition solutions.

3 Multiple Buoyant Jets in Stagnant Ambient

Previous sections have focused on the cases with long array of multiple pure jets and plumes, which are pure sources of momentum and buoyancy respectively. In most engineering applications, however, the discharges are governed by both momentum and buoyancy, and with more complex jet nozzles configurations. The resulting buoyant jets can interact kinematically due to the merging of jets as discussed in previous section; the asymmetric flow and pressure fields induced by neighbouring jets can also lead to dynamic interaction.

The dynamic interaction of multiple buoyant jets was first observed by Liseth (1970, 1976) and then Davidson (1989); both showed that for horizontal buoyant jets discharged alternately from both sides of the diffuser, the plumes tend to bend towards each other due to pressure interaction (Fig. 7). Roberts and Snyder (1993a, b) also studied the interaction of rosette buoyant jet groups, and observed the contraction of the multiple jets in the low pressure region above the source—affecting the mixing and merging of the discharge. In the absence of a theory to account for dynamic

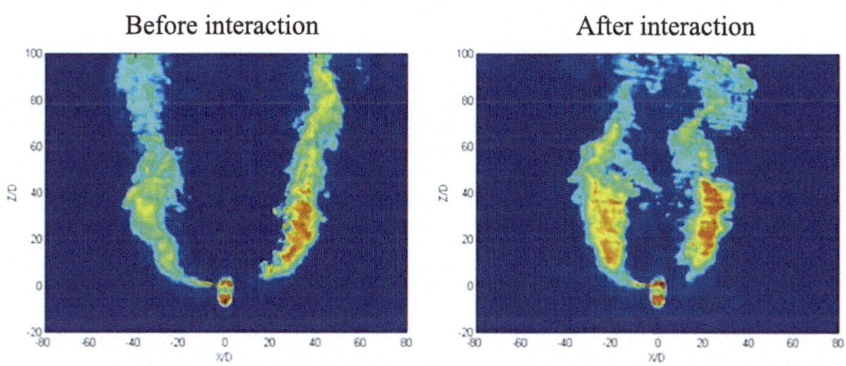

Fig. 7 Observed jet trajectory from an alternating diffuser ($D = 0.325$ cm; nozzle spacing $L = 2$ cm; $Fr \approx 10$) (Lai 2009)

interaction of buoyant jets, even in stagnant water, outfall design in engineering studies often totally rely on costly physical scale model studies.

3.1 Dynamic Interaction of Multiple Buoyant Jets

A Point Sink

To develop a model for the dynamic interaction of buoyant jets, it is useful to first consider the action of a 3D point sink. A 3D point sink draws flow radially to itself at a rate of m (m^3/s) ($m < 0$). The corresponding velocity potential can be expressed as:

$$\phi = \frac{m}{4\pi r} \quad (m < 0)$$

and the induced velocity is then given by:

$$u_r = -\nabla\phi = -\frac{\partial\phi}{\partial r} = \frac{m}{4\pi r^2}$$

Example 11.3 *Point sink induced velocity*
The flowrate into a point sink at (x_i, y_i, z_i) is given as $-\Delta Q$. Find the sink induced velocity at (x, y, z).

Solution: The induced velocity component in the x-, y-, and z-direction can be found by differentiation of the velocity potential in x-, y-, and z-direction respectively.

$$u_{xi} = -\frac{\partial\phi_i}{\partial r}\frac{\partial r}{\partial x}$$

$$= \frac{-\Delta Q}{4\pi}\frac{(x - x_i)[(x - x_i)^2 + (y - y_i)^2 + (z - z_i)^2]^{-1/2}}{[(x - x_i)^2 + (y - y_i)^2 + (z - z_i)^2]}$$

$$= \frac{-\Delta Q}{4\pi}\frac{(x - x_i)}{[(x - x_i)^2 + (y - y_i)^2 + (z - z_i)^2]^{3/2}}$$

$$u_{yi} = -\frac{\partial\phi_i}{\partial r}\frac{\partial r}{\partial y}$$

$$= \frac{-\Delta Q}{4\pi}\frac{(y - y_i)[(x - x_i)^2 + (y - y_i)^2 + (z - z_i)^2]^{-1/2}}{[(x - x_i)^2 + (y - y_i)^2 + (z - z_i)^2]}$$

$$= \frac{-\Delta Q}{4\pi}\frac{(y - y_i)}{[(x - x_i)^2 + (y - y_i)^2 + (z - z_i)^2]^{3/2}}$$

$$u_{zi} = -\frac{\partial \phi_i}{\partial r}\frac{\partial r}{\partial z}$$

$$= \frac{-\Delta Q}{4\pi}\frac{(z-z_i)[(x-x_i)^2+(y-y_i)^2+(z-z_i)^2]^{-1/2}}{[(x-x_i)^2+(y-y_i)^2+(z-z_i)^2]}$$

$$= \frac{-\Delta Q}{4\pi}\frac{(z-z_i)}{[(x-x_i)^2+(y-y_i)^2+(z-z_i)^2]^{3/2}}$$

Distributed Point Sinks

The external flow induced by a jet is irrotational and can be solved by potential flow theory, since the viscous effects or turbulent mixing is only important within the jet boundary layer. Taylor (1958) first showed that the jet induced flow can be solved by representing the jet as a line sink. Based on this concept, Lai and Lee (2012) showed that if the jet is discretized along its axis into N elements each of length Δs, the jet-induced flow can be obtained by summing the flow due to a distribution of N closely spaced point sinks with strength equal to the entrainment of the jet element $m = -\frac{dQ}{ds}$ (see Prob. 2.14). The distributed sink induced entrainment velocity field in the plane of symmetry is shown in Fig. 8, which has been shown to be equivalent to Taylor's line sink solution (Lai and Lee 2012).

Semi-analytical Model for Dynamic Interaction of Buoyant Jets

More generally, with the predicted buoyant jet trajectory (Chap. 4), and the entrainment flow of every "jet element", then the outside entrainment flow induced at a

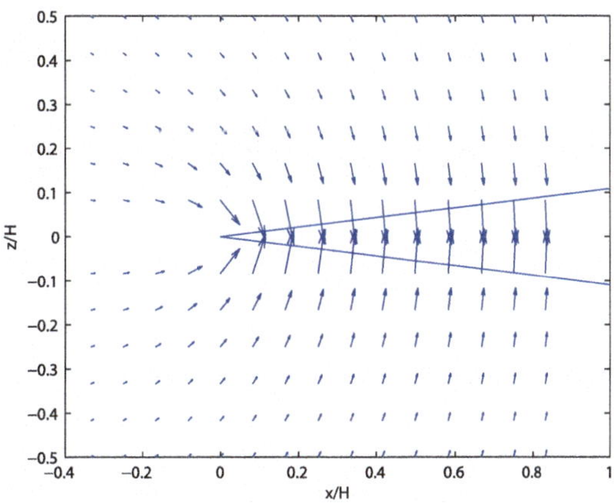

Fig. 8 External flow at centerline plane induced by a turbulent single free jet using distributed point sinks. $D = 0.1$ m; $u_o = 1.75$ m/s; $Q_o = 0.0138$ m^3/s; $M_o = 0.0242$ m^4/s^2; $\Delta s = 12D$

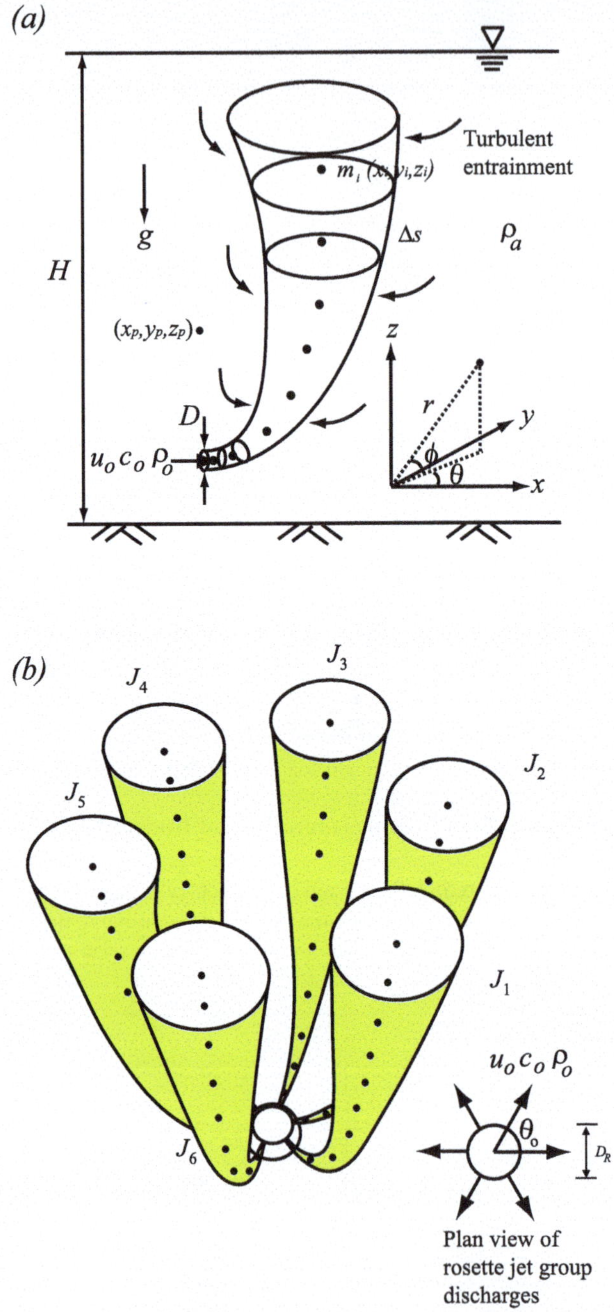

Fig. 9 Sketch of **a** an inclined buoyant jet—the action of the jet on the external irrotational flow is represented by a number of distributed point sinks; **b** a rosette buoyant jet group ($NJ = 6$)

given point (x_p, y_p, z_p) due to one jet can be predicted as the sum of the velocities induced by the point sinks with strength $m = -\frac{dQ}{ds}$ in all the jet elements (Fig. 9a); and the induced velocity due to all the jets can be similarly summed (Fig. 9b). Bottom and free surface boundaries can be accounted for using the method of images. Mathematically:

$$u_{xi}|_{x_p,y_p,z_p} = \sum_{j=1}^{NJ}\sum_{i=1}^{N} \frac{m_i}{4\pi} \frac{(x_p - x_i)\Delta s}{[(x_p - x_i)^2 + (y_p - y_i)^2 + (z_p - z_i)^2]^{3/2}} \tag{16}$$

$$u_{yi}|_{x_p,y_p,z_p} = \sum_{j=1}^{NJ}\sum_{i=1}^{N} \frac{m_i}{4\pi} \frac{(y_p - y_i)\Delta s}{[(x_p - x_i)^2 + (y_p - y_i)^2 + (z_p - z_i)^2]^{3/2}} \tag{17}$$

$$u_{zi}|_{x_p,y_p,z_p} = \sum_{j=1}^{NJ}\sum_{i=1}^{N} \frac{m_i}{4\pi} \frac{(z_p - z_i)\Delta s}{[(x_p - x_i)^2 + (y_p - y_i)^2 + (z_p - z_i)^2]^{3/2}} \tag{18}$$

where N is the number of jet elements for the j-th jet, and NJ is the number of jets.

Given that the outside entrainment flow of multiple jets in stagnant fluid can be simulated, the dynamic jet interaction can then be solved by choosing a suitable control volume (for example at the plane of symmetry where the net horizontal momentum flux is zero, so that only pressure needs to be accounted for), and performing an exact momentum balance (Lai 2009; Lai and Lee 2012). The jet properties can then be obtained by an iterative solution. Starting by assuming pressure $p = 0$, the jet trajectory is computed by solving the integral governing equations for each individual jet independent of other jets. The entrainment flow of each individual buoyant jet is computed using a well-accepted shear entrainment hypothesis (e.g. Chap. 4). Based on this, the outside entrainment flow is computed, which gives the pressure field via the Bernoulli equation $p_o - p = \rho(|\mathbf{q}|^2 - |\mathbf{q}_o|^2)/2$, where $|\mathbf{q}|^2 = u_x^2 + u_y^2 + u_z^2$, and $|\mathbf{q_o}|$ and p_o are reference velocity magnitude and pressure taken at infinity (assumed to be zero without loss of generality). Thus we have:

$$p = -\frac{1}{2}\rho|\mathbf{q}|^2 \tag{19}$$

Then the updated pressure field is incorporated in the jet model to compute the relevant pressure force on the control volume; the jet model is solved to give a revised jet trajectory, and so on until convergence (Fig. 10).

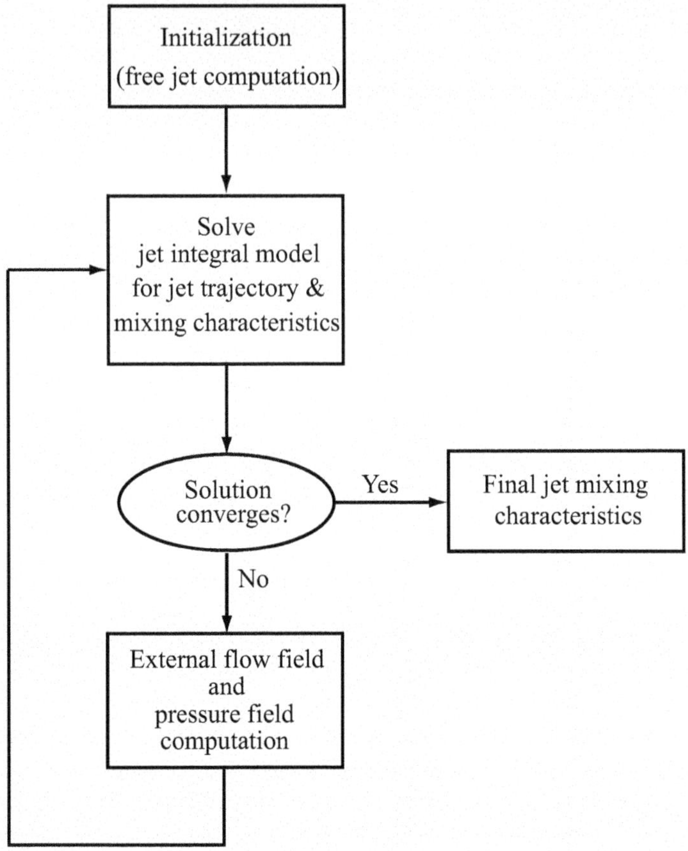

Fig. 10 Semi-analytical model for dynamic interaction of buoyant jets

Example 11.4 *Predicting jet interaction of twin buoyant jet group*
Consider a plume pair of equal buoyancy flux discharging vertically into a stagnant ambient in a laboratory water tank (Fig. 11). Each plume is discharged through a nozzle of diameter $D = 5$ *mm, flowrate* $Q_o = 0.5$ *cm*^{3}*/s, and buoyancy flux* $F_o =$ 30 *cm*^{4}*/s*$^{-3}$*. Spacing between the plume is* $s = 7.5$ *cm. Find the interacting plume trajectory. Assume the pressure at the plane of symmetry is constant and can be computed at* $(0, 0, 5D)$*. To simplify the problem, only consider two control volumes each with* $\Delta z = 15D$*,* $\Delta y = 20D$*, and first three entrainment sinks of each plume for the dynamic interaction computation.*

Solution: To determine the plume trajectory, first consider the governing equations for a vertical pure plume (Chap. 3), but also adding an interaction pressure term to the $x-$momentum equation:

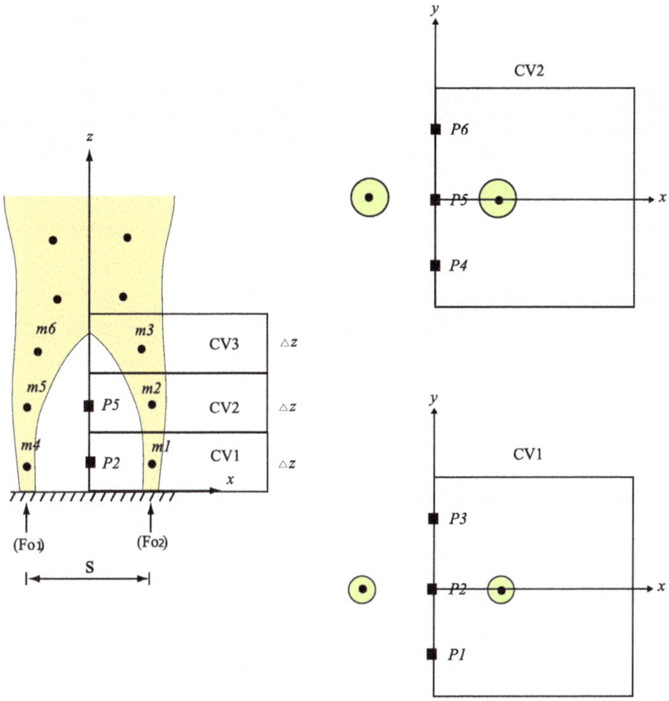

Fig. 11 Definition sketch of the plume pair for Example 11.4

$$\frac{dQ}{dz} = 2\pi\alpha(2M_z/\pi)^{1/2}$$

$$\frac{dM_z}{dz} = 0$$

$$\frac{dM_x}{dz} = 0 + \boxed{\sum \frac{P}{\rho} dy}$$

$$\frac{dx}{dz} = \frac{M_x}{\sqrt{M_x^2 + M_z^2}}$$

The crux of the model is to determine the interaction pressure term and compute the changes to the plume horizontal momentum due to the pressure. The pressure can be determined by a control volume (CV) approach. The CVs, each of height Δz, can be chosen such that they extends to infinity, where the horizontal momentum flux and pressure are practically zero. Since there must be no flow into the CVs at the plane of symmetry, only the plume induced pressure needs to be considered in the horizontal momentum equation. Given that the pressure can be assumed constant, we have

$P1 = P2 = P3 = P4 = P5 = P6 = P$, and P is to be computed at $P(0, 0, 5D)$ (Fig. 11).

The iterative solution procedures follow that of Fig. 10. Start by setting $P = 0$ with the plumes' trajectories being straight. The distributed sinks $m_1 - m_3, m_4 - m_6$ are then located at the each of the plume's centerline. The sink strength of each plume element can be computed by determining $\frac{dQ}{dz} = 2\pi\alpha(2M_z/\pi)^{1/2}$. Note that $M_z = (2\pi)^{1/3}[\frac{9\alpha(1+\lambda^2)}{20}]^{2/3} F_o^{2/3} z^{4/3}$, and using $\lambda = 1.19$ and $\alpha = 0.088$ for a round plume (Table 1 in Chap. 3), we have:

$$M_z = 0.386 F_o^{2/3} z^{4/3} = 1.73 \times 10^{-5} z^{4/3}$$

$$\frac{dQ}{dz} = 0.441 M_z^{1/2} = 0.441 \times [0.386 F_o^{2/3} z^{4/3}]^{1/2} = 0.274 F_o^{1/3} z^{2/3}$$

The sink strength of each element is then

$$m = -\Delta Q/\Delta z = -0.274 \times (3.0 \times 10^{-7})^{1/3} z^{2/3} = -0.001834 z^{2/3}$$

The assumed constant pressure at the plane of symmetry is to be evaluated at $P(0, 0, 5D) = P(0, 0, 0.025)$. The distributed sink induced velocity (u_x, u_y, u_z) at $(x_p, y_p, z_p) = (0, 0, 0.025)$ can be computed by Eqs. (16)–(18), where $\Delta s = \Delta z = 15D$. The results are tabulated in Table 1.

The net pressure is then calculated by Eq. (19) as $P/\rho = -1/2(u_x^2 + u_y^2 + u_z^2) = -1/2(0.001265^2) = -8.0 \times 10^{-7}\ \mathrm{m^2/s^2}$. The resulting horizontal momentum change is relate to the pressure by $\Delta M_x = (P/\rho)\Delta y \Delta z = -8.0 \times 10^{-7} \times 20 \times (5/1000) \times 15 \times (5/1000) = -6.0 \times 10^{-9}\ \mathrm{m^4/s^2}$. The trajectory change can then be computed with $\Delta x = \frac{M_x}{\sqrt{M_x^2 + M_z^2}} \Delta z$ and is tabulated in Table 2.

The trajectory of the other plume can be determined by symmetry. Repeating the above computation but noting that the plume sink locations (x_i, y_i, z_i) have now changed due to the updated trajectory, e.g. m_2 now located at $(0.0265, 0)$ instead of $(0.0375, 0)$, we can carry out the second iteration. The iteration can be continued until the solution converges.

As an example, a plot of the resulting trajectory with three iterations is given in Fig. 12, which shows the trajectory converges quickly to the final solution. The corresponding plume group velocity and concentration can then be determined by superposition of u^3 and uc. For more accurate computation, the pressure variation in the plane of symmetry should be considered; the entrainment sinks along the whole plume should be considered (and even accounting for boundary effects using method of images); control volume boundary should be extended further until pressure is practically zero; and with a finer computation step Δz and higher order integration scheme such as a 4th order Runge-Kutta solver. The plumes are assumed straight after merged. All these can be easily implemented into a computer program based on the concept illustrated in this example.

Table 1 Example 11.4: Velocity components at P(0, 0, 0.025) induced by the distributed point sinks $m_1 - m_6$ with coordinates (x_i, y_i, z_i) for the first iteration

Sink	x_i (m)	y_i (m)	z_i (m)	$m = -\frac{dQ}{dz}$ (m³/s/m)	$u_x(0, 0, 0.025)$ (m/s)	$u_y(0, 0, 0.025)$ (m/s)	$u_z(0, 0, 0.025)$ (m/s)
m_1	0.0375	0	0.0375	-2.05×10^{-4}	7.45×10^{-4}	0	2.48×10^{-4}
m_2	0.0375	0	0.1125	-4.27×10^{-4}	1.11×10^{-4}	0	2.59×10^{-4}
m_3	0.0375	0	0.1875	-6.01×10^{-4}	2.90×10^{-5}	0	1.26×10^{-4}
m_4	−0.0375	0	0.0375	-2.05×10^{-4}	-7.45×10^{-4}	0	2.48×10^{-4}
m_5	−0.0375	0	0.1125	-4.27×10^{-4}	-1.11×10^{-4}	0	2.59×10^{-4}
m_6	−0.0375	0	0.1875	-6.01×10^{-4}	-2.90×10^{-5}	0	1.26×10^{-4}
				SUM	0	0	0.001265

Table 2 Example 11.4: Computed plume's vertical and horizontal momentum fluxes and trajectory after accounting for the induced pressure for the first iteration

z (m)	M_x (m^4/s^2)	ΔM_x (m^4/s^2)	M_z (m^4/s^2)	Δx (m)	x (m)
0	0	-6.00×10^{-9}	0	0	0.0375
0.075	-6.00×10^{-9}	-6.00×10^{-9}	5.47121×10^{-7}	-1.10×10^{-2}	0.0265
0.150	-1.20×10^{-8}	-6.00×10^{-9}	1.37866×10^{-6}	-8.00×10^{-8}	0.0178

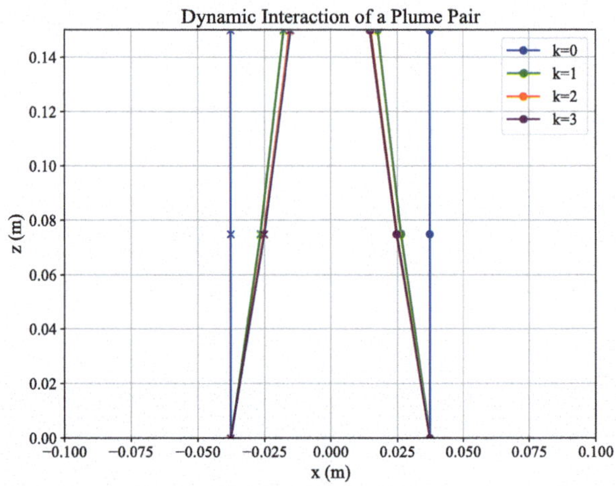

Fig. 12 Computed plume pair trajectory at different iteration k for Example 11.4

3.2 *Theory Versus Experiments*

The model developed for the dynamic interaction has been compared extensively with past experimental data of several complex cases.

Coalescing Plumes

The model has been tested on recent experiments on the merging of two axisymmetric plumes in connection with ventilation problems in the built environment. Kaye and Linden (2004) performed experiments with two plumes of equal and non-equal buoyancy flux; they studied the drawing together of two plumes caused by the respective entrainment fields. It is seen from the predicted concentration fields that due to the interaction of the entrainment fields the plumes are deflected towards each other; it is seen that the predicted merging heights agree well with data. When one plume has more buoyancy flux than the other, the weaker plume is drawn much more to the stronger plume. This is also in accord with observations (Fig. 13).

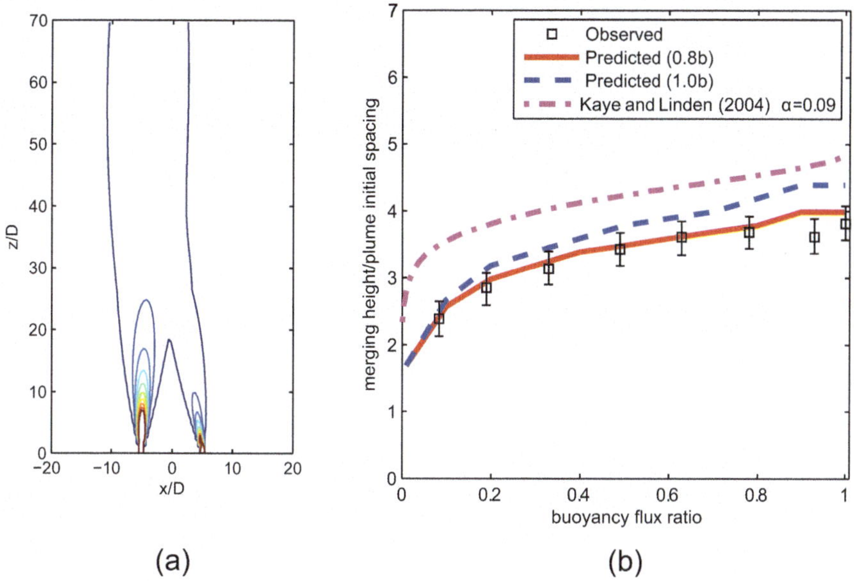

Fig. 13 **a** Predicted concentration field (buoyancy flux ratio $\Phi = 0.1$; c/c_o = 0.005–0.365 with interval 0.04); and **b** Merging height as a function of buoyancy flux ratio of unequal plume pairs (Symbols represent measurement data and error bar limits reported in Kaye and Linden 2004)

Rosette Buoyant Jet Group

For a rosette jet group discharging in stagnant ambient conditions, previous studies have observed that the jets tend to deflect towards each other, merging earlier than if they were independent, and form a single rising column. Lai and Lee (2012) carried experiments for a rosette buoyant group to study their dynamic interaction. The jet trajectory is computed (typically 3–4 iterations are required) until the outer edges of adjacent jets (defined by the Gaussian velocity half-width) intersect. Both the free surface and bottom have been accounted for using the method of images. The comparison of predicted versus observed jet trajectory shows that for the 2-jet case, the jet interaction appears rather weak; with 6 jets the interaction is seen to be stronger (Fig. 14a). In general the predicted jet trajectory is in good agreement with data (Fig. 14b). Figure 14c and d show the concentration field of the jet group at $z = 20D$ and $80D$ respectively. At $z = 20D$ the jets are still distinguishable, but become fully merged at $z = 80D$, which forms a single rising column as observed experimentally. The model has also been successfully applied to other complex jet group configuration such as clustered jet group, and alternating diffuser discharge (Lai 2009), demonstrating the generality of the approach.

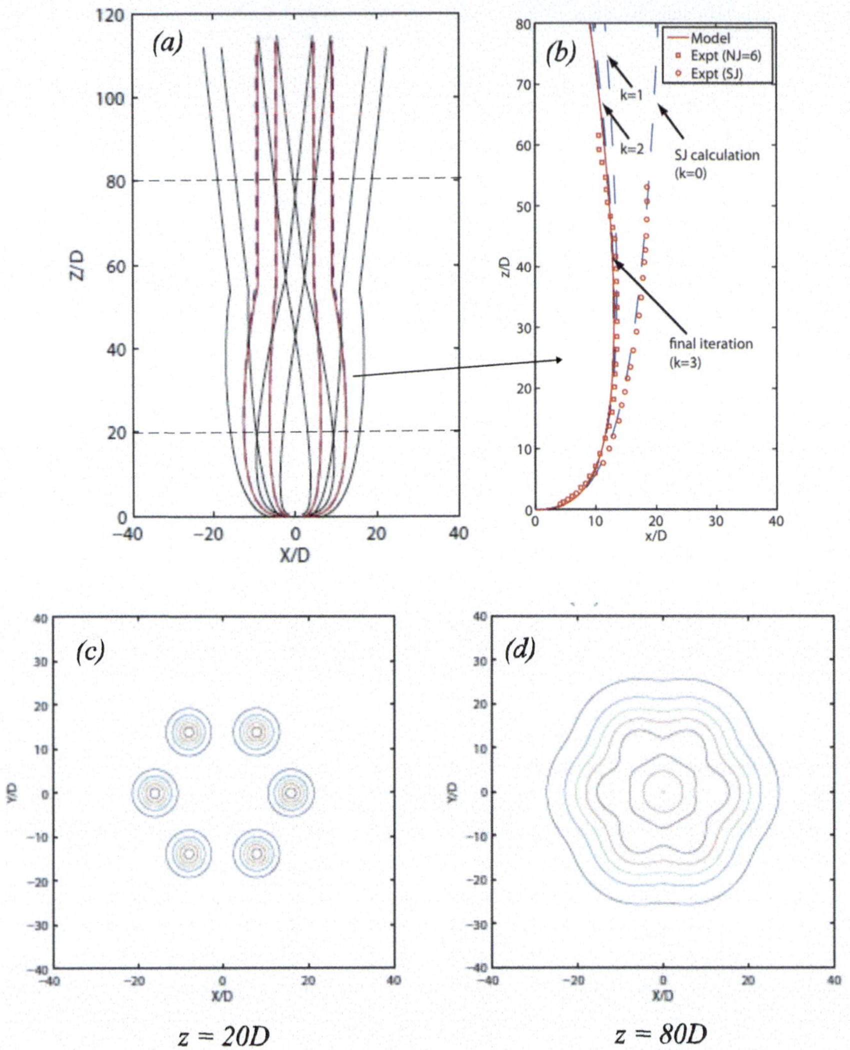

Fig. 14 Predicted trajectory and concentration field of a 6-jet group ($D = 0.0044$ m; $u_o = 0.365$ m/s; $Fr = 4.9$) by semi-analytical model; **a** and **b** Predicted and observed jet trajectories; the iterative calculations start from a single free jet (SJ); **c** and **d** Predicted concentration field at $z = 20D$ and $z = 80D$ respectively by superposition of uc

4 Multiple Buoyant Jets in a Crossflow

4.1 Effects of a Crossflow on Dynamic Interaction

In many cases, a buoyant jet group will be affected by an ambient crossflow, such as a rosette buoyant jet group discharged from a submarine outfall in a tidal current (Fig. 15). The effect of a weak ambient current can be incorporated into the dynamic interaction model by adding the ambient crossflow velocity to the jet group induced velocity field. The jet integral model remains valid if the current is weak so that entrainment is close to that of the stagnant ambient (Gaskin 1995 and Davidson et al. 2002).

To illustrate, a six-jet rosette jet group ($NJ = 6$, $D = 0.0044$ m) is discharged horizontally from a riser (horizontal angle 0–2π; vertical angle 0). $u_o = 0.365$ m/s and $\Delta\rho/\rho_a = 0.1266$, giving $Fr = 4.9$. Three ambient current speed (U_a) are tested: $U_a = 0$, 0.03, and 0.08 m/s, giving a jet-to-ambient velocity ratio of $K = u_o/U_a =$ 4.6–12.2. Water depth is 0.4 m. When an ambient current is added the external flow field is asymmetric (Fig. 16b). Figure 17 shows the comparison of the interacted jet trajectory and that without interaction for $U_a = 0.03$ and 0.08 m/s. Deflection of the interacted jets indicates the degree of interaction. Compared to the corresponding discharge in stagnant ambient (Fig. 14b), it can be seen with the addition of a current the degree of jet interaction tends to be less; the interaction is further reduced with increase of U_a—with a jet-to-ambient velocity ratio of $K = 4.6$ the interaction is so weak that the coflow jet (jet 1) is not affected by the adjacent jets at all as seen in Fig. 17b; the interaction also decreases with increasing Fr.

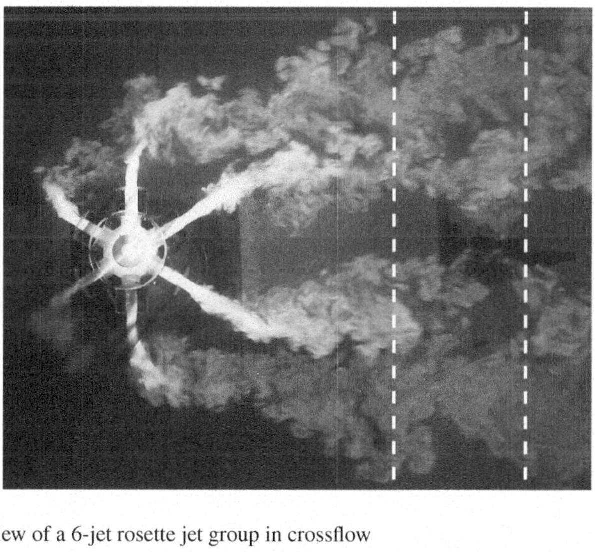

Fig. 15 Top view of a 6-jet rosette jet group in crossflow

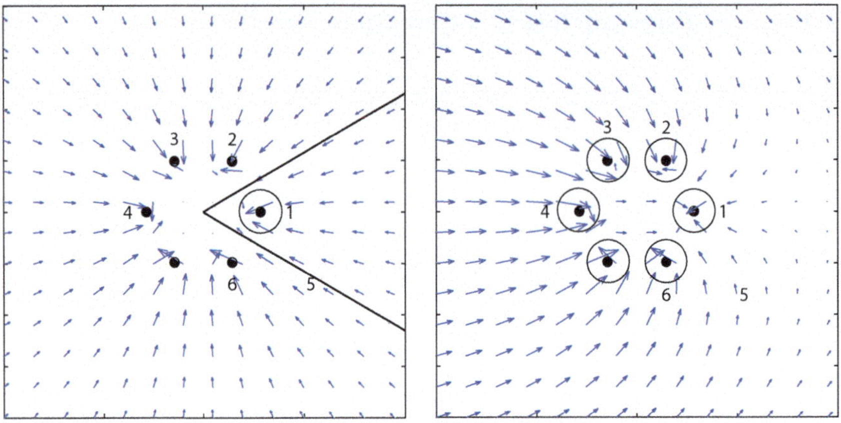

(a) stagnant ambient condition (b) crossflowing ambient

Fig. 16 Computed external induced by a rosette jet group in weak crossflow. The local control volumes chosen for horizontal momentum balance are also shown

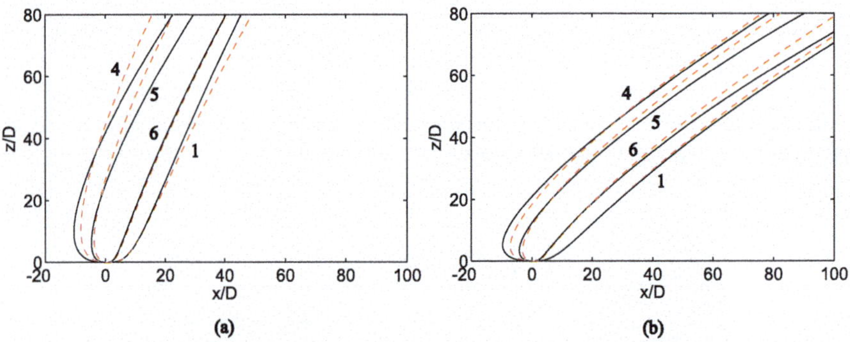

Fig. 17 Comparison of the interacted (solid line) and non-interacted jet (dash line) trajectory of a rosette jet group for two jet-to-ambient velocity ratios: $K = u_o/U_a$ **a** $K = 12.2$ and **b** $K = 4.6$. The numbers next to the jet trajectory refer to the individual jets labelled as in Fig. 16

Dynamic jet interaction in a stronger crossflow has been studied in the laboratory; by looking at single jets discharging separately one at a time, and then the multiple jets discharging together. It turns out that for typical jet to riser diameter of 0.1 or less, the fanned out jet arrangement does not result in significant dynamic interaction; the observed individual jet trajectories are almost the same as the multiple jet group. Figure 18 shows the measured individual jet trajectories for a 2-jet group (60° and 120°), and for a 3-jet group (30°, 90°, 150°) for $K = u_o/U_a = 5$–15. The circles are the measured trajectory for the single jet discharging separately, and the crosses are the observed corresponding jet trajectory in the multiple jet group. It is seen they are

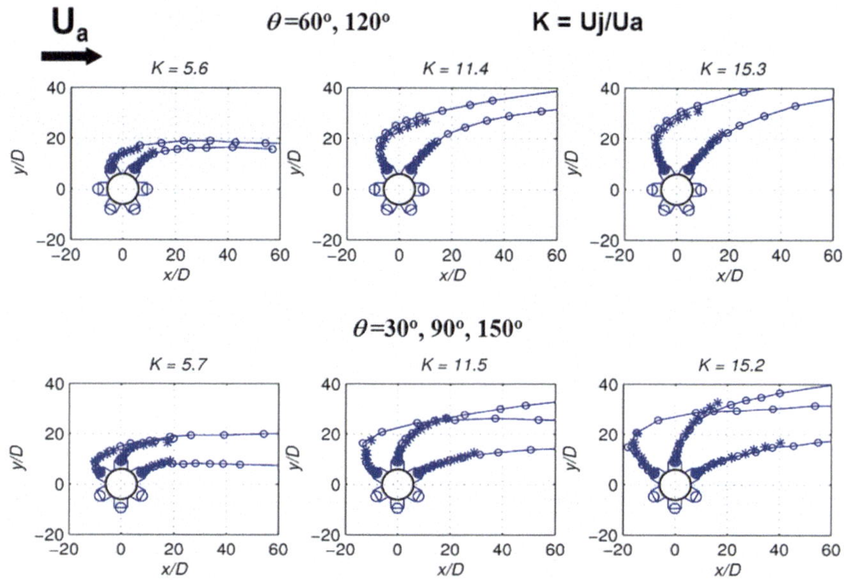

Fig. 18 Measured centerline trajectories of various discharge configurations (top view). Multiple jets are indicated by asterisks and single jets by circles joined by solid line (Lai et al. 2011)

very similar, and almost indistinguishable, suggesting that the dynamic interaction is negligible with the addition of an ambient crossflow.

4.2 *Composite Dilution of a Buoyant Jet Group*

Both theory and experiments showed that even in a weak crossflow the starved plume condition does not exist except in very confined surroundings, and hence dynamic interaction can be neglected. Each jet can then be treated independently without taking into account the dynamic interaction (i.e. only kinematic interaction of plume overlap needs to be considered). In a crossflow, in the bent-over phase, $u = U_a$; the plumes will merge, and the overlapping of plumes will lead to a reduction of cross section area and hence entrainment (Fig. 19).

In order to compute the kinematic interaction, it is useful to define a composite dilution that measures the mixing of a multiple buoyant jet group. In the bent-over phase of a jet in crossflow, experiments have shown that the jet horizontal excess-momentum is dissipated within a short distance, and the x velocity of the jet group will be approximately the same as the crossflow velocity U_a. Consider two plumes with cross-section area A_1 and A_2; if they overlap the same mass flux will be advected through a reduced area: $A_m = A_1 + A_2 - A_o$ (Fig. 20), and hence the concentration

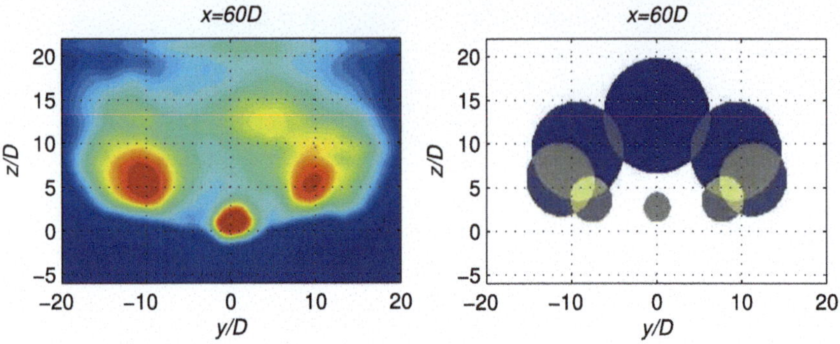

Fig. 19 Cross-sectional concentration profile of 8-jet rosette jet group of Lai et al. (2011) (left: LIF experiment; right: VISJET simulation)

is higher than without merging. The average concentration would then be given by the total mass flux divided by the merged area of the bent-over jets, and generally:

$$(u_m A_m)\, C_m = \sum_{i=1}^{N} (u_i A_i)\, C_i \tag{20}$$

where A_m, C_m, u_m = cross-sectional area, cross-sectional averaged concentration and velocity of the merged jets respectively; A_i, C_i, u_i = are the cross-sectional area, average concentration and velocity of an individual jet, respectively, and N = the N-th individual jets.

$$C_m = \sum_{i=1}^{N} (u_i\, A_i\, C_i\,)/(U_a\, A_m\,) \approx \sum_{i=1}^{N} A_i\, C_i / A_m \tag{21}$$

The comparison between the experimentally determined composite dilution and the observed dilution is shown in Fig. 21. It can be seen that they agree well with each other, showing that dilution of a jet group can be predicted by superimposing average tracer concentration of its composing individual jet and accounting for the overlapped area. The general model has also been tested against independent experimental data from physical model studies of ocean outfalls (Isaacson et al. 1983; Roberts et al. 1989a, b, c; Roberts and Snyder 1993a, b). For example, the agreement of predicted versus observed dilution for the 8-jet rosette jet group is most satisfactory when viewed against the complexity of the phenomenon (Fig. 22).

Example 11.5 *Composite dilution of a jet group in a crossflow*
Consider the two buoyant jet discharges in Fig. 20 in the bent-over phase (b) with initial discharge concentration 100 mg/L, if the individual jet area $A_1 = 80\ m^2$ *and* $A_2 = 50\ m^2$*, and average concentration* $C_1 = 5$ *mg/L and* $C_2 = 6$ *mg/L. Given the overlapped area of the jet group is* $30\ m^2$*, find the composite dilution of the jet group.*

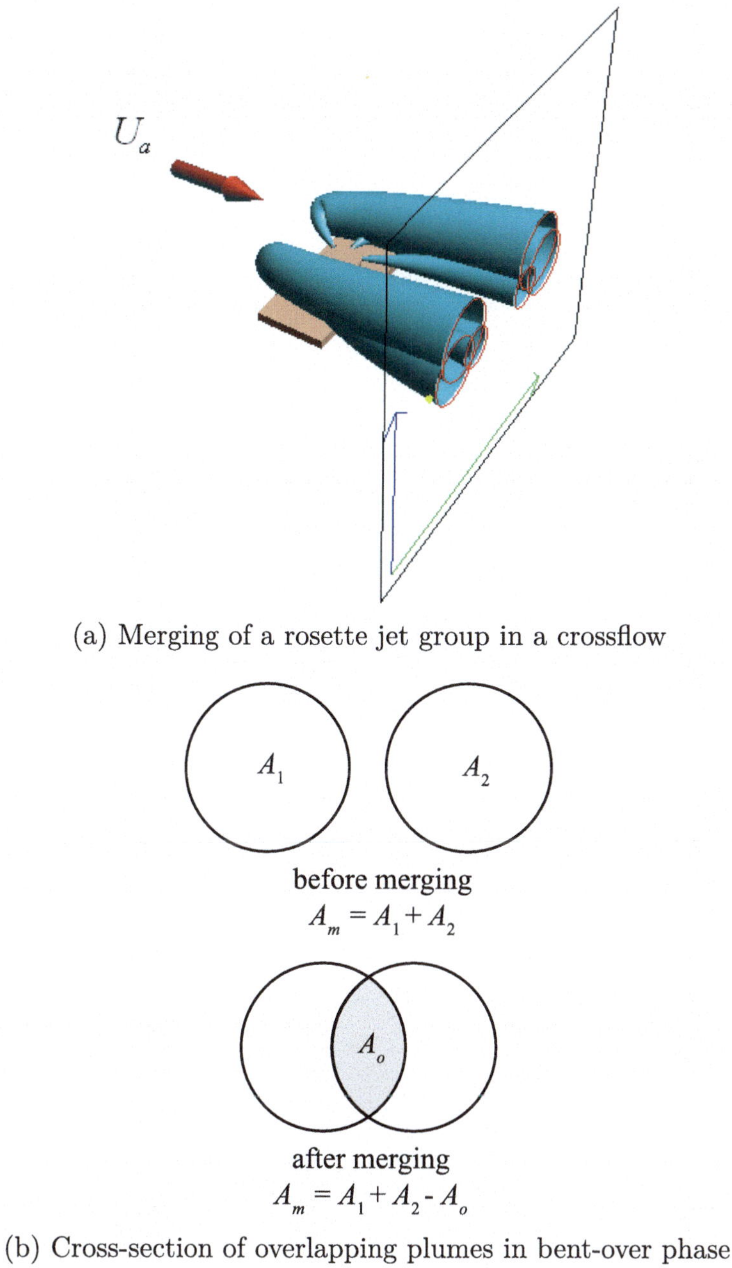

Fig. 20 Composite dilution of rosette jet group in crossflow

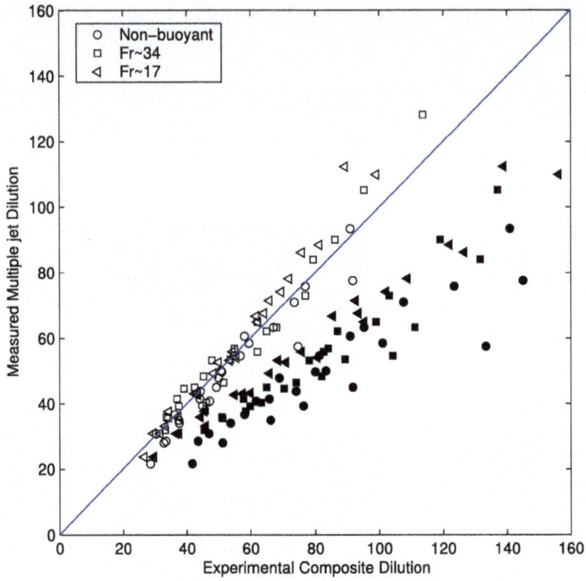

Fig. 21 Comparison between observed dilution and the experimentally determined composite dilution (Lai et al. 2011) (open and closed symbols represent the dilution determined with and without accounting for the overlapped area respectively)

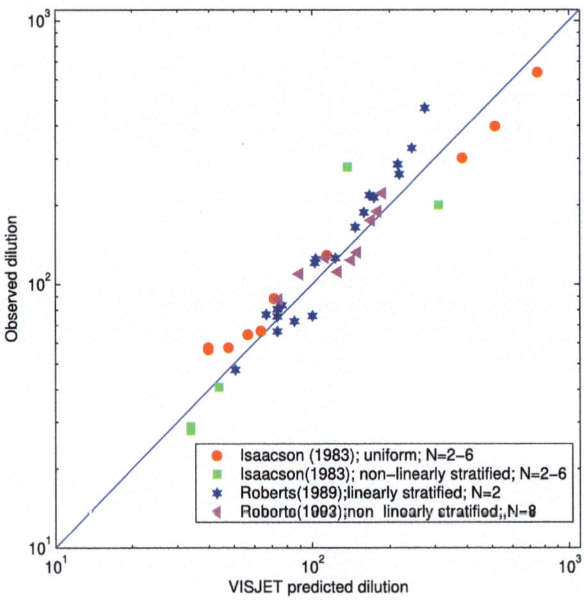

Fig. 22 Comparison of overall minimum dilution data obtained by various investigators in physical model tests of ocean outfalls

Solution: The merged area of the jet group is computed as $A_m = A_1 + A_2 - A_o = 80 + 50 - 30 = 100\,\text{m}^2$. The average concentration of the jet group can be computed using Eq. (21) as:

$$C_m = \frac{A_1C_1 + A_2C_2}{A_m} = \frac{80 \times 5 + 50 \times 6}{100} = 7\ \text{mg/L}$$

The composite dilution $S = 100/7 = \underline{14.3}$

4.3 VISJET Application in Multiple Buoyant Jet Groups

The composite dilution concept has been incorporated into the VISJET model (Chap. 10), and validated by extensive laboratory data (Lai et al. 2011). It can also be applied in predicting the mixing of submarine outfalls with complex jet configurations and ambient conditions. It should be noted that the VISJET virtual reality software facilitates easy graphical determination of the merged area A_m (Eq. 21) in any cross-section downstream of the jet group.

Boston Outfall—Non-linearly Stratified Crossflow

Roberts and Snyder (1993a, b) studied the mixing of the Boston outfall experimentally. The discharge is complex, with multiple rosette risers discharged into a non-linearly stratified crossflow (Fig. 23). The outfall design can be optimized by using VISJET. First, keep the outfall and jet velocity the same so that the discharge and momentum flux/diffuser length are kept constant. Then, increase the number of jets on a riser and observed the plume dilution for each case. It is interesting to note that the dilution increases initially as the number of jets per riser increases up to $NJ = 6$; beyond this point, the extensive plume merging with a larger number of jets results in decrease of plume dilution (Fig. 24). The optimal number of jets to be mounted on each riser for the outfall design can then be determined based on the maximum dilution that can be achieved.

VISJET can also be applied to study outfall discharges with complex atypical configuration, such as a combination of rosette risers with T-risers (twin jets discharging in opposite directions) (Fig. 25). For shallow water diffusers (e.g. discharges into rivers), the boundary effect can be significant; in such cases the method of images can provide a first order estimation (Pani et al. 2009, also see Problem 11.6).

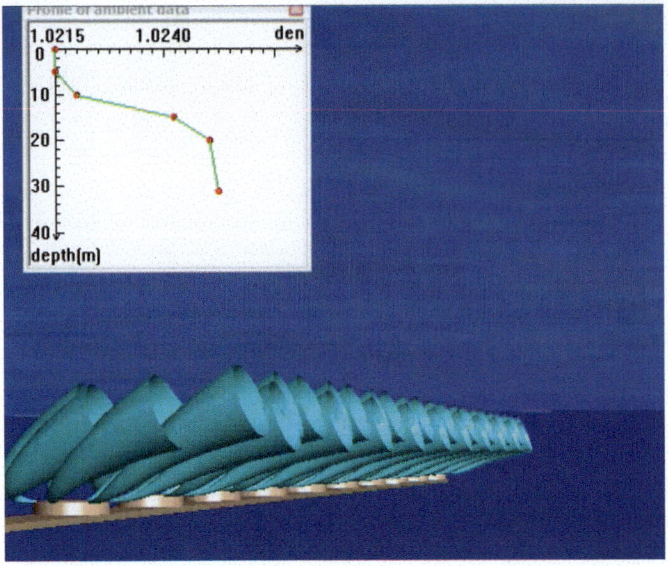

Fig. 23 Boston outfall discharge: multiple rosette buoyant jet groups into non-linearly stratified crossflow

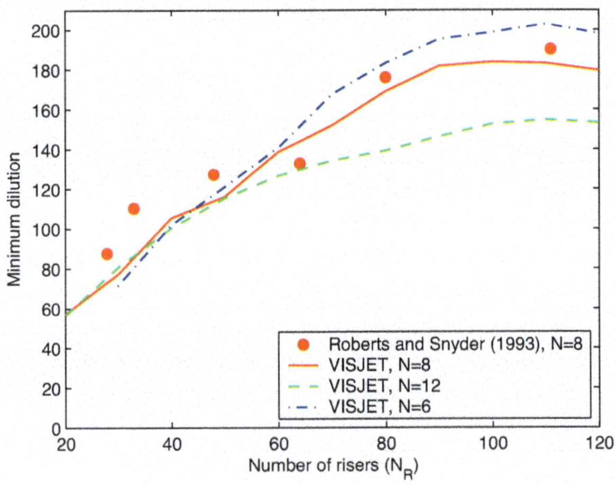

Fig. 24 Effect of riser spacing on minimum dilution

4.4 Remark on Dynamic Interaction of Jets in Crossflow

While the dynamic interaction of jets in crossflow has been found to be insignificant in the 'fanned-out' jet configuration commonly seen in outfall designs, it can be significant for the specific case of jet arranged "in tandem". For such a tandem-jet

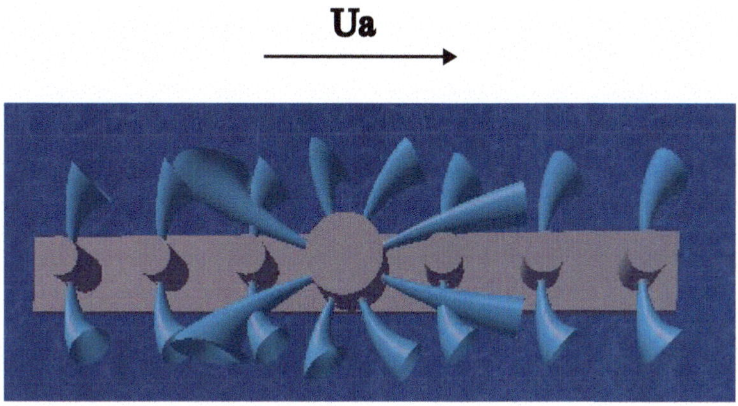

Fig. 25 River diffuser discharge in shallow water (hypothetical design)

configuration, the rear jets are sheltered from the ambient crossflow by the leading edge jet (most upstream jet); It was found that the effective ambient velocity of the rear jets is significantly reduced due to the sheltering effect as well as the entrainment demand in the momentum-dominated near field (MDNF—see Chap. 7, Sect. 6) of the leading jet (Yu et al. 2006). All the rear jets are hence less bent over (and interestingly to the same extent) than the leading jet as shown in Fig. 26.

The dynamic interaction model in Sect. 3.1 can be extended to model multiple tandem jets in a crossflow. The action of the jet on the external flow is modeled by a distribution of point sinks and doublets; the flow is treated like the flow past a

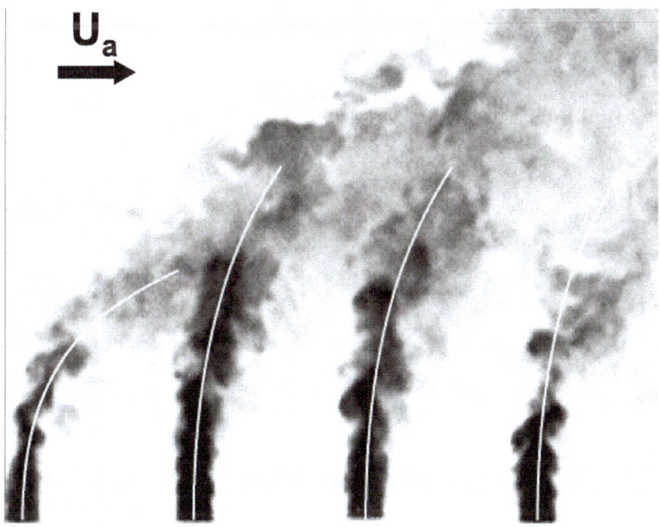

Fig. 26 The dynamic interaction of tandem jets in crossflow (Yu et al. 2006)

cylinder with uniform suction. The predicted jet trajectories are in good agreement with data; details of the model can be found in Lai (2009) or Lai and Lee (2010).

5 Dynamic Coupling of Near and Far Field Models in Predicting Mixing and Transport

5.1 Motivation

Climate change, urbanization, and population growth in recent decades are posing new challenges to water management in achieving the Sustainable Development Goals (SDGs) set by the United Nations. Fast growing population in many coastal cities put additional stress on the water supply system and drive innovations in alternative sources of water supply, such as membrane technology in desalination. For cities that rely on inland water bodies such as lakes for water supply, it is important to ensure the water quality is up to standard. Wastewater volume increases with population growth; however land scarcity and high cost may limit the use centralized advanced wastewater treatment technologies. Cost effective sustainable solutions and appropriate technologies for water and wastewater management can benefit from making best use of the natural assimilation capacity of nearby waters. In many engineering feasibility and environmental impact assessment (EIA) studies, a key assessment tool is a general near-field buoyant jet model coupled with a far-field circulation model; near-far field coupling is an important issue (Sect. 5.2 of Chap. 10).

In densely populated coastal cities, wastewater outfalls are often located not far from sensitive areas such as beaches or shellfisheries. The impact and risk assessment of effluent discharges poses particular technical challenges, as pollutant concentration needs to be accurately predicted across multiple scales—from the the near field to the far field with the intermediate field as transition (Fig. 27). The active mixing close to the discharge can be modelled by proven plume models (e.g. Chaps. 4 and 10), while the fate and transport far beyond the mixing zone can be well-predicted by 3D circulation models based on the hydrostatic pressure approximation (e.g. the Environmental Fluid Dynamics Code EFDC or Delft3D model). These models are usually applied separately with essentially one-way coupling; the action of the plume mixing on the external flow is neglected. Important phenomena such as surface buoyant spread or source-induced changes in ambient stratification cannot be satisfactorily addressed by such an approach.

Near-far field coupling of buoyant jets has been a long standing challenge because of the difficulty in representing the jet dynamics in a far-field model with grid size typically much larger than the size of the jet. Incorrect representation of the near field jet mixing in the far field model may result in drastic under-prediction of pollutant concentration and size of mixing zones.

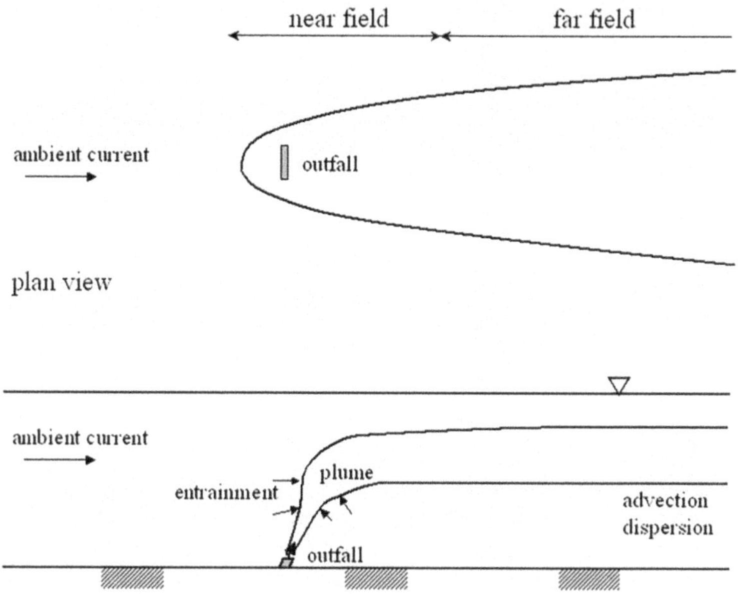

Fig. 27 Mixing and transport of coastal outfall discharge

In the following, a dynamic two-way coupling approach of embedding the near field model prediction within a far field circulation model computation will be introduced. Predictions are compared with representative laboratory experiments on the interaction of the near field jet mixing with ambient changes in the intermediate field, under both stratified and crossflow conditions. Case studies of prototype applications are also given.

5.2 *The Distributed Entrainment Sink Approach (DESA)*

From the viewpoint of the surrounding water, the important near field flows are the bulk sink flows ("loss") due to the turbulent jet entrainment and the bulk source flows ("gain") due to the diluted discharges. By treating the action of the jets on the outside flow as distributed sinks, a dynamic coupling of the near field and intermediate field can be made. The jet action—as entrainment sinks and sources—can be embedded seamlessly into a typical 3D shallow water circulation model based on the hydrostatic approximation (Choi and Lee 2007), with the continuity equation:

$$\frac{\partial U}{\partial x} + \frac{\partial V}{\partial y} + \frac{\partial W}{\partial z} = Q_s \tag{22}$$

and the solute mass transport equation (for both salinity and tracer):

$$\frac{\partial(HC)}{\partial t} + \mathbf{V} \cdot \nabla(HC) = \frac{\partial}{\partial x}\left[A_H H \frac{\partial C}{\partial x}\right] + \frac{\partial}{\partial y}\left[A_H H \frac{\partial C}{\partial y}\right] + \frac{\partial}{\partial z}\left[K_H H \frac{\partial C}{\partial z}\right] + Q_c \tag{23}$$

where U, V and W are the turbulent-mean velocity components in the x, y, z-directions respectively;

$$\mathbf{V} \cdot \nabla(\ ) \equiv U\frac{\partial(\ )}{\partial x} + V\frac{\partial(\ )}{\partial y} + W\frac{\partial(\ )}{\partial z};$$

H = water depth; C = solute concentration; A_H is the horizontal eddy diffusivity, K_H is the vertical eddy diffusivity, Q_s is the volumetric source/sink term representing the jet entrainment or diluted flows and Q_c is the source/sink term for the solute or tracer mass arising from the jet mixing. Both of these terms can be obtained at "grid cell" level from the near field model.

As the ambient conditions provided by the far field model (vertical profile of upstream velocity, solute concentration and density) change, the near field plume behaviour changes correspondingly. The changes in jet trajectory and dilution will be reflected in the positions and strengths of the sources and sinks. It should be noted that the time scale of plume rise to the surface or trap level ($\sim$1 min) is typically much smaller than that of far field tidal flows ($\sim$1 h); hence the near field jet mixing can be computed assuming quasi-steady ambient conditions—and all the theory of a single buoyant jet in the preceding chapters can be applied.

In order to have a true two-way coupling, the discharge is introduced into the far field model by considering: (i) a series of entrainment sinks along the predicted plume trajectory (similar to that described in the dynamic jet interaction model); and (ii) the diluted flow and pollution loading at the predicted terminal level of plume rise (Fig. 28). The sources and sinks are determined from JETLAG/VISJET embedded within the far field model. The key difference between the Distributed Entrainment Sink Approach (DESA) and the commonly employed 'source only' methods is that the near field mixing is fully resolved by the sub-grid plume model (Fig. 29); the bulk fluid and tracer mass transport induced by the effluent discharge as well as the influence of background concentrations or recirculation on the near field can be fully modelled in a dynamic manner (Fig. 30). It ensures mass conservation in the system and models the interaction of plume entrainment and ambient flow so that the pollutant built-up in a enclosed environmental can be accounted for. More details can be found in Choi and Lee (2007).

We begin with a simple example that illustrates the information passing between the plume model prediction and the sources/sinks in the transport equation of the far field circulation model.

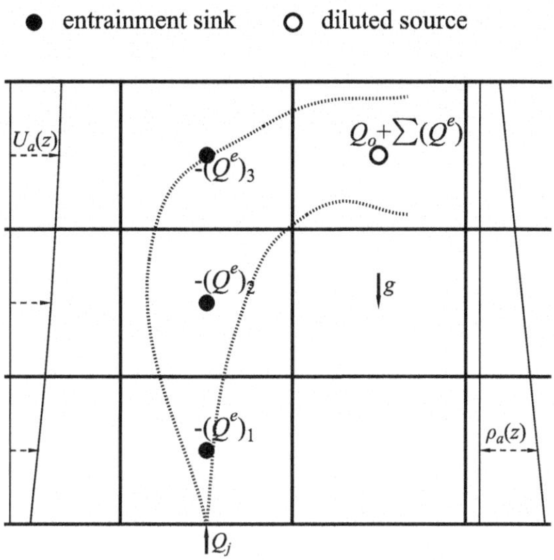

Fig. 28 Representation of plume mixing by diluted source flow and entrainment sinks along the jet trajectory (DESA)

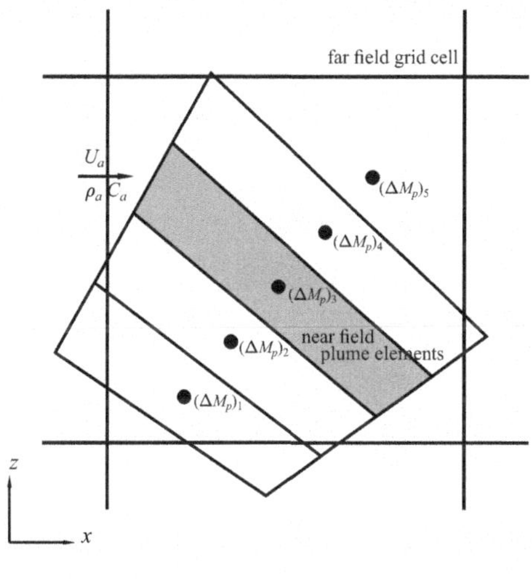

Fig. 29 Linking near field plume elements to far field grid cell via entrainment sinks

Example 11.6 *Distributed Entrainment Sink Approach*

Consider a buoyant jet with volume flux $Q_o = 0.2$ *m*^{3}*/s, relatively density difference* $\Delta\rho/\rho = 0.025$*, and buoyancy flux* $F_o = g'Q_o = 0.049$ *m*^{4}*/s*^{2}*, discharging into a stagnant ambient with depth* $H = 15$ *m. Initial concentration of the discharge* $C_o = 0.1$ *kg/m*^{3}*. A far-field model with grid cells of equal size 5 m × 5 m × 5 m has been set up to study the intermediate field mixing of the jet (Fig. 31). Background concentration of the ambient are* $C_{a1} = 0.02$ *kg/m*^{3} *and* $C_{a2} = 0.01$ *kg/m*^{3}*. Determine the entrainment sinks and the source of the jet in the far field grid using the DESA, and the tracer mass flux released at the surface layer.*

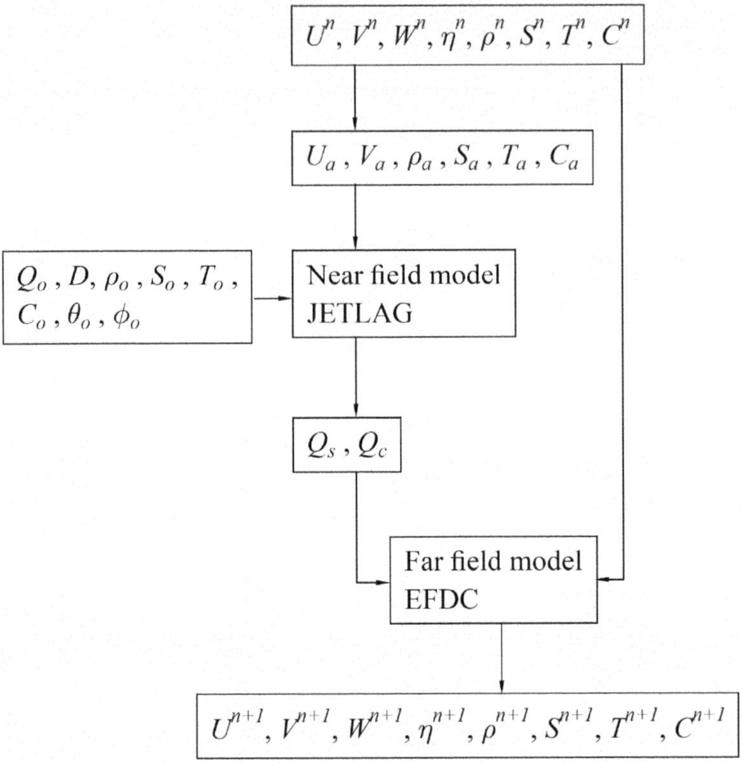

Fig. 30 Flow chart showing the numerical implementation of the coupling between the near field and far field models (η = free surface level above a reference datum; ρ = fluid density; S = salinity; T = temperature)

Solution: The buoyant jet discharge is highly buoyant and plume-dominated. Using Table 1 in Chap. 3, the volume flux of the plume as a function of height z is given as $Q = 0.163 F_o^{1/3} z^{5/3} = 0.06 z^{5/3}$ m^3/s. Hence:

$$-Q_{sink1} = Q_1 - Q_o = (0.06)(5^{5/3}) - 0.2 = 0.64 \text{ m}^3/\text{s}$$

$$-Q_{sink2} = Q_2 - Q_1 = (0.06)(10^{5/3}) - 0.64 = 2.7 \text{ m}^3/\text{s}$$

Note the sinks are negative sources. The source released at the surface layer of the far-field model grid will be the sum of the initial discharge and entrained water:

$$Q_{source} = Q_o + \sum |Q_{sink}| = 0.2 + 0.64 + 2.7 = 3.54 \text{ m}^3/\text{s}$$

Define tracer mass flux as $\Gamma = QC, then$:

$$-\Gamma_{sink1} = -Q_{sink1} C_{a1} = (0.64)(0.02) = 0.0128 \text{ kg/s}$$

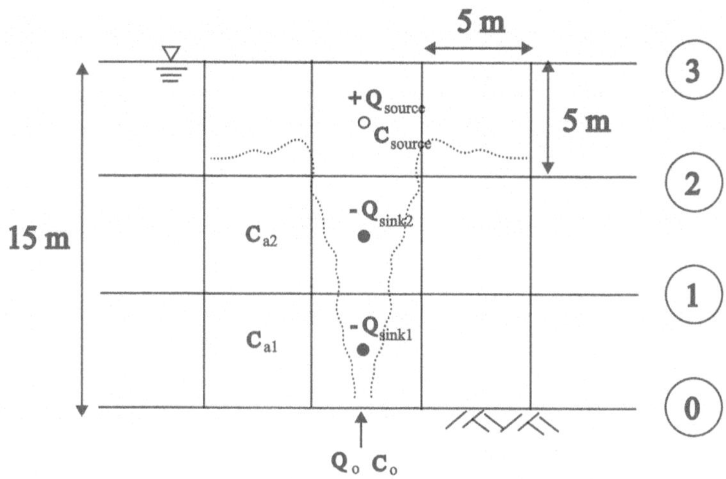

Fig. 31 Example 11.6: A buoyant jet discharge in the 3D far-field circulation model grid; connection between the jet model computed flows and sources/sinks in the far field model)

$$-\Gamma_{sink2} = -Q_{sink2}C_{a2} = (2.7)(0.01) = 0.027 \text{ kg/s}$$

$$\Gamma_{source} = Q_{source}C_{source} = \sum |\Gamma_{sink}| + \Gamma_o$$

$$= |Q_{sink1}C_{a1}| + |Q_{sink2}C_{a2}| + Q_oC_o$$

$$= (0.64)(0.02) + (2.7)(0.01) + (0.2)(0.1) = 0.0598 \text{ kg/s}$$

Note that $Q_{source} + \sum Q_{sink} = Q_o$, and $\Gamma_{source} + \sum \Gamma_{sink} = \Gamma_o$, so mass and tracer mass conservation are exact.

Various alternative near-far field coupling approaches have been proposed with their respective limitations and drawbacks (Choi and Lee 2007). A most common one-way coupling method is to introduce pollution source load (mass flux) at the predicted trap level.

5.3 Validation of DESA by Laboratory Experiments

DESA is particularly important when the near-field 'interacts' strongly with the far-field—as in many practical situations with strongly buoyant discharges (large source flow or buoyancy flux) in confined environment (shallow water, semi-enclosed bay). In the following the DESA approach is validated by comparing the prediction of the scalar concentration field with data of representative well-controlled laboratory experiments with practical relevance. All the results shown are from the 3D EFDC model computation dynamically coupled with the JETLAG model (Chap. 10). In every case, the results cannot be obtained by either the near field plume model or the 3D model alone.

Turbulent Buoyant Plume in a Confined Region

Baines and Turner (1969) carried out laboratory experiments to study the effect of continuous convection from small sources of (negative) buoyancy on the properties of the environment when the region of interest is bounded; the plume discharges in a small tank of length L = 57 cm, width W = 45 cm, and depth H = 45 cm, as shown in Fig. 32. A 19 × 15 × 20 model grid with horizontal grid size 3 cm by 3 cm and 20 uniform vertical layers is used. The buoyant plume mixes with the ambient fluid by turbulent entrainment; when it sinks to the bottom the mixed fluid spreads sideways as a bottom layer, with negligible vertical mixing (damped by density difference). The thickness of this layer of mixed fluid increases in depth with time, and the front of this bottom layer advances with a velocity dz/dt. As shown in Fig. 33, the model predicts ambient density profiles that change from an initially uniform concentration to a stable stratification due to the action of the plume. This computation can only be successful when the background concentration or salinity change is fed back to the near-field model. More details can be found in Choi and Lee (2007).

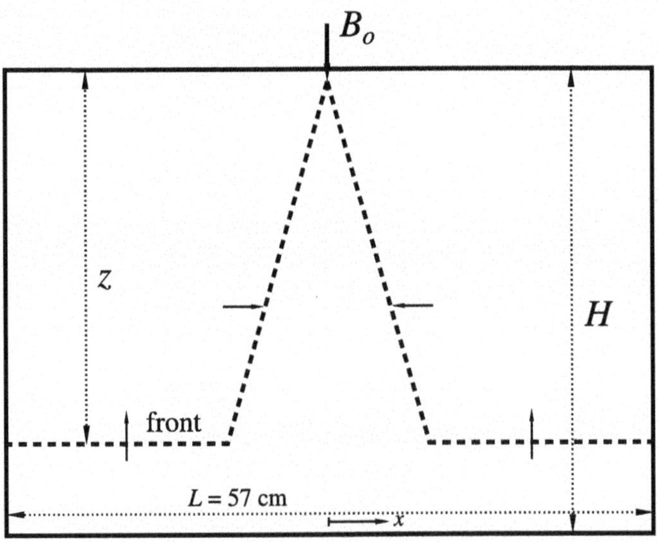

Fig. 32 The schematic setup of the turbulent buoyant jet in a confined region experiment

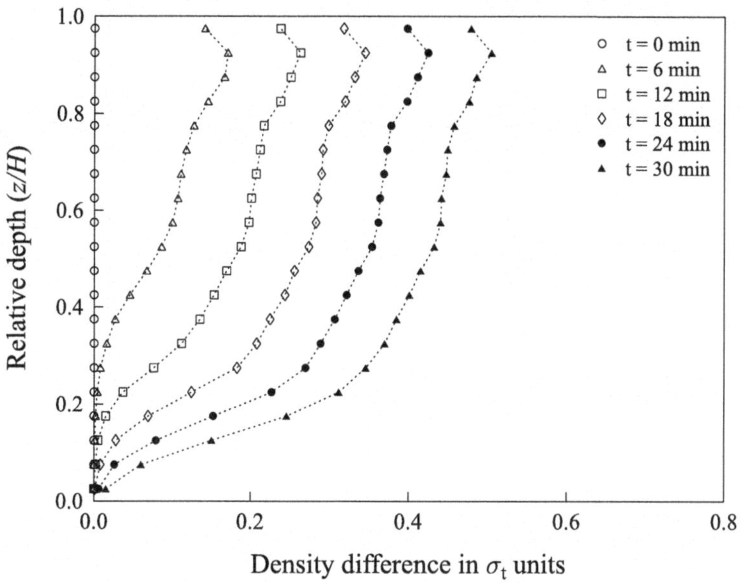

Fig. 33 Predicted ambient density profiles at $x = 5\Delta x$ from the jet at different times after release of a buoyant source with $B_o = 154\ \text{cm}^4/\text{s}^3$

Vertical Round Buoyant Jet in Linearly-Stratified Ambient Fluid

The interaction between a buoyant jet and the ambient fluid can be illustrated for a round jet discharged into an otherwise stagnant linearly stratified fluid. In Wong (1986)'s experiment, the (negatively) buoyant jet was created by discharging saline fluid downwards from a circular orifice into a linearly stratified receiving water (in a 457 cm × 61 cm × 91 cm deep tank). In this confined surrounding, the jet continuously modifies the ambient stratification. A 21 × 3 × 20 model grid with 20 uniform local layers is employed (Choi and Lee 2007). Figure 34 shows section views of the unsteady buoyant spreading at selected times. It is seen that the mixed jet fluid finds its level of density and spreads at the trapping level as a buoyant intrusion layer, with a weak counterflow both above and below; both the tracer concentration and velocity profiles are relatively uniform within the layer, and the predicted change in ambient stratification is in excellent agreement with experimental observation (Choi and Lee 2007).

Line Plume in Perpendicular Crossflow

As the last example, the DESA approach is demonstrated for the practically important problem of a finite "line plume" in a crossflow. Roberts (1979) performed experiments in a 6.1 m wide by 11 m long rectangular shallow water basin to study the mixing

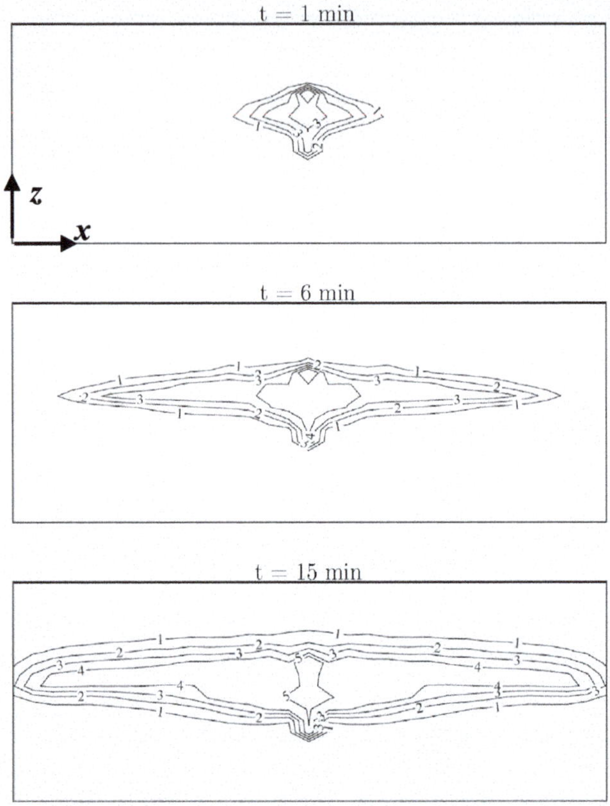

Fig. 34 Computed tracer concentration distribution (in 0.01Co units) in the vertical section (through the discharge point) for a round buoyant jet in stratified ambient

of multiport diffuser discharges. Experiments were carried out for a wide range of current speed Ua and discharge buoyancy flux per unit length $b = F_o$. DESA has been used to predict the intermediate field and buoyant spreading of the line plume; the finite diffuser (line plume) is represented by 6 non-interfering jets. Figure 35 shows the extent of the buoyant spreading layer is well predicted; From Fig. 36, it can be seen that the minimum surface dilution (computed from the maximum concentrations in the surface layer) obtained from DESA is in good agreement with the asymptotic dilution equation (best fit of data) given in Roberts (1979), while an alternative "actual source" coupling method (inputing mass flux at source directly) overpredicts dilution significantly.

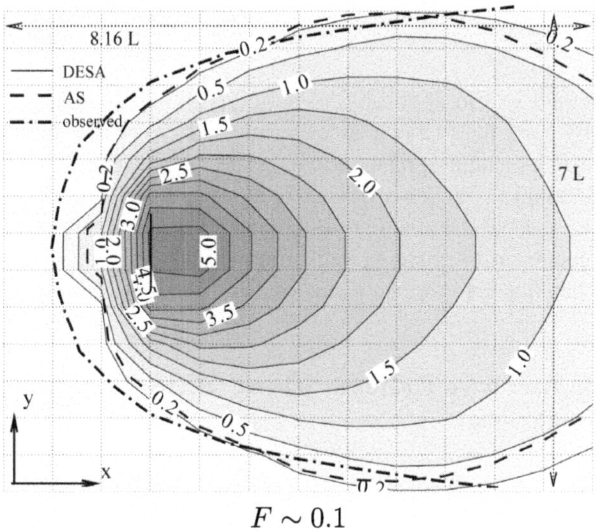

Fig. 35 Computed surface tracer concentration field (in units of $0.01C_o$) and observed surface field for a finite line plume in a perpendicular crossflow

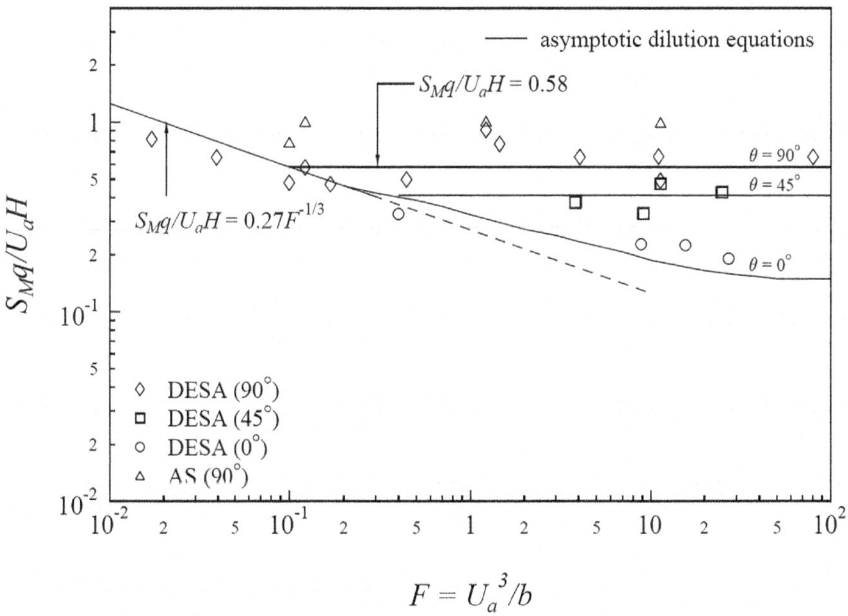

Fig. 36 Computed minimum surface dilution as a function of crossflow Froude number F (90° = perpendicular crossflow; AS = actual source method)

5.4 Case Studies

The DESA method is also applicable to prototype flow situations (Choi and Lee 2007). Carefully controlled laboratory experiments have confirmed that numerical predictions are in excellent agreement with experimental data—without the need to use non-hydrostatic circulation models to resolve the plume mixing, or ad hoc adjustment of far field model mixing coefficients. The general and robust method can be readily implemented in existing circulation models to yield accurate predictions of the intermediate/far field for many environmental transport problems.

Hong Kong Harbour Area Treatment Scheme (HATS)-Beach Water Quality Forecast System

As the key component of the Harbour Area Treatment Scheme (HATS), the Stonecutters Island Sewage Treatment Works (SCISTW) is the largest treatment plant of its kind (Chemically Enhanced Primary Treatment, CEPT) in the world. An average of 1.4 million m^3/d of partially treated wastewater is discharged into a stratified ambient water with depth $H \approx 12$ m via a 1.1 km long submarine outfall. The HATS outfall consists of 24 risers, each with 8 nozzles to discharge the effluent (Fig. 38). Nearby sensitive receptors, such as the beaches in Tsuen Wan, are only 8 km from the outfall and can be adversely impacted by the faecal pollution discharged from the outfall (Fig. 37).

To safeguard the health of the beach goers, a real-time beach water quality forecasting system (WATERMAN) has been developed to forecast the compliance and exceedance of the beach water quality objective (Chan et al. 2013). In view of the large flow and buoyancy of the HATS outfall, and because of the proximity to the sensitive receivers, it is vital to use a robust and accurate near-far coupling method to give realistic predictions of the buoyant spreading and dispersion of the *E. coli* from the outfall.

Figure 39 shows the the model grid, boundary conditions, major river discharges, and the meteorological forcings of the 3D hydrodynamic model used in the forecast system. Further details can be found in Chan et al. (2013). An example output from the system is shown in Fig. 40, where the DESA predicted near-field jet trajectories from the rosette risers of the HATS outfall, the far-field ambient velocity and salinity profile, and the *E. coli* concentration profile are shown. Figure 41 shows the comparison of the system predicted and the observed *E. coli* level at the Germini Beach (GEM in Fig. 37) during the testing of the HATS disinfection facility. The variation of the *E. coli* level is generally captured very well. The system has also been extensively validated with field data of *E. coli*, and an overall accuracy of 81–91% for compliance and exceedance has been achieved.

To compare the predicted intermediate field dilution based on the DESA, and the traditional 'mass source at trap level' approach, numerical experiments using a conservative tracer is released from the HATS outfall. Figure 42 shows a snapshot

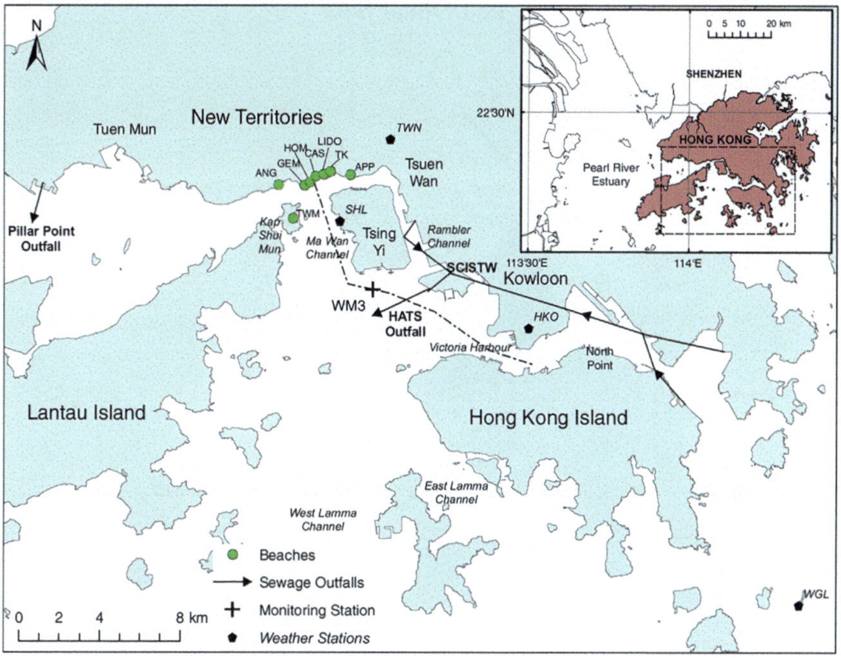

Fig. 37 Location of Tsuen Wan beaches and the HATS outfall in western Hong Kong coastal waters; the indicative path of main tidal flow is also shown as the dash-dot line

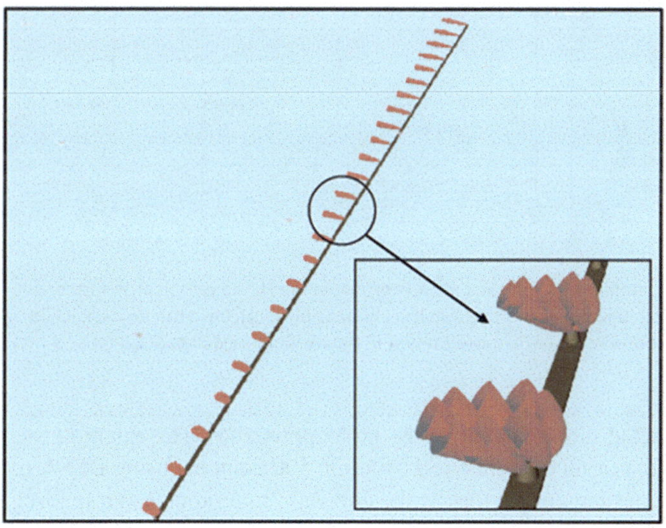

Fig. 38 A close up view of the diffuser of HATS outfall with two of the rosette risers shown in detail

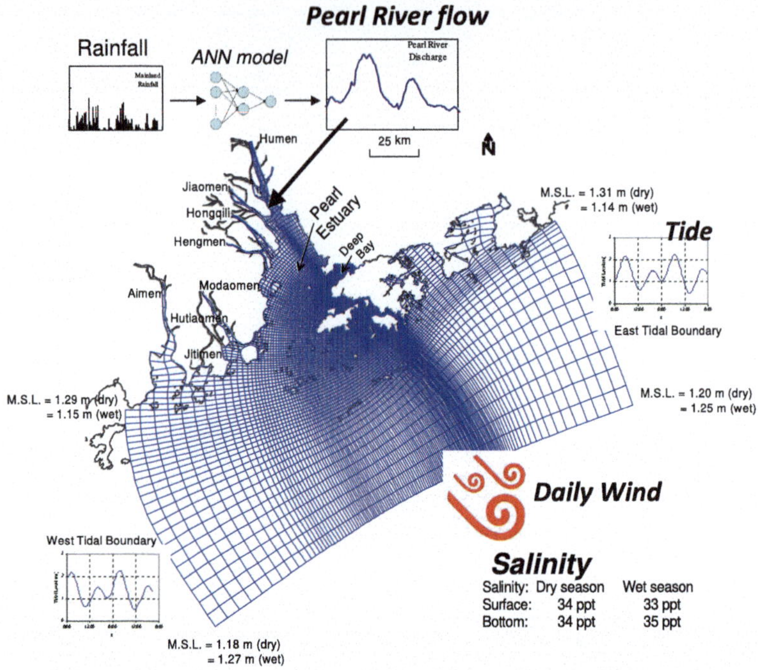

Fig. 39 Model grid and boundary conditions for the beach water quality forecast system

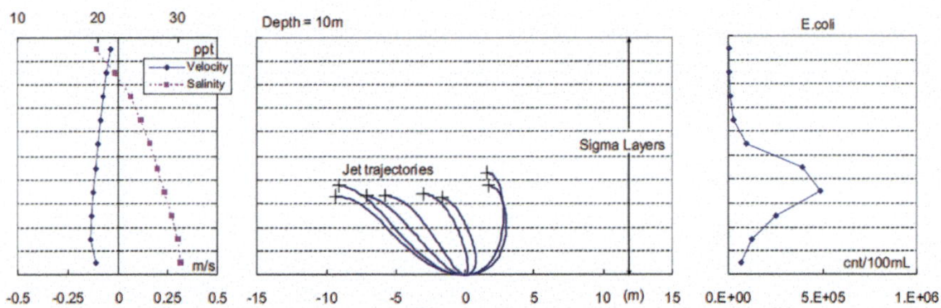

Fig. 40 Left: ambient velocity and salinity profile at the HATS outfall at an instant of a tidal cycle. Middle: predicted jet trajectory and trap level of multiple jets from a rosette riser. Right: the resultant vertical distribution of *E. coli* concentration in the far-field model grid cell

of the predicted dilution contours by both approaches. The extent of the predicted initial dilution or mixing zone (e.g. dilution 500 contour) using DESA can be seen to be 3–4 times larger than that by the one-way 'mass source at trap level' coupling method (Table 3). The importance of accounting for the intermediate field is amply demonstrated; incorrect near-far field coupling can lead to grossly inaccurate predictions, overly optimistic environmental impact assessment and under-protection of public health.

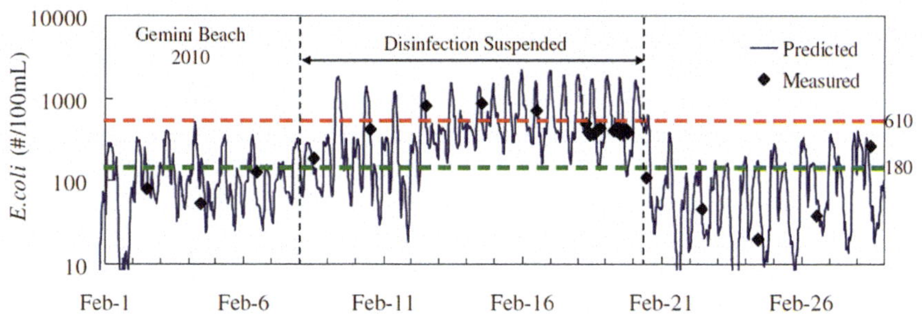

Fig. 41 Variation of beach *E. coli* level during the test of the HATS disinfection system in Feb 2010

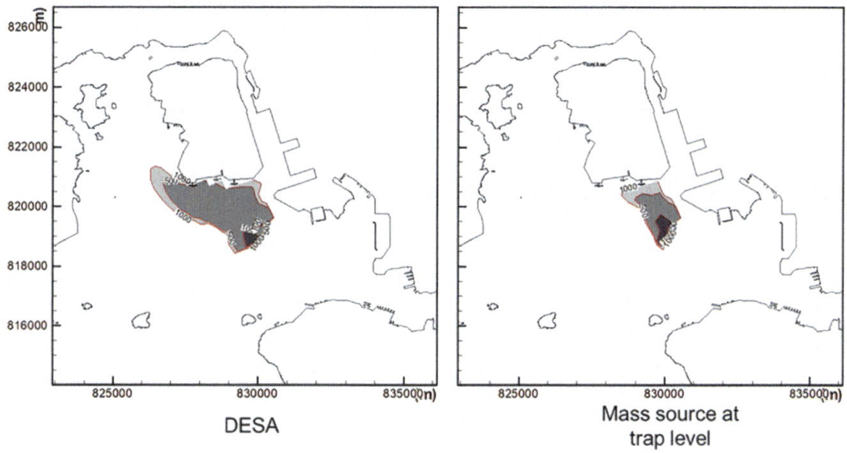

Fig. 42 Predicted surface dilution contours for HATS Outfall with DESA near-far field coupling during flood tide (left); and using one-way mass source at trap level coupling (right)

Table 3 Dilution contour areas (km^2) of the HATAS outfall during flood tide (4 h after dye release). Note that different near-far field coupling results in large differences of predicted mixing zones

Area of dilution contour of the HATS outfall (km^2)		
Dilution	DESA	Mass source at trap level
<50	0	0
<100	0.24	0.3
<500	4.71	1.49
<1000	6.36	2.25

Impact of Temporary Diversion of HATS Effluent Using the Northwest Kowloon (NWK) Outfall

To facilitate the construction of a new effluent tunnel for the HATS2A scheme, it was proposed to temporarily divert the HATS effluent to be discharged via an emergency bypass outfall—the Northwest Kowloon (NWK) outfall—for about 8 weeks in the non-bathing season. The NWK outfall was the original outfall for discharging sewage from the Northwest Kowloon Screening Plant before the commissioning of HATS in December 2001. The impact of the temporary diversion on the nearby beaches was assessed.

The NWK outfall is located close to the Stonecutters Island with a distance of only about 100 m offshore and at a depth of 11 m (Fig. 43). It consists of two risers each with 3 equally spaced jets of 1 m diameter and vertical inclination of 22.5° (Fig. 44). During the temporary diversion, all the effluent flow from HATS outfall $Q = 15.8\ m^3/s$ will be discharged through the NWK outfall. Considering the significantly less riser and nozzles to spread the flow comparing to the HATS outfall, the dilution is expected to be less during flow diversion. It is necessary to determine the disinfection dosage needed to mitigate the potential deterioration of water quality due to the diversion.

The model results showed that the area of mixing zone defined by the contour of 500 and 1000 surface dilution for NWK outfall discharge are significantly larger when DESA is used than that of the 'mass source at trap level' approach (Fig. 45 and Table 4), because of the buoyant spread of the low-dilution sewage field. The low

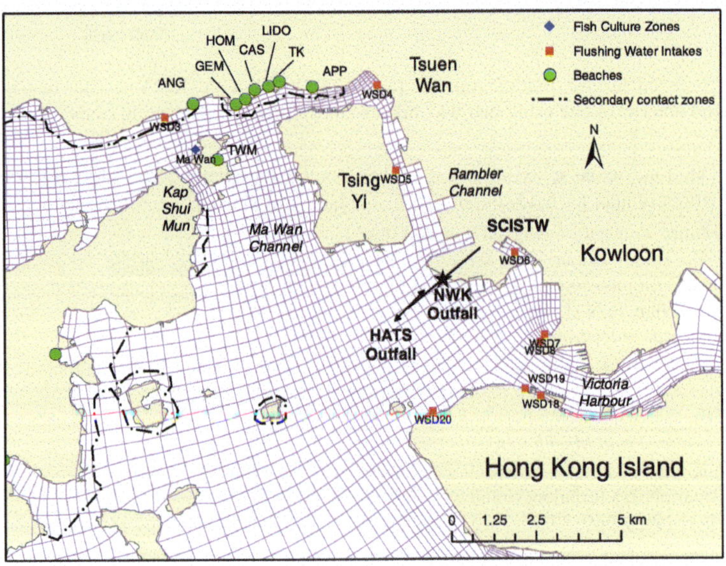

Fig. 43 Location of the HATS and NWK outfalls, and Tsuen Wan beaches

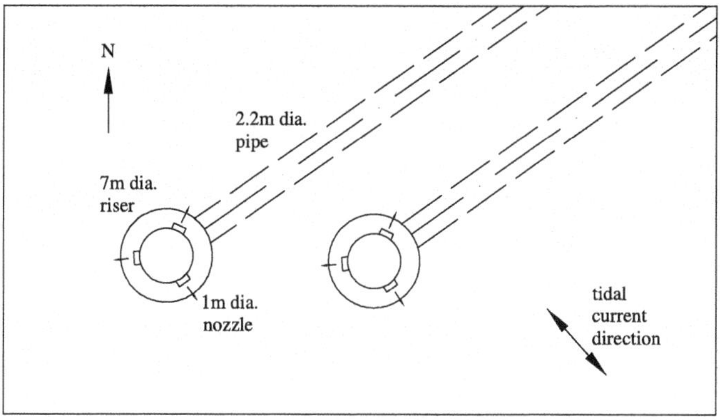

Fig. 44 Configuration of the NWK outfall with two risers (3-jet rosette)

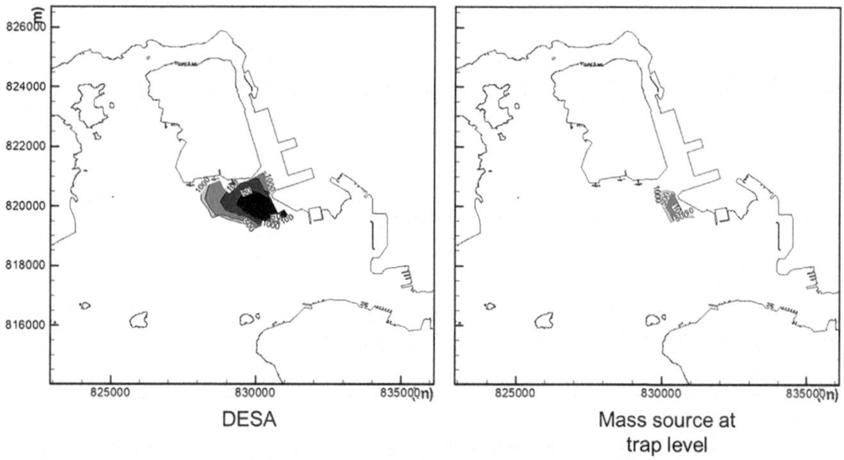

Fig. 45 Predicted surface dilution contours for NWK outfall with and without DESA during flood tide (mass source at trap level)

dilution of NWK outfall results in a much larger area of mixing zone with poor water quality and high concentration of *E. coli* at the sensitive receivers. For example, the area within the dilution = 50 contour can be under-predicted by an order of magnitude (1.19 vs. 0.09 km^2) if the coupling only involves mass source input at trap level. It was then concluded a stringent *E. coli* requirements should be maintained (by suitable disinfectant dosage) during the diversion period. If DESA is not used in this case, the dilution from the outfall will be over-predicted, leading to inadequate disinfection and under-protection of nearby beaches.

Table 4 Dilution contour areas (km^2) of the NWK outfall during flood tide (4 hours after dye release) for NWK outfall

Area of dilution contour of the NWK outfall (km^2)		
Dilution	DESA	Mass source at trap level
<50	1.19	0.09
<100	2.24	0.18
<500	3.49	0.57
<1000	3.91	0.84

6 Summary

Multiple buoyant jets can interact dynamically under 'starve plume' condition when the flows is being confined, and kinematically due to the jet group merging.

Dynamic interaction of multiple buoyant jets in stagnant ambient can be predicted by a semi-analytical model. The jet-induced external flow is computed by a distribution of point sinks along multiple free jet trajectories obtained using integral jet models based on a well-established shear entrainment hypothesis. The dynamic interaction is incorporated into the governing equations of a buoyant jet in an exact horizontal momentum balance using a control volume approach. The deflected jet trajectory is determined through the iterative solution of the external entrainment flow and the multiple buoyant jet trajectories. The mixing characteristics of the multiple jet group after merging are determined by superposition of local momentum (u^2 for jets) or kinetic energy (u^3 for plumes), and tracer mass (uc or buoyancy) flux.

Under typical ocean outfall configurations and ambient crossflows, the dynamic interaction among jets can be shown to be insignificant theoretically and experimentally. Each jet can then be treated as independent to each other. Based on this observation a general approach in predicting dilution of rosette jet groups is presented—the composite dilution concept. The Lagrangian jet model VISJET, in which the composite dilution has been incorporated, is used to interpret the experimental results. It is found that the VISJET can satisfactorily predict the trajectories, and the dilution of the rosette jet group; the predicted dilution are also in agreement with various previous studies under unstratified, uniformly stratified, non-linearly stratified ambient conditions. The implication on optimizing outfall designs using the present and previous studies have also been demonstrated using the VISJET model.

By treating the action of the jets on the outside flow as distributed sinks, a dynamic coupling approach of the near field and intermediate field has also been developed. The DESA method is generally applicable to both laboratory and prototype flows and has been validated extensively with both complex benchmark flow cases and field data (Choi and Lee 2007; Chan et al. 2013).

Problems

11.1 *Superposition approach for multiple coflow jets*
(a) Following Problem 6.1, the excess x-momentum equation for a coflow jet is conserved. By using similar argument as Sect. 1, show that uu_e can be superimposed to determine the mixing characteristics of multiple coflow jets, where $u = u_e + u_a$ is the coflow jet velocity, u_e is the velocity excess and u_a is the ambient coflow.
(b) It has been found experimentally that the excess momentum flux of a coflow jet is self-similar and follows a Gaussian profile (Pani et al. 2009), which can be expressed as:

$$\frac{uu_e}{u_m u_{em}} = e^{-\frac{1}{2}\left(\frac{r}{b_{me}}\right)^2}$$

where b_{me} is the spreading rate of the excess momentum flux. It should be noted that b_{me} is different from the Gaussian width b_g of a coflow jet.

Express the total jet excess momentum flux at a cross-section (i) using Gaussian distribution for jet excess velocity, and (ii) using Gaussian distribution for jet excess momentum flux. By equating (i) and (ii), show that b_{me} is relates to b_g as $b_{me} = k_l b_g$, where k_l is the link factor that can be expressed as:

$$k_l = \sqrt{\frac{\frac{u_{em}}{u_a} + 2}{4(\frac{u_{em}}{u_a} + 1)}}$$

(c) Consider three same coflow jets in Example 11.1. The jet spacing is $s = 3$ m, nozzle diameter $D = 0.4$ m, initial jet velocity $u_o = 3.0$ m/s. They are discharged into a river with a ambient coflow velocity of $u_a = 0.2$ m/s. Determine the centerline velocity of the jet group at $x = 100$ m from the diffuser.

11.2 *Superposition approach for buoyant jets*
A heuristics approach is proposed to predict the velocity field of multiple buoyant jets. Given a buoyant jet with velocity u, it was established that the velocity field induced by a group of the buoyant jet can be predicted by superposition of u^2 if it is jet-like ($Fr \to \infty$), i.e. $u_{Nj} = (\sum u^2)^{1/2}$, and superposition of u^3 if it is plume-like ($Fr \to Fr_p = 4.5$), i.e. $u_{Np} = (\sum u^3)^{1/3}$. It is then sensible to suggest the velocity of the buoyant jet group will be a function of (u_{Nj}, u_{Np}).

(a) Proposed a function that relates u to u_{Nj} and u_{Np} [Hint: recall that the entrainment coefficient of a buoyant jet is also dependent on that of the jet and plume $\alpha = \alpha_j + (\alpha_p - \alpha_j)(\frac{Fr_p}{Fr_L})^2$].

(b) Consider a group of buoyant jets with the following discharge parameters:
Number of jet $N = 11$
Jet diameter $D = 0.01$ m
Jet spacing $s = 5D$
Initial velocity $u_o = 0.127$ m/s
Initial relative density difference $\Delta\rho_o/\rho_a = 0.002$

Determine the jet group centerline velocity u_c/u_o at $z = 80D$ using superposition of (i) u^2, (ii) u^3, and (iii) your proposed function in (a). It is given that the free single buoyant jet centerline velocity at $z = 80D$ is $u_c/u_o = 0.115$, and $Fr_L = 5.25$.

(c) The equivalent slot buoyant jet of the discharge can be solved by governing equations (e.g. Lee and Cheung 1986; Choi and Lee 2007):

$$\frac{dq}{ds} = 2\sqrt{2}\alpha\frac{m}{q}$$

$$\frac{d(m^2)}{ds} = \sqrt{2}\lambda bq$$

$$\frac{db}{ds} = 0$$

$$\alpha = 0.055 + 0.16875b\left(\frac{q}{m}\right)^3$$

where q, m and b are the unit-width volume, momentum and buoyancy fluxes respectively, α is the entrainment coefficient and $\lambda = 1.35$ (Table 3 in Chap. 2 and Table 1 in Chap. 3). The equivalent slot buoyant jet centerline velocity at $z = 80D$ is given as $u_c/u_o = 0.153$ by solving the above equations. Compare your results in (b) with the slot jet solution and comment.

11.3 *Dynamic interaction of twin jets*
Consider two jets with equal momentum flux discharging vertically into a stagnant ambient in a laboratory water tank. Each plume is discharged through a nozzle of diameter $D = 5$ mm and initial velocity $u_o = 0.5$ m/s. Spacing between the plume is $s = 7.5$ cm. Find the interacting jet trajectory following the approach and assumptions in Example 11.4.

11.4 *Composite dilution of a rosette buoyant jet group*
Consider a jet group clustered around the circumference of an outfall riser, discharging into a horizontal ambient current U_a. Downstream of the riser, the bent-over jets merge and the plume cross sections overlap. For practical ocean outfall designs, the port diameter (D) is typically an order of magnitude smaller than the riser diameter D_R; for example, $D/D_R \approx 0:1$ and 0.05 for the Hong Kong Harbour Area Treatment Scheme and the Boston Outfall, respectively (e.g., Roberts and Snyder 1993a, b). Provided an adequate ambient flow is available for jet entrainment, it is reasonable to assume that the dynamic interaction of the widely separated jets can be neglected as a first approximation; the jets can be treated as independent of each other.

(a) Show by mass conservation that the merged jet concentration $C_m = \sum_{i=1}^{N} A_i C_i / A_m$, where A_i is the area of each individual jet, and A_m is the area of the merged jets.

(b) An ocean wastewater outfall system has been designed for a densely populated coastal city. The design total flow of the system is 19 m^3/s, and is discharged through 24 evenly spaced risers (radius 1 m) on a 1200 m diffuser. There are eight horizontal jets on each riser (D = 0.25 m) discharging at a depth of 12.0 m, with effluent temperature 20.0 °C, and salinity 1.0 ppt. Assume the following ambient profile:

Depth (m)	S (ppt)	T (°C)	U_a (m/s)
0.0	25.0	28.0	0.3
12.0	25.0	28.0	0.3

(i) Simulate a few risers of the outfall system using VISJET; do the jets of neighbouring risers interfere with each other?
(ii) Consider one of the risers. Use VISJET to visualize a vertical cross-section at a downstream distance of 20 m. What are the area and concentration of each individual jet? Also find the merged area and use the result in (a) to determine the jet group dilution. Compare with the VISJET composite dilution output. How is dilution affected by the jet interference?

11.5 *Optimization of outfall design*
For an outfall system with total flow Q and diffuser length L_D, the dilution initially increases with an increasing number of risers because the total jet surface area for ambient entrainment increases. However, the dilution can be adversely affected when too many risers are installed because jets of the neighbouring risers start to interfere with each other. It is of interest to determine the maximum number of risers to be placed before dilution begins to level off. This can be illustrated using the Boston outfall example (Roberts and Snyder 1993a, b).

(a) Assume $L_D = 2000$ m, $Q_T = 17.1$ m^3/s, effluent density $\rho_o = 1000$ kg/m^3. Use VISJET to simulate a portion (≈500 m) of the diffuser if 33 ($D = 0.198$ m), 48 ($D = 0.165$ m), 80 ($D = 0.127$ m), and 111 ($D = 0.108$ m) risers are placed. Note the nozzle diameter are varied to maintain the same diffuser momentum and buoyancy fluxes. Eight horizontal jets are discharging at a depth of 31.0 m for all cases. Determine the composite dilution of each case at 25 m downstream of the diffuser (by turning on the VISJET 'continuous mode'). The following ambient profile observed typically in late summer can be assumed:

Depth (m)	Density (kg/m^3)	U_a (m/s)
0.0	1021.5	0.12
5.0	1021.5	0.12
10.0	1022.0	0.12
15.0	1024.2	0.12
20.0	1025.0	0.12
31.0	1025.2	0.12

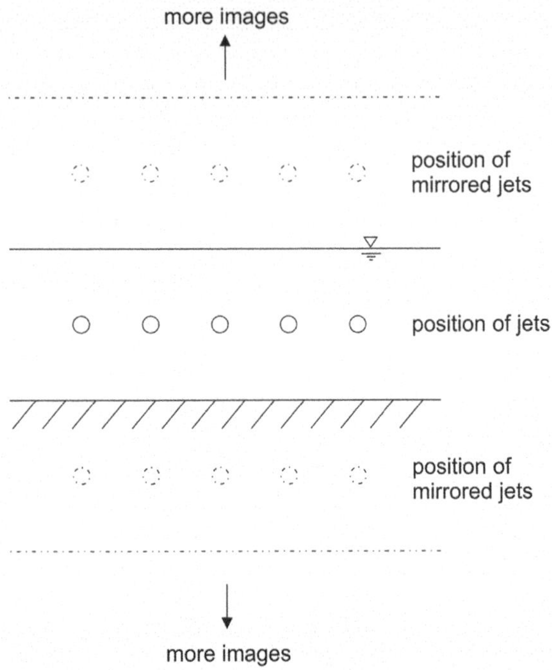

Fig. 46 Method of images to account for water surface and river bed

(b) Plot the composite dilution in (a) with the number of risers. Comment on your results.

11.6 *Method of images*
Outfall discharges in rivers are usually affected by boundaries because of the shallow water depth. An approach to provide a first order estimation accounting for the boundary effects is by the method of images (e.g. Pani et al. 2009), i.e. placing virtual jet sources by treating the boundaries as the planes of symmetry (Fig. 46). Consider the three round jet discharge in Example 11.1 again. The diffuser is now placed at mid-depth of a river with depth of $H = 3$ m, find the centerline velocity of the jet group at $x = 50m$ using the method of images. Use only one row of images each for the water surface and river bed. Assume the river flow is small comparing to the jet group such that the coflow velocity can be neglected.

11.7 *Mixing of a complex diffuser discharge in rivers*
Consider the complex diffuser discharge in Fig. 47. The river diffuser has a central rosette riser of 8 nozzles, three pairs of T-riser upstream, and three pairs of T-riser downstream, making a total of 20 jets. The total discharge is significant with $Q = 12.65\ \mathrm{m^3/s}$, giving each horizontal jet an initial velocity of $u_o = 3.2$ m/s. The river has a depth of $H = 11.5$ m, salinity $S_o = 0.1$ ppt, temperature $T_o = 15\ °\mathrm{C}$ and ambient velocity $U_a = 0.5$ m/s in the $+x$-direction. To better understand the problem, VISJET can be used for visualizing the discharges and getting more insights from the model outputs.

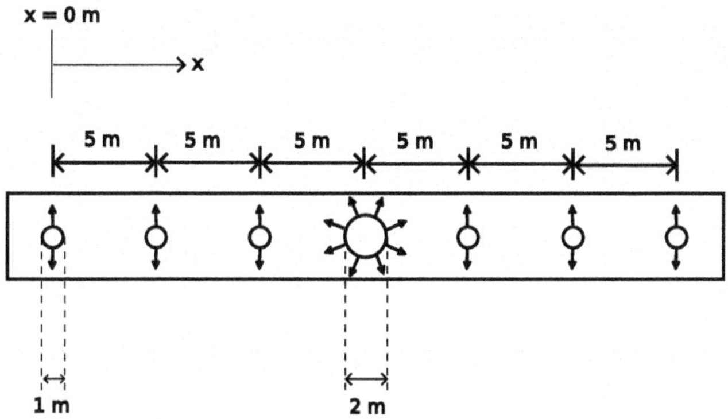

Fig. 47 The diffuser configuration in Problem 11.7

The riser and jet discharge parameters are given as:

<u>Riser Parameters</u>:

- The 7 risers are separated at equal distance $s = 5$ m
- The top and bottom radius of the T-risers is 1 m, and for the rosette-riser, the radius is 2 m, with all risers having heights of 2 m.
- The first three T-risers are identical and positioned at distances $x = 0, 5, 10$ m.
- The central rosette riser is positioned at $x = 15$ m.
- The last three T-risers are identical and positioned at distances $x = 20, 25, 30$ m.

<u>Jet Parameters</u>:

- All horizontal jets have a vertical angle of 0° and are identical but have different horizontal angles.
- Flow rate of each jet is $Q_o = 0.6325$ m^3/s.
- Nozzle diameter is $D = 0.5$ m.
- Port height is 1.05 m.
- Horizontal angles for T-risers: 90°, −90°.
- Horizontal angles for rosette-riser: ±22.5°, ±67.5°, ±112.5°, ±157.5°.

Find the composite dilution at $x = 200$ m assuming the discharge is unbounded.

Appendix A
Density of Seawater

Sea water density ρ is a function of salinity (S in ppt) and temperature (T in °C) and is often expressed in terms of the sigma unit σ_t, which is the density of water above the reference value of 1000 kg/m^3:

$$\rho\ (\mathrm{kg/m^3}) = 1000 + \sigma_t \tag{A.1}$$

The density of sea water under standard atmospheric pressure can be determined by one of the following two formulae:

(1) The FOFONOFF'S (Fofonoff, 1962) formula for σ_t is calculated as follows:

$$\sigma_t = (\sigma_o + 0.1324) \times [1 - A + B(\sigma_o - 0.1324)] - \Sigma(T) \tag{A.2}$$

where the coefficients $\sigma_o, \ A, \ B, \ \Sigma$ are functions of either S or T:

$$\sigma_o(S) = [(6.8 \times 10^{-6} S - 4.82 \times 10^{-4})S + .8149]S - 0.093$$

$$A(T) = .001T[(.0010843T - .09818)T + 4.7867]$$

$$B(T) = 1. \times 10^{-6} T[(.01667\ T - .8164)T + 18.03]$$

$$\Sigma(T) = (T - 3.98)^2 \frac{(T + 283.)}{503.57(T + 67.26)}$$

By the nature of the empirical equation derived from field data, Eq. A.2 does not reduce to the pure water formula (σ_t (freshwater) = $\Sigma(T)$) for $S = 0$. For determination of freshwater density the pure water density formula $\Sigma(T)$ can be used.

J. H. W. Lee et al., *Turbulent Jets and Plumes*,
https://doi.org/10.1007/978-3-032-01187-9

(2) The UNESCO formula (Milero & Poisson, 1981) is based on density measurements from 0 to 40°C and $S = 0.5 - 43$ ppt. Density is calculated in two steps. First, the density of fresh water:

$$\begin{aligned}\rho_w = &+ 999.842594 + 6.793952 \times 10^{-2} T - 9.095290 \times 10^{-3} T^2 \\ &+ 1.001685 \times 10^{-4} T^3 - 1.120083 \times 10^{-6} T^4 \\ &+ 6.536332 \times 10^{-9} T^5 \end{aligned} \tag{A.3}$$

Then, the addition due to salinity:

$$\rho = \rho_w + AS + BS^{\frac{3}{2}} + CS^2 \tag{A.4}$$

where

$$\begin{aligned} A = &+ 8.24493 \times 10^{-1} - 4.0899 \times 10^{-3} T + 7.6438 \times 10^{-5} T^2 \\ &- 8.2467 \times 10^{-7} T^3 + 5.3875 \times 10^{-9} T^4 \\ B = &- 5.72466 \times 10^{-3} + 1.0227 \times 10^{-4} T - 1.6546 \times 10^{-6} T^2 \\ C = &+ 4.8314 \times 10^{-4} \end{aligned}$$

Finally the sigma unit,

$$\sigma_t = \rho - 1000. \tag{A.5}$$

Equation A.4, the one-atmosphere international equation of state of seawater, has been suggested for use by the UNESCO joint panel on oceanographic tables and standards (UNESCO, 1981). The two formulae (FOFONOFF versus UNESCO) give results that differ by less than 0.1 σ_t in the range of $T = 0 - 40$ °C and $S = 0 - 40$ ppt, well within the accuracy required for typical engineering calculations. For density determination at near zero temperatures, the direction of buoyancy may be crucial. The table below compares the density computed by the two formulae at temperature $T = 0$ °C and 4 °C.

Temperature	FOFONOFF	UNESCO
σ_t at 0 °C	σ_t = - 0.092594	σ_t = - 0.157406
σ_t at 4 °C	σ_t = +0.038701	σ_t = - 0.025042

Figure A.1 shows the density contours in σ_t units as a function S (in ppt) and T (in ^oC). For a given temperature, the density varies approximately linearly with salinity. On the other hand, for a given salinity, the density-temperature variation can be quite nonlinear for temperature changes greater than a few degrees. Figure A.2 shows the nonlinear density-temperature variation for salinity $S = 0$, 10, 20, 30 and

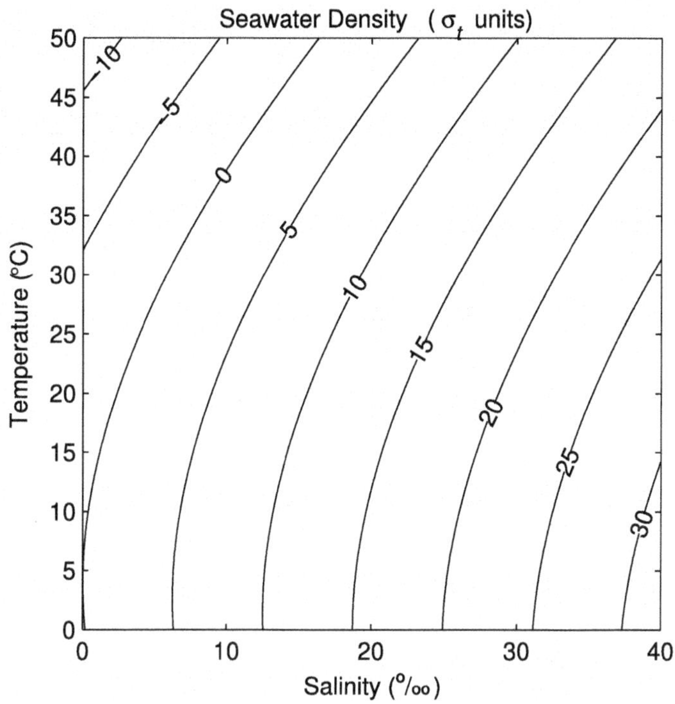

Fig. A.1 Density of sea water (in σ_t units) as a function of salinity (part per thousand) and temperature (°C) obtained using the FOFONOFF formula

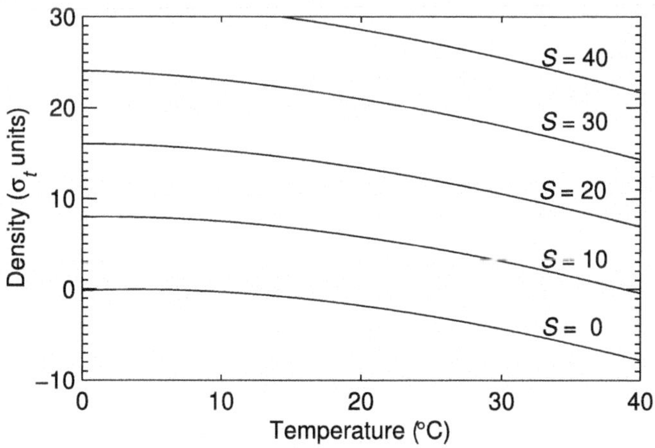

Fig. A.2 σ_t as a function of temperature using the UNESCO formula

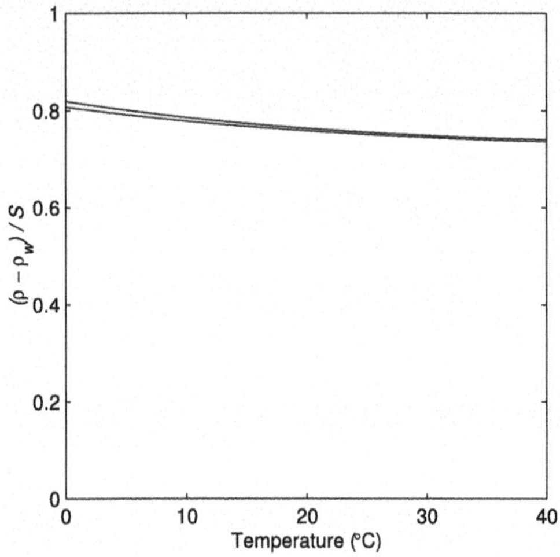

Fig. A.3 Density-to-salinity ratio, $(\rho - \rho_w)/S$, as a function of temperature (T in °C), for $S = 1$ and $S = 35$ (upper and lower curve respectively)

40 ppt. On average, the relative density above fresh water in sigma units, $(\rho - \rho_w)$, is equal to about 80% of the salinity.

The UNESCO formula shows the separate effect of salinity and temperature more clearly. Figure A.3 gives the relative density-to-salinity ratio, $(\rho - \rho_w)/S$, as a function of temperature for $S = 1$ and $S = 35$ ppt computed by the UNESCO equation. The ratio decreases slightly with increase of temperature. The linearized form of Eq. A.4 is $\rho = \rho_w + AS$; for $T = 0 - 30°C$, this linear approximation gives densities that differ from the UNESCO value by less than 0.2 σ_t units for $S = 0 - 10$ ppt. The error increases to around 0.8 σ_t when S increases to 40 ppt.

Example A.7 Nonlinear $\rho - T$ relation
A buoyant jet is produced in the laboratory by discharging 40 °C warm water at a rate of 3 cm^3/s into a tank of water at 20°C. (a) Find the buoyancy flux of the discharge at the source. (b) Estimate the buoyancy flux after the jet is diluted 100 times by entrainment of the ambient colder water into the jet.

Solution: The density of the $T_o = 40$°C water at the source is $\rho_o = 992.2204$ kg/m^3. The density of the surrounding $T_a = 20$°C water in the tank is $\rho_a = 998.2063$ kg/m^3. The specific buoyancy flux at the source

$$F_o = g[\frac{\rho_a - \rho}{\rho}]Q = 981 \times \frac{998.2063 - 992.2204}{992.2204} \times 3 = \underline{17.8} \text{ cm}^4/\text{s}^3$$

Entrainment of ambient fluid into the jet reduces the temperature difference. The reduction is determined by the heat conservation equation:

$$(T - T_a)Q = (T_o - T_a)Q_o$$

that gives the average temperature

$$T = T_a + (T_o - T_a)\frac{Q_o}{Q} = 20 + (40 - 20) \times \frac{1}{100} = 20.2\,°\text{C}$$

and the corresponding average density $\rho = 998.1649$ kg/m^3. Therefore, the specific buoyancy flux becomes

$$F = g[\frac{\rho_a - \rho}{\rho}](100Q_o) = 981 \times \frac{998.2063 - 998.1649}{998.1649} \times (100 \times 3)$$

$$= \underline{12.2}\ \text{cm}^4/\text{s}^3$$

after the jet is diluted by 100 times. The heat flux is conserved in the buoyant jet but the buoyancy flux is reduced by as much as 31 %.

A closely related example is the problem of negative buoyancy at 4 °C water. Thermal effluent from power plants into the Great Lakes has been observed to sink in the winter as heated water in the thermal plume is diluted by the freezing (0 °C) water from the lake. Pockets of water in a thermal effluent, with a source temperature of say 12 °C, will be reduced to 4 °C in region of the plume where a dilution of 3 is attained. Such pockets of water at 4 °C temperature is heavier than its surrounding water and have been observed to sink and impact on the winter ecosystem of the lake (Griffiths, 1980).

Appendix B
Notation

α	entrainment coefficient
β	spreading rate
γ	intermittency factor, thermal expansion coefficient
Γ	mass flux
ϵ	dissipation rate of turbulent kinetic energy
θ	horizontal angle for buoyant jet or vertical angle for momentum jet
ϕ	vertical angle for buoyant jet
λ	ratio of concentration to velocity width
ν	kinematic viscosity
ρ	fluid density
σ_t	sigma unit
τ_{ij}	stress tensor
Ψ_m^*	normalized stream function
ψ_k	angle between jet axial vector and crossflow direction

A	cross-sectional area, projection area
b	half-width of jet/plume
B	half-width of top-hat profile
c, C	concentration
c_p	specific heat at constant pressure
d_o	dimension for line source, slot width
D	dimension for point source, initial jet diameter
E	kinetic energy flux
F	buoyancy flux, specific buoyancy flux
Fr	source densimetric Froude number
$\mathrm{Fr_L}$	local densimetric Froude number
Fr_p	plume Froude number
g	acceleration due to gravity
g'	reduced gravity, buoyancy force per unit mass of fluid
H	depth, heat flux of source

J. H. W. Lee et al., *Turbulent Jets and Plumes*,
https://doi.org/10.1007/978-3-032-01187-9

$\hat{\mathbf{k}}$	unit vector in the opposite direction of the gravity
k, k_n	turbulent kinetic energy, added mass coefficient
K	jet to ambient velocity ratio
l_Q	source geometry length scale
l_s	momentum length scale
L_b	crossflow buoyancy length scale
L_d	diffuser length
L_m^*	excess momentum length scale
L_{mv}	crossflow momentum length scale
M	momentum flux, specific momentum flux
N	Brunt-Väisälä frequency
p	pressure
Q	volume flux, volume
s, r	streamwise and radial co-ordinate
R	radius, gas constant
Re	Reynolds number
S	dilution, salinity
Sc_t	turbulent Schmidt number
t	time
T	temperature, time scale
u, v, w	velocity
U_i	velocity
U, W	velocity of top-hat profile
V	volume
x, y, z	Cartesian co-ordinate

Subscript/Superscripts

a	ambient
c	concentration, centerline
e	excess, entrainment, end of potential core
G, g	Gaussian profile
h	horizontal
j	jet
l	line source
m	centerline maximum
o	source
p	plume
s	shear, stagnant ambient
t	thermal, turbulent
v	vertical, velocity
$\bar{}$	time average of turbulent fluctuation
$\tilde{}$	associated with dominant eddies
*	dimensionless
$'$	turbulent fluctuation

References

Abraham, G. (1963a). *Jet diffusion in stagnant ambient fluid* (p. 29). No: Delft Hydraulics Laboratory Publications.

Abraham, G. (1963b). Horizontal jets in stagnant fluid of other density. *Proceedings of the ASCE, Journal of the Hydraulics Division,91*(HY1), 63–68.

Abraham, G. (1970). The flow of round buoyant jets issuing vertically into ambient fluid flowing in a horizontal direction. In *Proceedings of the 5th International Conference on Water Pollution Research*, San Francisco, pp. III–15/1–7.

Abramovich, G. N. (1963). *The theory of turbulent jets* (p. 684). MIT Press.

Agg, A. R., & Wakeford, A. C. (1972). Field studies of jet dilution of sewage at sea outfalls. *Journal of the Institute of Public Health Engineers, 71*, 126–149.

Albertson, M. L., Dai, Y. B., Jensen, R. A., & Rouse, H. (1950). Diffusion of submerged jets. *Transactions of the ASCE, 115*, 639–644.

Altai, W., & Chu, V. H. (2001). Large eddy simulation by Lagrangian block method. In E. Satofuka (Ed.), *Computational fluid dynamics 2000* (pp. 467–472). Springer-Verlag.

Andreopoulos, J., & Rodi, W. (1984). Experimental investigation of jets in a cross-flow. *Journal of Fluid Mechanics, 138*, 93–127.

Antonia, R. A., & Bilger, R. W. (1973). An experimental investigation of an axisymmetric jet in a co-flowing air stream. *Journal of Fluid Mechanics, 61*, 805–822.

Antonia, R. A., & Bilger, R. W. (1974). The prediction of the axisymmetric turbulent jet issuing into a co-flowing stream. *Aeronautical Quarterly, 251*, 69–80.

Anwar, H. O. (1969). Behaviour of buoyant jet in calm fluid. *Journal of the Hydraulics Division, ASCE, 95*(HY4), 1289–1303.

Ayoub, G. M. (1971). Dispersion of buoyant jets in a flowing ambient fluid. Ph.D. thesis, Imperial College, the University of London.

Baines, W. D. (1975). Entrainment by a plume or jet at a density interface. *Journal of Fluid Mechanics, 68*, 309 320.

Baines, W. D. (1983). A technique for direct measurement of volume flux of a plume. *Journal of Fluid Mechanics, 132*, 247–256.

Baines, W. D., & Chu, V. H. (1996). Jets and plumes. In V. P. Singh and I. Hager (eds.) *Environmental hydraulics*, Chap. 2 (pp. 7–61). Kulwar Academic Publishers.

Baines, W. D., & Keffer, K. F. (1974). Entrainment by a multiple source turbulent jet. *Advances in Geophysics, 18B*, 289–298.

J. H. W. Lee et al., *Turbulent Jets and Plumes*,
https://doi.org/10.1007/978-3-032-01187-9

Baines, W. D., & Murphy, T. M. (1986). The temperature distribution in an enclosed space produced by a forced plume. In *Proceedings of the 8th International Heat Transfer Conference* (pp. 1507–1512). San Francisco, Hemisphere.

Baines, W. D., & Turner, J. S. (1969). Turbulent buoyant convection from a source in a confined region. *Journal of Fluid Mechanics, 37*, 51–81.

Baines, W. D., Corriveau, A. F., & Reedman, T. J. (1993). Turbulent fountains in a closed chamber. *Journal of Fluid Mechanics, 255*, 621–646.

Baines, W. D., Turner, J. S., & Campbell, I. H. (1990). Turbulent fountains in an open chamber. *Journal of Fluid Mechanics, 212*, 557–592.

Belatos, S., & Rajaratnam, N. (1973). Circular turbulent jet in an opposing infinite stream. In *Proceedings of the 1st Canadian Hydrotechnical Conference, Edmonton, Canada* (pp. 220–237).

Bennett, N. J. (1981). Initial dilution: A practical study on the Hastings long sea outfall. *Proceedings of the Institution of Civil Engineers, Part 1, 70*, 113–122.

Bennett, N. J. (1983). Design of sea outfalls–the lower limit concept of initial dilution. *Proceedings of the Institution of Civil Engineers, Part 2, 75*, 113–121.

Birch, A. D., Brown, D. R., Dodson, M. R., & Thomas, J. R. (1978). The turbulent concentration field of a methane jet. *Journal of Fluid Mechanics, 88*, 431–449.

Bourouiba, L., Dehandschoewercker, E., and Bush, J. W. M. (2014). Violent expiratory events: on coughing and sneezing. Journal of Fluid Mechanics, 745, 537–563.

Bradbury, L. J. S. (1965). The structure of a self-preserving turbulent plane jet. *Journal of Fluid Mechanics, 23*, 31–64.

Briggs, G. A. (1965). A plume rise model compared with observations. *Journal of the Air Pollution Control Association, 15*(9), 433–438.

Briggs, G. A. (1969). *Plume rise: A critical survey* (p. 81). U.S: Atomic Energy Commission Review.

Brooks, N. H. (1960). Diffusion of sewage effluent in an ocean current. In *Proceedings of the First International Conference on Waste Disposal in the Marine Environment* (pp. 246–267). Pergamon, New York.

Brooks, N. H. (1972). Dispersion in hydrologic and coastal environments. Report No. KH-R-29, W. M. Keck Laboratory of Hydraulics and Water Resources, California Institute of Technology, Pasadena, California.

Brunton, S. L., & Kutz, J. M. (2019). *Data-driven science and engineering: Machine learning, dynamical systems, and control*. Cambridge University Press.

Cathers, B., & Peirson, W. L. (1991). Verification of plume models applied to deepwater outfalls. In J. H. W. Lee, & Y. K. Cheung (eds.) *Proceedings of the International Symposium on Environment Hydraulics* (Vol. 1, pp. 261–266). Hong Kong, Balkema.

Cederwall, K. (1968). *Hydraulics of marine wastewater disposal* (p. 42). Chalmers Institute of Technology, Sweden, Report No: Hydraulics Division.

Cederwall, K. (1971). Buoyant slot jets into stagnant or flowing environments. Report No. KH-R-25, W. M. Kech Laboratory, California Institute of Technology, Pasedena, California.

Chan, D. T. L., Lin, J. T., & Kennedy, J. F. (1976). Entrainment and drag forces of deflected jets. *Journal of the Hydraulics Division, ASCE, 102*(HY5), 615–635.

Chan, C.H.C, & Lam, K.M. (2018). Centerline velocity decay of a circular jet in a counterflowing stream, Physics of Fluids, *10*(3), 637–644.

Chassaing, P., George, J., Claria, A., & Sananes (1974). Physical characteristics of subsonic jets in a cross-stream. *Journal of Fluid Mechanics,62*(part 1), 41–64.

Chan, S. N. (2013). Mixing and deposition of sediment-laden buoyant jets. Ph.D. thesis, The University of Hong Kong, Hong Kong S.A.R.

Chan, S. N., & Lee, J. H. W. (2014). Particle tracking modeling of sediment-laden jets. *Advances in Geosciences, 39*, 107–114.

Chan, S. N., Thoe, W., & Lee, J. H. W. (2013). Real-time forecasting of Hong Kong beach water quality by 3D deterministic model. *Water Research, 47*, 1631–1647.

Chan, S. N., Qiao, Q. S., Lee, J. H. W., Choi, K. W., & Huang, J. C. (2017). Modeling of mixing and rapid chlorine demand in sewage disinfection with dense chlorine jets. *Journal of Environmental Engineering, ASCE, 143*(11), 04017074.

Chan, Y. M., Thoe, W., & Lee, J. H. W. (2015). Field and laboratory studies of Escherichia coli decay rate in subtropical coastal water. *Journal of Hydro-Environment Research, 9*(1), 1–14.

Chen, G. Q., & Lee, J. H. W. (2002). Advected turbulent line thermal driven by concentration difference. *Communications in Nonlinear Science and Numerical Simulation, 7*, 175–195.

Cheung, V. (1991). Mixing of a round buoyant jet in a current. Ph.D. thesis, Department of Civil and Structural Engineering, The University of Hong Kong.

Cheung, V., & Lee, J. H. W. (1991). Heated jets with three-dimensional trajectories. In J. H. W. Lee, & Y. K. Cheung (eds.) *Proceedings of the International Symposium on Environmental Hydraulics*, Hong Kong (Vol. 1, pp. 71–75). Balkema.

Cheung, V., & Lee, J. H. W. (1996). Discussion of improved prediction of bending plumes. *Journal of Hydraulic Research, 34*(2), 260–262.

Cheung, V., & Lee, J. H. W. (1999). Discussion of simulation of oil spills from underwater accidents I: Model development. *Journal of Hydraulic Research, 37*, 425–429.

Chien, C. J., & Schetz, J. A. (1975). Numerical solution of the three-dimensional Navier-Stokes equations with application to channel flows and a buoyant jet in a cross-flow. *Transactions of the ASME E: Journal of Applied Mechanics, 42*, 575–579.

Choi, K. W., Lai, C. C. K., & Lee, J. H. W. (2016). Mixing in the intermediate field of dense jets in cross currents. *Journal of Hydraulic Engineering, ASCE, 142*(1), 04015041.

Choi, K. W., & Lee, J. H. W. (2007). Distributed entrainment sink approach for modeling mixing and transport in the intermediate field. *Journal of Hydraulic Engineering, 133*(7), 804–815.

Chu, P. C. K. (1996). Mixing of turbulent advected line puffs. Ph.D. thesis, Department of Civil and Structural Engineering, The University of Hong Kong.

Chu, P. C. K., & Lee, J. H. W. (1996). Vorticity dynamics of an inverted jet in crossflow. In *Proceedings of the 2nd International Conference on Hydrodynamics* (Vol. 2, pp. 743–748). Hong Kong.

Chu, P. C. K., Lee, J. H. W., & Chu, V. H. (1999). Spreading of a turbulent round jet in coflow. *Journal of Hydraulic Engineering, ASCE, 125*(2), 193–204.

Chu, V. H. (1975). Three-dimensional forced plumes in a laminar crossflow. In *Proceedings of the 15th International Association for Hydraulic Research Congress*(Vol. 4, pp. 31–39). Delft, The Netherlands.

Chu, V. H. (1977). A line-impulse model for buoyant jet in a cross-flow. In B. D. Spalding & N. H. Afgan (Eds.), *Heat transfer and turbulent buoyant convection* (pp. 325–337). Hemisphere Publishing Co.

Chu, V. H. (1979). L. N. Fan's data on buoyant jets in crossflow. *Journal of the Hydraulics Division, ASCE, 105*(HY5), 612–617.

Chu, V. H. (1985). Oblique turbulent jets in a crossflow. *Journal of Engineering Mechanics, 111*(11), 1343–1359.

Chu, V. H. (1994). Lagrangian scalings of jets and plumes with dominant eddies. In P. A. Davies & M. J. Valente Neves (Eds.), *Recent research advances in the fluid mechanics of turbulent jets and plumes, NATO ASI series E: Applied sciences* (Vol. 255, pp. 45–72). Kluwer Academic Publishers.

Chu, V. H., & Altai, W. (2001a). Simulation of shallow transverse shear flow by generalized second moment method. *Journal of Hydraulic Research, 39*(6), 575–582.

Chu, V. H., & Altai, W. (2001b). Lagrangian block method. In E. Satofuka (Ed.), *Computational fluid dynamics 2000* (pp. 771–772). Springer-Verlag.

Chu, V. H., & Baddour, R. E. (1984). Turbulent stratified shear flows. *Journal of Fluid Mechanics, 132*, 353–378.

Chu, V. H., & Baines, W. D. (1989). Entrainment by a buoyant jet between confined walls. *Journal of Hydraulic Engineering, ASCE, 115*(4), 475–492.

Chu, V. H., & Baines, W. D. (1991). Starting jets between confined walls. In *Proceedings of the 1st International Symposium on Environmental Hydraulics* (Vol. 1), pp. 53–58.

Chu, V. H., & Goldberg, M. B. (1974). Buoyant forced plumes in crossflow. *Journal of the Hydraulics Division, ASCE, 100*(HY9), 1203–1214.

Chu, V. H., & Lee, J. H. W. (1996). A general integral formulation of turbulent buoyant jets in crossflow. *Journal of Hydraulic Engineering, ASCE, 122*(1), 27–34.

Chu, V. H., Senior, C., & List, J. (1981). Transition from a turbulent jet into a turbulent plume. ASME Publication 81-FE-29, Boulder, Colorado, 8 pp.

Corrsin, S., & Kistler, A. L. (1954). The free stream boundaries of turbulent flows. NACA Technical Note 3133.

Csanady, G. T. (1965). The buoyant motion within a hot gas plume in a horizontal wind. *Journal of Fluid Mechanics, 22*, 225–239.

Davidson, M. J. (1989). The behaviour of single and multiple horizontally discharged buoyant jets in a non-turbulent ambient fluid. Ph. D. thesis, University of Canterbury, Christchurch, New Zealand.

Davidson, M. J., Gaskin, S., & Wood, I. R. (2002). A study of a buoyant axisymmetric jet in a small coflow. *Journal of Hydraulic Research, 40*(4), 477–489.

Davidson, M. J., Knudsen, M., & Wood, I. R. (1991). The behaviour of a single, horizontally discharged, buoyant flow in a non-turbulent coflowing ambient fluid. *Journal of Hydraulic Research, 29*(4), 545–566.

Davidson, M. J., & Pun, K. L. (1999). Weakly advected jets in crossflow. *Journal of Hydraulic Research, ASCE, 125*(1), 47–58.

Dehghani, M. R. (1996). Line thermals and buoyant jets in cross flow. Ph. D. thesis, Department of Civil Engineering, McGill University, 181 pp.

Demuren, A. O. (1983). Numerical calculations of steady three-dimensional turbulent jets in cross flow. *Computer Methods in Applied Mechanics and Engineering, 37*(3), 309–328.

Dimotakis, P. E., Miake-Lye, R. C., & Papantoniou, D. A. (1983). Structure and dynamics of round turbulent jets. *Physics of Fluids, 26*(11), 3185–3192.

Escudier, M. P., & Maxworthy, T. (1973). On the motion of turbulent thermals. *Journal of Fluid Mechanics, 61*, 541–552.

Everitt, K. W., & Robins, A. G. (1978). The development and structure of turbulent plane jets. *Journal of Fluid Mechanics, 88*, 563–583.

Fage, A., & Falkner, V. M. (1935). Note on experiments on the temperature and velocity in the wake of a heated cylindrical obstacle. *Proceedings of the Royal Society of London A, 135*, 702–705.

Fan, L. N. (1967). Turbulent buoyant jets into stratified or flowing ambient fluids. Report No. KH-R-15, W. M. Keck Laboratory of Hydraulics and Water Resources, California Institute of Technology, Pasadena, California.

Fan, L. N., & Brooks, N. H. (1969). Numerical solutions of turbulent buoyant jet problems. Report No. KH-R-18, W. M. Keck Laboratory, California Institute of Technology, Pasadena.

Fischer, H. B., List, E. J., Koh, R. C. Y., Imberger, J., & Brooks, N. H. (1979). *Mixing in inland and coastal waters*. Academic Press.

Fofonoff, N. P. (1962). Physical properties of sea-water. In M.N. Hill (Ed.), *The Sea* (Vol. 1, pp. 3–30). Interscience.

Fox, D. G. (1970). Forced plume in a stably stratified fluid. *Journal of Geophysical Research, 75*, 6818–6835.

Fric, T. F., & Roshko, A. (1994). Vortical structure in the wake of a transverse jet. *Journal of Fluid Mechanics, 279*, 1–47.

Frick, W. E. (1984). Non-empirical closure of the plume equations. *Atmospheric Environment, 18*(4), 653–662.

Gartshore, I. S. (1966). An experimental examination of the large-eddy equilibrium hypothesis. *Journal of Fluid Mechanics, 24*, 89–98.

Gartshore, I. S. (1967). Two-dimensional turbulent wakes. *Journal of Fluid Mechanics, 30*, 547–560.

Gaskin, S. J. (1995). Single buoyant jet in a crossflow and the advected line thermal. Ph.D. thesis, University of Canterbury, Christchurch, New Zealand.

Gaskin, S. J., Papps, D. A., & Wood, I. R. (1995). The axisymmetric equations for a buoyant jet in a crossflow. In *Proceedings of the 12th Australasian Fluid Mechanics Conference*, Sydney.

Gaskin, S. J., & Wood, I. R. (2001). Flow structure and mixing dynamics of the advected line thermal. *Journal of Hydraulic Research, 39*(5), 459–468.

George, W. K., Alphert, R. L., & Tamanini, F. (1977). Turbulent measurements in an axisymmetrical buoyant plume. *International Journal of Heat and Mass Transfer, 20*, 1145–1154.

Germeles, A. E. (1975). Forced plumes and mixing of liquids in tanks. *Journal of Fluid Mechanics, 71*, 601–625.

Glezer, A., & Coles, D. (1990). An experimental study of a turbulent vortex ring. *Journal of Fluid Mechanics, 211*, 243–283.

Gordon, A. D., & Fagan, P. W. (1991). Ocean outfall performance monitoring. In J. H. W. Lee, & Y. K. Cheung (eds.) *Proceedings of the International Symposium on Environment Hydraulics*, Hong Kong, Vol. 1, Balkema, pp. 243–248.

Griffiths, J. S. (1980). Potential effects of unstable thermal discharges on incubation of round whitefish eggs. Ontario Hydro Research Division Report No. 80-140-K.

Gutmark, E., & Wygnanski, I. (1976). The planar turbulent jet. *Journal of Fluid Mechanics, 73*, 465–495.

Harleman, D. R. F. (1982). Hydrothermal analysis of lakes and reservoirs. *Journal of the Hydraulics Division, ASCE, 108*(HY3), 302–325.

Heskestad, G. (1984). Engineering relations for fire plumes. *Fire Safety Journal, 7*, 25–32.

Hewett, T. A., Fay, J. A., & Hoult, D. P. (1971). Laboratory experiments of smokestack plumes in a stable atmosphere. *Atmospheric Environment, 5*, 767–789.

Hinze, J. O. (1975). *Turbulence* (p. 790). McGraw-Hill.

Hirst, E. (1972). Buoyant jets with three-dimensional trajectories. *Journal of the Hydraulics Division, ASCE, 98*(HY11), 1999–2014.

Hodgson, J. E., & Rajaratnam, N. (1992). An experimental study of jet dilution in crossflows. *Canadian Journal of Civil Engineering, 19*(5), 733–743.

Hodgson, J. E., Moaward, A. K., & Rajaratnam, N. (1999). Concentration field of multiple circular turbulent jets. *Journal of Hydraulic Research, 37*(2), 249–256.

Horton, P. R., Lee, J. H. W., & Wilson J. R. (1997). Near-field JETLAG modelling of the Northwest new territories sewage outfall, Urmston Road, Hong Kong. In *Proceedings of the 13th Australasian Coastal and Ocean Engineering Conference* (Pacific Coasts and Ports '97), Christchurch, New Zealand, Vol. 2, pp. 561–566.

Hoult, D. P., Fay, J. A., & Forney, L. J. (1969). A theory of plume rise compared with field observations. *Journal of the Air Pollution Control Association, 19*, 585–590.

Hoult, D. P., & Weil, J. C. (1972). Turbulent plume in a laminar crossflow. *Atmospheric Environment, 6*, 513–531.

Howard, R. A., Charlton, J. A., & Webb, M. B. (1991). The design of the stage I outfall for the Hong Kong strategic sewage disposal scheme. In J. H. W. Lee, & Y. K. Cheung, Y.K. (eds.), *Proceedings of the International Symposium on Environmental Hydraulics*, Hong Kong, Vol. 1, Balkema, pp. 335–339.

Hydraulics Research Station. (1977). *Horizontal outfalls in flowing water*. Hydraulics Research Station Report EX763. Wallingford, United Kingdom.

Imberger, J., & Patterson, J. C. (1990). Physical limnology. In J. W. Hutchinson, & T. Y. Wu (eds.), *Advances in Applied Mechanics* (Vol. 27). Academic Press, pp. 303–475.

Iribarne, J. V., & Godson, W. L. (1981). *Atmospheric thermodynamics* (p. 259). Reidel Publishing Co.

Isaacson, M. S., Koh, R. C. Y., & Brooks, N. H. (1983). Plume dilution for diffusers with multiport risers. *Journal of Hydraulic Engineering, ASCE, 109*(2), 199–220.

Jenkins, P. E., & Goldschmidt, V. W. (1973). Mean temperature and velocity in a plane turbulent jet. *Journal of Fluids Engineering, ASME, 95*, 581–584.

Jiang, M., Law, A. W. K., & Lai, A. C. H. (2019). Turbulence characteristics of 45° inclined dense jets. *Environmental Fluid Mechanics, 19*, 27–54.

Jirka, G. H. (1982). Turbulent buoyant jets in shallow fluid layers. In W. Rodi (ed.) *Turbulent buoyant jets and plumes: HMT: The science and applications of heat and mass transfer* (Vol. 6, pp. 69–119), Pergamon.

Jirka, G. H., Abraham, G., & Harleman, D. R. F. (1975). *An assessment of techniques for hydrothermal prediction* (p. 203). M.I: T. Parsons Lab. for Water Resources & Hydrodyn Report.

Jirka, G. H., & Doneker, R. L. (1991). Hydrodynamic classification of submerged single-port discharges. *Journal of Hydraulic Engineering, ASCE, 117*(9), 1095–1112.

Jirka, G. H., & Harleman, D. R. F. (1973). The mechanics of submerged multiport diffusers for buoyant discharges in shallow water. Technical Report 169, R. Parson Lab. for Water Resources and Hydrodynamics, MIT.

Jirka, G. H., & Harleman, D. R. F. (1979). Stability and mixing of a vertical plane buoyant jet in confined depth. *Journal of Fluid Mechanics, 94*(2), 275–304.

Kaminski, E., Tait, S., & Carazzo, G. (2005). Turbulent Entrainment in Jets with Arbitrary Buoyancy. *Journal of Fluid Mechanics*, 526, 361–376.

Kamotani, Y., & Greber, I. (1972). Experiments on a turbulent jet in a cross-flow. *AIAA Journal, 10*, 1425–1429.

Kaye, N. B., & Linden, P. F. (2004). Coalescing axisymmetric turbulent plumes. *Journal of Fluid Mechanics, 502*, 41–63.

Keffer, J. F., & Baines, W. D. (1963). The round turbulent jet in a cross wind. *Journal of Fluid Mechanics, 15*, 481–496.

Killworth, P. D., & Turner, J. S. (1982). Plumes with time varying buoyancy in a confined region. *Geophysical and Astrophysical Fluid Dynamics, 20*, 265–291.

Kiya, M., Ohyama, M., & Hunt, J. C. R. (1986). Vortex pairs and rings interacting with shear layer vortices. *Journal of Fluid Mechanics, 172*, 1–15.

Knudsen, M. (1988). Buoyant horizontal jets in an ambient flow. Ph. D. thesis, University of Canterbury, Christchurch, New Zealand.

Knudsen, M., & Wood, I. R. (1985). An axisymmetric jet in a moving fluid. In *Proceedings of the 9th Australasian Fluid Mechanics Conference*, Auckland, New Zealand, pp. 484–487.

Knystautas, R. (1964). The turbulent jet from a series of holes in line. *Aeronautical Quarterly, 15*, 1–28.

Koh, R. C. Y., & Brooks, N. H. (1975). Fluid mechanics of waste disposal in the ocean. *Annual Review of Fluid Mechanics, 7*, 187–211.

Kotsovinos, N. E. (1975). A study of entrainment and turbulence in a plane buoyant jet. Report No. KH-R-32, W. M. Keck Laboratory of Hydraulics and Water Resources, California Institute of Technology, Pasadena, California.

Kotsovinos, N. E. (1976). A note on the spreading rate and virtual origin of a plane turbulent jet. *Journal of Fluid Mechanics, 77*, 305–311.

Kotsovinos, N. E. (1977). Plane turbulent buoyant jet: Part 2 turbulence structure. *Journal of Fluid Mechanics, 81*, 55–62.

Kotsovinos, N. E., & List, E. J. (1977). Plane turbulent buoyant jet: Part I integral properties. *Journal of Fluid Mechanics, 81*, 25–44.

Kuang, C.P., Lee, J.H.W., Liu, S.G., & Gu, J. (2006). Numerical study on plume interaction above an alternating diffuser in stagnant water, China Ocean Engineering, *20*(2), 289–302.

Lai, A. C. H. (2009). Mixing of a rosette buoyant jet group. Ph. D. thesis, The University of Hong Kong, Hong Kong S.A.R.

Lai, A. C. H., & Lee, J. H. W. (2010). Multiple tandem jet interaction in a crossflow. *Journal of Hydrodynamics, Series B, 22*(5), 639–643.

Lai, A. C. H., & Lee, J. H. W. (2012). Dynamic interaction of multiple buoyant jets. *Journal of Fluid Mechanics, 708*, 539–575.

Lai, A. C. H., Law, A. W. K., & Adams, E. E. (2019). A second-order integral model for buoyant jets with background homogeneous and isotropic turbulence. *Journal of Fluid Mechanics, 871*, 271–304.

Lai, A. C. H., Yu, D., & Lee, J. H. W. (2011). Mixing of a rosette jet group in a crossflow. *Journal of Hydraulic Engineering, ASCE, 137*(8), 787–803.

Lai, A. C. H., Chan, S. N., Law, A. W. K., & Adams, E. E. (2016). Spreading hypothesis of a particle plume. *Journal of Hydraulic Engineering, ASCE, 142*(12), 04016065.

Lai, C. C. K., & Lee, J. H. W. (2012). Mixing of inclined dense jets in stationary ambient. *Journal of Hydro-Environment Research, 6*(1), 9–28.

Lai, C. C. K., & Lee, J. H. W. (2014). Initial mixing of inclined dense jet in perpendicular crossflow. *Environmental Fluid Mechanics, 14*(1), 25–49.

Lai, C. C. K., & Socolofsky, S. A. (2018). Budgets of turbulent kinetic energy, Reynolds stresses, and dissipation in a turbulent round jet discharged into a stagnant ambient. *Environmental Fluid Mechanics, 19*(2), 349–377.

Lai, C. C. K., & Socolofsky, S. A. (2019). The turbulent kinetic energy budget in a bubble plume. *Journal of Fluid Mechanics, 865*, 993–1041.

Lai, C.C.K., Lee, J.H.W., and Lam, K.M. (2009). "Digital Simulation of Dominant Eddies of A Co-Flowing Jet." In *Advances in Water Resources and Hydraulic Engineering 2*, Springer, pp. 618-623.

Lam, K.M. and Chan, H.C. (1997). "Round jet in ambient counterflowing stream," *J. Hydr. Engrg.*, Vol.123, pp. 895-903. (Discussions: Vol.125, 1999, pp. 428-432)

Lamb, H. (1932). *Hydrodynamics*. U.K.: Cambridge University Press.

Launder, B. E., & Spalding, D. B. (1974). The numerical computation of turbulent flows. *Comp. Meth. in Appl. Mech. Engrg., 3*, 269–289.

Launder, B.E. (1978). "Heat and Mass Transport," in Chapter 6, *Turbulence*, P. Bradshaw editor, Springer-Verlag Series on Topics in Applied Physics, Vol. 12, pp. 228-331.

Law, S., Davidson, M. J., McConnochie, C., & Lagrava, D. (2023). Improved numerical predictions of inclined negatively buoyant jet behavior. *Environmental Fluid Mechanics, 23*, 879–906.

Lawrence, G. A., Burke, J. M., Murphy, T. P., & Prepas, E. E. (1997). Exchange of water and oxygen between the two basin of Amisk Lake. *Canadian Journal of Fisheries and Aquatic Sciences, 54*, 2121–2132.

Lee, J.H.W. (1989). "Note on Ayoub's data of horizontal round buoyant jet in current," *J. Hydr. Engrg., ASCE*, Vol. 115, No. 7, pp. 969-975.

Lee, J. H. W. (1993). Discussion of "An experimental study of jet dilution in crossflows, Canadian. *J. Civil Engrg., 20*, 1073–1076.

Lee, J. H. W., & Chen, G. Q. (1998). A numerical study of turbulent line puffs via the renormalization group (RNG) k-ϵ model. *International Journal for Numerical Methods in Fluids, 26*, 217–234.

Lee, J. H. W., & Chen, G. Q. (2002). Numerical experiment on two-dimensional line thermal. *China Ocean Engineering, 16*, 453–467.

Lee, J. H. W., & Cheung, V. W. L. (1986). An inclined plane buoyant jet in stratified fluid. *Journal of Hydraulic Engineering, ASCE, 112*, 580–589.

Lee, J. H. W., & Cheung, V. (1990). Generalized Lagrangian model for buoyant jets in current. *Journal of Environmental Engineering, ASCE, 116*(6), 1085–1106.

Lee, J. H. W., & Cheung, V. (1991). Mixing of buoyancy-dominated jets in a weak current. *Proceedings of the Institution of Civil Engineers, 91*(2), 113–129.

Lee, J. H. W., & Chu, P. C. K. (1995). Application of video image processing in the study of environmental flows. In *Proceedings of 10th ASCE Engineering Mechanical Conference*, Boulder, Colorado, pp. 1014–1017.

Lee, J. H. W., & Jirka, G. H. (1980). Multiport diffuser as line source of momentum in shallow water. *Water Resources Research, 16*(4), 695–708.

Lee, J. H. W., & Jirka, G. H. (1981). Vertical round buoyant jet in shallow water. *Journal of the Hydraulics Division, ASCE, 107*(HY12), 1651–1675.

Lee, J. H. W., & Neville-Jones, P. (1987). Initial dilution of horizontal jet in crossflow. *Journal of Hydraulic Engineering, ASCE, 113*(5), 615–629.

Lee, J. H. W., & Neville-Jones, P. (1987). Sea outfall design–prediction of initial dilution. *Proceedings of the Institution of Civil Engineers, 82*(1), 981–994.

Lee, J. H. W., & Wong, C. F. (1993). Experiments on advected line momentum puffs. In *Proceedings of the 25th IAHR Congress*. Tokyo, Vol. 5, pp. 250–257.

Lee, J. H. W., Chen, G. Q., & Kuang, C. P. (2002). Mixing of a turbulent jet in crossflow—the advected line puff. In h. H. Shen et al. (Eds.) *Environmental fluid mechanics: Theories and applications* ASCE, Chap. 3, pp. 47–83.

Lee, J. H. W., Cheung, Y. K., & Cheung, V. (1987c). Mathematical modelling of a round buoyant jet in a current: An assessment. In *Proceedings of the International Symposium on River Pollution, Control and Management*, Shanghai, China, pp. S62–S82.

Lee, J. H. W., Cheung, V., & Lai, C. C. K. (2008). JETLAG2008—An update. Technical Note, Croucher Laboratory of Environmental Hydraulics, The University of Hong Kong, Hong Kong China.

Lee, J. H. W., Kuang, C. P., & Chen, G. Q. (2002). The structure of a turbulent jet in crossflow–effect of jet-to-crossflow velocity. *China Ocean Engineering,16*, 1–20.

Lee, J. H. W., Li, L., & Cheung, V. (1999). A semi-analytical self-similar solution of a bent-over jet in crossflow. *Journal of Engineering Mechanics, ASCE, 125*, 733–746.

Lee, J. H. W., Rodi, W., & Wong, C. F. (1996). Turbulent line momentum puffs. *Journal of Engineering Mechanics, ASCE, 122*(1), 19–29.

Lee, J. H. W., Cheung, V., Wang, W. P., & Cheung, S. K. B. (2000). Lagrangian modeling and visualization of rosette outfall plumes. In *Proceedings of the Hydroinformatics 2000 (CDROM)*, University of Iowa, 23–27 July 2000.

Lee, J. H. W., Qiao, Q. S., Chan, S. N., Choi, K. W., & Huang, C. H. (2017). Field experiments of chlorine demand in disinfection of treated primary effluent. *Journal of Environmental Engineering, ASCE, 143*(11), 04017073.

Lee, J. H. W., Qiao, Q. S., Chan, S. N., Choi, K. W., & Chau, H. K. M. (2023). Reduction of chlorine disinfection dosage through optimal jet design in the Hong Kong Harbor area treatment scheme. *Journal of Environmental Engineering, ASCE, 149*(2), 05022009.

Lilly, D. K. (1964). Numerical solution for the shape preserving 2-D thermal convection element. *Journal of the Atmospheric Sciences, 21*, 83–98.

Linden, P. F. (1999). The fluid mechanics of natural ventilation. *Annual Review of Fluid Mechanics, 31*, 201–238.

Liseth, P. (1970). Mixing of merging buoyant jets from a manifold in stagnant receiving water of uniform density. Ph. D. Thesis, University of California, Berkeley, California.

Liseth, P. (1976). Wastewater disposal by submerged manifolds. *Journal of the Hydraulics Division, ASCE, 102*(1), 1–14.

List, E. J. (1979). Turbulent jets and plumes. *In Mixing in inland and coastal waters, Academic Press, Chap., 9*, 315–389.

List, E. J. (1982). Turbulent jets and plumes. *Annual Review of Fluid Mechanics, 14*, 189–212.

List, E. J., & Imberger, J. (1973). Turbulent entrainment in buoyant jets and plumes. *Journal of the Hydraulics Division, ASCE, 99*(HY9), 1461–1474.

Maczynski, J. F. J. (1962). A round jet in an ambient co-axial stream. *Journal of Fluid Mechanics, 13*, 597–608.

Mancini, J. L. (1978). Numerical estimates of coliform mortality rates. *Journal (Water Pollution Control Federation), 50*(11), 2477–2484.

McClimans, T., & Eidnes, G. (2000). Forcing nutrients to the upper layer of a fjord by a buoyant plume. In G. Lawrence (ed.) *Proceedings of the 5th International Symposium on Stratified Flows*, University of British Columbia, Vol. 1, pp. 199–204.

McQuivey, R. S., Keefer, T. N., & Shirazi, M. A. (1971). *Basic data report on the turbulent spread of heat and matter*. U. S. Department of the Interior Geological Survey and U. S: Environmental Protection Agency.

Metzger, P. T., Li, X., Immer, C. D., & Lane, J. E. (2009). ISRU implications for lunar and martian plume effects. In *47th AIAA Aerospace Sciences Meeting including The New Horizons Forum and Aerospace Exposition*. Orlando, Florida

Mih, W. C., & Hoopes, J. A. (1972). Mean and turbulent velocities for plane jet. *Journal of the Hydraulics Division, ASCE, 98*(HY7), 1275–1294.

Milgram, J. H. (1983). Mean flow in round bubble plumes. *Journal of Fluid Mechanics, 133*, 345–376.

Miller, D., & Comings, E. (1957). Static pressure distribution in the free turbulent jet. *Journal of Fluid Mechanics, 3*, 1–16.

Millero, F. J. (1978). Freezing point of sea water. In Eighth Report of the Joint Panel on Oceanographic Tables and Standards. UNESCO Technical Paper Marine Science, No. 28, Annex 6, UNESCO, Paris.

Millero, F. J., & Poisson, A. (1981). International one-atmosphere equation of state for seawater. *Deep Sea Research, 28A*, 625–629.

Milne-Thomson, L.M. (1968). *Theoretical hydrodynamics*, Macmillan, New York

Mohammadian, A., Gildeh, H. K., & Yan, X. (2023). *Numerical simulation of effluent discharges: Applications with OpenFOAM*. CRC Press.

Monismith, S. G., Koseff, J. R., Thompson, J. K., O'Riordan, C. A., & Nepf, H. M. (1990). A study of model bivalve siphonal currents. *Limnology and Oceanography, 35*(3), 680–696.

Morton, B. R., Taylor, G. I., & Turner, J. S. (1956). Turbulent gravitational convection from maintained and instantaneous sources. *Proceedings of the Royal Society of London A, 234*, 1–23.

Moussa, Z. M., Trischka, J. W., & Eskinazi, S. (1977). The near field in the mixing of a round jet with a cross-stream. *Journal of Fluid Mechanics, 80*, 49–80.

Muellenhoff, W. P., Soldate, A. M., Baumgartner, D. J., Schuldt, M. D., Davis, L. R., & Frick, W. E. (1985). Initial mixing characteristics of municipal ocean discharges. Report EPA-600/3-85-073, United States Environmental Protection Agency, Newport, Oregon.

Newman, B. G. (1967). Turbulent jets and wakes in a pressure gradient. In G. Sovran (Ed.), *Fluid mechanics of internal flow* (pp. 176–209). Elsevier.

Nickels, T. B., & Perry, A. E. (1996). An experimental and theoretical study of the turbulent coflowing jet. *Journal of Fluid Mechanics, 309*, 157–182.

O'Riordan, C. A., Monismith, S. G., & Koseff, J. R. (1995). The effect of bivalve excurrent jet dynamics on mass transfer in a benthic boundary layer. *Limnology and Oceanography, 40*(2), 330–344.

Pani, B. S., Lee, J. H. W., & Lai, A. C. H. (2009). Application of Reichardt's hypothesis for multiple coflowing jets. *Journal of Hydro-Environment Research, 3*(3), 121–128.

Pani, B. S., & Dash, R. (1983). Three dimensional single and multiple jets. *Journal of Hydraulic Engineering, ASCE, 109*(2), 254–269.

Pani,B.S., & Dugad, S.B. (2002). Turbulent jets: Application of point source concept, in Research perspectives in hydraulics and water resources engineering, World Scientific, (pp. 1–37).

Papantoniou, D. A. (1986). Observations in turbulent buoyant jets by use of Laser-induced fluorescence. Ph.D. thesis, California Institute of Technology, Pasadena, California.

Papanicolaou, P., & List, E. J. (1987). Statistical and spectral properties of tracer concentration in round buoyant jets. *International Journal of Heat and Mass Transfer, 30*(10), 2059–2071.

Papanicolaou, P., & List, E. J. (1988). Investigations of round vertical turbulent buoyant jets. *Journal of Fluid Mechanics, 195*, 341–391.

Papantoniou, D., & List, E. J. (1989). Large-scale structure in the far field of buoyant jets. *Journal of Fluid Mechanics, 209*, 151–190.

Patankar, S. V. (1980). *Numerical heat transfer and fluid flow*. Hemisphere Publishing Co.

Patankar, S. V., Basu, D. K., & Alpay, S. A. (1977). Prediction of the three-dimensional velocity field of a deflected turbulent jet. *Transactions of the ASME I: Journal of Fluids Engineering, 99*, 758–762.

Patel, R. P. (1971). *Turbulent jets and wall jets in uniform streaming flow* (pp. 311–326). XXII: Aeronautical Quarterly.

Phillips, O. M. (1966). *The dynamics of the upper oceans* (p. 261). Cambridge University Press

Pratte, B. D., & Baines, W. D. (1967). Profiles of the round turbulent jet in a cross flow. *Journal of the Hydraulics Division, ASCE,92*(HY6), 53–64; see also 1968 correction, *93*(HY3), 815–816.

Pratte, B. D., & Keffer, J. F. (1972). The swirling turbulent jet. *Journal of Basic Engineering, Transactions of the ASME, 94*(4), 739–748.

Proni, J. R., Huang, H., & Dammann, W. P. (1994). Initial dilution of Southeast Florida ocean outfalls. *Journal of Hydraulic Engineering, ASCE, 120*(12), 1409–1425.

Pun, K. L., & Davidson, M. J. (1999). On the behaviour of advected plumes and thermals. *Journal of Hydraulic Research, 37*(4), 519–540.

Rajaratnam, N. (1976). *Turbulent jets*. Amsterdam, Netherlands: Elsevier.

Richards, J. M. (1961). Experiments on the penetration of an interface by buoyant thermals. *Journal of Fluid Mechanics, 11*, 369–384.

Richards, J. M. (1963). Experiments on the motion of isolated cylindrical thermals through unstratified surroundings. *International Journal of Air, Water and Pollution, 7*, 17–34.

Richards, J. M. (1965). Puff motion in unstratified surroundings. *Journal of Fluid Mechanics, 21*, 97–106.

Ricou, F. P., & Spalding, D. B. (1961). Measurements of entrainment by axisymmetrical turbulent jets. *Journal of Fluid Mechanics, 11*, 21–32.

Roberts, P. J. W. (1979). Line plume and ocean outfall dispersion. *Journal of the Hydraulics Division, ASCE,105*(HY4), 313–331 (discussion by G. H. Jirka, HY12, 1573–1575).

Roberts, P. J. W. (1980). Ocean outfall dilution: Effect of currents. *Journal of the Hydraulics Division, ASCE, 106*(HY5), 769–782.

Roberts, P. J. W., & Snyder, W. H. (1993a). Hydraulic model study for Boston outfall I: Riser configuration. *Journal of Hydraulic Engineering, ASCE, 119*(9), 970–987.

Roberts, P. J. W., & Snyder, W. H. (1993b). Hydraulic model study for Boston outfall II: Environmental performance. *Journal of Hydraulic Engineering, ASCE, 119*(9), 988–1002.

Roberts, P. J. W., Snyder, W. H., & Baumgartner, D. J. (1989). Ocean outfalls I, submerged wastefield formation. *Journal of Hydraulic Engineering, ASCE, 115*(1), 1–25.

Roberts, P. J. W., Snyder, W. H., & Baumgartner, D. J. (1989). Ocean outfalls II, spatial evolution of submerged wastefield. *Journal of Hydraulic Engineering, ASCE, 115*(1), 26–48.

Roberts, P. J. W., Snyder, W. H., & Baumgartner, D. J. (1989). Ocean outfalls III, effect of diffuser design on submerged wastefield. *Journal of Hydraulic Engineering, ASCE, 115*(1), 49–70.

Rodi, W. (1980). *Turbulence models and their application in hydraulics*. International Association for Hydraulic Research.

Roshko, A. (1976). Structure of turbulent shear flows: A new look. *AIAA Journal, 14*, 1349–1357; *15*, 768 (1977).

Rouse, H., Baines, W. D., & Humphreys, H. W. (1953). Free convection over parallel sources of heat. *Proceedings of the Physical Society Section B, 56*, 393–399.

Rouse, H., Yih, C. S., & Humphreys, H. W. (1952). Gravity convection from a boundary source. *Tellus, 4*, 201–210.

Schatzmann, M. (1979). An integral model of plume rise. *Atmospheric Environment, 13*, 721–731.

Schatzmann, M. (1981). Mathematical modeling of submerged discharges into coastal waters. In *Proceedings of the 19th IAHR Congress*. New Delhi, Vol. 3, pp. 239–246.

Schladow, S. G., & Fisher, I. H. (1995). The physical response of temperate lakes to artificial destratification. *Limnology and Oceanography, 40*, 359–373.

Schmid, P. T. (2010). Dynamic mode decomposition of numerical and experimental data. *Journal of Fluid Mechanics, 656*, 5–28.

Scorer, R. S. (1957). Experiments on convection of isolated mass of buoyant fluid. *Journal of Fluid Mechanics, 2*, 583–594.

Scorer, R. S. (1959). The behaviour of chimney plumes. *International Journal of Air and Pollution, 1*, 198–220.

Scorer, R. S. (1978). *Environmental Aerodynamics*. Ellis Horwood.

Sharp, J. J., & Vyas, B. D. (1977). The buoyant wall jet. *Proceedings of the Institution of Civil Engineers, Part 2, 63*, 593–611.

Shaughnessy, E. J., & Morton, J. B. (1977). Laser light-scattering measurements of particle concentration in a turbulent jet. *Journal of Fluid Mechanics, 80*, 120–148.

Shirazi, M. A., McQuivey, R. S., & Keefer, T. N. (1974). Heated water jet in coflowing turbulent stream. *Journal of the Hydraulics Division, ASCE, 100*(HY7), 919–934.

Sobey, R. J., Johnston, A. J., & Keane, R. D. (1988). Horizontal round buoyant jet in shallow water. *Journal of Hydraulic Engineering, ASCE, 114*(8), 910–929.

Sonar, P., & Katsuragi, H. (2024). Air-jet impact craters on granular surfaces: A universal scaling. *Journal of Fluid Mechanics, 998*, A29.

Socolofsky, S. A., & Adams, E. E. (2002). Multi-phase plumes in uniform and stratified crossflow. *Journal of Hydraulic Research, 40*(6), 661–672.

Sun, T. Y., & Faeth, G. M. (1986). Structure of turbulent bubbly jets-I: Methods and centerline properties. *International Journal of Multiphase Flow, 12*(1), 99–114.

Sykes, R. I., Lewellen, W. S., & Parker, S. F. (1986). On the vorticity dynamics of a turbulent jet in a crossflow. *Journal of Fluid Mechanics, 168*, 393–413.

Syrbin, A. N., & Lyakhovskiy, D. N. (1936). *Aerodynamics of an elementary flame Soobshch.* Kotloturoinnyi Inst: Nauchn-Issled.

Taylor, G. I. (1932). The transport of vorticity and heat through fluids in turbulent motion. *Proceedings of the Royal Society of London, Series A, Containing Papers of a Mathematical and Physical Character, A135*, 685–702.

Taylor, G. I. (1958). Flow induced by jets. *Journal of the Aeronautical Sciences, 25*, 464–465.

Thoe, W. (2010). A daily forecasting system of marine beach water quality in Hong Kong. Ph. D. thesis, The University of Hong Kong, Hong Kong S.A.R.

Thomas, R. M. (1973). Conditional sampling and other measurements in a plane turbulent wake. *Journal of Fluid Mechanics, 57*, 549–581.

Toledo, J. C., Singleton, V. L., Little, J. C., Lawrence, G. A., Ramon, C. L., & Rueda, F. J. (2021). Fate of artificially injected oxygen in the hypolimnion of a two-basin lake: Amisk Lake, revisited. *Water Resources Research,57*(6), e2020WR028840.

Townsend, A. A. (1976). *The structure of turbulent shear flow* (2nd ed.). Cambridge University Press.

Tsang, G. (1971). Laboratory study of line thermals. *Atmospheric Environment, 5*, 445–471.

Turner, J. S. (1960). A comparison between buoyant vortex rings and vortex pairs. *Journal of Fluid Mechanics, 7*, 419–432.

Turner, J. S. (1966). Jets and plumes with negative and reversing buoyancy. *Journal of Fluid Mechanics, 26*, 779–792.

Turner, J. S. (1973). *Buoyancy effects in fluids* (p. 367). Cambridge University Press.

Turner, J. S. (1986). Turbulent entrainment: The development of the entrainment assumption, and its application to geophysical flows. *Journal of Fluid Mechanics, 173*, 431–471.

UNESCO. (1981). Background papers and supporting data on the International Equation of State of Seawater 1980. UNESCO Technical Paper Marine Science, No. 38, UNESCO, Paris.

Van der Hegge Zijnen, B. G. (1957). Measurements of the distribution of heat and matter in a plane turbulent jet of air. *Applied Scientific Research, 7A*, 277–292.

Wallace, R. B., & Wright, S. J. (1984). Spreading layer of two-dimensional buoyant jet. *Journal of Hydraulic Engineering, ASCE, 110*(6), 813–828.

Wang, H. J., & Davidson, M. J. (2001). A profile tracking system for investigating the behaviour of discharges in moving environments. *Experiments in Fluids, 31*(5), 533–541.

Wang, H. J., & Davidson, M. J. (2003). Jet interaction in a still ambient fluid. *Journal of Hydraulic Engineering, ASCE, 129*(5), 349–357.

Wang, H., & Law, A. W. K. (2002). Second-order integral model for a round turbulent buoyant jet. *Journal of Fluid Mechanics, 459*, 397–428.

Wilkinson, D. L. (1988). Avoidance of seawater intrusion into ports of ocean outfalls. *Journal of Hydraulic Engineering, ASCE, 114*(2), 218–228.

Winiarski, L. D., & Frick, W. E. (1976). Cooling tower plume model. Report EPA-600/3-76-100. U.S. Environment Protection Agency, Corvallis, Oregon

Wong, D. R. (1986). Buoyant jet entrainment in stratified fluids. Report No. UMCE 85-9, The University of Michigan, Michigan, USA.

Wong, C. F. (1991). Advected line thermals and puffs. M. Phil. thesis, Department of Civil and Structural Engineering, University of Hong Kong, Hong Kong.

Wong, C. F., & Lee, J. H. W. (1991). Experiments on advected line thermals. *In Proceedings of the International Symposium on Environmental Hydraulics, Hong Kong, 1*, 153–157.

Wong, D. R., & Wright, S. J. (1988). Submerged turbulent jets in stagnant linearly stratified fluids. *Journal of Hydraulic Research, 26*(1), 199–223.

Wood, I. R. (1993). Asymptotic solutions and behaviour of outfall plumes. *Journal of Hydraulic Engineering, ASCE, 119*, 555–580.

Wood, I. R., Bell, R. G., & Wilkinson, D. L. (1993). *Ocean disposal of wastewater*. World Scientific.

Worster, M. G., & Leith, A. M. (1985). Laminar free convection in confined regions. *Journal of Fluid Mechanics, 156*, 301–319.

Wright, S. J. (1977). Mean behavior of buoyant jet in a crossflow. *Journal of the Hydraulics Division, ASCE, 103*(HY5), 499–513.

Wright, S. J. (1977b). Effects of ambient crossflow and density stratification on the characteristic behaviour of round turbulent buoyant jets. Report No. KH-R-36, W. M. Keck Lab. of Hydraulics and Water Resources, California Institute of Technology, Pasadena, California.

Wright, S. J., & Wallace, R. B. (1979). Two-dimensional buoyant jets in stratified fluid. *Journal of the Hydraulics Division, ASCE, 105*(HY11), 1393–1406.

Wright, S. J., Wong, D. R., Zimmerman, K. E., & Wallace, R. B. (1982). Outfall diffuser behavior in stratified ambient fluid. *Journal of the Hydraulics Division, ASCE, 108*(HY4), 483–501.

Wright, S. J. (1984). Buoyant jets in density-stratified crossflow. *Journal of Hydraulic Engineering, ASCE, 110*(5), 643–656.

Wright, S. J., & Wallace, R. B. (1991). Surface dilution of round submerged buoyant jets. *Journal of Hydraulic Research, 29*(1), 67–89.

Wright, S. J. (1994). The effect of ambient turbulence on jet mixing. In P. A. Davies & M. J. Valente Neves (Eds.), *Recent research advances in the fluid mechanics of turbulent jets and plumes, NATO ASI series E: Applied sciences* (Vol. 255, pp. 13–27). Kluwer Academic Publishers.

Wygnanski, I. J., & Fiedler, H. (1969). Some measurement in the self-preserving jet. *Journal of Fluid Mechanics, 38*(3), 577–612.

Wygnanski, I. J., & Fiedler, H. (1970). The two-dimensional mixing region. *Journal of Fluid Mechanics, 41*, 327–361.

Yang, A. J. K., Chan, S. N., Lee, J. H. W., & Zhang, X. (2019). Integral model of a reacting chlorine jet in ammonia nitrogen and treated primary effluent. *Journal of Hydro-Environment Research, 27*, 50–64.

Yannopoulos, P. C., & Noutsopoulos, G. C. (2006). Interaction of vertical round turbulent buoyant jets–part II: Superposition method. *Journal of Hydraulic Research, 44*(2), 233–248.

Yapa, P. D., Zheng, L., & Nakata, K. (1999). Modeling underwater oil/gas jets and plumes. *Journal of Hydraulic Engineering, ASCE, 125*(5), 481–491.

Yih, C. S. (1981). Similarity solutions for turbulent jets and plumes. *Journal of the Engineering Mechanics Division, ASCE, 107*, 455–478.

Yoda, M., Hesselink, L., & Mungal, M. G. (1992). The evolution and nature of large-scale structures in the turbulent jet. *Physics of Fluids A: Fluid Dynamics, 4*, 803–811.

Yu, D., Ali, M. S., & Lee, J. H. W. (2006). Multiple tandem jets in cross-flow. *Journal of Hydraulic Engineering, ASCE, 132*(9), 971–982.

Zhang, H., & Baddour, R. E. (1997). Maximum vertical penetration of plane turbulent negatively buoyant jets. *Journal of Engineering Mechanics, 123*(10), 973–977.

Zhang, H., & Baddour, R. E. (1998). Maximum penetration of vertical round dense jets at small and large Froude number. *Journal of Hydraulic Engineering ASCE, 124*(5), 550–552.

Zhang, S., Jiang, B., Law, A. W. K., & Zhao, B. (2016). Large eddy simulations of 45° inclined dense jets. *Environmental Fluid Mechanics, 16*, 101–121.

Zhang, H., Li, C., You, J., Zhang, X., Wang, Y., Chen, L., Fu, Q., Zhang, B., & Wang, Y. (2022). The investigation of plume-regolith interaction and dust dispersal during Chang'E-5 descent stage. *Aerospace, 9*(7), 358–375.

Zic, K., & Stefan, H. G. (1988). Lake aerator effect on temperature stratification analyzed by MINLAKE model. *Lake and Reservoir Management, 4*, 73–83.

Index

A
Added mass, 9, 311
  line puff, 243
  line thermal, 275
  sphere, 156
Added mass coefficient
  line puff, 226, 252, 291
  line thermal, 261, 290
Advected line puff, 219–252
  dilution, 242
  laser-induced fluorescence image, 240
  summary of experiments, 241
  summary of properties , 235
  time-averaged properties, 240
  trajectory, 242
  turbulence properties, 245
  width, 242
Advected line thermal, 253–290
  laser-induced fluorescence image, 253
  shadowgraph image, 273
  summary of experiments, 283
  summary of properties, 270
Analogy
  bent-over jet in crossflow, 223
  bent-over plume in crossflow, 256
Archimedes principle, 9, 146
Aspect ratio
  line puff, 234
  line thermal, 272
Average dilution, 84, 99, 140
  line puff, 227
  line thermal, 262

B
Bent-over buoyant jet, 292
Bent-over jet, 219
Bifurcation, 275
Biochemical oxygen demand, 143
Boundary layer, 69
Boundary layer approximation, 31, 142
Boussinesq approximation, 72
Brunt-Väisälä frequency, 151, 164, 174
Buoyancy, 64, 146
  flux, 14
  frequency, 185
  thermal, 155
Buoyancy-dominated far field (BDFF), 161, 256, 285, 322
Buoyancy-dominated near field (BDNF), 256, 284, 322
Buoyancy flux, 69
  specific, 72, 167
Buoyant jet
  bottom attachment, 141
  boundary effect, 128
  horizontal, 125
  inclined, 111–137
  in crossflow, 221
  in stratified crossflow, 345, 346
  plume, 97
  round, 83
  vertical, 121
Buoyant jet in crossflow, 282, 308
Buoyant plume in stratified crossflow, 359

J. H. W. Lee et al., *Turbulent Jets and Plumes*,
https://doi.org/10.1007/978-3-032-01187-9

C

Cederwall equation, 128, 186
Centerline dilution
  coflow jet, 203
  line puff, 235
  line thermal, 275
Coanda effect, 128, 141
Coflow jet, 189–214
  summary of properties, 214
Composite dilution, 414, 416
Concentration, 46
  coflow jet, 192, 193
  line puff, 230
Concentration distribution, 4
Concentration fluctuation
  coflow jet, 213
  line puff, 245
Conservation
  of momentum, 41
  of momentum flux, 33, 40
  of tracer-mass flux, 34
Continuity equation, 32, 39
Continuous source, 5, 25
Control volume, 27
Cooling water, 98
Counterflowing jet, 346

D

Dense jet, 93, 360
Densimetric Froude number, 19, 103, 112
  local, 78, 104
Density
  difference, 5, 86
  nonlinear temperature dependence, 86, 422
  seawater, 419
  stratification, 5, 180
Density stratification, 145–180
  oscillation, 155
Depth of penetration, 156
Desalination plant, 103
Dilution, 4, 15, 124
  minimum, 126, 167
Dimensional analysis, 14, 18, 69
Dissipation rate, 228, 263
Dominant eddy, 25, 226, 248, 297, 309, 313
Double vortex, 219
Dredger, 8
Dynamic interaction, 414, 416

E

Eddy-viscosity, 227
Elemental volume, 297–298, 308
Energy flux, 29
Entrainment, 334
  projected area entrainment, 302, 330, 335
  shear entrainment, 305, 334, 363
  surface area entrainment, 304
  velocity, 73, 362
  vortex entrainment, 334
Entrainment coefficient, 74, 106, 196
  coflow jet, 206, 216
  line puff, 235
  line thermal, 268, 292
  plume, 75
Entrainment hypothesis, 37, 73, 75
Environmental impact, 1
Environmental impact assessment, 349
Environmental monitoring, 342
Epilimnion, 150
Equation of motion, 4, 72
Equation of state, 86, 419
Equilibrium depth, 158
Equilibrium level, 158, 162
Eulerian Integral model, 72
Eulerian integral model, 34
Excess jet velocity, 197
Excess momentum, 189
Excess velocity, 194, 211
External flow, 363

F

Finite volume method, 229
Fjords, 132
Froude number, 103

G

Galilean transformation
  bent-over jet in crossflow, 227
  bent-over plume in crossflow, 262
Gaussian profile, 70, 114, 191
Gradient transport, 263
Gravity
  acceleration, 147

H
Half-width
  jet, 31, 39
  line puff, 235
  line thermal, 273
  plume, 70
High Island reservoir, 146
Horizontal round buoyant jet, 128
Hypolimnion, 145, 150

I
Initial dilution
  field data, 323–326
Initial dilution, 322
Instantaneous image, 238
Instantaneous source, 5
Integral analysis, 33
Integral model
  coflow jet, 193
  jet, 34
  line puff, 226
  line thermal, 260
  plume, 72, 88
Intermittency
  coflow jet, 207
  jet, 48
  line puff, 245
  line thermal, 282
  plume, 81
Irrotational flow
  jet in crossflow, 363
  line puff, 226
  line thermal, 261

J
Jet
  in coflow, 189–214
  in counterflow, 216, 346
  in crossflow, 219–252
  interference, 59
  negatively buoyant, 93, 423
  round, 38
  summary of properties, 45
  trajectory, 121
  turbulent, 1, 27–55
  two dimensional, 30
Jet boundary, 209
Jet in crossflow, 291
  3D model, 233
  initial dilution, 219
  laser-induced fluorescence image, 238
  summary of properties, 235
  tracer concentration, 237
JETLAG, 343, 361
  entrainment formulation, 337
  field verification, 349
  model validation, 342
Jet mixing
  in river, 210, 249
JETLAG, 329
  field verification, 342
  model validation, 341
  theory, 331

K
Kinematic momentum flux, 40
Kinetic energy, 58
Kinetic energy flux, 35

L
Lagrangian method, 167, 297, 318
Lagrangian model, 88
  general formulation, 297–318
  inclined buoyant jet, 114
  JETLAG, 329, 361
  line plume, 106
  spreading hypothesis, 116, 301
Large Eddy Simulation, 170, 172, 175
Laser doppler anemometry, 191
Laser-induced fluorescence, 46, 124, 190
Length scale, 15
  in stratification, 164, 187
  jet in crossflow, 221
  momentum, 84
  plume in crossflow, 256
Linear stratification, 183
Line puff, 237
  characteristics, 283
  1D model, 226
  2D model, 227
Line thermal
  characteristics, 283
  1D model, 260
  2D model, 262

M
Maximum concentration ratio
  line puff, 234
  line thermal, 272
Method of excesses, 299–301
Mixing layer, 54, 114, 312
Momentum

flux, 14
line puff, 245
line thermal, 265
method of superposition, 59
Momentum-Dominated Far Field (MDFF), 223, 248
Momentum-Dominated Near Field (MDNF), 223, 247
Momentum equation, 32, 39, 59
round thermal, 154
Momentum flux, 28
specific, 35
Multiple buoyant jets, 365, 413, 415
Multiple coflow jets, 413, 415
Multiport diffuser, 132

N
Near-Far field transition, 337
Numerical experiment
line puff, 228
line thermal, 263
Numerical modelling, 321
model validation, 341–342
Nutrient, 132

O
Ocean outfall, 295
Hong Kong SSDS ocean outfall, 349
Orange County outfall, 175
Sydney ocean outfall, 342
Oscillation
in stratification, 165
Outfall mixing in weak current, 358

P
Plume
black smoker, 185
buoyant gas, 102
2D linear stratification, 167
fire plume, 1, 68, 76
gas, 107
in arbitrary density stratification, 174
in container, 176
in crossflow, 161, 253–290
in linear stratification, 164
in stratified fluid, 162
line, 96, 100, 105, 106, 175
maximum rise, 164
maximum rise in stratified fluid, 186
round, 69
sandslurry, 10
summary of properties, 101
thermal, 5, 87
turbulent, 1, 63–100
waste gas, 85, 95
wastewater, 185
Plume element, 63–100, 330, 332
Plume in crossflow
3D model, 270
laser-induced fluorescence image, 254
stratified, 293
summary of properties, 270
tracer concentration, 277
Plume in crossflow
initial dilution, 322
Pollution, 143
Potential core, 26, 27, 31, 54, 112, 114, 117, 123, 124, 311
coflow jet, 200
development, 121
Power plant, 98, 139
Puff, 5, 306
round, 19

R
Reduced gravity, 64
Reynolds-averaged equation, 227, 262
Reynolds number
jet, 19, 25
Rhodamine dye, 25
Ring vortices, 219
Risk assessment, 1
Hong Kong SSDS ocean outfall, 349
River discharge, 416, 418

S
Salinity, 86, 147, 419
Salt balance, 184
Sand dumping, 8
Scalar field
advected line puff, 238
jet in crossflow, 233, 238
plume in crossflow, 272, 277
Schmidt number, 228, 263
Self-similarity
jet, 31, 46
line puff, 224, 230
line thermal, 258
plume, 69, 80, 96
Separation angle, 337
Sewage discharge, 63, 131
Sewage disposal, 13
Sewage outfall, 1, 131, 143

Urmston Road, 152
Sewage outfall
North-West New Territories outfall, 347
Shear entrainment, 219
Sigma unit, 152, 419
Spreading coefficient, 13, 116, 310
jet, 55
line puff, 226
line thermal, 261
plume, 100
Spreading hypothesis, 226, 301–306
Spreading rate
coflow jet, 203
line puff, 252
line thermal, 290
Stagnation point, 339
Stratification frequency, 151
Stream function
line puff, 232
line thermal, 264, 265, 270
Submarine diffuser, 108

T
Temperature, 148, 419
excess, 86
Thermal, 5, 8, 306
in stratified fluid, 153
line, 10, 160
round, 153, 184
Thermal diffuser, 142
Thermal discharge, 98, 99
Thermocline, 151, 158, 159, 176
Time scale
buoyant jet, 121
line puff, 224
line thermal, 259
Top-hat profile
coflow jet, 197, 215
jet, 29, 51
plume, 89
Toxic spill, 156, 158
Tracer concentration
jet, 46
line puff, 220, 233
line thermal, 266
plume, 80
Tracer-mass conservation equation, 32
Trajectory
horizontal buoyant jet, 124
inclined buoyant jet, 122
line puff, 235
line thermal, 273
Turbulence intensity
coflow jet, 213
jet, 82
plume, 82
Turbulence model
line puff, 227
line thermal, 262
Turbulent closure, 33
Turbulent entrainment, 25
Turbulent kinetic energy, 263
line puff, 230
line thermal, 266
Turbulent viscosity, 228

U
Upwelling, 133

V
Velocity gradient, 229
Vertical momentum
line puff, 229
line thermal, 275
Virtual reality, 8, 350–353
VISJET, 8, 350–353
Visual boundary
coflow jet, 207
line puff, 238
line thermal, 277
Vortex flow, 219
Vortex pair, 219, 253, 264
Vorticity
line puff, 228
line thermal, 266

Z
Zone of established flow, 31, 39, 114, 115, 309
Zone of flow establishment, 31, 38, 114, 309